# OPTIMIZATION MODELS
# USING FUZZY SETS AND POSSIBILITY THEORY

## THEORY AND DECISION LIBRARY

General Editors: W. Leinfellner and G. Eberlein

Series A: Philosophy and Methodology of the Social Sciences
Editors: W. Leinfellner (Technical University of Vienna)
G. Eberlein (Technical University of Munich)

Series B: Mathematical and Statistical Methods
Editor: H. Skala (University of Paderborn)

Series C: Game Theory, Mathematical Programming and Mathematical Economics
Editor: S. Tijs (University of Nijmegen)

Series D: System Theory, Knowledge Engineering and Problem Solving
Editor: W. Janko (University of Vienna)

---

## SERIES B: MATHEMATICAL AND STATISTICAL METHODS

Editor: H. Skala (Paderborn)

**Editorial Board**

J. Aczel (Waterloo), G. Bamberg (Augsburg), W. Eichhorn (Karlsruhe), P. Fishburn (New Jersey), D. Fraser (Toronto), B. Fuchssteiner (Paderborn), W. Janko (Vienna), P. de Jong (Vancouver), M. Machina (San Diego), A. Rapoport (Toronto), M. Richter (Aachen), D. Sprott (Waterloo), P. Suppes (Stanford), H. Theil (Florida), E. Trillas (Madrid), L. Zadeh (Berkeley).

**Scope**

The series focuses on the application of methods and ideas of logic, mathematics and statistics to the social sciences. In particular, formal treatment of social phenomena, the analysis of decision making, information theory and problems ofinference will be central themes of this part of the library. Besides theoretical results, empirical investigations and the testing of theoretical models of real world problems will be subjects of interest. In addition to emphasizing interdisciplinary communication, the series will seek to support the rapid dissemination of recent results.

# OPTIMIZATION MODELS USING FUZZY SETS AND POSSIBILITY THEORY

*Edited by*

J. KACPRZYK

*Systems Research Institute,*
*Polish Academy of Sciences, Warsaw, Poland*

and

S. A. ORLOVSKI

*International Institute for Applied Systems Analysis,*
*Laxenburg, Austria*

Springer-Science+Business Media, B.V.

**Library of Congress Cataloging in Publication Data**

CIP

Optimization models using fuzzy sets and possibility theory

(Theory and decision library. Series B, Mathematical and statistical methods)
Includes index.
1. Mathematical optimization. 2. Fuzzy sets. 3. Decision-making.
I. Kacprzyk, Janusz. II. Orlovski, S. A. (Sergeĭ A.) III. Series.
QA402.5.0653 1987 511.3′2 87–9459

ISBN 978-94-010-8220-4 ISBN 978-94-009-3869-4 (eBook)
DOI 10.1007/978-94-009-3869-4

---

**THE INTERNATIONAL INSTITUTE FOR APPLIED SYSTEMS ANALYSIS**

is a nongovernmental research institution, bringing together scientists from around the world to work on problems of common concern. Situated in Laxenburg, Austria. IIASA was founded in October 1972 by the academies of science and equivalent organizations of twelve countries. Its founders gave IIASA a unique position outside national, disciplinary, and institutional boundaries so that it might take the broadest possible view in pursuing its objectives:

*To promote international cooperation* in solving problems arising from social, economic, technological, and environmental change
*To create a network of institutions* in the national member organization countries and elsewhere for joint scientific research
*To develop and formalize systems analysis* and the sciences contributing to it, and promote the use of analytical techniques needed to evaluate and address complex problems
*To inform policy advisors and decision makers* about the potential application of the Institute's work to such problems

The Institute now has national member organizations in the following countries:

**Austria**
The Austrian Academy of Sciences

**Bulgaria**
The National Committee for Applied Systems Analysis and Management

**Canada**
The Canadian Committee for IIASA

**Czechoslovakia**
The Committee for IIASA of the Czechoslovak Socialist Republic

**Finland**
The Finnish Committee for IIASA

**France**
The French Association for the Development of Systems Analysis

**German Democratic Republic**
The Academy of Sciences of the German Democratic Republic

**Federal Republic of Germany**
Association for the Advancement of IIASA

**Hungary**
The Hungarian Committee for Applied Systems Analysis

**Italy**
The National Research Council

**Japan**
The Japan Committee for IIASA

**Netherlands**
The Foundation IIASA – Netherlands

**Poland**
The Polish Academy of Sciences

**Sweden**
The Swedish Council for Planning and Coordination of Research

**Union of Soviet Socialist Republics**
The Academy of Sciences of the Union of Soviet Socialist Republics

**United States of America**
The American Academy of Arts and Sciences

CONTENTS

## PREFACE

Optimization is of central concern to a number of disciplines. Operations Research and Decision Theory are often considered to be identical with optimization. But also in other areas such as engineering design, regional policy, logistics and many others, the search for optimal solutions is one of the prime goals. The methods and models which have been used over the last decades in these areas have primarily been "hard" or "crisp", i.e. the solutions were considered to be either feasible or unfeasible, either above a certain aspiration level or below. This dichotomous structure of methods very often forced the modeller to approximate real problem situations of the more-or-less type by yes-or-no-type models, the solutions of which might turn out not to be the solutions to the real problems. This is particularly true if the problem under consideration includes vaguely defined relationships, human evaluations, uncertainty due to inconsistent or incomplete evidence, if natural language has to be modelled or if state variables can only be described approximately.

Until recently, everything which was not known with certainty, i.e. which was not known to be either true or false or which was not known to either happen with certainty or to be impossible to occur, was modelled by means of probabilities. This holds in particular for uncertainties concerning the occurrence of events. Probability theory was used irrespective of whether its axioms (such as, for instance, the law of large numbers) were satisfied or not, or whether the "events" could really be described unequivocally and crisply.

In the meantime one has become aware of the fact that uncertainties concerning the occurrence as well as concerning the description of events ought to be modelled in a much more differentiated way. New concepts and theories have been developed to do this: the theory of evidence, possibility theory, the theory of fuzzy sets have been advanced to a stage of remarkable maturity and have already been applied successfully in numerous cases and in many areas. Unluckily, the progress in these areas has been so fast in the last years that it has not been documented in a way which makes these results easily accessible and understandable for newcomers to these areas: text-books have not been able to keep up with the speed of new developments; edited volumes have been published which are very useful for specialists in these areas, but which are of very little use to nonspecialists because they assume too much of a background in fuzzy set theory. To a certain degree the same is true for the existing professional journals in the area of fuzzy set theory.

The editors of this book have succeeded to avoid this weakness by starting with an introductory section which provides - even for the newcomer to this area - the necessary background to understand the contributions of the following sections.

Section II of this volume focuses on methodological advances in the areas of optimization and decision making. Three of the most relevant topics in this area have been chosen to illustrate modern tools and techniques using fuzzy sets and possibility theory: preference theory, decision theory and multicriteria decision analysis.

The editors can be congratulated on the selection of authors they have succeeded to convince to contribute to these sections. They are all internationally well-reputed and leading scientists in their respective areas.

The same is true for the two subsections of this chapter which are of a slightly different character. One treats fuzzy approaches to location and distribution problems. This will certainly be of particular interest to people working in logistics. The second subsection introduces the reader into the most modern area of knowledge-based decision support systems which links past experience and available optimization models to future developments as they will be needed, for instance, in the 5th Generation Computer Technology.

One of the frequently asked questions is: "Can fuzzy sets be used in practice?" At the start of any new theory this question is particularly hard to answer. Nevertheless the editors of this book give an answer by presenting in the third chapter of the book five descriptions of the use of fuzzy sets in solving real world problems in quite diverse areas such as regional policy, water resource allocation and hydrocracking processes. They could not have thought of a better and more convincing conclusion of their book.

Altogether this volume is a very important and appreciable contribution to the literature on fuzzy set theory. The editors have succeeded in presenting a well composed selection of contributions by leading scientists from all over the world. They have also provided enough background information to make the book selfcontained and valuable to newcomers to this area as well as to specialists. It can only be hoped that it will be read in all parts of the world. It really deserves it!

H.-J. Zimmermann
President
International Fuzzy Systems Association

Aachen, June 1986

# I. INTRODUCTORY SECTIONS

# NEW PARADIGMS IN SYSTEMS ENGINEERING: FROM "HARD" TO "SOFT" APPROACHES

Brian R. Gaines

Department of Computer Science
University of Calgary
Calgary, Alberta, Canada T2N 1N4

Abstract. Developments in fuzzy sets theory are considered in relation to those in expert systems. It is suggested that these are not just mathematical and technological advances but also represent major paradigm shifts in system theory. The main shift is away from the normative application of technology to change the world to be theoretically tractable, and towards increasing model realism. The limitations of classical "hard" system theory when applied to natural systems are the impetus behind the development of modern "soft" system theory, its foundations in fuzzy sets theory and its application in expert systems.

Keywords: systems theory, systems engineering, fuzzy sets, expert systems.

## 1. INTRODUCTION

Lotfi A. Zadeh first discussed the need for a "mathematics of fuzzy or cloudy quantities" in a paper entitled "From circuit theory to system theory" published in 1962. This led to his publishing his seminar paper, "Fuzzy sets", proposing such a mathematics in 1965. A comprehensive bibliography for the first decade shows an increase from 2 papers published in 1965 to over 227 in 1975 with a cumulative total in 1975 of some 620 items (Gaines and Kohout, 1977) and in 1979 of some 1400 items (Kandel and Yager, 1979). The number of papers a year and cumulative total fit well to exponential growth at 60% a year for the first decade. However, it is now almost impossible to track the growth of a literature which has grown from the output of a small group of specialists to that of an international community involving almost every nation and discipline. The growth rate of the dissemination of knowledge about, interest in, and work on fuzzy sets theory (FST) and its applications has been spectacular.

Why has there been this tremendous growth of interest in the past twenty years in the mathematics of fuzzy or cloudy quantities? What changes have occurred in systems engineering? This paper suggests that there has been a shift in the modes of thinking and problem-solving for a significant community of theoretical and applied scientists and technologists. The "hard" systems approach that has proved so powerful in the development of man-made systems is far less useful in developing models for the management of natural systems, or coupled man-made and natural systems. A "soft" systems approach has become necessary to extend our science and technology to systems engineering for major ecological, social and economic processes.

Nowhere is the significance of this change more apparent than in the development of expert systems (ESs) where a soft systems approach has been taken to the encoding of human expertise for computer-based systems (e.g. Michie, 1979; Gevarter, 1983; and Reitman, 1984). This is an interesting area, not only for its high intrinsic value, but also because it enables us to contrast differing aspects of the role of FST in modern information science. Expert systems development leads to requirements for reasoning with imprecise data where FST provides an alternative paradigm to those of classical logic and probability theory. The most well-recognized breakthroughs in ESs such as MYCIN (Shortliffe, 1976) were not based on FST, but on heuristic methods that turn out to closely resemble FST. Other early breakthroughs such as linguistic process controllers (Mamdani and Assilian, 1981) were based directly on FST.

The next section outlines the development of ESs and the role of FST, and the following section considers the paradigm changes involved.

## 2. EXPERT SYSTEMS AND FUZZY SETS THEORY

The computer simulation of people in the roles of experts on some topic has become an important application of interactive computer systems. It has generated a new industry based on creating expert systems to make the practical working knowledge of a human expert in a specific subject area such as medicine or geology widely available to those without direct access to the original expert (Reitman, 1984). Programs now exist that have made practical achievements in medical diagnosis, interpretation of mass spectroscopy results, analysis of geological survey data, and other problems where one would normally go to a human expert for advice.

One of the first ES developments was the fuzzy logic control system developed in 1974 by Mamdani and Assilian. The system accepted human knowledge of control strategies expressed verbally and encoded it directly as computer programs which acted on the environment (Mamdani and Assilian, 1981). This work was undertaken as part of a study of machine learning in process control and the system controlled was a small steam engine. The verbal rules were of the form shown in Fig. 1.

> IF the pressure error is positive and big and the change in pressure error is not negative medium or big
> THEN make the heat change negative and big

Fig. 1. Rule from a fuzzy logic controller

What was surprising at the time and made the 1974 results

a recognized breakthrough was that the control rules derived from the verbal statements were extremely effective. They compared favorably with those derived by tuning a standard PID (proportional-integral-derivative) controller for optimum performance. It was also found that the learning machine then proceeded only to learn less effective strategies. Hence interest switched to the process whereby human expression of verbal rules that appear vague can lead to highly effective control strategies. In the past ten years Mamdani and Assilian's results have been replicated in many different countries for many different control processes, including a number of significant industrial processes such as pig iron smelting where effective automatic control had been thought to be impossible (Mamdani, Østergaard and Lembessis, 1984).

The concept of an ES was not prevalent at the time of the initial fuzzy control studies and their significance as examples of early ES development was noted only later. In parallel with the controller development, other rule-based ESs were being developed for completely different domains. The system widely recognized as an early breakthrough, MYCIN, is a medical diagnosis ES which aids a clinician to act as a consultant on infectious diseases (Shortliffe, 1976). It uses rules of the form shown in Fig. 2.

```
RULE 50
  IF 1) the infection is primary-bacteremia, and
     2) the site of the culture is one of the sterile
        sites, and
     3) the suspected portal of entry of the organism is
        the gastro-intestinal tract,
  THEN there is suggestive evidence (.7) that the identity
       of the organism is bacteroides
```

Fig. 2. A MYCIN rule

These rules are obtained from specialists in microbial infections and their application to particular data is fairly simple data processing. The rules are validated through their application to many cases and revised when they fail to give the correct diagnosis. MYCIN is designed to interact with a clinician in order to make a diagnosis and suggest therapy for a particular patient with suspected microbial infections. It first gathers data about the patient and then uses this to make inferences about the infections and their treatment.

Note that the MYCIN rule of Fig. 2 involves an assertion that is evidential rather than true. Shortliffe found it necessary to encode rules of inference that were imprecise and could not be encoded simply in terms of truth and falsity. He ascribed a verbal label, "suggestive evidence", and a numerical truth value, "0.7", to a rule and developed a calculus for combining such truth values in chains of logical inference. Thus, linguistic reasoning and multivalued logics were key components of early ES developments although the MYCIN developers were initially unaware of FST and the linguistic controller developers were initially unaware of ES concepts.

It was also discovered that the rule-based approach to knowledge encoding could be applied to high-level capabilities such as learning processes through the use of metarules. Mamdani, Procyk and Baaklini (1976) found that learning could be introduced effectively in the steam engine controller through metarules that expressed the way in which the basic rules should be changed as a result of performance feedback. The learning level of their controller operated on rules of the form shown in Fig. 3. This was sufficient for the fuzzy controller to acquire a control strategy similar to that induced through verbal rules from a human expert.

```
IF time is small and error is negative big
   THEN desired change is big
IF time is big and error is positive zero
   THEN desired change is zero
```

Fig. 3. Metarules from a fuzzy learning controller

Metarules were also introduced independently by Davis to aid the debugging of MYCIN. It was difficult to set up the MYCIN rules initially and also difficult to trace errors in the deductions. To overcome these problems TEIRESIAS (Davis and Lenat, 1982) was added as an auxiliary ES with expertise about MYCIN to explain MYCIN's decisions and help the clinician amend the rules when they lead to incorrect conclusions. TEIRESIAS uses a similar rule-based approach to reasoning as does MYCIN but the rules are now rules about the forms of rules and the use of rules. A typical such metarule is shown in Fig. 4. Whereas MYCIN's rules are specific to microbial infections, those of TEIRESIAS are more general and can be used in other domains. Davis (1983), for example, shows TEIRESIAS being used as an investment decision system for clients of a stockbroker.

```
METARULE003:
IF 1) there are rules which do not mention the
      current goal in their premise
   2) there are rules which mention the current
      goal in their premise
   THEN it is definite that the former should be
      done before the latter
```

Fig. 4. A TEIRESIAS metarule

One of the important features of MYCIN/TEIRESIAS that has become an essential characteristic of ES is their capacility to provide explanations of the deductions given. "Why?" questions are accepted as responses when data is requested and are interpreted as a request for the rule to be shown that requires the data requested. A "why?" question may also be asked when conclusions are drawn and this is interpreted as a request for the complete chain of logic used in arriving at that conclusion to be shown. The facility to answer such questions make ES accountable for their behavior and conclusions. This is itself a major new feature of systems programmed for computers.

Another important feature of ES is that they are not static representations of knowledge but can continue to acquire knowledge as they are used. Essentially, the use of metarules allows ES to be programmed interactively by their users. From one perspective the metarules of the fuzzy learning controller and TEIRESIAS can be seen as an important development in automatic programming. From another they can be seen as a way in which a machine acquires knowledge through interaction with its environment or with a person. These are analogous to the fundamental ways in which people acquire knowledge (Gaines and Shaw, 1984a). For such applications computational logics capable of dealing with the uncertainties of imprecise data and fallible hypotheses are essential.

Since the early success with fuzzy logic control and MYCIN, a very wide range of ESs have been developed. Gevarter (1983) has summarized some well-known expert systems and their applications but the numbers and domains have since increased so rapidly that it is now impossible to make any accurate count. Most universities have some activity in this field and many industrial ESs are regarded as highly proprietary. ESs are a pragmatic example of the success of a soft systems approach based on linguistic reasoning with uncertain rules and data (Gaines and Shaw, 1984b). They contrast with previous unsuccessful approaches to similar problems based on the development of precise mathematical models and their use in the development of optimum control and decision algorithms.

The next section considers the significance of this change of approach.

## 3. SHIFTS IN SYSTEMS PARADIGMS

The previous section has shown how the early applications of FST to control and decision systems paralleled the development of early expert systems in the use of linguistic rules, fuzzy reasoning and metarules. This role of FST, significant as it is in itself, is only an indication of the deeper paradigm shifts from which FST and ES both stem. The classical approach in decision and control system design is shown in Fig. 5. This positivistic paradigm underlies the methodologies of the physical sciences and technologies based on them. It has the merit that it has been extremely successful in engineering much of the technological infrastructure of our current civilization.

However, this paradigm is successful only to the extent that the systems under consideration are amenable to instrumentation and modeling. Its greatest successes have been where this amenability can be achieved normatively, that is in cases where the system to be controlled is itself a human artifact. For example, linear system theory has not become a major tool in systems engineering because most natural systems are linear - they are not. The implication is in the opposite direction: that linear systems are mathematically tractable and that we design artificial systems to be linear so that we may model them readily.

The application of "a linear model with quadratic peformance criterion" to natural systems is often attempted but, in general, it does not work. We have done so not because the tool was appropriate, but because it was the only one we had. How-

STEP 1
Thoroughly instrument the system to be controlled or about which decisions are to be made

STEP 2
Use the instrumentation to gather data about the system behavior under a wide variety of circumstances

STEP 3
From this data build a model of the system that accounts for this behavior

STEP 4
From this model derive algorithms for decision or control that are optimal in terms of prescribed performance parameters

Fig. 5. The hard systems approach

ever, the use of a hammer to insert screws, although partially effective, tends to distort, destroy, and generally defeat the purpose of using a screw. Similarly, the use of an inappropriate system theory to model a system may give useful, but limited, results when we have no other, but it distorts reality, destroys information and generally defeats the purpose of modeling that system.

Much of our current technology succeeds to the extent that it is normative. In agriculture we reduce the complexity of a natural ecology to a comprehensible simplicity by the use of pesticides, herbicides and chemical fertilizers (Gaines and Shaw, 1984c). We reduce the system to one which is amenable to our modeling techniques. That simpler is not necessarily better and that re-engineering nature to impose uniformity destroys variety which is itself valuable have only been realised in recent years.

The four shifts in perspective that we see in FST and ES are shown in Fig. 6. The last three perspectives all stem from the first. The importance of this first perspective to Zadeh is apparent in his 1962 paper where he discusses the fundamental inadequacy of conventional mathematics for coping with the analysis of biological systems, noting also that the need for a new mathematics was becoming increasingly apparent even in the realm of inanimate systems.

The second perspective is that which lead to FST. Optimal control theory was regarded as the peak achievement of system theory in the 1950s and 1960s. However, it proved limited in application because it demanded precision in system modeling that was impossible in practice. It was too sensitive to the nuances of system structure expressed through over-precise system definition.

The third perspective is that which led to the success of linguistic fuzzy controllers and later ESs. Hayes-Roth (1984) has noted the many problems that have been felt to require human management are now amenable to ESs. Modeling the way the expert

PROBLEM 1
The models available are inadequate to capture the system

OLD APPROACH: PROCRUSTEAN DESIGN
Change the world to fit the model - normative technology

PARADIGM SHIFT: MODEL REALISM
Use system methodologies and information technology that enable the natural world to be modeled without distortion and destruction

PROBLEM 2
Optimal control is over-sensitive to system uncertainties

OLD APPROACH: SUB-OPTIMALITY
Use a sub-optimal controller that is robust

PARADIGM SHIFT: MODEL UNCERTAINTY
Model the uncertainty as part of the system

PROBLEM 3
Data is unavailable or inadequate for modeling

OLD APPROACH: MANAGE
Do not automate - leave to human decision/control

PARADIGM SHIFT: EXPERT SYSTEMS
Model the person as a decision-maker or controller

PROBLEM 4
Neither a human nor an automatic system alone is adequate

OLD APPROACH: AD HOC SYSTEM DESIGN
Use a mixture of automatic and human decision/control

PARADIGM SHIFT: ACCOUNTABLE INTEGRATION
Integrate automatic and human activity - make the automation accountable ("Why?" in ES)

Fig. 6. Paradigm shifts in systems engineering - from the hard to the soft systems approach

performs the task rather than modeling the task itself is the primary characteristic of an ES.

The fourth perspective is an important one for both ES and FST. They are knowledge-based systems because they make provision for explaining the decisions reached in terms of the data and inferences used. It is interesting to note that logics of uncertainty that aggregate evidence, such as probabilistic logics, do not provide a simple mechanism for explanation. Explicable logics have to be truth-functional and non-aggregative; fuzzy logic satisfies these requirements (uniquely among those logics satisfying the weak axioms of a standard uncertainty logic (Gaines, 1983)). It is also interesting to note that the capability to give explanations is seen by some philosophers of science as a key difference between models that make accurate predictions and scientific theories that in addition provide causal explanations (Salmon, 1984). The "why?" question has important implications for both the practical and theoretical significance of ESs and the logics on which they are based.

## 4. CONCLUSIONS

FST was from the outset an attempt to create a new mathematical system theory that corresponds to Paradigm Shift 1 and fits the realities of the world without distorting them. It was created by a person who had extended the boundaries of current system theory, attempted to encompass in generality the key concepts of applied systems engineering (Zadeh, 1956, 1957, 1963 and 1964), and recognized the failure of that theory in this task. Developments in ES have shown the practical significance of this paradigm shift in enabling systems to be engineered for problems previously considered intractable. FST and ESs, and the application of one to the other, are not just mathematical and technological advances but also represent major paradigm shifts in system theory. This has involved fundamental changes in system philosophy and technology, shifts from a positivistic, normative approach to a more realistic and naturalistic approach. These shifts are apparent throughout science and technology and its application to our world and society.

Fuzzy sets theory cannot be either right or wrong. It is applicable mathematics tested by its uses. However, the rationale behind it, the systemic principles involved, can be right or wrong. They are right for our time, for the objectives of dealing adequately with a complex universe and extending the capabilities of the person with computer enhancements. The soft systems principles involved do not replace hard systems principles but extend the domain of systems theory to encompass both. The re-development of system theory is not yet complete and the seminal notions of stability, adaptivity, modeling, and so on, still need adequate expression. However, we now have the foundations on which to build a system theory that combines realism with power and provides applicable mathematics for our knowledge-based society.

The papers in this volume present the state-of-the-art of soft systems engineering based on FST, mainly in a decision making and optimization context, and its application to a wide variety of practical problems.

## ACKNOWLEDGEMENT

Financial assistance for this work has been made available by the National Sciences and Engineering Research Council of Canada.

## REFERENCES

Davis, R. (1983). TEIRESIAS: experiments in communicating with a knowledge-based system. In M.E. Sime, and M.J. Coombs (eds.), Designing for Human-Computer Communication. Academic Press, London, 87-137.

Davis, R,. and D.B. Lenat (1982). Knowledge-Based Systems in Artificial Intelligence. McGraw-Hill, New York.

Gaines, B.R. (1983). Precise past - fuzzy future. Int. J. Man-Machine Studies 19, 117-134.

Gaines, B.R., and L.J. Kohout (1977). The fuzzy decade: a bibliography of fuzzy systems and closely related topics. Int. J. Man-Machine Studies 9, 1-68.

Gaines, B.R., and M.L.G. Shaw (1984a). The Art of Computer Conversation: A New Medium for Communication. Prentice Hall, Englewood Cliffs, N.J.

Gaines, B.R., and M.L.G. Shaw (1984b). Logical foundations of expert systems. Proc. of IEEE Int. Conference on Systems, Man and Cybernetics. Halifax, Nova Scotia, 238-247.

Gaines, B.R. and M.L.G. Shaw (1984c). Expert systems: the substitution of knowledge for materials in world food production. Possible Worlds, 5-8.

Gevarter, W.B. (1983). Expert systems: limited but powerful. IEEE Spectrum 18, 39-45.

Hayes-Roth, F. (1984). The industrialization of knowledge engineering. In W. Reitman (ed.), Artificial Intelligence Applications for Business. Ablex, Norwood, N.J., 159-177.

Kandel, A., and R.R. Yager (1979). A 1979 bibliography of fuzzy sets, their applications and related topics. In M.M. Gupta, R.K. Ragade, and R.R. Yager (eds.), Advances in Fuzzy Set Theory and Applications. North-Holland, Amsterdam, 621-744.

Mamdani, E.H., and S. Assilian (1981). An experiment in linguistic synthesis with a fuzzy logic controller. In E.H. Mamdani and B.R. Gaines (eds.), Fuzzy Reasoning and its Applications. Academic Press, London, 311-323.

Mamdani, E.H., J.J. Ostergaard, and E. Lembessis (1984). Use of fuzzy logic for implementing rule-based control of industrial processes. In H.J. Zimmermann, L.A. Zadeh, and B.R. Gaines (eds.), Fuzzy Sets and Decision Analysis. North-Holland, Amsterdam.

Mamdani, E.H., T. Procyk and N. Baaklini (1976). Application of fuzzy logic to controller design based on linguistic protocol. In E.H. Mamdani and B.R. Gaines (eds.), Discrete Systems and Fuzzy Reasoning. Queen Mary College, London, 125-149.

Michie, D., (ed.) (1979). Expert Systems in the Micro Electronic Age. Edinburgh University Press, Edinburg.

Reitman, W., (ed.) (1984). Artificial Intelligence Applications for Business. Ablex, Norwood, N.J.

Salmon, W. (1984). Scientific Explanation and the Causal Structure of the World. Princeton University Press, Princeton, N.J.

Shortliffe, E.H. (1976). Computer-Based Medical Consultations: MYCIN. Elsevier, New York.

Zadeh, L.A. (1956). On the identification problem. IRE Trans. Circuit Theory CT-3, 277-281.

Zadeh, L.A. (1957). What is optimal? IRE Trans. Circuit Theory. CT-4, 3.

Zadeh, L.A. (1962). From circuit theory to system theory. Proc. Institute of Radio Engineers 50, 856-865.

Zadeh, L.A. (1963). On the definition of adaptivity. Proc. IEEE 51, 469-470.

Zadeh, L.A. (1964). The concept of state in system theory. Proc. of the Second Systems Symposium. John Wiley, New York.

Zadeh, L.A. (1965). Fuzzy sets. Inf. and Control 8, 338-353.

# INTRODUCTION TO FUZZY SETS AND POSSIBILITY THEORY

Mario Fedrizzi

Institute of Statistics and Operations Research,
University of Trento,
Via Verdi 26, 38100 Trento, Italy

Abstract. The purpose of this introduction is to discuss the essence of fuzzy sets and possibility theory in order to make the interested reader familiar with the basic elements of these growing fields of research. Thus such issues as basic definitions and properties of a fuzzy set, fundamental operations, fuzzy relations, the extension principle, fuzzy numbers, linguistic variables and some basic elements of possibility theory are briefly reviewed.

Keywords: fuzzy set, fuzzy relation, fuzzy number, linguistic variable, possibility distribution

## 1. INTRODUCTION

The purpose of this paper is to provide the reader with an introduction to fuzzy sets and possibility theory. Mathematics is used only instrumentally, that is, as a tool; no proof appears explicitly in the paper and no specific mathematical prerequisite is required.

To see fuzzy sets in a proper perspective, let us notice that analysis and modelling of any real world phenomenon or process must take into account an inherent uncertainty. In many cases this uncertainty is not due to randomness but to some imprecision whose formal treatment cannot be performed inside the mathematical framework of probability theory. We could say that such an imprecision may be: ambiguity, i.e. the association with a given object of a number of alternative meanings, generality, i.e. the application of the symbol's meaning to a multiplicity of objects, and vagueness, i.e. a lack of clear-cut boundaries of the set of objects to which the symbol (meaning) is applied. Notice that all the above imprecisions, and more particularly vagueness, may be viewed as an effect of natural languages used by humans.

In 1965 Zadeh (1965) provided the first tools, i.e. fuzzy sets, specially devised for dealing with this last form of imprecision, vagueness, and by now more then two thousand works dealing with this topic have been published, and hundreds of researchers all over the world are still working on the theory itself or on its application.

The theory of what Zadeh called fuzziness also stimulated a constructive debate on the several forms of uncertainty (Gaines, 1976), and on their mathematical representation (among others, Shafer, 1976; Höhle and Klement, 1984).

## 2. BASIC DEFINITIONS

In mathematics, sets are used to formally represent a concept. For instance the "integer numbers which are greater than 4 and smaller than 12" may be represented by the set $A = \{5, 6, 7, 8, 9, 10, 11\}$ or by its characteristic function $\emptyset_A : X \to \{0, 1\}$. Here $X$ is the universe of discourse (the set of integer numbers), $\emptyset_A(x) = 0$ means that $x$ does not belong to set $A$, while $\emptyset_A(x) = 1$ means that $x$ belongs to it.

Some difficulty arises when we want to use set theory to characterize vague concepts, say "numbers more or less equal to 8" which do not present a clear-cut differentiation between the elements belonging and not belonging to the set.

Zadeh (1965) suggested the replacement of the characteristic function by the so-called membership function $\mu_A: X \to [0,1]$ which associates with each element of the universe $X$ its grade of membership in a fuzzy set $A$, belonging to the interval $[0,1]$. Thus, $\mu_A(x) = 0$ means that $x$ does not belong to $A$, $\mu(x) = 1$ means that $x$ belongs to $A$, while $0 < \mu(x) < 1$ means that $x$ partially belongs to $A$.

For example, a fuzzy set $A$ = "numbers more or less equal to 8" may be represented by the membership function $\mu_A(x)$ shown in Fig. 1.

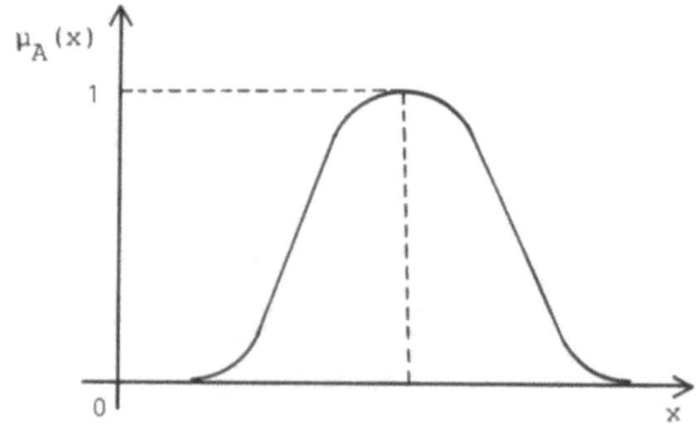


Fig. 1

Let us notice that the membership function is in this case in fact discrete but it is represented in the figure in a continuous form to make it more illustrative. We could also notice that the form of membership function is subjective due to the fact that a statement such as "number more or less equal to 8" contains some inherent subjectivity.

For instance, as another example, consider the set X of roses growing in a garden, and the subset A of X including only red roses. Of course, some roses of X will be definitely red, others definitely not red, but there will be borderline cases. The more an element of X belongs to A, the closer to 1 is its membership grade. The membership grades are obviously subjectively assessed and reflect an ordering of the universe X with respect to the vague predicate, i.e. the fuzzy set A.

Formally, we can now give the following definition: a fuzzy set A in a universe of discourse $X = \{x\}$, $A \subseteq X$, is defined as the set of pairs

$$A = \{(\mu_A(x), x)\}, \quad x \in X \tag{1}$$

where $\mu_A : X \to [0,1]$ is the membership function of A and $\mu_A(x)$ is called the grade of membership of $x \in X$ in A.

Generally, for the sake of brevity, fuzzy sets are equated with their membership functions and so we can say "fuzzy set $\mu_A(x)$" instead of "fuzzy set A characterized by membership function $\mu_A(x)$".

Usually the pair $(\mu_A(x), x)$ is also denoted by $\mu_A(x)/x$ and the following notations are introduced:

$$A = \mu_A(x_1)/x_1 + \ldots + \mu_A(x_n)/x_n = \sum_{i=1}^{n} \mu_A(x_i)/x_i, \quad \text{when } |X| = n$$

$$A = \int_X \mu_A(x)/x, \quad \text{when } X \text{ is a continuum}$$

and "+" and "Σ" are in the set-theoretic sense.

For example, the fuzzy set whose membership function is shown in Fig. 1 may be written as

$$A = 0.1/5 + 0.3/6 + 0.8/7 + 1/8 + 0.8/9 + 0.3/10 + 0.1/11,$$

In order to simplify the exposition, only finite universes of discourse will be used, even if membership functions will be graphed in a continuous form to make them more illustrative.

We must underline some facts here: first, the range of values of the membership function may be generalized, for instance to some lattice (see, e.g., Goguen, 1967); second, the exposition of fuzzy sets theory could be axiomatized even if some attempts of axiomatization, e.g., Chapin, 1971; or Novak, 1980, are not generally accepted.

Before concluding this section we will introduce other useful definitions.

The <u>support</u> of a fuzzy set $A \subseteq X$ is defined as

$$\text{supp } A = \{x \in X : \mu_A(x) > 0\}$$

Example. If $X = \{2,4,6,8,10\}$ and $A = 0.2/2 + 0.1/4 + 0/6 + 0/8 + 0.3/10$, then $\text{supp } A = \{2,4,10\}$.

The α-level set (α-cut) of $A \subseteq X$ is defined as

$$A_\alpha = \{x \in X : \mu_A(x) \geqslant \alpha\}$$

Example. If $X = \{1,3,5,7,9\}$ and $A = 0/1 + 0.6/3 + 0.7/5 + 0.9/7 + 1/9$, then $A_{0.2} = \{3,5,7,9\}$, $A_{0.5} = \{3,5,7,9\}$, $A_{0.9} = \{7,9\}$.

Let us observe that supp A and $A_\alpha$ are conventional (nonfuzzy) sets.

The height of a fuzzy set $A \subseteq X$ is defined by

$$\text{hgt}(A) = \sup_{x \in X} \mu_A(x)$$

Example. If $X = \{1,2,3,4,5\}$ and $A = 0.3/1 + 0.5/2 + 0.6/3 + 1/4 + 0/5$, then $\text{hgt}(A) = 1$.

Following Dubois and Prade (1979b) we could say "hgt(A) evaluates the possibility of finding in X at least one element which fits the predicate A exactly".

We close this section defining two fundamental relations between fuzzy sets, i.e. equality and containment.

The fuzzy sets $A,B \subseteq X$ are said to be equal, written $A = B$, if and only if $\mu_A(x) = \mu_B(x)$, for each $x \in X$.

This definition seems, however, to contradict to some extent a "soft" character that the equality of two fuzzy sets should have. Bandler and Kohout (1980) introduce a degree of equality of two fuzzy sets suggesting some "indexes" for the measurement of such a degree.

We say that A is a fuzzy subset of B or, alternatively, that A is contained in B, written $A \subseteq B$, if and only if $\mu_A(x) \leqslant \mu_B(x)$, for each $x \in X$.

Bandler and Kohout (1980) have suggested the use of a degree of containment in this case too (see also Dubois and Prade, 1980).

## 3. SET OPERATIONS AND THEIR PROPERTIES

All the definitions of the basic operations of the algebra of fuzzy sets will be given, as usual, in terms of the respective membership functions.

Given the fuzzy sets $A, B \subseteq X$ the following basic operations are defined:

union: $\mu_{A \cup B}(x) = \mu_A(x) \vee \mu_B(x)$ for each $x \in X$ (2)

intersection: $\mu_{A \cap B}(x) = \mu_A(x) \wedge \mu_B(x)$ for each $x \in X$ (3)

where "v" and "∧" are the maximum and minimum operators, respectively.

Example. If $X = \{1,2,3,4,5\}$, $A = 0.2/1 + 0.4/2 + 0.2/3 + 0.5/4 + 0.8/5$ and $B = 0.3/1 + 0.6/2 + 1/3 + 0.1/4 + 0/5$, then:

$$A \cup B = 0.3/1 + 0.6/2 + 1/3 + 0.5/4 + 0.8/5$$

$$A \cap B = 0.2/1 + 0.4/2 + 0.2/3 + 0.1/4 + 0/5.$$

Graphically the union and intersection may be portrayed as in Figs. 2 and 3.

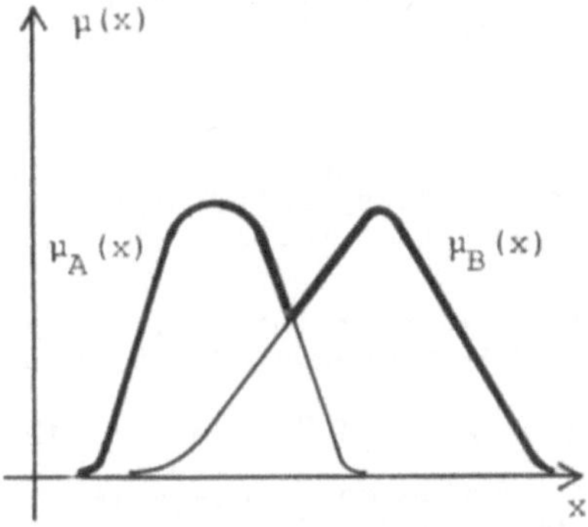


Fig. 2. Union

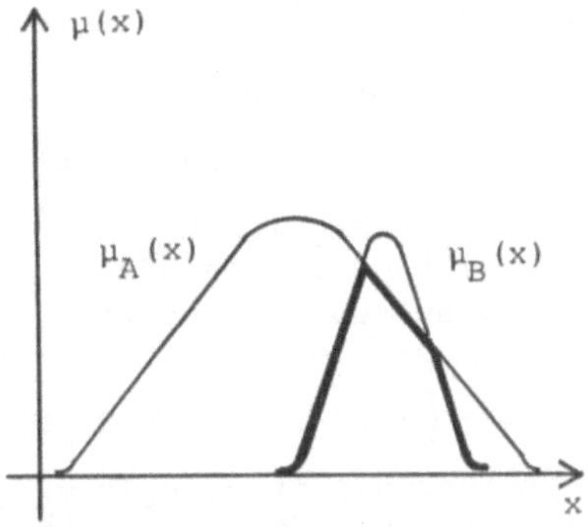


Fig. 3. Intersection

The complement of a fuzzy set $\bar{A} \subseteq X$ is defined as

$$\mu_{\bar{A}}(x) = 1 - \mu_A(x) \qquad \text{for each } x \in X \tag{4}$$

Example: If $X = \{1,2,3,4\}$ and $A = 0.8/1 + 0.6/2 + 0.3/3 + 0.1/4$, then $\bar{A} = 0.2/1 + 0.4/2 + 0.7/3 + 0.9/4$.

Graphically, the complement may be illustrated as in Fig. 4.

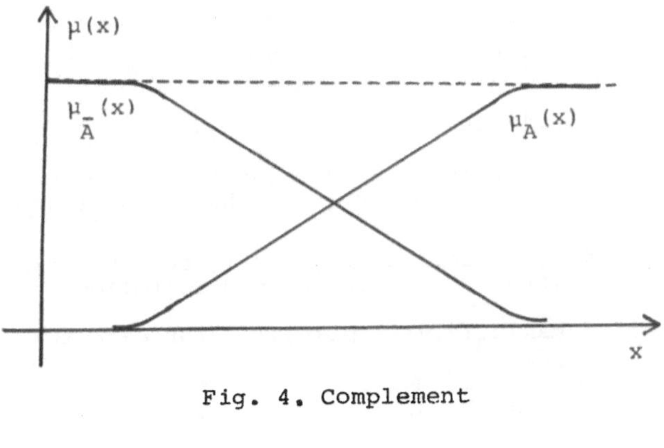


Fig. 4. Complement

These operations were originally defined by Zadeh, and many more were proposed later. Their definitions and properties may be found by the interested reader in, i.e., Dubois and Prade (1980), Kaufmann (1975), Mizumoto and Tanaka (1981) and Mizumoto (1981).

Let us conclude this section by pointing out a problem crucial for the theory of fuzzy sets, i.e. the adequacy of basic operations on fuzzy sets. This problem is still somewhat open and several approaches were proposed for studying it. Some exposition of these approaches may be found by the interested reader in, e.g., Bellman and Giertz (1973), Fung and Fu (1975), Rödder (1975), Yager (1980), and Kacprzyk (1983).

## 4. FUZZY RELATIONS

The concept of a relation plays a key role in mathematics. The same is true for a fuzzy relation in fuzzy mathematics. For the sake of simplicity the exposition will be restricted here to fuzzy binary relations.

Given two (nonfuzzy) universes X and Y, a (binary) <u>fuzzy relation</u> R is a fuzzy set in the Cartesian product X × Y, hence

$$R = \left\{ (\mu_R(x,y)/(x,y) \right\}, \text{ for each } (x,y) \in X \times Y \qquad (5)$$

The membership grade $\mu_R(x,y)$ may be considered an estimated value of the strength of the link between x and y.

Example. If $X = \{\text{John, Paul, Ronald}\}$ and $Y = \{\text{Richard, Jim}\}$, the fuzzy relation R labelled "resemblance" may be, e.g., defined as follows:

R = 0.5/(John, Richard) + 0.4/(John, Jim) + 0.7/(Paul, Richard) + 0.3/(Paul, Jim) + 0.9/(Ronald, Richard) + + 0.1/(Ronald, Jim).

Any fuzzy relation (in a finite X × Y) may be represented in a matrix form. The following matrix corresponds to the above relation "resemblance"

| | Richard | Jim |
|---|---|---|
| John | 0.5 | 0.4 |
| Paul | 0.7 | 0.3 |
| Ronald | 0.9 | 0.1 |

As ordinary relations, fuzzy relations can be composed, and the most important composition is the so-called <u>max-min composition</u>.

Given two fuzzy relations $R \subsetneq X \times Y$ and $S \subsetneq Y \times Z$, such a composition, written R o S, is defined as follows

$$\mu_{RoS}(x,z) = \max_{y \in Y}(\mu_R(x,y) \wedge \mu_S(y,z)), \text{ for each } x \in X,\ z \in Z \qquad (6)$$

Example. If $X = \{1,3\}$, $Y = \{2,4,6\}$, $Z = \{1,2,3\}$,

$$R = \begin{bmatrix} 0.2 & 0.3 & 0.1 \\ 0.5 & 0.9 & 0.6 \end{bmatrix} \quad \text{and} \quad S = \begin{bmatrix} 0.7 & 0.5 & 0.8 \\ 0.4 & 0.2 & 0.6 \\ 0.1 & 1 & 0.3 \end{bmatrix}$$

then

$$R \circ S = \begin{bmatrix} 0.3 & 0.2 & 0.3 \\ 0.5 & 0.6 & 0.6 \end{bmatrix}$$

Fuzzy binary relations satisfy a set of properties in an analogous way as nonfuzzy finary relations do. For example, a fuzzy binary relation $R \subseteq X \times X$ is said to be:

(i) reflexive iff $\mu_R(x,x) = 1$, for each $x \in X$;

(ii) symmetric iff $\mu_R(x_1,x_2) = \mu_R(x_2,x_1)$, for each $x_1,x_2 \in X$;

(iii) min-transitive iff $\mu_R(x_1,x_3) \geq \mu_R(x_1,x_2) \wedge \mu_R(x_2,x_3)$, for each $x_1,x_2,x_3 \in X$.

A fuzzy binary relation satisfying (i) - (ii) is called a <u>proximity relation</u>. A proximity relation satisfying (iii) is called a <u>similarity</u>.

Fuzzy (binary) relations play the same fundamental role in decision analysis under fuzziness as nonfuzzy (binary) relations in decision making in the conventional (nonfuzzy) settings, e.g., for preference modelling. Hence fuzzy partial ordering, preordering, etc. are defined and used to find, e.g., nondominated sets of elements in ordered structures. The interested reader may find an exhaustive exposition of such topics in, e.g., Ovchinnikov (1981) and Orlovsky (1978). For a detailed discussion of fuzzy relations, see Kaufmann (1973, 1975).

## 5. THE EXTENSION PRINCIPLE AND FUZZY NUMBERS

The extension principle, introduced by Zadeh (1975), is one of the most important and powerful tools in fuzzy sets theory. It addresses the following fundamental problem: if there is some relationship between nonfuzzy entities, what is its equivalent between fuzzy entities? Owing to this principle, models and algorithms involving nonfuzzy variables can be extended to the case of fuzzy variables.

The principle may be so stated: given some fuzzy sets $A_1 \subseteq X_1,\ldots,A_n \subseteq X_n$ and a (nonfuzzy) function $f\colon X_1\times\ldots\times X_n \to Y$, $y = f(x_1,\ldots,x_n)$, the fuzzy image $B \subseteq Y$ of $A_1,\ldots,A_n$, through $f$, has the following membership function

$$\mu_B(y) = \max_{\substack{(x_1,\ldots,x_n)\in A_1\times\ldots\times A_n:\\ y = f(x_1,\ldots,x_n)}} \bigwedge_{i=1}^{n} \mu_{A_i}(x_i), \text{ for each } y\in Y \qquad (7)$$

where the Cartesian product $A_1\times\ldots\times A_n$ is defined as

$$A_1 \times \ldots \times A_n = \int_{X_1 \times \ldots \times X_2} \min(\mu_{A_1}(x_1), \ldots, \mu_{A_n}(x_n)/(x_1, \ldots, x_n)$$

Example. If $X_1 = \{1,2,3\}$, $X_2 = \{1,2,3,4\}$, f is addition, i.e. $y = x_1+x_2$, $A_1=0.1/1+0.6/2+1/3$ and $A_2=0.6/1+1/2+0.5/3+0.1/4$, then

$$B = A_1+A_2 = 0.1/2+0.6/3+0.6/4+1/5+0.5/6+0.1/7$$

Notice that we use here "+" both in the arithmetic and set-theoretic sense.

The proposed example pertains to a very important application area of the extension principle, i.e. to real algebra. In fact a composition law "*" in the set of real numbers R can be extended, according to (7) to a composition law "$\circledast$" in the set of fuzzy numbers.

<u>A fuzzy number</u> is defined as a normal and convex fuzzy set $A \subsetneq R$, i.e. a fuzzy set satisfying the two following properties:

i) $\mu_A(x) = 1$, for at least one $x \in R$

ii) $\mu_A[\lambda x_1 + (1-\lambda)x_2] \geqslant \mu_A(x_1) \wedge \mu_A(x_2)$, for each $x_1, x_2 \in R$, and $\lambda \in [0,1]$

Then, if A and B are two fuzzy numbers, $A \circledast B$ is defined as

$$\mu_{A \circledast B}(z) = \max_{x * y=z} (\mu_A(x) \wedge \mu_B(x)), \text{ for each } z \in R \tag{8}$$

Thus, for example, if $A,B \subsetneq R$ are two fuzzy numbers with respective membership functions $\mu_A(x)$ and $\mu_B(x)$, the four basic extended arithmetic operations, i.e., addition, subtraction, multiplication and division give, for each $x,y,z \in R$, the following results:

$$\mu_{A \oplus B}(z) = \max_{x+y=z} (\mu_A(x) \wedge \mu_B(y)) \tag{9}$$

$$\mu_{A \ominus B}(z) = \max_{x-y=z} (\mu_A(x) \wedge \mu_B(y)) \tag{10}$$

$$\mu_{A \odot B}(z) = \max_{x \cdot y=z} (\mu_A(x) \wedge \mu_B\ y)) \tag{11}$$

$$\mu_{A \ominus B}(z) = \max_{\substack{x/y=z \\ y \neq 0}} (\mu_A(x) \wedge \mu_B(y)) \tag{12}$$

An exhaustive treatment of all extended operations and their properties may be found in Dubois and Prade (1978, 1979, 1980) who also suggest an efficient approach for computing the membership grades of the resulting fuzzy members. It consists in assuming the fuzzy numbers to be given in a standard form, i.e. the so-called L-R representation, characterized by three parameters. All the extended operations are performed only on these parameters.

Formally, a fuzzy number A is said to be an <u>L-R type</u> fuzzy number iff:

$$\mu_A(x) = \begin{cases} L((a - x)/\alpha) & \text{for } x \leqslant a,\ \alpha > 0 \\ R((x - a)/\beta) & \text{for } x > a,\ \beta > 0. \end{cases} \tag{13}$$

L and R are the so-called left and right reference, respectively, while a is the mean value of A and $\alpha$ and $\beta$ are called the left and right spreads, respectively.

Symbolically, we write

$$A = (a, \alpha, \beta)_{LR}$$

and graphically A may look like in Fig. 5.

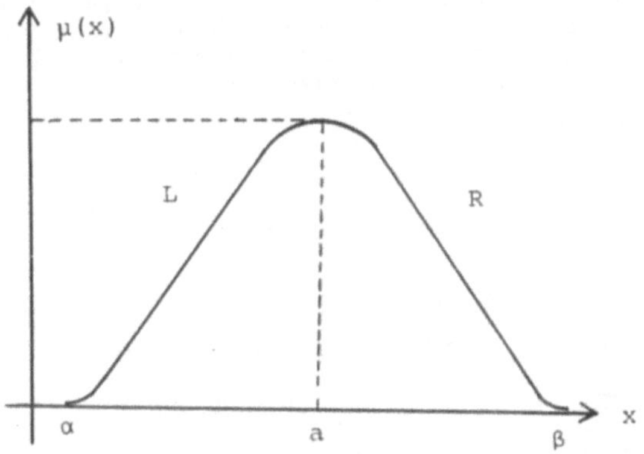


Fig. 5. An L-R type fuzzy number

## 6. LINGUISTIC VARIABLES

As the complexity of a problem increases, the ability of the conventional mathematical tools to precisely yet significantly represent it diminishes, says the <u>principle of incompatibility</u> formulated by Zadeh (1973).

The solution suggested by Zadeh leads to the use of a linguistic description in order to provide a simple but often adequate tool to describe even the most complex situations and to provide an extraordinary information aggregation.

Such an approach, called the <u>linguistic approach</u>, was developed by Zadeh (1973, 1975) starting from the notion of a <u>linguistic variable</u>, i.e. a variable whose values are not numbers but words or sentences in a natural or artificial language. For example "small", "high", "around 4", "a little less than high", are values of linguistic variables.

In fuzzy set theory, the values of linguistic variables are equated with appropriate fuzzy sets. For example, a variable such as "Age" may be viewed as a linguistic variable which takes its values in a so-called universe of discourse, e.g., U = {young, not young, very young, not very young, quite young, old, etc.}, and each of these values may be represented by a fuzzy subset of the universe of discourse $X = [0,150]$.

Generally, as suggested by Zadeh (1981) the values of a linguistic variable such as "Age" can be built up applying the so-called <u>fuzzy hedges</u> together with conjunctions and disjunctions to a set of primary terms, e.g.,"young" and "old". The interested reader may see Zadeh (1975, 1981).

A fundamental problem to be solved in the linguistic approach is how to characterize a relationship (dependence) between linguistic variables.

Usually a <u>fuzzy conditional statement</u> is used. For instance, if L and K are linguistic variables taking on fuzzy values $A \subsetapprox X$ and $B \subsetapprox Y$, respectively, then a dependence between L and K may be given by a fuzzy conditional statement IF(L = A) THEN (K = B), or, shorter, IF A THEN B.

It is usually assumed that

$$\text{IF } A \text{ THEN } B = A \times B \tag{14}$$

i.e. is equated with the Cartesian product A × B being a fuzzy relation.

And similarly, for more complex fuzzy conditional statements

$$\text{IF } A \text{ THEN } B \text{ ELSE } C = A \times B \times A^{-} \times C \tag{15}$$

$$\text{IF } A_1 \text{ THEN } B_1 \text{ ELSE IF } A_2 \text{ THEN } B_2 \text{ ELSE } \dots \text{ ELSE IF}$$

$$A_n \text{ THEN } B_n = A_1 \times B_1 + A_2 \times B_2 + \dots + A_n \times B_n \tag{16}$$

An immediate problem associated with the use of fuzzy conditional statements is: if L takes on a value, what is the value of K implied by the dependence between L and K?

The answer gives the <u>compositional rule of inference</u>: if $R \subsetapprox X \times Y$ is a fuzzy relation representing a dependence between L and K (a fuzzy conditional statement), L is taking on a value A, then the induced value of K is

$$\mu_B(y) = \max_{x \in X} (\mu_A(x) \wedge \mu_R(x,y)) \text{ for each } y \in Y \tag{17}$$

which evidently corresponds to the max-min composition (7).

Example. Let the fuzzy conditional statement be

IF (L is "low") THEN(K is "high") = IF("low") THEN("high")

= ("low") × ("high")

which, for "low" = 1/1 + 0.7/2 + 0.3/3 and "high" = 0.2/1 + 0.5/2 + 0.8/3 + 1/4, corresponds to

$$R = (\text{"low"}) \times (\text{"high"}) = X \begin{array}{c} \\ 1 \\ 2 \\ 3 \end{array} \overset{\displaystyle y}{\begin{array}{cccc} 1 & 2 & 3 & 4 \\ \end{array}} \begin{bmatrix} 0.2 & 0.5 & 0.8 & 1 \\ 0.2 & 0.5 & 0.7 & 0.7 \\ 0.2 & 0.3 & 0.3 & 0.3 \end{bmatrix}$$

If now L = "medium" = 0.5/1 + 1/2 + 0.5/3, then

K = ("medium") o R = 0.2/1 + 0.5/2 + 0.7/3 + 0.7/4.

For more details, see the source papers by Zadeh (1973, 1975).

## 7. POSSIBILITY DISTRIBUTIONS

Zadeh (1978) in his seminal paper wrote: "... the mathematical apparatus of the theory of fuzzy sets provides a natural basis for the theory of possibility, playing a role which is similar to that of measure theory in relation to the theory of probability...".

Even if Zadeh's concept of possibility does not state a clear difference between probability theory and fuzzy sets theory, it undoubtly offers some tools to represent most of the imprecision intrinsic in human decision processes.

Let us start with the following non-fuzzy proposition p:

$p \triangleq u$ is an integer in the interval [1,6],

which asserts that it is possible for any integer in the interval [1,6] to be a value of u.

Without any other information it seems natural to say that p induces a possibility distribution which is uniform, i.e. the possibility values are defined as

$$\text{Poss}\{u=x\} = \begin{cases} 1 & \text{for} \quad 1 \leqslant x \leqslant 6 \\ 0 & \text{for} \quad x < 1 \text{ or } x > 6 \end{cases}$$

Here Poss$\{u=x\}$ means "the possibility that u may assume the value x" and usually it is also written as

$$\text{Poss}\{u = x\} \triangleq \pi_u(x) \tag{18}$$

Now, let us fuzzify the proposition p in this way:

$q \triangleq X$ is a small integer,

where "small integer" may be considered as a label of a fuzzy set, for example

"small integer" = 1/1 + 0.9/2 + 0.7/3 + 0.5/4 + 0.3/5+0.1/6.

In this case we may write:

$$\mathrm{Poss}\{u = 1\} = 1$$
$$\mathrm{Poss}\{u = 2\} = 0.9$$
$$\mathrm{Poss}\{u = 3\} = 0.7$$
$$\mathrm{Poss}\{u = 4\} = 0.5$$
$$\mathrm{Poss}\{u = 5\} = 0.3$$
$$\mathrm{Poss}\{u = 6\} = 0.1$$
$$\mathrm{Poss}\{u = x\} = 0 \quad \text{for} \quad x < 1 \text{ or } x > 6.$$

More formally, if $u$ is a variable which takes values in $X$ and $A$ is a fuzzy subset of $X$, the proposition

$$q \triangleq X \text{ is } A \qquad (19)$$

induces a possibility distribution $\Pi(X = u)$ which is equal to $A$, i.e.

$$\Pi(u{=}x) = \pi_u(x) = \mu_A(x), \quad \text{for each } x \in X \qquad (20)$$

Given $\pi_u$, the possibility for $x$ to belong to a non-fuzzy set $E$ is defined as

$$\Pi(u \in E) = \sup_{x \in X} \pi_u(x) \qquad (21)$$

According to Dubois and Prade (1979) we could say that such a definition is consistent with our intuition of the possibility of any one of several events ($\exists x \in E,\ u = x$) as the possibility of the most possible one.

Starting from the concept of possibility distribution, we can define, in the framework of possibility theory, something analogous to the probability measure in the framework of probability theory, i.e. a possibility measure.

In fact, if the normality condition $\sup_{x \in X} \pi_u(x) = 1$ is satisfied, a possibility measure is defined as a function $\Pi : P(X) \to [0,1]$, such that:

i) $\pi(\emptyset) = 0, \quad \pi(X) = 1$

ii) $\pi(\bigcup_i A_i) = \sup_i \pi(A_i)$, for any collection $A_i$ of subsets of $X$

Zadeh (1978) points out that it seems natural to think that "what is possible may not be probable and what is improbable need not to be impossible". This principle, known as the <u>consistency principle</u>, gives rise to a deep discussion and a consequent portion of papers which aimed at developing a better understanding of the interplay between possibility and probability.

Details on this problem may be found, e.g., in Dubois and Prade (1980), Hisdal (1982), Höhle and Klement (1984), Lindley (1982), Nahmias (1978), Prade (1979).

## 8. CONCLUDING REMARKS

The aim of this introduction was to make the interested readers, and especially those not yet exposed to fuzzy sets, familiar with the relevant elements of fuzzy sets and possibility theory. We hope that the exposition was facilitated by the illustrative examples which should also help the readers find possible applications of fuzzy tools in their specific areas of work.

## REFERENCES


Bandler, W., and L. Kohout (1980). Fuzzy power sets and fuzzy implication operators. Fuzzy Sets and Syst. 4, 13-30.

Bellman, R.E., and M. Giertz (1973). On the analytic formalism of the theory of fuzzy sets. Inf. Sci. 5, 149-157.

Chapin, E.W. (1971). An axiomatization of the set theory of Zadeh. Not. Amer. Math. Soc. 68, 702-704.

Dubois, D., and H. Prade (1978). Operations on fuzzy numbers. Int. J. Syst. Sci. 9, 613-626.

Dubois, D., and H. Prade (1979a). Fuzzy real algebra: some results. Fuzzy Sets and Syst. 2, 327-348.

Dubois, D., and H. Prade (1979b). Outline of fuzzy set theory. In M.M. Gupta, R.K. Ragade, and R.R. Yager (eds.), Advances in Fuzzy Set Theory and Applications. North-Holland, Amsterdam.

Dubois, D., and H. Prade (1980). Fuzzy Sets: Theory and Applications. Academic Press, New York.

Fung, L.W., and K.S. Fu (1975). An axiomatic approach to rational decision-making in a fuzzy environment. In L.A. Zadeh, K.S. Fu, K. Tanaka, M. Shimura (eds.); Fuzzy Sets and their Applications to Cognitive and Decision Processes. Academic Press, New York, 227-256.

Gaines, B.R. (1976). Foundations of fuzzy reasoning. Int. J. Man-Machine Stud. 8, 623-668.

Goguen, J.A. (1967). L - fuzzy sets. J. Math. Anal. and Appl. 18, 145-174.

Hisdal, E. (1982). Possibilities and probabilities. In Proc. of the Second World Conference on Mathematics at the Service of Man. Las Palmas, The Canaries.

Höhle, U., and E.P. Klement (1984). Plausibility measures. A general framework for possibility and fuzzy probability measures. In H.J. Skala, S. Termini, and E. Trillas (eds.), Aspects of Vagueness. D. Reidel, Dordrecht.

Kacprzyk, J. (1983). Multistage Decision-Making under Fuzziness. Verlag TÜV Rheinland, Cologne.

Kaufmann, A. (1973). Introduction a la Theorie des Sous-ensembles Floues. Vol. 1: Elements Theoriques de Base. Masson, Paris.

Kaufmann, A. (1975). Introduction to the Theory of Fuzzy Subsets. Vol. 1: Fundamental Theoretical Elements. Academic Press, New York.

Lindley, D. (1982). Scoring rules and inevitability of probability. Int. Statist. Review 50, 1-26. (With discussion among others by L.A. Zadeh).

Mizumoto, M., and K. Tanaka (1981). Fuzzy sets and their operations, Part I. Inf. and Control 48, 30-38.

Mizumoto, M. (1981). Fuzzy sets and their operations, Part II. Inf. and Control 50, 160-174.

Nahmias, S. (1978). Fuzzy variables. Fuzzy Sets and Syst., 97-110.

Novak, V. (1980). An attempt at Gödel - Bernays - like axiomatization of fuzzy sets. Fuzzy Sets and Syst. 3, 323-326.

Ovchinnikov, S.V. (1981). Structure of fuzzy binary relations. Fuzzy Sets and Syst. 6, 169-195.

Orlovsky, S.A. (1978). Decision-making with a fuzzy preference relation. Fuzzy Sets and Syst. 1, 155-168.

Prade, H. (1979). Nomenclature of fuzzy measures. In E.P. Klement (ed.), Proc. of International Seminar on Fuzzy Set Theory. Johannes Kepler Universität, Linz, Austria.

Rödder, W. (1975). On "and" and "or" Connectives in Fuzzy Set Theory. Working Pap. 75/07. RWTH Aachen (FRG).

Shafer, G. (1976). A Mathematical Theory of Evidence. University Press, Princeton.

Yager, R.R. (1980). On a general class of fuzzy connectives. Fuzzy Sets and Syst. 4, 235-242.

Zadeh, L.A. (1965). Fuzzy sets. Inf. and Control. 8, 338-353.

Zadeh, L.A. (1973). Outline of a new approach to the analysis of complex systems and decision processes. IEEE Trans. on Syst., Man and Cybern. SMC - 3, 28-44.

Zadeh, L.A. (1975). The concept of a linguistic variable and its application to approximate reasoning. Inf. Sci. Part I: 8, 199-249; Part II: 8, 301-357; Part III: 9, 43-80.

Zadeh, L.A. (1978). Fuzzy sets as a basis for a theory of possibility. Fuzzy Sets and Syst. 1, 3-28.

Zadeh, L.A. (1981). PRUF - a meaning representation language for natural languages. In E.H. Mamdani, and B.R. Gaines (eds.); Fuzzy Reasoning and its Applications. Academic Press, London, 1-66.

Zadeh, L.A. (1983). A computational approach to fuzzy quantifiers in natural languages. Comput. and Math. with Appl. 9, 149-184.

# INTRODUCTION TO DECISION MAKING UNDER VARIOUS KINDS OF UNCERTAINTY

Thomas Whalen
Decision Sciences Department
Georgia State University
Atlanta, GA 30303, USA

Abstract. All interest in decision making processes stems from uncertainty: when we are certain what course of action is best, we simply perform it without further consideration. Unfortunately, many obstacles exist which can prevent us from having this ideal certainty. Section 1 of this chapter surveys some of these obstacles, together with the paradigmatic problems that arise from each obstacle in its pure form. Section 2 presents a unifying framework, the general multiple facet optimization problem, which exploits some important isomorphisms among these problems. In Section 3, a fairly simple two-stage decision problem is viewed at several different levels of information, ranging from a mere incomplete ordering to an approximate statistical specification, in order to illustrate a number of different techniques for decision making that have been developed for the various levels of information. Finally, Section 4 summarizes the results presented in this chapter and discusses some promising areas for future research.

Keywords: uncertainty, decision making under uncertainty, optimization

## 1. OBSTACLES TO CERTAINTY

In order to know for certain what to do, we must satisfy three conditions. First, we must comprehend all of the alternative courses of action from which we can choose. Second, we must know all the consequences of each alternative course of action. Third, we must know which set of consequences is preferable to any other achievable set.

### 1.1. Uncertainty About Alternative Courses of Action

Comprehension of the set of alternative courses of action can be limited in three ways: failure of imagination, immensity of choice, and imprecision of specification.

Failure of imagination simply means that relevant alternative courses of action exist which we are unaware of. The advance of technology provides a simple example of this. Engineering design choices that not so long ago were limited to choosing between metal and wood construction are now enriched, but also complicated, by the availability of many new plastics and ceramics. Unless we can be certain that all possible alternative courses of action have been enumerated, we cannot be certain that the one we select is indeed optimal.

Sometimes it is possible to specify all available choices in an abstract (intensive) way, but the resulting set is too large to be extensively listed, or at least too large to be exhaustively evaluated. When this immense set of alternative courses of action can be represented as a continuum of real numbers or vectors, there are many well-known tools such as mathematical programming to proceed more or less efficiently to an optimal solution. However, in other cases the large number of alternative courses of action is due to a combinatoric explosion rather than a real continuum. To take an example widely advertised to be unsolvable in the remaining lifetime of the universe, cracking a public-key cryptogram requires selecting correctly from a set of pairs of very large prime numbers. If such combinatoric problems are to be solved at all, heuristic methods of search must be used. These heuristics typically do not afford proofs of optimality, so a decision made in this manner is uncertain.

When a foreman on a job site decides which order to give to a laborer, the foreman knows what the demanded course of action is; the laborer may perform well or badly, but the foreman's decision alternatives are clear. On the other hand, when a senior executive chooses a policy directive for a large business, government, or voluntary organization, the policy must be interpreted and fleshed out by successive layers of intermediate decision makers before it is eventually carried out by the operative personnel (Dimitrov and Driankova, 1977). Thus, the senior executive does not really know exactly what it means to choose one policy rather than another; the policy decision is by nature fuzzy, and thus uncertain.

The process of limiting and coping with uncertainty in the set of alternative courses of action has not generally been the focus of paradigms for decision making under uncertainty. As part of Simon's (1977) "intelligence" phase of problem solving, the requirements for this process vary more significantly from problem to problem than uncertainty about consequences or about preferences.

## 1.2. Uncertainty About Consequences

When we cannot predict with certainty what outcome will follow from a given course of action, we usually model this situation using the concept of "states of the world". (For an alternative approach see Fishburn, 1960). We hypothesize that the outcomes of our actions depend on two things: on which course of action we select and on the current values of one or more variables called "state variables". If we knew the values of the state variables, we would know the outcomes of each alternative course of action; if we do not know these values for certain, we must make an uncertain choice.

Much analysis and specific background knowledge of the domain of the decision in question are necessary to enumerate the relevant set of states of the world. Once these are enumerated, the next step is to marshal whatever information is available regarding the relative likelihood of these states. Several levels of information have been studied. The lowest level of information we shall consider is when the states of the world are specified but no information about their relative degree of possibility or probability is known. With more information we reach the second level, in which some states of the world are known to be more possible than others (incomplete order); a third level is reached when states can be put in a complete weak order from most to least possible, so that for any two states we can either say which one is more possible than the other or else we can say that they are of exactly equal possibility. The fourth level of information is when we can specify approximate statistical probabilities for all states of the world using fuzzy real numbers, and the fifth level is when we can specify the probability distribution over states exactly, using (crisp) real numbers. Game theory can be viewed as a sixth level of information, in which our opponent's actions, while unknown in advance, will be determined by our own actions and the payoff structure of the game.

Because the nature and amount of information about the relative possibilities of states of the world that can be usefully applied to decision making depends strongly on the nature and amount of information about preferences that is available, the paradigmatic problems for each of the above levels will be discussed in the context of uncertainty about preferences.

## 1.3. Uncertainty About Preferences

The most generally accepted view of preferences among economists is that utilities are measurable by a complete weak order. In other words, an individual will always either be able to specify one of a pair of outcomes as better than the other, or else be strictly indifferent between the two. In this view, it is meaningless to assign numbers to the utilities of outcomes, and hence no arithmetic can be performed on them. From an information content point of view, it is clearly equivalent to talk about ordinal gains, in which the best outcome ranks first, and ordinal losses, in which the worst outcome ranks first. A more sophisticated view of ordinal utilities postulates that it is not a static position that is valued, but rather the gain or loss between a prior position and a subsequent position. On this basis, well substantiated by studies of human behavior, it is possible to talk about an ordinal theory of regrets in the context of decision making under uncertainty. The regret associated with a particular (state-action) pair is defined by the difference between the outcome of that particular (state - action) pair and the outcome of the best possible action for that particular state.

A well-established minority view, however, holds that meaningful numeric measures of an individual's utilities for outcomes can be generated. The most sophisticated varieties of this theory derive from the work of von Neumann and Morgenstern (1947). In these approaches, utility is measured on an interval scale anchored by specific, context-dependent "best" and "worst" outcomes, and utilities for intermediate outcomes are determined by betting preferences. More recently, work has been done using fuzzy numbers rather than crisp numbers to represent these utilities (Watson, Weiss and Donnell, 1979; Freeling, 1980); this can be a very valuable way to handle the fact that some of the hypothetical choices between bets are much easier than others in the von Neumann - Morgenstern methodology. Fuzzy utilities come into play even more directly when the outcomes themselves are only vaguely known in advance. When utilities are measured by crisp or fuzzy real numbers, it is possible to compute regrets by subtracting the utility of the outcome of each (state - action) pair from the utility of the best possible action for that particular state.

In the following two subsections, we examine in detail some of the paradigmatic problems which arise from specific combinations of information about states of the world and about utility; first we examine cases that arise when utility is ordinal, then cases which require crisp or fuzzy real numbers to measure utility.

### 1.3.1. Ordinal Utilities

NO RELATIVE POSSIBILITY INFORMATION. When we have no information about the relative likelihood of the various states of the world, we must make our decision on the basis of the utilities of the outcomes of the various (state - action) pairs together with a fundamental choice of philosophies. The "optimistic" philosophy in such a situation is to choose the course of action whose best possible outcome is better than that of any other (maximax algorithm). The "pessimistic" philosophy, on the contrary, seeks to cut losses by choosing the course of action whose worst possible outcome is better (or less bad) than the worst possible outcome of any other course of action (minimax loss algorithm).

The minimax regret approach steers a course between the extremes of optimism and pessimism. Outcomes are ordered in terms of regret rather than actual gains or losses, and that course of action is selected for which the worst possible regret is less bad than the worst possible regret for any other course of action. This approach has the effect of focusing our attention primarily on those states of the world for which our choices have the greatest effect, whereas minimax focuses on the most dangerous states of the world and maximax on the most promising ones.

ORDINAL POSSIBILITIES. If only ordinal information about utilities is available, then whatever information is available about the relative possibility or probability of the various possible states of the world is also most appropriately expressed in an ordinal manner. The Commensurate Ordinal Decision Analysis algorithm (Whalen, 1984a) uses two distinct ordinal scales, one for disutility (loss or regret) and the other for possibility. These scales define three L-fuzzy sets (Goguen, 1967): the set of poor outcomes, the set of possible states of the world, and the set of risky exposures. An "exposure" is an ordered pair consisting of an outcome and the state of the world in which that outcome occurs; its membership in the set of risky exposures is also defined by an ordered pair consisting of the poorness of the outcome and the possibility of the state.

The inputs to the algorithm are a (complete or incomplete) rank ordering of the poorness of all possible outcomes, and a separate rank ordering of the possibilities of all possible states of the world. It is also necessary to specify, by means of a decision tree in normal form, which outcomes go with which states of the world and which strategies. The algorithm then automatically determines the fuzzy set of risky exposures and uses this to eliminate suboptimal strategies using a series of dominance criteria which are successively more powerful but less robust. These criteria, discussed below in turn, are complete dominance, global riskiest-states dominance, and pairwise riskiest-states dominance. Typically, commensurate ordinal risk minimization alone will not be sufficient to narrow the range of alternative strategies to just one, but it can be very useful as a preliminary screen. Given the results of a commensurate ordinal decision analysis, we are better prepared to seek additional information about those states and actions identified as critical or to use informal/intuitive methods to pick a final course of action from the "short list".

Complete Dominance. A strategy $\alpha$ is completely dominated by another strategy $\alpha'$ if for all possible states of the world , the disutility $D(\alpha|\sigma)$ arising from strategy $\alpha$ when $\sigma$ is the actual state of the world is worse than or equal to the disutility $D(\alpha'|\sigma)$ arising from strategy $\alpha'$ in the same state of the world $\sigma$, and the inequality is strict for at least one $\sigma$. This is essentially the Pareto rule; a strategy is dominated by another if it is possible to improve one criterion of the outcome without worsening any other criterion. The different criteria in this case are the conditional outcomes given the different possible states of the world. Note, however, that mixed strategies are undefined when utilities are ordinal; a strategy can only be dominated by a specific other strategy, not by a convex combination of two or more as is possible in numerical utility theory.

Global Riskiest-States Dominance. For each alternative strategy $\alpha$, let $R_\alpha$ equal (The set of all $\sigma$ such that $(p(\sigma') > p(\sigma))$ implies $(D(\alpha|\sigma') < D(\alpha|\sigma))$ . For any strategy $\alpha$, $R_\alpha$ is the (nonfuzzy) set of states $\sigma$ such that, if another state $\sigma'$ is more possible, then the outcome of $\alpha$ when state $\sigma'$ is in effect is less poor than the outcome of $\alpha$ under state $\sigma$. $R_\alpha$ is referred to as the set of riskiest states for strategy $\alpha$, since any state not in $R_\alpha$ is either less likely or leads to a less poor outcome for $\alpha$ than any state in $R_\alpha$.

Let $R = \cup_\alpha(R_\alpha)$, the set of states of the world which belong to the set of riskiest states for any strategy. $R^c$, the set of states not in R, is thus the set of states which are neither very possible nor ever very poor regardless of what strategy is selected.

The "global riskiest-states dominance criterion" is evaluated by deleting the states in $R^c$ from consideration and eliminating any strategies which are completely dominated on just those states in R. A strategy $\alpha$ is global-riskiest-states dominated by another strategy $\alpha'$ if $D(\alpha|\sigma) > D(\alpha'|\sigma)$ for all $\alpha$ in R and the inequality is strict for at least one $\sigma$ in R. In effect, we are saying that $\alpha$ completely dominates $\alpha$ if we ignore the "unimportant" states of the world in $R^c$.

Pairwise Riskiest-States Dominance. For each pair of alternative strategies $\alpha$ and $\alpha'$, let $R_{\alpha\alpha'}$ be the set union of $R_\alpha$ and $R_{\alpha'}$, the set of states of the world which are in the riskiest set for either of the two strategies $\alpha$ and $\alpha'$. Then strategy $\alpha$ is pairwise riskiest-states dominated by strategy $\alpha'$ if $D(\alpha|\sigma) > D(\alpha'|\sigma)$ for all $\sigma$ in $R_{\alpha\alpha'}$, and the inequality is strict for at least one $\sigma$ in $R_{\alpha\alpha'}$. The argument in this case is that two strategies can be compared taking into consideration only those states which are risky ones for one or the other action, ignoring any states which may be risky for some extraneous third alternative as well as the unimportant states in $R^c$.

Clearly, any strategy which is completely dominated is also dominated according to the global riskiest-states criterion and any strategy which is dominated according to the latter is also dominated according to the pairwise riskiest-states criterion. Nevertheless, it is useful to know the most robust criterion under which a specific strategy can be eliminated, since each of the three criteria differs from its predecessor by making stronger assumptions and discarding more information as "unimportant".

The assumptions of the L-Fuzzy Risk Minimization algorithm (Whalen, 1980) differ from those of the Commensurate Ordinal Decision Analysis algorithm by allowing, on one hand, direct comparisons between the grade of membership of an outcome in the set of bad outcomes and, on the other, the grade of membership of a state of the world in the set of possible states. The riskiness of an exposure is equal to the minimum of the poorness of the relevant outcome and the possibility of the corresponding state of the world. As in the Commensurate Ordinal Decision Analysis algorithm, the incompletely ordered lattice structure of the L-fuzzy risk minimization algorithm allows many comparisons to remain undefined, concentrating our attention on just those few comparisons which actually affect the course of the decision making process. Furthermore, the user has the option of refusing to make any given requested comparison. In this case, the algorithm continues to pass through the decision tree, and in many instances, the difficult comparison which the user has declined to make can be rendered most by further analysis. If the user's refusals to make a final solution impossible, the algorithm will identify several alternative unresolved pairs of memberships such that at least one of these difficulties must be resolved by the user before analysis can continue. Symbolically, the strategy selected by L-Fuzzy risk minimization is the one for which $\max[\min[D(\alpha|\sigma), p(\sigma)]$ is least.

FUZZY OR CRISP NUMERIC PROBABILITIES. When utilities are known on an ordinal scale only, there is no meaningful way to weight them with numeric probabilities. If probabilities are known, they should be rescaled to ordinal possibilities, perhaps after applying a cutoff to eliminate extremely unlikely states of the world.

GAME THEORY. The literature of game theory will not be reviewed in detail here; in brief, ordinal payoff information allows us to evaluate only pure strategies. A pure strategy is one in which, if identical circumstances occur repeatedly, we will predictably take the same course of action each time rather than attempting to keep our opponent guessing.

### 1.3.2. Fuzzy or Crisp Numeric Utilities

When the utility of the outcome of each alternative course of action under each possible state of the world is specified by a real number, we can combine these utilities with numeric probability measures to compute expected values and choose of action for which the expected value is best. If the utilities and/or the probabilities are only known aproximately, we can represent them as fuzzy numbers and calculate fuzzy expected utilities by the extension principle of fuzzy mathematics; this process reduces to ordinary arithmetic when the operands are crisp.

NO RELATIVE POSSIBILITY INFORMATION. With numeric utilities, minimax loss and maximax gain approaches are simple matters of numeric comparisons, while the regret measures needed for the minimax regret approach can be found by subtracting the utility of each outcome from the best utility obtainable in the relevant state of the world. Another approach, unique to the situation with numeric utilities and no information about relative possibilities, is the maximum entropy approach. In this approach, we treat all possible states of the world as equally probable in the absence of information to the contrary; operationally, this means simply taking the average across states of the world of the utilities which might arise from each alternative course of action, and choosing that course of action for which this average is best.

ORDINAL POSSIBILITIES. If by "ordinal possibilities" we mean only that some possibilities are known to be greater than others, there is little advantage to combining this information with fuzzy or crisp numeric measures of utilities. However, if we also know just a little more, for instance that one state has a probability of more than .5 or that state $\sigma'$ is more than three times as likely as state $\sigma$, these constraints allow meaningful bounds to be placed on the expected value of the outcome of each alternative course of action. Smith's (1980) "textured sets" approach demonstrates how linear programming techniques can be used to find the maximum and minimum possible expected values of each alternative course of action subject to linear constraints on the probabilities of the possible states of the world. Any course of action whose maximum expected utility is less than the minimum expected utility of another can then be eliminated from further consideration.

FUZZY OR CRISP NUMERIC UTILITIES. The most commonly discussed technique for decision making under uncertainty is statistical decision analysis (Raiffa, 1968). In this technique, the imperfect information about the state of the world is represented by a probability distribution over the set of such states, and the utility of each strategy given each state of the world is expressed on an interval scale after the manner of von Neumann and Morgenstern (1947). The expected value of each strategy is found by multiplying the corresponding utilities and probabilities and adding the products; the strategy whose utility is greatest (or, equivalently, whose disutility is least) is the one that is chosen.

Sometimes it is possible to specify the utility and probability information required by statistical decision analysis, but only in an approximate way. If the degree of imprecision in the estimates of probability or of utility is relatively small, statistical decision analysis provides for the use of sensitivity analysis, in which the numerical inputs are "perturbed" about their original values and the analysis re-done to see whether the final decision changes.

Fuzzy statistical decision analysis as presented by Watson, Weiss and Donnel (1979) and by Freeling (1980) can be viewed as an extension of sensitivity analysis to the case where the degree and qualitative shape of the imprecision need to be considered throughout the entire analysis of a decision. A major goal of this approach is to represent the imprecision of each value explicitly, and to manipulate these imprecise values in such a way as to determine the degree and nature of the resulting imprecision in the final decision.

In order to accomplish this, fuzzy decision analysis uses linguistic and graphical techniques to elicit probabilities and utilities in the form of fuzzy numbers (Dubois and Prade, 1979). A fuzzy number is a set of numbers, some of whose members are considered to have higher degrees of membership in the set than other members do, along a scale ranging from total membership to total nonmembership. For example, the number 11 has a lower membership in the fuzzy number "around a dozen" than the number 12 does, but a higher membership than the number 10 does.

The "extension principle of fuzzy mathematics" (Zadeh, 1965; Dubois and Prade, 1979) allows any mathematical operations that can be performed on real numbers to be performed on fuzzy numbers as well. Fuzzy statistical decision analysis makes extensive use of this principle to compute a fuzzy number representing the statistical expected value of each alternative course of action given the fuzzy probabilities and utilities in the input. The course of action for which this fuzzy number is highest is chosen; the method also specifies the degree of confidence that this action is actually the best, by measuring the degree to which the highest expected utility is clearly higher than the next-highest as opposed to the degree to which these two fuzzy numbers overlap. (It is in the assessment of confidence that Freeling differs from Watson, Weiss and Donnel).

GAME THEORY. When the payoff table is expressed in terms of numerical utilities, there exist situations in game theory in which our opponent could gain an advantage over us if our behavior were too predictable. In these situations, solving the appropriate linear equations for a maximum payoff to us results in a convex combination of strategies rather than a single strategy. The interpretation of this is that, on each "play" of the game, we should randomly select one of those strategies with a probability proportional to the weight given to that strategy in the mathematical solution to the game equations. A familiar example of this is seen in the strategy of bluffing in the game of poker; a player who is known never to bluff or one who is known always to bluff will do less well than one whose bluffs are random.

## 2. GENERAL MULTIPLE FACET OPTIMIZATION

The above discussion centered around problems which satisfy two important simplifying features: first, the amount of information about states of the world was fixed throughout the course of the decision making process rather than increasing at later decision stages as a result of what is learned at earlier ones; and second, the utility of any single possible outcome was viewed as a unit. We will now relax each of these simplifying assumptions, and state a unified theoretical framework for the resulting broader class of problems.

### 2.1. Multistage Decision Making

A very important and widely-studied class of problems arises when it is possible to break a decision process down into stages so that later decisions are made in the light of information gained in earlier stages of the process. In fact, we may often choose to perform experiments or otherwise take actions designed deliberately to obtain information about the states of the world; typically this information is both imperfect and costly, so that a major part of our burden as decision makers is knowing when to seek information and when to make a substantive decision on the basis of what we already know.

For analytic purposes, however, it is convenient to transform a multistage problem into an equivalent single-stage problem in "normal form" (Raiffa, 1968). A multistage problem can be diagrammed by a decision tree with alternating choice and chance nodes: at each choice node that we encounter in working through the tree we must pick one of several alternative action branches, while at each chance node that we encounter, the unknown state of the world will determine which one of several possible outcome branches we will observe.

The first step to convert the problem into normal form is to define all possible "strategies" for moving through the decision tree. To specify a strategy, begin by selecting one alternative action at the first decision node of the decision tree. This action branch will lead to a chance node, each of whose branches in turn will lead to another choice node. For each of these possible second choice nodes, we must specify what action branch our strategy would dictate, and so on through the tree. The normal form decision tree will have only one choice node, with one branch for each possible strategy derived from the original tree. (A multistage specification of a decision problem and its corresponding tree are called the "extended form" to distinguish them from the normal form specification of the same problem.)

The second step in normalization is to respecify the set of possible states of the world. To do this, we must enumerate all possible combinations that can be formed by selecting one outcome branch from each chance node. Knowledge of the background of the specific problem-situation is essential here to avoid a combinatoric explosion; while the total number of combinations is likely to be unmanageably large, many combinations will be physically impossible because of identity or dependency between the variables being observed at the corresponding chance nodes.

The last step in converting a problem into normal form is to determine the utility of each strategy defined in step 1 under each state of the world defined in step 2. This involves working through the extended form of the tree for each (strategy - state) pair, using the strategy to decide all choice branches and the state of the world to decide all chance branches, and accumulating all the gains and losses associated with the various partial actions and outcomes. The result is a shorter but wider tree; a satisfactory or optimal solution of the structurally simpler normal form of the problem is guaranteed to yield a satisfactory or optimal strategy for traversing the extended form of the problem.

## 2.2. Compound Measures of Utility

The current literature on utility theory devotes much concern to conditions which make numeric utility measurements or even ordinal utility comparisons difficult. These conditions include: multicriterion or multiattribute decision making, in which outcomes are valued along several dimensions; discount theory, in which costs and benefits occur over a long period of time after the decision is made; and social decision making, in which several different stakeholders interests must be respected.

These problems, along with the problem of uncertainty about the state of the world, can be subsumed in to a general mathematical structure, which I will call the general multiple facet decision problem. In this abstract problem, we have a number of possible courses of action to choose from; the value of each strategy depends on a number of different facets, some of which may be more important than others.

In multicriterion or multiattribute decision making, each facet is one of the criteria or attributes that different choices are being judged on, and the relative importance of each facet depends on the importance weight given to that attribute or criterion. The multiple facet approach can be viewed as an extension of multicriterion decision making to situations which have traditionally been viewed as distinct topics.

In discount theory, each facet is the net cost or benefit accruing at a particular point in time, and the relative importance of each facet is the degree of discount to be applied to events at that point in time; the further into the future an event is, the more it is discounted and thus the lower the relative importance of the facet.

In social decision making, the various facets of an alternative course of action are the utility assessments of that course of action by the various interested individuals and groups, and the relative importance of each facet may be associated with the "clout" of each interested party. In a pure democracy, the clout of a facet depends on the number of persons it represents; in other situations, it may mean rhetorical skill, financial resources, or political or military power, depending on the circumstances and mores surrounding the decision making process.

In the problems considered in Section 1 above, the different facets of a given course of action consist of the outcome of that course of action under the different possible states of the world, and the relative importance of each facet depends on the relative possibility or probability of the corresponding state of the world.

Obviously, treating these different decision making problems under a single theoretical framework closely resembling traditional views of multicriterion decision making does not remove the need for considerable situation-specific work in unraveling these and other difficulties in any specific situation. However, recognizing the structural commonalities between the problems will allow any methodological advance in one field to be readily transported to the others.

A fruitful area for future research will be to use the multiple facet approach for problems where two or more of the above sources of complexity interact; for example, many pressing problems in economic and energy policy revolve around social decisions with uncertain outcomes distributed over a long future. When we must combine such fundamentally different kinds of information as social, financial and engineering, the result tends to be less precise than the least precise individual class of information. Thus, the "soft optimization models" discussed here and elsewhere in this book can be of great usefulness.

## 3. EXAMPLE

The following example, adapted from a classic text in statistical decision analysis (Raiffa, 1968), will serve to illustrate the operation of some of the algorithms discussed in Section 1. The problem illustrates the normal form of analysis, with simple utilities dependent on the selected strategy and on the state of the world; as discussed in Subjection 2.2, these algorithms can be extended to other types of multiple facet problems. The algorithms illustrated are: the commensurate ordinal risk minimization algorithm, L-fuzzy risk minimization, and fuzzy statistical decision analysis.

The example concerns an oil wildcatter's decision whether to drill an exploratory well at a new site, and whether to perform a seismic experiment to get additional information about the site before making the drilling decision. The seismic structure of the area, revealed by the experiment (if it is performed) may be "No structure," "Open structure," or "Closed structure," (N,O and C, respectively, in Tables 1-11) and the oil content may be "Dry," "Wet,", or "Soaking." (D, W and S, respectively, in Tables 1-11). The three seismic structures together with the three oil contents generate 9 possible states of the world, ranging from dry-no structure (DN) to soaking-closed structure (SC). Ten strategies are possible: strategy 1 is to do nothing (NO) and strategy 2 is to drill without experimentation (DRILL), while strategies 3 through 10 prescribe drilling only if the experimental outcome is in a particular subset of the three seismic structures. Table 1 shows the dollar profits, in thousands, of each of the ninety possible outcomes generated by pairing each strategy with each state of the world, and the numeric probabilities of each state of the world, taken from the original text.

However, the numbers in Table 1 come from a specific example in a book published in 1968. Changes in prices and technology will have altered the dollar amounts, and the probabilities will be different at a different class of site. Even so, it is reasonable to assume that the ordinal structure of the problem remains the same; if one outcome was more profitable than another in 1968 it is probably more profitable than the other today although the exact ratio between them will have changed, and if a state of the world is more probable than another in one situation it will be more probable than the other in a broader class of similar situations than the class where the two probabilities remain unchanged. Thus, it is appropriate to see what can be deduced from only the ordinal data contained in Table 1, abstracting from its numerical details.

### 3.1. Commensurate Ordinal Decision Making

The first step in analyzing the problem ordinally is to view the outcomes as regrets rather than as profits and losses. In Table 2, this is done in terms of the numeric values given by Raiffa; in Table 3 the more stable ordinal relations among the regrets and among the probabilities are abstracted from the specific numbers appropriate to Raiffa's example. To help us in using the information on the relative possibilities of states, Table 4 shows the same information as Table 3 with the columns representing the more possible states listed before those representing the less possible states, and Table 5 shows the rows representing the ten strategies sorted by regrettability, using regret under the most possible state (Dry-No structure) as the primary sort key, regret under the second-likeliest state as the second key, etcetera.

Examination of Table 5 shows that strategy 10, which is to experiment but drill regardless of the outcome, is worse than strategy 2, to drill without experimenting, regardless of the state of the world. Similarly, strategy 3, to experiment but not drill regardless of the outcome, is always worse than strategy 1, to do nothing. Thus, strategies 10 and 3 are completely dominated. In Table 6, these two strategies are eliminated from further consideration.

Tables 7 through 9 demonstrate the global riskiest-states dominance criterion. In Table 7, the set of riskiest exposures for each strategy are marked with an asterisk. Thus, any exposure not so marked is either less regrettable or less possible (or both) than any marked exposure in its row. At the top of Table 7, states which are in the set of riskiest states for any strategy are also marked with an asterisk. These states form the global riskiest set R, and those not marked are the states in $R^c$. In Table 8, the states in $R^c$ are eliminated. Considering only the states in R, we can see that strategy 8 (experiment; drill if No or Open structure) is always worse than or equal to strategy 2 (drill without experimenting), that strategy 7 (experiment; drill if No structure) is always worse than or equal to strategy 9 (experiment; drill if No or Closed structure), and that strategy 5 (experiment; drill if Open structure) is always worse than strategy 6 (experiment; drill if Open or Closed structure). The only exceptions to these dominance relations are under the states of the world in $R^c$, which are neither very possible nor ever very regrettable and thus may be ignored. Table 9 shows the result of eliminating strategies 8, 7, and 5, which are dominated according to the global riskiest-states criterion.

Table 10 shows the pairwise comparisons involved in evaluating the pairwise riskiest-states criterion. Consider a pair of strategies $\alpha$ and $\alpha'$. $R_{\alpha\alpha'}$ is the set of all states which are in the riskiest set for either $\alpha$ or $\alpha'$ or both. A state is entered into Table 10 in the row corresponding to strategy $\alpha$ and the column corresponding to strategy $\alpha'$ if $\sigma$ is in $R_{\alpha\alpha'}$ and $D(\alpha|\sigma)$ is worse than $D(\alpha'|\sigma)$. States for which the outcomes of the two strategies are tied are ignored; thus, the main diagonal cells are automatically empty. The asterisks in the strategy 9 column of the rows for strategies 2 and 4 indicate that there were no states of the world satisfying the above conditions when $\alpha$ is 2 or 4 and $\alpha'$ is 9. Strategy 9 is worse than or equal to strategy 2 for every state in $R_{2,9}$ and worse than or equal to strategy 4 for every state in $R_{4,9}$, so we say that strategies 2 and 4 each dominate strategy 9 by the pairwise riskiest-states criterion.

Table 11 shows the four remaining nondominated strategies. No further reduction is possible using only commensurate ordinal comparisons, but we have reduced ten original strategies down to a "short list" of four reasonable candidates. This short list would then be subjected to some more information-intensive decision analysis technique to arrive at a final decision.

## 3.2. L-Fuzzy Risk Minimization

In order to apply the L-fuzzy risk minimization technique (Whalen 1980, 1984b) let us make the following assumptions:

(1) The truth value of the statement "State SC is very possible" is intermediate between the truth values of the statements "Outcome B is very regrettable" and "Outcome F is very re-

grettable."

(2) It is truer to say "State DO is very possible" than to say "Outcome C is very regrettable."

(3) The truth value of the statement "Outcome E is very regrettable" is intermediate between the truth values of the statements "State WO is very possible" and "State SO is very possible."

According to the L-fuzzy risk minimization algorithm, the disutility of each of the ten candidate strategies is:

$$D(2) = \max\{D \wedge P,\ D \wedge Q,\ H \wedge R,\ H \wedge S,\ H \wedge T,\ H \wedge T,\ H \wedge U,\ D \wedge V,\ H \wedge W\}$$
$$= \max\{D,\ D,\ H,\ H,\ H,\ H,\ H \wedge U,\ V,\ H \wedge W\} = D$$

$$D(6) = \max\{g \wedge P,\ C \wedge Q,\ g \wedge R,\ g \wedge S,\ E \wedge T,\ g \wedge T,\ g \wedge U,\ C \wedge V,\ A \wedge W\}$$
$$= \max\{g,\ C,\ g,\ g,\ E,\ g,\ g \wedge U,\ U,\ W\} = C$$

$$D(4) = \max\{g \wedge P,\ g \wedge Q,\ E \wedge R,\ g \wedge S,\ E \wedge T,\ g \wedge T,\ A \wedge U,\ C \wedge V,\ A \wedge W\}$$
$$= \max\{g,\ g,\ E,\ g,\ E,\ g,\ U,\ U,\ W\} = E$$

$$D(1) = \max\{H \wedge P,\ H \wedge Q,\ F \wedge R,\ B \wedge S,\ F \wedge T,\ F \wedge T,\ B \wedge U,\ H \wedge V,\ B \wedge W\}$$
$$= \max\{H,\ H,\ F,\ S,\ F,\ F,\ U,\ H\ V,\ W\} = S$$

Since outcomes C and D are more regrettable than outcome E, we can eliminate strategies 2 and 6. This allows us to conclude that strategy 4 is preferable if it is truer to say "Outcome E is very regrettable" than to say "State SC is very possible," and that strategy 1 is preferable otherwise. (Note that our assumptions do not allow us to conclude whether it is truer to say "g is regrettable" or to say "U is possible", and similarly for H and U, H and V, and H and W, but none of these comparisons are necessary to arrive at a final decision.) See Whalen (1984a) for an evaluation of these ordinal results, using a Monte Carlo simulation sampling from the space of all probability and utility distributions meeting the ordinal constraints.

## 3.3. Fuzzy Statistical Decision Analysis

To illustrate the use of fuzzy statistical decision analysis for this problem, let us make the following assumptions:

(1) The cost of the seismic study and the cost of drilling are fixed by contract at $10 000 and $70 000 each.

(2) The conditional probabilities of No structure, Open structure, and Closed structure given Dry, Wet or Soaking oil content are known to be as follows:

$$P(N|D)=.6,\ P(O|D)=.3,\ P(C|D)=.1,$$
$$P(N|W)=.3,\ P(O|W)=.4,\ P(C|W)=.3,$$
$$P(N|S)=.1,\ P(O|S)=.4,\ P(C|S)=.5.$$

(3) The probability of a Dry well is 1 minus the sum of the probabilities of Wet and Soaking.

(4) The revenue from a Dry well is zero.

(5) Our knowledge of the probability of Wet is given by a triangular fuzzy number with support running from .25 to

.35 with a peak at .30 .

(6) Our knowledge of the probability of Soaking is given by a triangular fuzzy number with support running from .25 to .35 with a peak at .20 .

(7) Our knowledge of the revenue from a Wet well is given by a triangular fuzzy number with support running from $70 000 to $170 000 with a peak at $120 000.

(8) Our knowledge of the revenue from a Soaking well is given by a triangular fuzzy number with support running from $170 000 to $370 000 with a peak at $270 000.

Assumptions 1 through 4 are exactly as given in Raiffa (1968), while assumptions 5 through 8 are fuzzifications of the Raiffa data.

The goal of fuzzy statistical decision analysis is to find a strategy whose expected profit is not less than any other strategy. We compute expected values by exactly the same formula as in statistical decision analysis, applying the extension principle of fuzzy mathematics to perform the required multiplications and additions of fuzzy numbers. Fig. 1 shows the graph of expected profit versus possibility for the four nondominated strategies.

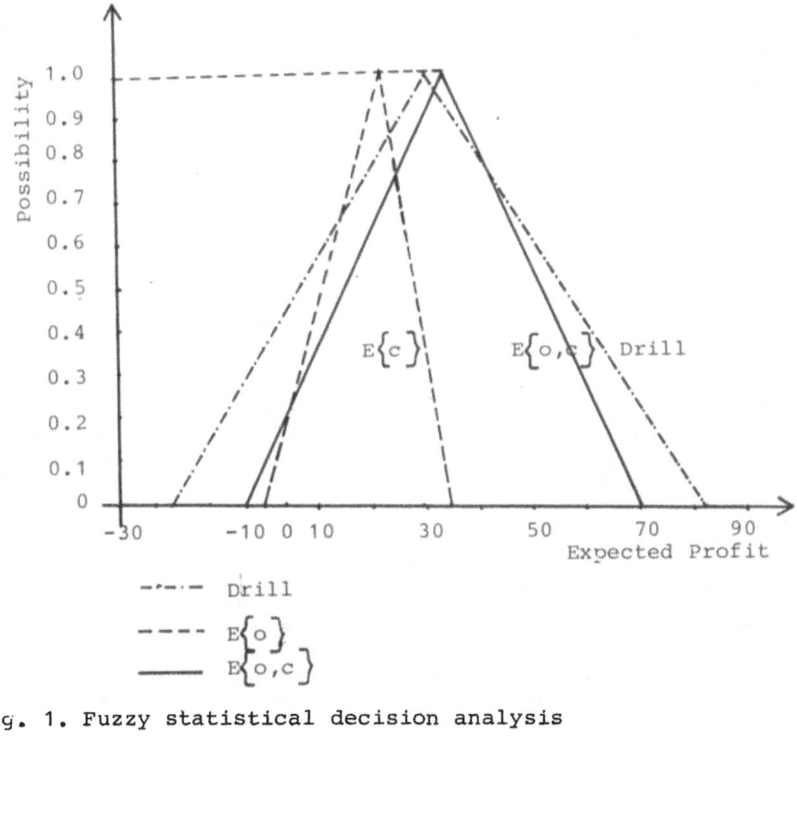


Fig. 1. Fuzzy statistical decision analysis

If we restrict our attention to only the top 2.6 percent of membership grades in Fig. 1 (memberships greater than or equal to .974), then the lowest expected value for strategy 6 (test; drill only if Open or Closed structure) is higher than the highest expected value for strategy 2 (drill without testing) or for any other strategy. Thus, by this very tight standard of possibility, we can say that the expected value of strategy 6 is strictly greater than the expected value of strategy 2. Watson, Weiss and Donnell (1979) suggest that the strength with which we can make this statement, its "truth value", is given by the complement of the lowest membership grade above which the statement is true; thus the statement "Strategy 6 is strictly better than strategy 2" has a truth value of .026.

For membership grades above .803, there may be overlap between the range of possible expected values of strategy 6 and strategy 2, but the highest possibility for strategy 6 is higher than the highest possibility for strategy 2 and the lowest possibility for strategy 6 is higher than the lowest possibility for strategy 2. Thus, the truth value of the statement "strategy 6 is at least as good as strategy 2" is 1-.803 or .197.

Strategy 6 also outperforms the other two nondominated strategies; the statement "strategy 6 is strictly better than strategy 4" has a truth value of .184, the statement "strategy 6 is at least as good as strategy 4" has a truth value of .595, and the statement "strategy 6 is strictly better than strategy 1" has a truth value of .586. Since no statement asserting that any of the strategies is better than strategy 6 has a positive truth value, the fuzzy statistical decision analysis algorithm advises us to select strategy 6; unlike the crisp statistical approach discussed in Raiffa, it also tells us how much confidence we are entitled to have in the superiority of the chosen strategy as a function of how much confidence we have in our data.

## 4. CONCLUSION

One of the most important parts of making a decision is the early choice of what formal model (if any) will be used to structure the remainder of the decision process. Different decision models make different assumptions about the nature of the alternative actions, goals (utilities) and other considerations for evaluating the alternative actions. Early choices among models, made on the basis of the general appropriateness of their assumptions to the case in point, further determine the way in which the relevant data will be collected and defined.

Because of this effect on the way a decision will be structured, it is important to have a wide variety of techniques with differing assumptions. Furthermore, these techniques need to be classified within an integrative framework according to the nature of their assumptions. Only then can we be confident of choosing a model which makes the most effective possible use of the available data without introducing the distortions which result from a mismatch between the data and the algorithm (e.g., treating nominal or ordinal scale data as if it were measured on a ratio scale).

The methodologies discussed in this chapter constitute a subset of the various possible assumptions about the kind of information that can be obtained and used. Once a new practical decision problem has been identified as belonging to the general class of decision-tree type problems, the quality and quantity of the data associated with the new problem can be compared with the information presented herein to select the best model around which to structure the processes of estimating numeric probabilities or relative possibilities, assessing utilities, and arriving at a final decision.

The goal is to maximize the efficient use of whatever information is actually available while minimizing the need for introducing arbitrary assumptions or spurious precision. For example, if the information actually available in a given problem situation were just sufficient to satisfy the requirements of the L-fuzzy risk minimization algorithm, then to use a less information-intensive algorithm such as minimax regret would require ignoring real information which might be critical to an optimal decision, while using a more information-intensive technique such as statistical decision analysis would require introducing arbitrary assumptions about cardinal measurement scales which might distort the solution enough to lead to a suboptimal decision. In general, a problem situation will not fit the assumptions of any one model exactly. In such a case, a good strategy might be to bracket the problem by comparing the results of using two techniques: the most information-intensive technique whose assumptions are completely satisfied by the situation (but which does not use all the available information); and the least information - intensive technique which uses all the available information (but which also requires some additional assumptions). If the two "bracketing" techniques agree on a single decision alternative, that alternative may be adopted with some confidence; if the two techniques disagree, their respective recommendations may be compared more intensively as a "short list" from which the final action is to be selected.

Further advances in the field of enriching and guiding the choice of methodologies for soft optimization can take three separate directions; development and refinement of individual techniques; systematic comparisons of their characteristics and, development of tools to aid in the selection of appropriate techniques for a particular problem.

One advantage of the conceptual framework used in this chapter is that it can suggest important gaps in the spectrum of techniques, and thus serve as a stimulus to the development of additional techniques which may fit some practical problems better than the ones currently in place. Examples of useful potential additions include hybrid systems combining information at different levels such as ordinal and real numbers, and an extension of the L-fuzzy risk minimization technique based on wholistic comparisons between commensurate pairs of (possibility, utility) tuples to eliminate the conceptually difficult comparisons between a possibility on one hand and a utility on the other. In addition to investigation of the technical efficiency of new and existing techniques, research is also needed regarding their potential for user acceptance; any decision making methodology which imposes major conceptual shifts on its intended users will be accepted only very slowly, regardless of its other merits, as witnessed by the histories of Bayesian statistics and, more recently, fuzzy mathematics.

The framework of this chapter provides a starting point for the systematic comparison of techniques in terms of their basic assumptions regarding uncertainty. However, in order to provide really effective guidance as to what techniques ought to be used in a particular situation, it is also necessary to have a body of knowledge comparing the difficulty of use and the quality of results using each technique in a variety of situations. Such a body of knowledge exists only in fragmentary form at present, and needs to be expanded and systematized using both axiomatic analysis and experimental studies with realistic problems and user populations.

As the number of techniques in the collection and the number of criteria for selection become large, the difficulty of choosing a technique using printed reports such as this one becomes greater. This suggests a third avenue of research: the development of an "intelligent index" to help a decisionmaker to find the technique which best matches his perception of his problem. Since the choice of technique must be made very early in the decision process, at a time when the problem is still relatively ill-structured, a fuzzy ordinal approach to such an index seems most appropriate.

TABLE 1: PROFITS AND PROBABILITIES

| | DN | DO | DC | WN | WO | WC | SN | SO | SC |
|---|---|---|---|---|---|---|---|---|---|
| 1 NO | 0 | 0 | 0 | 0 | 0 | 0 | 0 | 0 | 0 |
| 2 DRILL | -70 | -70 | -70 | 50 | 50 | 50 | 200 | 200 | 200 |
| 3 E{ } | -10 | -10 | -10 | -10 | -10 | -10 | -10 | -10 | -10 |
| 4 E{c} | -10 | -10 | -80 | -10 | -10 | 40 | -10 | -10 | 190 |
| 5 E{o} | -10 | -80 | -10 | -10 | -40 | -10 | -10 | 190 | -10 |
| 6 E{o,c} | -10 | -80 | -80 | -10 | 40 | 40 | -10 | 190 | 190 |
| 7 E{n} | -80 | -10 | -10 | 40 | -10 | -10 | 190 | -10 | -10 |
| 8 E{n,o} | -80 | -80 | -10 | 40 | 40 | -10 | 190 | 190 | -10 |
| 9 E{n,c} | -80 | -10 | -80 | 40 | -10 | 40 | 190 | -10 | 190 |
| 10 E{n,o,c} | -80 | -80 | -80 | 40 | 40 | 40 | 190 | 190 | 190 |
| PROBABILITY | .30 | .15 | .05 | .09 | .12 | .09 | .02 | .08 | .10 |

TABLE 2: REGRETS

| | DN | DO | DC | WN | WO | WC | SN | SO | SC |
|---|---|---|---|---|---|---|---|---|---|
| 1 NO | 0 | 0 | 0 | -50 | -50 | -50 | -200 | -200 | -200 |
| 2 DRILL | -70 | -70 | -70 | 0 | 0 | 0 | 0 | 0 | 0 |
| 3 E{ } | -10 | -10 | -10 | -60 | -60 | -60 | -210 | -210 | -210 |
| 4 E{c} | -10 | -10 | -80 | -60 | -60 | -10 | -210 | -210 | -10 |
| 5 E{o} | -10 | -80 | -10 | -60 | -10 | -60 | -210 | -10 | -210 |
| 6 E{o,c} | -10 | -80 | -80 | -60 | -10 | -10 | -210 | -10 | -10 |
| 7 E{n} | -80 | -10 | -10 | -10 | -60 | -60 | -10 | -210 | -210 |
| 8 E{n,o} | -80 | -80 | -10 | -10 | -10 | -60 | -10 | -10 | -210 |
| 9 E{n,c} | -80 | -10 | -80 | -10 | -60 | -10 | -10 | -210 | -10 |
| 10 E{n,o,c} | -80 | -80 | -80 | -10 | -10 | -10 | -10 | -10 | -10 |
| PROBABILITY | .30 | .15 | .05 | .09 | .12 | .09 | .02 | .08 | .10 |

TABLE 3: ORDINAL REGRETS

| | DN | DO | DC | WN | WO | WC | SN | SO | SC |
|---|---|---|---|---|---|---|---|---|---|
| 1 NO | H | H | H | F | F | F | B | B | B |
| 2 DRILL | D | D | D | H | H | H | H | H | H |
| 3 E{ } | g | g | g | E | E | E | A | A | A |
| 4 E{c} | g | g | C | E | E | g | A | A | g |
| 5 E{o} | g | C | g | E | g | E | A | g | A |
| 6 E{o,c} | g | C | C | E | g | g | A | g | g |
| 7 E{n} | C | g | g | g | E | E | g | A | A |
| 8 E{n,o} | C | C | g | g | g | E | g | g | A |
| 9 E{n,c} | C | g | C | g | E | g | g | A | g |
| 10 E{n,o,c} | C | C | C | g | g | g | g | g | g |
| POSSIBILITY | P | Q | V | T | R | T | W | U | S |

TABLE 4: SORT STATES BY POSSIBILITY

| | DN | DO | WO | SC | WN | WC | SO | DC | SN |
|---|---|---|---|---|---|---|---|---|---|
| 1 NO | H | H | F | B | F | F | B | H | B |
| 2 DRILL | D | D | H | H | H | H | H | D | H |
| 3 E{ } | g | g | E | A | E | E | A | g | A |
| 4 E{c} | g | g | E | g | E | g | A | C | A |
| 5 E{o} | g | C | g | A | E | E | g | g | A |
| 6 E{o,c} | g | C | g | g | E | g | g | C | A |
| 7 E{n} | C | g | E | A | g | E | A | g | g |
| 8 E{n,o} | C | C | g | A | g | E | g | g | g |
| 9 E{n,c} | C | g | E | g | g | g | A | C | g |
| 10 E{n,o,c} | C | C | g | g | g | g | g | C | g |
| POSSIBILITY | P | Q | R | S | T | T | U | V | W |

TABLE 5: SORT STRATEGIES BY REGRET (MOST POSSIBLE STATE = PRIMARY KEY)

| | DN | DO | WO | SC | WN | WC | SO | DC | SN | |
|---|---|---|---|---|---|---|---|---|---|---|
| 8 E{n,o} | C | C | g | A | g | E | g | g | g | |
| 10 E{n,o,c} | C | C | g | g | g | g | g | C | g | >2 |
| 7 E{n} | C | g | E | A | g | E | A | g | g | |
| 9 E{n,c} | C | g | E | g | g | g | A | C | g | |
| 2 DRILL | D | D | H | H | H | H | H | D | H | |
| 5 E{o} | g | C | g | A | E | E | g | g | A | |
| 6 E{o,c} | g | C | g | g | E | g | g | C | A | |
| 3 E{ } | g | g | E | A | E | E | A | g | A | >1 |
| 4 E{c} | g | g | E | g | E | g | A | C | A | |
| 1 NO | H | H | F | B | F | F | B | H | B | |
| POSSIBILITY | P | Q | R | S | T | T | U | V | W | |

TABLE 6: ELIMINATE COMPLETELY DOMINATED STRATEGIES

| | DN | DO | WO | SC | WN | WC | SO | DC | SN |
|---|---|---|---|---|---|---|---|---|---|
| 8 E{n,o} | C | C | g | A | g | E | g | g | g |
| 7 E{n} | C | g | E | A | g | E | A | g | g |
| 9 E{n,c} | C | g | E | g | g | g | A | C | g |
| 2 DRILL | D | D | H | H | H | H | H | D | H |
| 5 E{o} | g | C | g | A | E | E | g | g | A |
| 6 E{o,c} | g | C | g | g | E | g | g | C | A |
| 4 E{c} | g | g | E | g | E | g | A | C | A |
| 1 NO | H | H | F | B | F | F | B | H | B |
| POSSIBILITY | P | Q | R | S | T | T | U | V | W |

TABLE 7:
IDENTIFY WORST-SET EXPOSURES FOR EACH STRATEGY

| | * | * | * | * | | | * | | * |
|---|---|---|---|---|---|---|---|---|---|
| | DN | DO | WO | SC | WN | WC | SO | DC | SN |
| 8 E{n,o} | C* | C | g | A* | g | E | g | g | g |
| 7 E{n} | C* | g | E | A* | g | E | A | g | g |
| 9 E{n,c} | C* | g | E | g | g | g | A* | C | g |
| 2 DRILL | D* | D | H | H | H | H | H | D | H |
| 5 E{o} | g* | C* | g | A* | E | E | g | g | A |
| 6 E{o,c} | g* | C* | g | g | E | g | g | C | A* |
| 4 E{c} | g* | g | E* | g | E | g | A* | C | A |
| 1 NO | H* | H | F* | B* | F | F | B | H | B |
| POSSIBILITY | P | Q | R | S | T | T | U | V | W |

TABLE 8:
DELETE STATES WITH NO EXPOSURE IN WORST SET

| | DN | DO | WO | SC | SO | SN | |
|---|---|---|---|---|---|---|---|
| 8 E{n,o} | C* | C | g | A* | g | g | >2 |
| 7 E{n} | C* | g | E | A* | A | g | >9 |
| 9 E{n,c} | C* | g | E | g | A* | g | |
| 2 DRILL | D* | D | H | H | H | H | |
| 5 E{o} | g* | C* | g | A* | g | A | >6 |
| 6 E{o,c} | g* | C* | g | g | g | A* | |
| 4 E{c} | g* | g | E* | g | A* | A | |
| 1 NO | H* | H | F* | B* | B | B | |
| POSSIBILITY | P | Q | R | S | U | W | |

TABLE 9:
ELIMINATE STRATEGIES DOMINATED ON REMAINING STATES

| | DN | DO | WO | SC | SO | SN |
|---|---|---|---|---|---|---|
| 9 E{m,c} | C* | g | E | g | A* | g |
| 2 DRILL | D* | D | H | H | H | |
| 6 E{o,c} | g* | C* | g | g | g | A* |
| 4 E{c} | g* | g | E* | g | A* | A |
| 1 NO | H* | H | F* | B* | B | B |
| POSSIBILITY | P | Q | R | S | U | W |

TABLE 10: PAIRWISE COMPARISONS

(States listed are in riskiest set for either row strategy or column strategy, and have a worse outcome for row strategy than for column strategy.)

| | 9 | 2 | 6 | 4 | 1 |
|---|---|---|---|---|---|
| 9 E{n,o} | | DN<br>SO | DN<br>SO | DN | DN<br>WO<br>SO |
| 2 DRILL | * | | DN | DN | DN |
| 6 E{o,c} | DO | SN<br>DO | | DO | DN<br>DO<br>SN |
| 4 E{c} | * | WO<br>SO | WO<br>SO | | WO<br>SO<br>DN |
| 1 NO | SC | WO<br>SC | WO<br>SC | SC | |

TABLE 11:
ELIMINATE PAIRWISE WORST-SET DOMINATED STRATEGIES

| | DN | DO | WO | SC | SO | SN |
|---|---|---|---|---|---|---|
| 2 DRILL | D* | D | H | H | H | H |
| 6 E{o,c} | g* | C* | g | g | g | A* |
| 4 E{c} | g* | g | E* | g | A* | A |
| 1 NO | H* | H | F* | B* | B | B |
| POSSIBILITY | P | Q | R | S | U | W |

REFERENCES

Dimitrov, V., and L. Criankova (1977). Program system for social choice under fuzzy managing. Inf. Processing 77. North - Holland, Amsterdam.
Dubois, D., and H. Prade (1979). Fuzzy real algebra: some results. Fuzzy Sets and Syst. 2, 327 - 348.
Fishburn, P.C. (1969). Information analysis without states of the world. Op. Res. 17, 413 - 424.
Freeling, S. (1980). Fuzzy sets and decision analysis. IEEE Trans. Systems, Man, and Cyber. SMC - 10.
Goguen, J.A. (1967). L-fuzzy sets. J. Math. Analysis and Appl. 18, 145 - 174.
Raiffa, H. (1968). Decision Analysis: Introductory Lectures on Choices under Uncertainty. Addison-Wesley, Reading, Mass.
Simon, H. (1977). The New Science of Management Decision. Prentice-Hall, London.
Smith, G.R. (1980). Textured sets: an approach to aggregation problems with multiple concerns. IEEE Trans. Syst., Man and Cyber. SMC-10 .
Von Neumann, J., and O. Morgenstern (1947). Theory of Games and Economic Behavior. J. Wiley, New York.
Watson, S., J.Weiss, and M. Donnell (1972). Fuzzy decision analysis. IEEE Trans. Syst., Man, and Cyber. SMC-9.
Whalen, T. (1980). Risk minimization using L-fuzzy sets. Proc. of the Int. Conference on Cybernetics and Society.
Whalen, T. (1984a). Decision analysis with commensurate ordinal data. Proc. of 1984 IEEE Int. Conference on Systems, Man and Cybernetics.
Whalen, T. (1984b). Decision-making under uncertainty with various assumptions about available information. IEEE Trans. Systems, Man and Cyber. SMC-14, 880 - 900.
Zadeh, L.A. (1965). Fuzzy sets. Inf. and Control 8, 338 - 353 .

# FUZZY OPTIMIZATION AND MATHEMATICAL PROGRAMMING: A BRIEF INTRODUCTION AND SURVEY

Janusz Kacprzyk* and Sergei A. Orlovski**

* Systems Research Institute, Polish Academy of Sciences, Newelska 6, 01-447 Warsaw, Poland

** International Institute for Applied Systems Analysis (IIASA), A-2361 Laxenburg, Austria (On leave from the Computing Center of the USSR Academy of Sciences, Moscow, USSR)

Abstract. Some general concepts and ideas related to fuzzy optimization as, e.g., a fuzzy constraint, fuzzy goal (objective function), fuzzy optimum, etc. are introduced first. A general fuzzy optimization problem involving these elements is formulated and solved. The cases of single and multiple objective functions are dealt with. Secondly, basic classes of fuzzy mathematical programming are discussed, including: fuzzy linear programming (with single and multiple objective functions), fuzzy integer programming, fuzzy 0-1 programming and fuzzy dynamic programming. Finally some newer, knowledge-based approaches are mentioned. An extended list of literature is included.

Keywords: fuzzy decision making, fuzzy optimization, fuzzy mathematical programming.

## 1. INTRODUCTION

The book focuses on optimization problems which belong to a much wider class of decision making problems.

Decision making has always played, and is playing, a crucial role in human life. In fact, any human activity is a succession of decision-making-related acts. A growing complexity of social, economic, technical, military, etc. problems faced by human decisionmakers has finally led to a necessity of using some formal (scientific) tools. This has stimulated the development of modern mathematical tools and techniques for that purpose.

The analysis of a real decision making situation is virtually based on two types of information:

- information on feasible alternative decisions (options, choices, alternatives, variants, ...),
- information making possible the comparison of alternative decisions with each other in terms of "better", "worse",

"indifferent", etc.

To apply mathematical tools and techniques, these two types of information should be adequately quantified and formalized in the form of some mathematical models.

Having such models, diverse, more or less formal analytical methods may be used by the analyst to derive a rational choice (s) to be recommended to the decision maker. Evidently, the applicability of analytical methods at the analyst's disposal depends in a straightforward way upon the form of the model employed to represent a real decision making situation.

If a model of a decision making situation is not adequate enough, then the results of analysis may be misleading. This may also occur in case of unreliable or inaccurate data. Unfortunately, in many cases - above all in economic, social etc. systems where human judgments, preferences, etc. play a crucial role - information on a particular decision making situation must be elicited from human experts. It is therefore full of subjectivity and of ambiguity or vagueness which stem from the use of a natural language that is the only fully natural means of human communication. And it is our ability to adequately incorporate this type of information in an analytical mathematical framework that is crucial for enhancing the applicability of mathematical methods in real-world decision making situations.

Optimization problems constitute a wide class within decision making. Basically, information on the preferences among alternatives is in them described by some utility (objective, performance, ...) function that maps a given set of feasible alternatives into the real axis; this allows one to compare the alternatives with each other in a straightforward way through their numerical evaluations as, e.g., the greater the value of that function, the better the corresponding alternative.

The set of feasible alternatives in an optimization problem is frequently described by a system of equations and/or inequalities. In such a case the problem is referred to as one of mathematical programming.

Methods and techniques of optimization, or - more specifically - those of mathematical programming have been successfully used for years in various problems involving, and related to, technical systems of relatively well-defined structure and behavior, the so-called "hard" ones. This has allowed the formulation of corresponding optimization problems with precisely specified constraints and objective functions solvable by well-developed and quite efficient traditional analytical and computational means.

That success has motivated a direct application of the same traditional approaches to the modeling and analysis of what is often called the "soft" systems in which a key role is played by human judgments, preferences, etc. Unfortunately, the progress in this direction has been much less than expected, which has even raised doubts whether traditional mathematical tools are at all applicable to problems with relevant human-related elements.

It seems, however, that a more justified viewpoint is probably such that to be able to successfully use optimization methods in complex systems, which are "soft" in the above sense, a "technological change" towards "softer" approaches is needed, toward approaches that would make it possible to incorporate fuzziness (imprecision) of information into optimization models and into methods of dealing with them. This need for "softer" approaches in broadly perceived systems analysis and systems engineering has been articulated and advocated for a long time (e.g., Rapaport, 1970; or Checkland, 1972) and has recently gained impetus in view of advances in knowledge engineering (see Gaines' paper earlier in this volume).

As already mentioned, a major "obstacle" in the application of traditional modeling and optimization tools in "soft" economic, social, environmental, etc. systems is the subjective nature of available information and its predominantly imprecise (fuzzy) form due to the use of a natural language. A rapid development of fuzzy sets theory over the last two decades gives more and more evidence that this theory provides useful means for a more adequate modeling of "soft" information and for the development of analytical approaches that make possible an adequate processing of such information to finally arrive at a realistic decision. It is in this sense that we say fuzzy sets theory is a promising tool for "softening" traditional optimization models and techniques.

In this introductory paper we briefly review some basic developments in the field of "soft" optimization via fuzzy sets and, to a lesser extent, possibility theory. We present various existing approaches to the formulation of fuzzy optimization problems, and methods of their solution. In principle, in all of them a fuzzy optimization problem is transformed into some equivalent nonfuzzy problem which, in turn, can be solved by using some traditional techniques (e.g., mathematical programming) and widely available commercial software packages.

## 2. APPROACHES TO FUZZY OPTIMIZATION WITH AN EXPLICITLY SPECIFIED FEASIBLE SET

As it has been already mentioned, the formulation of an optimization problem contains two essential elements: (1) a set of feasible alternatives, and (2) an objective function whose values serve the purpose of comparing the alternatives with each other. The optimization problem itself lies in determining some "best" (in a sense) alternative(s).

The description of both the objective function and feasible set may be fuzzy. In this section we consider formulations in which the feasible set is explicitly specified by its corresponding membership function whose values indicate the degrees of feasibility of the particular alternatives. In further sections we also consider formulations in which the feasible sets are described by systems of fuzzy equations and/or inequalities, and refer to such formulations as to fuzzy mathematical programming. Paranthetically, let us mention that some of them are based on extracting an explicit specification of the feasible set in the form of a membership function.

The class of problems considered in this section can be

stated as follows. Let $X = \{x\}$ be a set of relevant alternatives (options, choices, decisions, ...).

The objective function is generally defined as a mapping $F : X \to L(R)$ where $L(R)$ is a class of fuzzy subsets of the real line $R$. The value of $F$ for $x \in X$, $F(x)$, is a fuzzy number which represents a "soft" evaluation of the alternative $x \in X$.

The feasible alternatives are those "belonging" to a fuzzy set $C \subseteq X$ described by its membership function $\mu_C : X \to [0,1]$. The alternatives may therefore differ in their degrees of feasibility represented by the values of $\mu_C(x)$.

In traditional terms, our "soft" optimization problem can be written as

$$f(x,r) \to \underset{x \in \tilde{C}}{\text{"max"}} \tag{1}$$

to be read as to "maximize" (the quotation marks mean that maximization is not understood in the "hard" traditional sense but in a "soft" one, i.e. to attain "possibly great" fuzzy values of $f(x,r)$) the objective function $f(x,r)$ with respect to $x$ "belonging" to the fuzzy constraint $C$; $r$ is a parameter.

We will outline now two approaches to solving the above general formulation of a fuzzy optimization problem.

### 2.1. Attainment of a fuzzy goal subject to a fuzzy constraint: Bellman and Zadeh's approach

In this approach by Bellman and Zadeh (1970) which forms the basis of an overwhelming majority of fuzzy decision-making -related models, the underlying assumption is that besides an explicitly formulated fuzzy set of feasible alternatives $C \subseteq X$, called a fuzzy constraint, we also have an explicitly specified fuzzy set of alternatives that attain a goal, denoted $G \subseteq X$ and called a fuzzy goal.

The value of $\mu_G(x)$, the membership function of $G$, indicates the degree to which an alternative $x \in X$ satisfies the fuzzy goal $G$. For example (see, e.g., Kacprzyk, 1983a), the membership function $\mu_G(x)$ may be defined as

$$\mu_G(x) = \begin{cases} 1 & \text{for} \quad f(x) \geqslant \bar{f} \\ g(x) & \text{for} \quad \underline{f} < f(x) < \bar{f} \\ 0 & \text{for} \quad \bar{f}(x) \leqslant \underline{f} \end{cases} \tag{2}$$

to be read as: we are fully satisfied ($\mu_G(x) = 1$) with the values of $x$ for which our objective function $f(x)$ is not below an aspiration level $\bar{f}$, we are less satisfied ($0<g(x)<1$) with $x$'s for which $\underline{f} < f(x) < \bar{f}$, and we are fully dissatisfied with $x$'s which do not exceed a lowest possible level $\underline{f}$, i.e. such $x$'s are unacceptable.

The problem is now generally stated as

$$\text{"satisfy } C \text{ and attain } G\text{"} \tag{3}$$

i.e. satisfy the fuzzy constraint G and attain the fuzzy goal G.

If we introduce a fuzzy set $D \subseteq X$ which solves this problem, and is called a fuzzy decision, then (3) can be written as

$$D = C \cap G \tag{4}$$

where " $\cap$" is an intersection operator corresponding to "and" in (3).

In terms of membership functions, we can write (4) as

$$\mu_D(x) = \mu_C(x) * \mu_G(x) \quad \text{for each } x \in X \tag{5}$$

where "*" is an operation corresponding to " $\cap$".

Most frequently, "*" is assumed to be a minimum denoted by "$\wedge$", i.e. $a \wedge b = \min(a,b)$, and then (5) is

$$\mu_D(x) = \mu_C(x) \wedge \mu_G(x) \quad \text{for each } x \in X \tag{6}$$

This form of fuzzy decision may be viewed as a fuzzily specified instruction (which x to choose), the execution of which ensures the achievement of the fuzzy goal subject to the fuzzy constraint. Evidently, the fuzziness of this instruction is implied by the fuzziness of the problem formulation.

In the above form of fuzzy decision there still remains some uncertainty as to its execution, i.e. the choice of a specific x. A number of approaches are possible here (see,e.g., Zadeh 1968; Bellman and Zadeh, 1970,or Sommer and Pollatschek 1978). One of the most popular is to choose an alternative "belonging" to D to the maximum extent, that is to choose $x^* \in X$ such that

$$\mu_D(x^*) = \max_{x \in X} \mu_D(x) \tag{7}$$

In case of multiple fuzzy goals, $G_1, \ldots, G_n \subseteq X$, and fuzzy constraints $C_1, \ldots, C_m \subseteq X$, (6) becomes

$$\mu_D(x) = \mu_{G_1}(x) \wedge \ldots \wedge \mu_{G_n}(x) \wedge \mu_{C_1}(x) \wedge \ldots \wedge \mu_{C_m}(x) \tag{8}$$

$$\text{for each } x \in X$$

and we seek $x^* \in X$ such that (7).

Moreover, if the fuzzy constraint is defined as a fuzzy set in X, $C \subseteq X$, and the fuzzy goal in Y, $G \subseteq X$, and a function $y = w(x)$ is known, then (6) becomes

$$\mu_D(x) = \mu_C(x) \wedge \mu_{G'}(w(x)), \quad \text{for each } x \in X \tag{9}$$

where $G' \subseteq X$ is a fuzzy goal in X induced by $G \subseteq Y$; (7) remains the same.

Finally, let us notice that within Bellman and Zadeh's approach other forms of trade-offs between the degrees of feasibility and goal satisfaction are possible (see, e.g., Bellman and Zadeh, 1970; Kacprzyk, 1983a; Kacprzyk and Straszak, 1984; Yager, 1978, 1979); the choice depends here on the specifics of the problem considered.

We will now outline the application of Bellman and Zadeh's approach to the analysis of our optimization problem (1). First, we assume that our goal is to attain, by choosing an appropriate feasible alternative, some fuzzily specified value of the objective function. We assume this desired fuzzy value to be described by the membership function of a fuzzy goal, $\mu_G : R \to [0,1]$.

Consider first the case when the objective function is precisely defined, i.e. $f: X \to R$. To use Bellman and Zadeh's framework, we first determine the subset of alternatives providing for the satisfaction of our fuzzy goal. Clearly, this subset is the inverse image $\bar{\mu}_G$ of $\mu_G$ under the mapping (objective function) $f : X \to R$, that is

$$\bar{\mu}_G(x) = \mu_G(f(x)) \quad \text{for each} \quad x \in X \tag{10}$$

Now, our problem is of type (6) with $\bar{\mu}_G(x)$ replacing $\mu_G(x)$ as the fuzzy goal, and with $\mu_C(x)$ as the fuzzy feasible set (fuzzy constraint).

The fuzzy solution of the problem is now of type (9), i.e.

$$\mu_D(x) = \mu_C(x) \wedge \mu_G(f(x)) \quad \text{for each} \quad x \in X \tag{11}$$

For determining $x^* \in X$, such that

$$\mu_D(x^*) = \max_{x \in X} \mu_D(x) \tag{12}$$

some well-known methods of mathematical programming can be used (see, e.g., Tanaka, Okuda and Asai, 1974, or Negoita and Ralescu, 1975).

Let us now consider a more general case with fuzzily specified values of the objective function, i.e. $f : X \times R \to [0,1]$. As before, we assume that $\mu_C(x)$ is the membership function of the fuzzy feasible set in $X$ (fuzzy constraint), and $\mu_G(r)$, $r \in R$, is the membership function of the set of satisfactory values of $f$ (fuzzy goal).

To apply in this case Bellman and Zadeh's framework, we can introduce (Orlovski, 1981) the following equivalent definition of a solution to the problem of fuzzy goal satisfaction.

A fuzzy decision $D \subsetneqq X$ in our problem is a maximal (with respect to the containment of two fuzzy sets - see Fedrizzi's paper earlier in this volume) fuzzy set satisfying:

1. $D \subset C$ (feasibility of the solution),

2. $D \circ f \subset G$ (attainment of the fuzzy goal),

where $D \circ f$ is the image of D under the fuzzy objective function $f : X \times R \to [0,1]$, and "$\subset$" is the containment of two fuzzy sets.

It can be easily verified that for the above case of a precisely defined objective function, this definition of D is equivalent to (11).

Following Orlovski (1981), we introduce now the following sets:

$$N = \left\{ (x,r) : (x,r) \in X \times R,\ f(x,r) > \mu_G(r) \right\} \tag{13}$$

$$N_x = \left\{ r : r \in R,\ (x,r) \in N \right\} \tag{14}$$

$$X^0 = \left\{ x : x \in X,\ N_x \neq \emptyset \right\} \tag{15}$$

Then, the membership function of the fuzzy solution to our problem is

$$\mu_D(x) = \begin{cases} \mu_C(x) \wedge \inf\limits_{r \in N_x} \mu_G(x) & \text{for } x \in X^0 \\ \mu_C(x) & \text{otherwise} \end{cases} \tag{16}$$

It can be easily seen that for the conventional (nonfuzzy) objective function, $f:X \to R$, this boils down to (11).

As before, we can seek an alternative yielding the maximum value of $\mu_D(x)$, and use for its determination computational methods of mathematical programming.

## 2.2. The use of α-cuts of the fuzzy feasible set

While the approaches outlined in the previous subsection use the concept of goal satisfaction, there also exist approaches that use in a more explicit manner the concept of maximization. As an illustrative example, we will outline here an approach by Orlovski (1977).

The problem is as (1), that is

$$f(x) \to \underset{x \tilde{\in} C}{\text{"max"}} \tag{17}$$

where $f:X \to R$ is an objective function and $C \subsetneq X$ is a fuzzy constraint characterized by its membership function $\mu_C : X \to [0,1]$.

The first problem is to introduce some concept of a solution. We will present two of them; both define the solutions as some fuzzy sets.

In the first solution concept, we employ the α-cuts of C, i.e. $C_\alpha = \left\{ x \in X : \mu_C(x) \geqslant \alpha \right\}$, $\alpha \in (0,1]$. For each α, such that $C_\alpha \neq \emptyset$, we introduce the (nonfuzzy) set

$$N(\alpha) = \left\{ x \in X : f(x) = \sup_{z \in C_\alpha} f(z) \right\}$$

By solution 1 to the problem (17) we now mean a fuzzy set $S_1 \subsetneq X$, such that

$$\mu_{S_i}(x) = \begin{cases} \sup\limits_{x \in N(\alpha)} \alpha & \text{for } x \in \bigcup\limits_{\alpha>0} N(\alpha) \\ 0 & \text{otherwise} \end{cases} =$$

$$= \begin{cases} \mu_C(x) & \text{for } x \in \bigcup\limits_{\alpha>0} N(\alpha) \\ 0 & \text{otherwise} \end{cases} \quad (18)$$

We say that solution 1 exists if an only if there exists $\alpha>0$, such that $N(\alpha) \neq \emptyset$.

Next, we define the fuzzy maximal value of $f(x)$ over $\mu_C(x)$ as

$$\mu_f(r) = \sup_{x \in f^{-1}(r)} \mu_{S_1}(x) = \sup_{x \in f^{-1}(r)} \sup_{x \in N(\alpha)} \alpha \quad \text{for each } r \in R \quad (19)$$

Notice that the choice of a single $x$ as a final solution is not simply based on taking the $x$ with the highest $\mu_{S_1}(x)$ but also on the value of $f(x)$ corresponding to that $x$. Namely, the greater $r_o$ the smaller the value of $\mu_{S_1}(x)$ for $x=x_o$ such that $f(x_o) = r_o$. A compromise is therefore needed.

Solution 2 to the problem (17) is based on the concept of the Pareto optimum. Namely, for the two functions $f(x)$ and $\mu_C(x)$ we first define $P$, the set of Pareto maximal elements, i.e. the (nonfuzzy) subset $P$ of $X$, such that $x \in P$ if there exists no $y \in X$ for which either:

$$f(y) > f(x) \quad \text{and} \quad \mu_C(y) \geq \mu_C(x)$$

or

$$f(y) \geq f(x) \quad \text{and} \quad \mu_C(y) > \mu_C(x).$$

Solution 2 is now defined as a fuzzy set $S_2 \subsetneq X$ such that

$$\mu_{S_2}(x) = \begin{cases} \mu_C(x) & \text{for } x \in P \\ 0 & \text{otherwise} \end{cases} \quad (20)$$

As shown in Orlovski (1977), this solution gives the same fuzzy maximal value of $f(x)$ over $\mu_C(x)$ as solution 1 in (18), i.e.

$$\mu_f(r) = \sup_{x \in f^{-1}(r)} \mu_{S_2}(x) = \sup_{x \in f^{-1}(r)} \sup_{x \in N(\alpha)} \alpha \quad \text{for each } r \in R \quad (21)$$

Solution 2 explicitly suggests therefore that we should consider as rational choices only those x's which cannot be simultaneously improved in the values of both f(x) and $\mu_C(x)$.

Moreover, notice that $P \subset \bigcup_{\alpha>0} N(\alpha)$ which implies $\mu_{S_2}(x) \leqslant \mu_{S_1}(x)$ for any $x \in X$, i.e. solution 2 is a subset of solution 1.

Among other approaches to fuzzy optimization with a separate treatment of f(x) and $\mu_C(x)$, some of which employ and extend Orlovski's (1977) ideas, we should mention, e.g., Negoita, and Ralescu (1977), Ralescu (1979, 1984) or Yager (1979).

### 2.3. The case of multiple objective functions and fuzzy constraints

The problem is now basically as follows. We have n objective functions, $f_i(x)$, $i=1,\ldots,n$, and m fuzzy constraints $\mu_{C_1}(x),\ldots,\mu_{C_n}(x)$. We seek $x^* \in X$, such that

$$(f_1(x),\ldots,f_n(x)) \rightarrow \overline{\max_x} \tag{22}$$

subject to $\mu_{C_1}(x),\ldots,\mu_{C_m}(x)$

where $\overline{\max}$ is maximization in the sense of Pareto.

Similarly as in case of a single objective function, we can apply the approaches outlined in subsections 2.1 and 2.2.

Here we will sketch another approach based on Orlovski (1980, 1981, 1983, 1984) which is intuitively appealing and efficient.

First, let us assume a more general case when the objective functions are not real valued as before but take on fuzzy values $F_i(x) \subsetapprox X$; the real valued functions are here evidently special cases. Thus, the membership grade of a value of $f_i(x) = r$ in $F_i(x)$ is $\mu_{F_i(x)}(r)$.

Through $\mu_{F_1}(x)$ we obtain, using the extension principle, n fuzzy nonstrict preference relations over the set of alternatives X, i.e. $p_i : X \times X \rightarrow [0,1]$, given by

$$p_i(x_1,x_2) = \sup_{z \geqslant y}(\mu_{F_i(x_1)}(z) \wedge \mu_{F_i(x_2)}(y)) \quad i=1,\ldots,n \tag{23}$$

The next step is to define a way of comparing alternatives using these n fuzzy preference relations. We define a fuzzy strict preference relation $p_i^S : X \times X \rightarrow [0,1]$ corresponding to $p_i$ as

$$p_i^s(x_1,x_2) = 0 \vee (p_i(x_1,x_2) - p_i(x_2,x_1)) \tag{24}$$

where "v" is maximum, i.e. $a \vee b = \max(a,b)$. Then

$$p^s(x_1,x_2) = \bigwedge_{i=1}^{n} p_i^s(x_1,x_2) \tag{25}$$

is the degree to which $x_1$ is strictly preferred to $x_2$, where $\bigwedge_{i=1}^{n} a_i = a_i \wedge \ldots \wedge a_n$.

Next, we introduce a fuzzy subset of nondominated alternatives

$$\mu_{ND}(x) = 1 - \sup_{y \in X} p^s(y,x) = 1 - \sup_{y \in X} \bigwedge_{i=1}^{n} p_i^s(y,x) =$$
$$= 1 - \sup_{y \in X} \bigwedge_{i=1}^{n} (p_i(y,x) - p_i(x,y)) \tag{26}$$

The value of $\mu_{ND}(x)$ is a nondominance degree of alternative x. Thus, if $\mu_{ND}(x) \geqslant \alpha$, then x may be strictly dominated by some other alternative to a degree smaller than $1-\alpha$.

As the second element of the approach, we define a degree of feasibility of alternative x with respect to the fuzzy constraints $C_1, \ldots, C_m$. This can be done for instance as follows:

$$\mu_{FS}(x) = \mu_{C_1}(x) \wedge \ldots \wedge \mu_{C_m}(x) \tag{27}$$

The solution of the optimization problem (22) is now meant to find an alternative $x^* \in X$ for which

$$\mu_{ND}(x^*) \geqslant \alpha \quad \text{and} \quad \mu_{FS}(x^*) \geqslant \beta \tag{28}$$

where $\alpha$ is a desired degree of nondominance and $\beta$ is a desired degree of feasibility. In fact, a compromise between $\alpha$ and $\beta$ is sought.

In Orlovski (1984) some conventional (nonfuzzy) optimization problems equivalent to (22) are described.

Among other approaches, most of them also being based on some degree of nondominance, we should mention, e.g., Takeda and Nishida (1980), Leung (1982, 1983, 1984), Yager (1980) and Carlsson (1982).

## 3. BRIEF INTRODUCTION TO FUZZY MATHEMATICAL PROGRAMMING

The point of departure is here a general mathematical programming problem written as

$$\begin{cases} f(x) \to \max_x \\ \text{subject to} \\ g_i(x) \leqslant b_i \quad i=1,\dots,m \end{cases} \tag{29}$$

where $x = (x_1,\dots,x_n) \in R^n$ is a vector of decision variables, $f:R^n \to R$ is an objective function, $g_i:R^n \to R$ are constraints, and $b_i \in R$ are the so-called right-hand-sides. Evidently, maximization and "$\leqslant$" can be easily transformed into minimization and "$\geqslant$".

Specific forms of the decision variables, objective function and constraints lead to specific types of mathematical programming as, e.g., linear, quadratic, nonlinear, integer, 0-1, dynamic, etc.

In the following we present some basic approaches to introducing fuzziness into the general mathematical programming problem (29). Emphasis is on fuzzy linear programming, which, as its nonfuzzy counterpart, is of particular relevance from the practical viewpoint.

### 3.1. Fuzzy linear programming

The problem of conventional nonfuzzy linear programming may be written as

$$\begin{cases} f(x) = \sum_{i=1}^{n} c_i x_i \to \max_x \\ \text{subject to:} \\ \sum_{i=1}^{n} a_{ij} x_i \leqslant b_j \quad j=1,\dots,m \end{cases} \tag{30}$$

$$x_i \geqslant 0 \quad i=1,\dots,n$$

"Softening" of this problem may proceed along two main lines. First, we may "soften" the rigid requirement to strictly maximize the objective function and to strictly satisfy the constraints. Second, we may allow the coefficients, i.e. $c_i$'s, $a_i$'s, and $b_j$'s, to be fuzzy numbers. We will sketch now the two approaches.

#### 3.1.1. Fuzzy linear programming in the setting of Zimmermann

The first attempt to fuzzify a linear program is due to Zimmermann (1975, 1976).
To show its essence, we first rewrite (30) as

$$f(x) = \sum_{i=1}^{n} e_i x_i \to \min_x$$

$$\begin{cases} \text{subject to} \\ \sum_{i=1}^{n} a_{ji}\, x_i \leqslant b_j \qquad j=1,\ldots,m \\ x_i \geqslant 0 \qquad i=1,\ldots,n \end{cases} \tag{31}$$

where, evidently, $e_i = - c_i$.

The fuzzy version of this problem is now written as

$$\begin{cases} \sum_{i=1}^{n} e_i x_i \lesssim Z \\ \sum_{i=1}^{n} a_{ji}\, x_i \lesssim b_j \qquad j=1,\ldots,m \\ x_i \geqslant 0 \qquad i=1,\ldots,n \end{cases} \tag{32}$$

which is read as: the objective function $f(x) = \sum_{i=1}^{n} e_i\, x_i$ should be "essentially smaller than or equal to" an aspiration level $Z$, and the constraints $\sum_{i=1}^{n} a_{ji}\, x_i$ should be "essentially smaller than or equal to" the right hand sides $b_j$ or, in another words, should be possibly well satisfied.

The above "essentially smaller than or equal to", written "$\lesssim$" is formalized as follows. First, we denote by $H = [h_{ki}]$, $k=1,\ldots,m+1$, $i=1,\ldots,n$, the matrix obtained by adding to the matrix $A = [a_{ji}]$ the row vector $[e_i]$ as the first row of A. We denote $(Hx)_k = \sum_{i=1}^{n} h_{ki}\, x_i$, and define the function

$$f_k((Hx)_k) = \begin{cases} 1 & \text{for} \quad (Hx)_k \leqslant w_k \\ 1 - \dfrac{(Hx)_k - w_k}{d_k} & \text{for} \quad w_k < (Hx)_k \leqslant w_k + d_k \\ 0 & \text{for} \quad (Hx)_k > w_k + d_k \end{cases} \tag{33}$$

where $w_k$'s are the original right-hand-sides $b_j$'s and the aspiration level $Z$, i.e. $w^T = (w_1,\ldots,w_{m+1})^T = (Z, b_1,\ldots,b_m)^T$, and $d_k$'s are some subjectively chosen admissible violations of the constraints.

We wish to satisfy all the constraints of (32), hence the new objective function of the fuzzified linear programming problem, i.e. the fuzzy decision (see (8)), is

$$\mu_D(x) = \bigwedge_{k=1}^{m+1} f_k((Hx)_k) \tag{34}$$

and we seek an optimal solution $x^* = (x_1^*, \ldots, x_n^*)$, such that

$$\bigwedge_{k=1}^{m+1} f_k((Hx)_k) \to \max_{x=(x_1,\ldots,x_n)} \qquad (35)$$

or, by substituting $w_k^- = w_k/d_k$ and $(Hx)_k^- = (Hx)_k/d_k$ and dropping 1 in (2).

$$\bigwedge_{k=1}^{m+1} (w_k^- - (Hx)_k^-) \to x = \max_{x=(x_1,\ldots,x_n)} \qquad (36)$$

It is easy to show (see, e.g. Negoita and Sularia, 1976) that this is equivalent to

$$\begin{cases} \lambda \to \max\limits_{\lambda\in[0,1]} \\ \text{subject to:} \\ \lambda \leqslant w_k^- - (Hx)_k^- \qquad k=1,\ldots,m+1 \\ x_i \geqslant 0 \qquad i=1,\ldots,n \end{cases} \qquad (37)$$

in the sense that an optimal solution to (37) is also optimal to (32).

While using a fuzzy linear program the decision maker is not forced to state the problem in precise terms, required by the mathematics involved but possibly strange from the practical point of view. This is a serious advantage.

Zimmerman's approach has found numerous applications in, e.g.: designing the size and structure of a truck fleet (Zimmermann, 1975), designing of a parking place (Rödder and Zimmermann, 1977), media selection in advertising (Wiedey and Zimmermann, 1978), air pollution regulation (Sommer and Pollatschek, 1978), determination of agricultural policies (Kacprzyk and Owsiński, 1984; and Owsiński, Zadrożny and Kacprzyk later in this volume).

The model was also a point of departure for some extensions, as, e.g., the transportation problem (Oheigeartaigh, 1984; Verdegay, 1983; but particularly Chanas and Kołodziejczyk, 1984; and Delgado, Verdegay and Vila later in this volume); fuzzy linear programming with constraints given as fuzzy relations (Nakamura, 1984), fuzzy stochastic linear programming (Luhandjula, 1983), etc.

Let us mention that for fuzzy linear programming in Zimmermann's setting there are some works on duality (Hamacher, Leberling and Zimmermann, 1978; but particularly Verdegay, 1984a and Llena, 1985) sensitivity analysis (Hamacher, Leberling and Zimmermann, 1978), derivation of the whole fuzzy decision (34) using parametric linear programming (Chanas, 1983; and Verdegay, 1982) etc.

The presented approach can also be employed for solving multiobjective linear programming problems (see, e.g., Zimmermann, 1978, Hannan, 1981a, 1981c).

Moreover, attempts to develop fuzzy goal programming (e.g., Hannan, 1981b; Narasimhan, 1984; Llena, 1985) are relevant. Interactive approaches to a practical solution of multiobjective problems, also in a nonlinear case, have been developed by Sakawa and collaborators (e.g., Sakawa, 1983, 1984a, 1984b; Sakawa and Seo, 1983; Sakawa and Yumine, 1983).

### 3.1.2. Fuzzy linear programming with fuzzy coefficients

The first attempt to solve linear programs with fuzzy parameters is due to Negoita, Minoiu and Stan (1976). Basically, they consider the problem

$$\begin{cases} f(x) = \sum_{i=1}^{n} c_i x_i \rightarrow \max_x \\ \text{subject to:} \\ x_1 K_1 + \ldots + x_n K_n \subset K \\ x_1 \geqslant 0 \qquad i=1,\ldots,n \end{cases} \tag{38}$$

where $K_i$'s are fuzzy sets. By using $\alpha$-cuts of $K_i$'s, i.e. $R_\alpha(K_i) = \left\{ x \in X : \mu_{K_i}(x) \geqslant \alpha \right\}$, they replace (38) by

$$\begin{cases} f(x) = \sum_{i=1}^{n} c_i x_i \rightarrow \max_x \\ \text{subject to:} \\ x_1 R_\alpha(K_1) + \ldots + x_n R_\alpha(K_n) \subset R_\alpha(K) \\ x_i \geqslant 0 \qquad i=1,\ldots,n \quad \text{for each } \alpha \in (0,1] \end{cases} \tag{39}$$

which, for $\mu_K(x) \in \left\{\alpha_1, \ldots, \alpha_p\right\}$, is a finite set of the so-called set inclusive linear programs solvable by conventional linear programming techniques (see, e.g., Soyster (1973) for details).

It should be noted that the above approach has some serious drawbacks which are often prohibitive in its practical use. First, "$\subset$" is the conventional fuzzy set inclusion (see Fedrizzi's paper earlier in the volume) of a "yes-no" character which makes the problem unnecessarily rigid. The use of a less rigid definition of containment of two fuzzy sets to a degree (cf. Kacprzyk, 1983a) could help, although presumably at the expense of analytical tractability. Moreover, even if $\mu_K(x)$ takes on a finite number of distinct values, this number is usually high so that we obtain a high dimensional equivalent conventional linear program. Several approaches have appeared to overcome this difficulty. In one of them, due to Orlovski (1984b) fuzzy information on coefficients in a fuzzy mathematical programming problem is used to extract a fuzzy preference relation over the set of alternatives, and then to use this relation for determining nondominated alternatives as solutions to the problem.

Many approaches assume fuzzy coefficients to be some specific numbers. For instance, Dubois and Prade (1980) use the so-called L - R representation of fuzzy numbers, and Tanaka and Asai (1984a) and Tanaka, Ichihashi and Asai (1984) use fuzzy numbers with triangular (pyramidal) membership functions. The latter approach makes it also possible to obtain a fuzzy or nonfuzzy optimal solution. It seems to be quite promising as it has been successfully applied to designing agricultural policies (Owsiński, Zadrożny and Kacprzyk's paper later in this volume). Some more information on approaches to using some specific representations of fuzzy coefficient numbers can be found in Słowiński's paper later in this volume.

### 3.2. Fuzzy integer and 0-1 programming

Although mathematical programming problems in which decision variables are required to take on discrete values, integer or 0-1, as opposed to real values in the previous problems, are of utmost importance in many fields, e.g., in all operations - research - and management-related ones, not much work has been done on fuzzification of those models. We will sketch below some attempts.

#### 3.2.1. Fuzzy integer programming

Almost all of the progress in the field is due to Fabian and Stoica (1984). They start from the conventional nonfuzzy integer program

$$\begin{cases} f(x) \to \max_x \\ \text{subject to:} \\ g(x) \leqslant 0 \\ x = (x_1, \ldots, x_n), \; x_i \geqslant 0, \quad x_i \text{ - integer} \end{cases} \tag{40}$$

where $f(x)$ and $g(x)$ are real-valued functions. This problem (40) is now fuzzified as follows:

$$\begin{cases} f(x) \to \widetilde{\max}_x \\ \text{subject to:} \\ g(x) \lesssim 0 \\ x_i \geqslant 0, \quad x_i \text{ - } \widetilde{\text{integer}} \end{cases} \tag{41}$$

to be read as: find a "possibly maximal" ($\widetilde{\max}$) solution $x^*$ which satisfies the constraints to a "possibly high" degree, and whose components, $x_i$'s, are "almost" integers. Let us notice that the last requirement does not force the solutions to be exactly integers which may be a source of serious numerical difficulties in conventional large integer programs.

Basically, by choosing appropriate fuzzy sets to represent fuzziness in (40) concerning an approximate optimization, constraint satisfaction and integral values of the decision variables, an equivalent nonlinear mixed integer program is derived, and a procedure for its solution is given.

The model has found application in some production scheduling problems.

A solution technique for solving fuzzy integer programming models with multiple criteria appeared in Ignizio and Daniels (1983).

### 3.2.2. Fuzzy 0-1 programming

Practically, the only works on fuzzy 0-1 programming are those of Zimmermann and Pollatschek (1979, 1984). They extend Zimmermann's fuzzy linear programming model (cf. Section 3.1.1) by adding the requirements $x_i \in \{0,1\}$, i.e.

$$\begin{cases} \sum_{i=1}^{n} e_i x_i \gtrsim Z \\ \sum_{i=1}^{n} a_{ji} x_i \lesssim b_j & j=1,\ldots,m \\ x_i \in \{0,1\} & i=1,\ldots,n \end{cases} \tag{42}$$

Then, following in principle the line of reasoning (33) - (36), a conventional (nonfuzzy) equivalent of (42) is derived. A branch-and-bound procedure for its solution is developed.

## 3.3. Fuzzy dynamic programming

Dynamic programming is an effective approach to solving a variety of optimization (decision making) problems of multistage (dynamic) character. The first attempts at the fuzzification of dynamic programming appeared very early (Chang, 1969; Bellman and Zadeh, 1970). Their essence may be best seen by using the following framework, Let: $X = \{x\} = \{s_1,\ldots,s_n\}$ be a state space, $U = \{u\} = \{c_1,\ldots,c_m\}$ be a control space, the temporal evolution of a dynamic system under control be described by its state equation $x_{t+1} = f(x_t,u_t)$, where $x_t$, $x_{t+1} \in X$ are states at times $t$ and $t+1$, respectively, and $u_t \in U$ is a control at time $t$. $x_0 \in X$ is an initial state and $N$ is a fixed and specified termination time.

For simplicity, we assume that for each $t = 0,1,\ldots,N-1$, a fuzzy constraint $\mu_{C^t}(u_t)$, and only for $t = N$ a fuzzy goal $\mu_{G^N}(x_N)$ are defined. The problem is to find an optimal sequence of controls $u_0^*,\ldots,u_{N-1}^*$, such that

$$\mu_D(u_0^*,\ldots,u_{N-1}^* | x_0) = \max_{u_0,\ldots,u_{N-1}} \mu_D(u_0,\ldots,u_{N-1} | x_0) =$$

$$= \max_{u_0,\ldots,u_{N-1}} (\mu_{C^0}(u_0) \wedge \ldots \wedge \mu_{C^{N-1}}(u_{N-1}) \wedge \mu_{G^N}(x_N)) \qquad (43)$$

where, $x_{t+1} = f(x_t, u_t)$, $t=0,1,\ldots,N-1$.

It is easy to see that this problem can be solved by dynamic programming through the following set of recurrence equations

$$\begin{cases} \mu_{G^{N-i}}(x_{N-i}) = \max\limits_{u_{N-1}} (\mu_{C^{N-1}}(u_{N-i}) \wedge \mu_{G^{N-i+1}}(x_{N-i+1})) \\ x_{N-1+1} = f(x_{N-1}, u_{N-1}) \qquad i=1,\ldots,N \end{cases} \qquad (44)$$

This basic formulation can be considerably extended, mainly with respect to: (1) the type of termination time: implicity given by entering a termination set of states, fuzzy, and infinite, and (2) the type of system under control: stochastic or fuzzy. For an excellent short review, see Esogbue and Bellman (1984), and for a detailed analysis - Kacprzyk (1983a).

Among numerous applications of fuzzy dynamic programming we should mention those for: research and development control (Esogbue, 1983), health care systems, clustering, water systems (for all, see Esogbue and Bellman, 1984), and regional development (Kacprzyk and Straszak, 1984).

## 4. REMARKS ON SOME RECENT KNOWLEDGE-BASED APPROACHES

Recently, some newer approaches to fuzzy optimization have appeared. Basically, they try to further "soften" the models presented in the previous sections by representing some commonsense perceptions. In fuzzy multicriteria optimization an optimal solution is sought that best satisfies, e.g., most of the important objectives (Yager, 1983, Kacprzyk and Yager, 1984a, 1984b) as opposed to that satisfying all the objectives in the conventional models. In the multistage case an optimal sequence of controls is sought that best satisfies the goals and contraints at, e.g., most of the earlier control stages (Kacprzyk, 1983b; Kacprzyk and Yager, 1984a, 1984b). The approach may also be used in other problems, as, e.g., in group decision making (see Kacprzyk (1985a) for a review). The above approaches employ Zadeh's (1983a, 1983b) representation of commonsense knowledge equated with a collection of dispositions (propositions with implicit linguistic quantifiers) handled by using fuzzy logic.

The above may be seen as attempts to develop what might be called knowledge-based optimization and mathematical programming models as opposed to the data-based conventional ones. This should eventually lead to an expert-system-based decision support for optimization which should greatly enhance implementability of optimization tools and techniques in real world problems.

## 5. CONCLUDING REMARKS

This paper is a brief introduction to fuzzy optimization and mathematical programming and a survey of the main contributions in these fields. An interested reader, who has not yet been exposed to the subject, will here find a body of basic knowledge needed to be able to read both the following articles in the volume and other literature. For other readers, the paper can be a source of basic contributions in the field.

## REFERENCES

Bellman, R.E., and L.A. Zadeh (1970). Decision-making in a fuzzy environment. Mang. Sci. 17, 151-169.

Buckley, J.J. (1983). Fuzzy programming and the Pareto optimal set. Fuzzy Sets and Syst. 10, 56-64.

Carlsson, C. (1981). Solving ill-structured problems through well-structured fuzzy programming. In J.P. Brans (ed.), Op. Res. '81. North-Holland, Amsterdam.

Carlsson, C. (1982). Fuzzy multiobjective programming with composite compromises. In M. Grauer, A. Lewandowski, and A.P. Wierzbicki (eds.), Multiobjective and Stochastic Optimization. CP-82-S12, IIASA, Laxenburg.

Chanas, S. (1983). The use of parametric programming in fuzzy linear programming. Fuzzy Sets and Syst. 11, 243-251.

Chanas, S., W. Kołodziejczyk, and A. Machaj (1984). A fuzzy approach to the transportation problem. Fuzzy Sets and Syst. 13, 211-222.

Chanas, S., and M. Kulej (1984). A fuzzy linear programming problem with equality constraints. In Kacprzyk (1984a), 195-202.

Chang, S.S.L. (1969). Fuzzy dynamic programming and the decision making process. In Proc. 3rd Princeton Conf. on Inf. Sci and Syst., 200-203.

Checkland, P.B. (1973). Towards a system-based methodology for real - world problem solving. J. Syst. Mang. 3.

Delgado, M. (1983). A resolution method for multiobjective problems. Eur. J. Op. Res. 13, 165-172.

Dubois D., and H. Prade (1980). Systems of linear fuzzy constraints. Fuzzy Sets and Syst. 3, 37-48.

Dyson, R.G. (1980). Maxmin programming, fuzzy linear programming and multicriteria decision making. J. Op. Res. Soc. 31, 263-267.

Esogbue, A.O. (1983). Dynamic programming, fuzzy sets, and the modeling of R & D management control systems. IEEE Trans. on Syst. Man and Cyber. SMC-13, 18-30.

Esogbue, A.O., and R.E. Bellman (1984). Fuzzy dynamic programming. In Zimmermann, Zadeh and Gaines (1984), 147-167.

Esogbue, A.O. (1986). Bellman memorial issue. Fuzzy Sets and Syst. To appear.

Fabian, C., and M. Stoica (1984). Fuzzy integer programming. In Zimmermann, Zadeh and Gaines (1984), 123-131.

Feng, Y.J. (1983). A method using fuzzy mathematics to solve vector maximum problem. Fuzzy Sets and Syst. 9, 129-136.

Flachs, J., and M.A. Pollatschek (1978). Further results on fuzzy mathematical programming. Inf. and Control 38, 241-257.

Flachs, J., and M.A. Pollatschek (1979). Duality theorems for certain problems involving minimum and maximum operation. Math. Prog. 16, 348-370.

Fung, L.W., and K.S. Fu (1977). Characterization of a class of fuzzy optimal control problems. In B.R. Gaines, M.M. Gupta, and G.N. Saridis (eds.), Fuzzy Information and Decision Processes. North-Holland, Amsterdam.

Hamacher, H., H. Leberling, and H.J. Zimmermann (1978). Sensitivity analysis in fuzzy linear programming. Fuzzy Sets and Syst. 1, 269-281.

Hannan, E.L. (1979). On the efficiency of the product operator in fuzzy programming with multiple objectives. Fuzzy Sets and Syst. 2, 259-262.

Hannan, E.L. (1981a).Linear programming with multiple fuzzy goals. Fuzzy Sets and Syst. 6, 235-248.

Hannan, E.L. (1981b). On fuzzy goal programming. Decision Sci. 12, 522-531.

Hannan, E.L. (1981c). Fuzzy programming with multiple fuzzy goals. Fuzzy Sets and Syst. 6, 235-248.

Hannan, E.L. (1982). Contrasting fuzzy goal programming and "fuzzy" multicriteria programming. Decision Sci. 13, 337-339.

Ignizio, J.P. (1982). On the(re) discovery of fuzzy goal programming. Decision Sci. 13, 331-336.

Ignizio, J.P., and S.C. Daniels (1983). Fuzzy multicriteria integer programming via fuzzy generalized networks. Fuzzy Sets and Syst. 10, 261-270.

Kabbara, G. (1982). New utilization of fuzzy optimization method. In M.M. Gupta, and E. Sanchez (eds.), Fuzzy Information and Decision Processes. North-Holland, Amsterdam.

Kacprzyk, J. (1982). Multistage decision processes in a fuzzy environment: a survey. In M.M. Gupta, and E. Sanchez (eds.), Fuzzy Information and Decision Processes. North-Holland, Amsterdam.

Kacprzyk, J. (1983a). Multistage Decision-Making under Fuzziness: Theory and Applications. ISR Series. Verlag TÜV Rheinland, Cologne.

Kacprzyk, J. (1983b). A generalization of fuzzy multistage decision-making and control via linguistic quantifiers. Int. J. Control 38, 1249-1270.

Kacprzyk, J. (Guest ed.) (1984a). Special issue on fuzzy sets and possibility theory in optimization models. Control and Cyber. 4, No. 3.

Kacprzyk, J. (1985). Zadeh's commonsense knowledge and its use in multicriteria, multistage and multiperson decision making, In M.M. Gupta et al. (eds.), Approximate Reasoning in Expert Systems. North-Holland, Amsterdam.

Kacprzyk, J. (1986a). Group decision making with a fuzzy linguistic majority. Fuzzy Sets and Syst. To appear.

Kacprzyk, J. (1986b). Towards "human-consistent" multistage decision making and control models using fuzzy sets and fuzzy logic. In A.O. Esogbue (1986). To appear.

Kacprzyk, J., and J.W. Owsiński (1984). Nonstandard mathematical programming models including imprecision as a planning tool in an agricultural enterprise operating in varying conditions (in Polish) In Proc. Conf. on Organization of Agricultural Enterprises. Kołobrzeg, 1984.

Kacprzyk, J., and A. Straszak (1984). Determination of "stable" trajectories of integrated regional development using fuzzy decision models. IEEE Trans. on Syst. Man and Cyber. SMC-14, 310-313.

Kacprzyk, J., and R.R. Yager (1984a). "Softer" optimization and control models via fuzzy linguistic quantifiers. Inf. Sci. 34, 157-178.

Kacprzyk, J., and R.R. Yager (1984b). Linguistic quantifiers and belief qualification in fuzzy multicriteria and multistage decision making. In Kacprzyk (1984a), 155-174.

Kacprzyk, J., and R.R. Yager (eds.) (1985). Management Decision Support Systems Using Fuzzy Sets and Possibility Theory. Verlag TÜV Rheinland, Cologne.

Leberling, H. (1981). On finding compromise solutions in multicriteria problems using the fuzzy min operator. Fuzzy Sets and Syst. 6, 105-118.

Leung, Y. (1982). Multicriteria conflict resolution through a theory of displaced fuzzy ideal. In M.M. Gupta, and E. Sanchez (eds.), Approximate Reasoning in Decision Analysis. North Holland, Amsterdam.

Leung, Y. (1983). Concept of a fuzzy ideal for multicriteria conflict resolution. In P.P. Wang (ed.), Fuzzy Sets Theory and Applications. Plenum, New York.

Leung, Y. (1984). Compromise programming under fuzziness. In Kacprzyk (1984a), 203-216.

Llena, J. (1985). On fuzzy linear programming. Eur. J. Op. Res. 22, 216-223.

Luhandjula, M.K. (1982). Compensatory operators in fuzzy linear programming with multiple objectives. Fuzzy Sets and Syst. 8, 245-252.

Luhandjula, M.K. (1983). Linear programming under randomness and fuzziness. Fuzzy Sets and Syst. 10, 57-63.

Luhandjula, M.K. (1984). Fuzzy approaches for multiple objective linear fractional optimization. Fuzzy Sets and Syst. 13, 11-24.

Nakamura, K. (1984). Some extensions of fuzzy linear programming. Fuzzy Sets and Syst. 14, 211-229.

Narasimhan, R. (1980). Goal programming in a fuzzy environment. Decision Sci. 11, 325-336.

Narasimhan, R. (1981). On fuzzy goal programming - some comments. Decision Sci. 12, 532-538.

Negoita, C.V. (1981). The current interest in fuzzy optimization. Fuzzy Sets and Syst. 6, 261-269.

Negoita, C.V. (1984). Structure and logic in optimization. In: Kacprzyk (1984a), 121-128.

Negoita, C.V., P. Flondor, and M. Sularia (1977). On fuzzy environment in optimization problems. Econ. Comp. and Econ. Cybern. Stud. and Res. 1, 3-12.

Negoita, C.V., S. Minoiu, and E. Stan (1976). On considering imprecision in dynamic linear programming. Econ, Comp. and Econ. Cybern. Stud. and Res. 3, 83-95.
Negoita, C.V., and D. Ralescu (1975). Applications of Fuzzy Sets to Systems Analysis. Birkhauser, Basel.
Negoita, C.V., and D. Ralescu (1977). On fuzzy optimization. Kybernetes 6, 193-195.
Negoita, C.V., and A.C. Stefanescu (1982). On fuzzy optimization. In M.M. Gupta, and E. Sanchez (eds.), Fuzzy Information and Decision Processes. North Holland, Amsterdam.
Negoita, C.V., and M. Sularia (1976). On fuzzy programming and tolerances in planning. Econ. Comp. and Econ. Cybern. Studies and Res. 1, 3-15.
Oheigertaigh, M.A. (1982). A fuzzy transportation algorithm. Fuzzy Sets and Syst. 8, 235-245.
Ollero, A., J. Aracil, and E.F. Carmacho (1984). Optimization of dynamic regional models: an interactive multiobjective approach. Large Scale Syst. 6, 1-12.
Orlovski, S.A. (1977). On programming with fuzzy constraint sets. Kybernetes 6, 197-201.
Orlovski, S.A. (1978). Decision making with a fuzzy preference relation. Fuzzy Sets and Syst. 1, 155-167.
Orlovski, S.A. (1980). On formalization of a general fuzzy mathematical programming problem. Fuzzy Sets and Syst. 3, 311-321.
Orlovski, S.A. (1981). Problems of Decision-Making with Fuzzy Information (in Russian). Nauka, Moscow.
Orlovski, S.A. (1982). Effective alternatives for multiple fuzzy preference relations. In R. Trappl (ed.), Cybernetics and Systems Research. North Holland, Amsterdam.
Orlovski, S.A. (1983). Problems of Decision-Making with Fuzzy Information. WP-83-28, IIASA, Laxenburg.
Orlovski, S.A. (1984a). Multiobjective programming problems with fuzzy parameters. In: Kacprzyk (1984a), 175-184.
Orlovski, S.A. (1984b). Mathematical Programming Problems with Fuzzy Parameters. WP-84-38. IIASA, Laxenburg.
Ralescu, D. (1978). The interfaces between orderings and fuzzy optimization. ORSA/TIMS Meeting. Los Angeles.
Ralescu, D. (1979). A survey of representations of fuzzy concepts and its application. In M.M. Gupta, R.K. Ragade, and R.R. Yager (eds.), Advances in Fuzzy Set Theory and Applications. North-Holland, Amsterdam.
Ralescu, D. (1984). Optimization in a fuzzy environment. In M.M. Gupta, and E. Sanchez (eds.), Fuzzy Information, Knowledge Representation and Decision Analysis. Proc. of IFAC Workshop, Pergamon Press, Oxford.
Ramik, J. (1983). Extension principle and fuzzy-mathematical programming. Kybernetika 19, 516-525.
Ramik, J., and J. Rimanek (1985). Inequality relation between fuzzy numbers and its use in fuzzy optimization. Fuzzy Sets. and Syst. 16, 123-138.
Rapaport, A. (1970). Modern Systems theory - an outlook for coping with change. Gen. Syst. Yearbook XV, Soc. for Gen. Syst.
Rödder, W., and H.J. Zimmermann (1977). Duality in fuzzy programming. Int. Symp. on Extremal Methods and Syst. Anal. Austin, Texas.
Rubin, P.A., and B. Narasimhan (1984). Fuzzy goal programming with nested priorities. Fuzzy Sets and Syst. 14, 115-129.

Sakawa, M. (1983). Interactive computer programs for fuzzy linear programming with multiple objectives. Int. J. Man-Machine Stud. 18, 489-503.
Sakawa, M. (1984a). Interactive fuzzy goal programming for multiobjective nonlinear problems and its application to water quality management. In: Kacprzyk (1984a),217-228.
Sakawa, M. (1984b). Interactive multiobjective decision making by the fuzzy sequential proxy optimization technique: FSPOT. In Zimmermann, Zadeh and Gaines (1984),241-260.
Sakawa, M., and F. Seo (1983). Interactive multiobjective decision - making in environmental - systems using the fuzzy sequential proxy optimization technique. Large Scale Syst. 4, 223-243.
Sakawa, M., and T. Yumine (1983). Interactive fuzzy decision making for multiobjective linear fractional programming problems. Large Scale Syst. 5, 105-114.
Sher, A.P. (1980). Solving a mathematical programming problem with a linear goal function in fuzzy constraints (in Russian). Aut. and Remote Control 40, 137-143.
Soyster, A.L. (1973). Convex programming with set-inclusive constraints. Application to inexact linear programming. Op. Res. 21, 1154-1157.
Sommer, G., and M.A. Pollatschek (1978). A fuzzy programming approach to an air pollution regulation problem. Eur. J. Op. Res. 10, 303-313.
Takeda, E., and T.N. Nishida (1980). Multiple criteria decision making with fuzzy domination structures. Fuzzy Sets and Syst. 3, 123-136.
Tanaka, H., and K. Asai (1981). Fuzzy linear programming based on fuzzy functions. Proc. 8th IFAC World Congress (Kyoto). Pergamon Press, Oxford.
Tanaka, H., and K. Asai (1984a). Fuzzy linear programming problems with fuzzy numbers. Fuzzy Sets and Syst. 13, 1-10.
Tanaka, H., and K. Asai (1984b). Fuzzy solution in fuzzy linear programming problems. IEEE Trans. on Syst. Man and Cyber. SMC-14, 285-388.
Tanaka, H., H. Ichihashi, and K. Asai (1984). A formulation of fuzzy linear programming problems based on comparison of fuzzy numbers. In: Kacprzyk (1984a), 185-194.
Tanaka, H., T. Okuda, and K. Asai (1974). On fuzzy mathematical programming. J. Cyber. 3-4, 37-46.
Verdegay, J.L. (1982). Fuzzy mathematical programming. In M.M. Gupta, and E. Sanchez (eds.), Fuzzy Information and Decision Processes. North-Holland, Amsterdam.
Verdegay, J.L. (1983). Transportation problem with fuzzy parameters (in Spanish). Rev. Acad. Cien. Mat. Fis. Quim. y Nat. de Granada 2, 47-56.
Verdegay, J.L. (1984a). A dual approach to solve the fuzzy linear programming problem. Fuzzy Sets and Syst. 14, 131-141.
Verdegay, J.L. (1984b). Applications of fuzzy optimization in operational research. In: Kacprzyk (1984a), 229-240.
Wagenknecht, M., and K. Hartmann (1983). On fuzzy rank ordering in polyoptimization. Fuzzy Sets and Syst. 11, 253-264.
Wiedey, G., and H.J. Zimmermann (1979). Media selection and fuzzy linear programming. J. Op. Res. Soc. 29, 1071-1084.

Yager, R.R. (1977). Multiple objective decision making using fuzzy sets. Int. J. Man-Machine Stud. 9, 375-382.
Yager, R.R. (1978). Fuzzy decision making including unequal objectives. Fuzzy Sets and Syst. 1, 87-95.
Yager, R.R. (1979). Mathematical programming with fuzzy constraints and a preference on the objective. Kybernetes 9, 109-114.
Yager, R.R. (1983). Quantifiers in the formulation of multiple objective decision functions. Inf. Sci. 31, 107-139.
Zadeh, L.A. (1968). Fuzzy algorithms. Inform. and Control 12, 94-102.
Zadeh, L.A. (1983a). A computation approach to fuzzy quantifiers in natural languages. Comp. and Math. with Appls. 9, 149-184.
Zadeh, L.A. (1983b). A Theory of Commonsense Knowledge. Memo. UCB/ERL M83/27, University of California, Berkeley.
Zimmermann, H.J. (1975). Optimal decisions in problems with fuzzy description (in German). Z. f. Betriebswirtschaftliche Forschung 12, 785-795.
Zimmermann, H.J. (1976). Description and optimization of fuzzy systems. Int. J. Gen. Syst. 2, 209-215.
Zimmermann, H.J. (1978). Fuzzy programming and linear programming with several objective functions. Fuzzy Sets and Syst. 1, 45-55.
Zimmermann, H.J. (1983). Using fuzzy sets in operational research. Eur. J. Op. Res. 13, 201-216.
Zimmermann, H.J., and M.A. Pollatschek (1979). A Unified Approach to Three Problems in Fuzzy 0-1 Linear Programs. Working Paper, RWTH Aachen.
Zimmermann, H.J., and M.A. Pollatschek (1984). Fuzzy 0-1 linear programs. In Zimmermann, Zadeh and Gaines (1984).
Zimmermann, H.J., L.A. Zadeh, and B.R. Gaines (eds.) (1984). Fuzzy Sets and Decision Analysis. TIMS Studies in the Management Sciences, Vol. 20, North-Holland, Amsterdam.

# II. ADVANCES IN FUZZY DECISION MAKING, FUZZY OPTIMIZATION, AND FUZZY MATHEMATICAL PROGRAMMING

## II.1. Fuzzy Preferences and Choice

FUZZY PREFERENCES IN AN OPTIMIZATION PERSPECTIVE

Marc Roubens* and Philippe Vincke**

* State University of Mons, B-7000 Mons, Belgium

** Free University of Brussels, B-1050 Brussels, Belgium

Abstract. This paper should be considered as an introduction to the fundamental properties of binary fuzzy relations. It summarizes some of the proposed definitions of a fuzzy preference relation, compares them and introduces the reader to the difficult problems of ranking and choice on the basis of a preference relation. The last part points out an important role of fuzzy relations in multicriteria analysis.

Keywords: fuzzy relation, fuzzy preference relation, ranking, choice, multiple criteria, decision making.

## 1. INTRODUCTION

In many decision making problems, the preference relations in the set of alternatives are of a fuzzy nature, reflecting the imprecision of experts' estimates or uncertain aspects of preferences.

The literature on non-fuzzy preference relations is rather rich and deals with structures called complete or partial orders and preorders, semiorders, interval orders, etc. These definitions can be extended to the fuzzy case in different ways.

Sections 2 and 3 deal with the basic properties of binary fuzzy relations using the min and max operators. Related $\alpha$-cuts and nested families of crisp relations are emphasized.

Section 4 investigates three different strict preference concepts and determines the logical relationships between the transitivity properties of these preference relations.

Sections 5 and 6 introduce different tools to solve the ranking and choice problems: utility functions, domination concepts, etc.

Section 7 presents a short survey on multiple criteria decision making methods using fuzzy outranking relations.

## 2. SOME PROPERTIES FOR A BINARY FUZZY RELATION

We consider a binary fuzzy relation S in a finite set A,

that is a mapping $\mu_S$ from $A \times A$ to $[0,1]$; $\mu_S(a,b)$ will denote the image of the ordered pair (a,b) by this mapping. A binary fuzzy relation S is said to be:

| | |
|---|---|
| reflextive | if $\mu_S(a,a)=1,\ \forall a\in A$; |
| irreflexive | if $\mu_S(a,a)=0,\ \forall a\in A$; |
| symmetric | if $\mu_S(a,b)=\mu_S(b,a),\ \forall a,b\in A$; |
| weakly antisymmetric | if $\min[\mu_S(a,b),\mu_S(b,a)]\leq\frac{1}{2}, \forall a,b\in A, a\neq b$; |
| weakly complete | if $\max[\mu_S(a,b),\mu_S(b,a)]\geq\frac{1}{2}, \forall a,b\in A, a\neq b$; |
| antisymmetric | if $\min[\mu_S(a,b),\mu_S(b,a)]=0, \forall a,b\in A, a\neq b$; |
| complete | if $\max[\mu_S(a,b),\mu_S(b,a)]=1, \forall a,b\in A, a\neq b$; |
| saturated | if $\min[\mu_S(a,b),\mu_S(b,a)]>0, \forall a,b\in A, a\neq b$; |
| transitive | if $\mu_S(a,c)\geq\min[\mu_S(a,b),\mu_S(b,c)], \forall a,b,c\in A$; |
| negatively transitive | if $\mu_S(a,c)\leq\max[\mu_S(a,b),\mu_S(b,c)], \forall a,b,c\in A$; |
| linear | if $\mu_S(a,b)>\mu_S(b,c)\rightarrow\mu_S(a,d)\geq\mu_S(b,d), \forall a,b,c,d\in A$; |
| probabilistic | if $\mu_S(a,b)+\mu_S(b,a)=1, \forall a,b\in A,\ a\neq b$. |

Some of these properties naturally follow from the equivalent crisp relations if we adopt the following usual concepts, S and T being two fuzzy relations on A:

$S\subset T$ iff $\mu_S(a,b) \leq \mu_T(a,b), \forall a,b\in A$;

$\mu_{S\cap T}(a,b) = \min[\mu_S(a,b),\mu_T(a,b)]$;

$\mu_{S\cup T}(a,b) = \max[\mu_S(a,b),\mu_T(a,b)]$;

$\mu_{S.T}(a,b) = \max_c \min[\mu_S(a,c),\mu_T(c,b)]$;

$\mu_{S\setminus T}(a,b) = \max[0,\mu_S(a,b)-\mu_T(a,b)]$;

$\mu_{S^-}(a,b) = \mu_S(b,a)$ : $S^-$ is the converse relation of S;

$\mu_{S^c}(a,b) = 1-\mu_S(a,b)$ : $S^c$ is the complementary relation of S;

$\mu_{S^d}(a,b) = 1-\mu_S(b,a)$ : $S^d$ is the dual relation of S.

For example, the crisp antisymmetry is defined as

$$S \cap S^- \cap\{(a,b),\ a\neq b\} = \emptyset.$$

Dealing with a fuzzy relation, we obtain the definition presented here. It is also the case of transitivity and negative transitivity which, for a crisp relation, are defined as $S.S\subset S$ and $S^c.S^c \subset S^c$, respectively.

On the other hand, the reader will easily verify the following proposition, generalizing the crisp situation.

Proposition 1. The dual relation of a reflexive (resp. irreflexive, symmetric, weakly antisymmetric, weakly complete, antisymmetric, complete, saturated, transitive, negatively transitive and probabilistic) relation is irreflexive (resp. reflexive, symmetric, weakly complete, weakly antisymmetric, complete, antisymmetric, non-saturated, negatively transitive, transitive and probabilistic).

The previous properties are common and were used for instance by Zadeh (1971) to define concepts like fuzzy orderings, preorderings, partial orderings, weak orderings and linear orderings.

## 3. $\alpha$ -CUTS OF A BINARY FUZZY RELATION AND NESTED FAMILIES OF CRISP RELATIONS

For each fuzzy relation S, a nested sequence of crisp relations $\{S_\alpha, \alpha \in (0,1]\}$, called $\alpha$-cuts, can be defined as follows:

$$a\, S_\alpha b \quad \text{iff} \quad \mu_S(a,b) \geqslant \alpha \; .$$

A natural way of defining a property P of fuzzy relation consists in asking that all its $\alpha$-cuts have this property P. As an exercise we leave the proof of the following proposition to the reader.

Proposition 2. A fuzzy relation S is reflexive (respectively: irreflexive, symmetric, antisymmetric, complete, transitive and negatively transitive) iff every $\alpha$-cut of S has the corresponding crisp property.

Conversely, given a family $F = \{S_\lambda, \lambda \in (0,1]\}$ of crisp relations on A such that

$$\lambda_1 > \lambda_2 \rightarrow S_{\lambda_1} \subseteq S_{\lambda_2} \; ,$$

we obtain a fuzzy relation S given by

$$\mu_S(a,b) = \max_\lambda \; \mu_{S_\lambda}(a,b) .$$

It is clear that the family of $\alpha$-cuts of S coincides with the initial family F. So, it is equivalent to study properties of fuzzy relations and of nested families of crisp relations (for a more complete and rigorous proof, see DOIGNON, MONJARDET, ROUBENS and VINCKE, submitted).

Now, giving a fuzzy relation S, it may be interesting to determine the crisp relation which is, in some sense, the nearest to S. Using the Hamming distance, we have:

Proposition 3. Given a fuzzy relation S on A, the 0.5-cut of S minimizes

$$d(S,T) = \sum_{a,b \in A} |\mu_S(a,b) - \mu_T(a,b)|$$

among the set of all possible crisp relations T on A.

Proof. $d(S,T)$ will be minimum if $\mu_T(a,b)=1$ when $\mu_S(a,b)>.5$, and $\mu_T(a,b)=0$ when $\mu_S(a,b)<.5$.

4. FUZZY PREFERENCES

Suppose that $\mu_S(a,b)$ represents the degree to which the proposition "a is not worse than b" is true so that S may be considered a fuzzy preference relation. It is reasonable to define fuzzy indifference I and fuzzy incomparability R as in the crisp situation, i.e.

$$I = S \cap S^-,$$

$$R = (S \cup S^-)^c,$$

that is

$$\mu_I(a,b) = \min[\mu_S(a,b),\ \mu_S(b,a)],$$

$$\mu_R(a,b) = 1-\max[\mu_S(a,b),\mu_S(b,a)].$$

Now, several possible expressions exist to define the strict preference P. We give three of them here. The first one is an extension of the crisp definition

$$P_1 = S \cap S^d,$$

that is

$$\mu_{P_1}(a,b) = \min[\mu_S(a,b),\ \mu_{S^d}(a,b)].$$

The second one is from ORLOVSKY (1978) and consists in considering the so-called antisymmetrized relation of S given by

$$\mu_{P_2}(a,b) = \max[0,\ \mu_S(a,b) - \mu_S(b,a)]$$

corresponding to the crisp definition $P_2 = S\setminus S^-$.

The third one is by OVCHINNIKOV (1981):

$$\mu_{P_3}(a,b) = \begin{cases} \mu_S(a,b) & \text{if } \mu_S(a,b) > \mu_S(b,a), \\ 0 & \text{if } \mu_S(a,b) \leqslant \mu_S(b,a). \end{cases}$$

The last definition is also obtained in considering the nested family of crisp preferences $\{S_\alpha \cap S^d_\alpha,\ \alpha \in (0,1]\}$, where $S_\alpha$ are the $\alpha$-cuts of S.

The following proposition summarizes the main properties of all these fuzzy relations.

Proposition 4. We have:

- (i) S reflexive → I reflexive, R, $P_1$, $P_2$, $P_3$ irreflexive;
- (ii) I and R are symmetric;
- (iii) S antisymmetric or complete → $P_1$ antisymmetric;
- (iv) S weakly antisymmetric or weakly complete → $P_1$ weakly antisymmetric;
- (v) $P_2$ and $P_3$ are antisymmetric;
- (vi) $S=P_1$ iff $\mu_S(a,b)+\mu_S(b,a)\leq 1$, $\forall a,b$;
- (vii) $S^d=P_1$ iff $\mu_S(a,b)+\mu_S(b,a)\geq 1$, $\forall a,b$;
- (viii) $S=P_2=P_3$ iff S is antisymmetric;
- (ix) S crisp → $P_1=P_2=P_3$;
- (x) S transitive → I transitive;
- (xi) $S^d$ transitive (i.e. S negatively transitive) → R transitive;
- (xii) S and $S^d$ transitive → $P_1$ transitive;
- (xiii) S or $S^d$ transitive → $P_2$ transitive;
- (xiv) $P_1$ transitive → $P_2$ transitive;
- (xv) S transitive → $P_3$ transitive;
- (xvi) S transitive ↛ $P_1$ transitive;
- (xvii) $P_1$ transitive ↛ $P_2$ or S transitive;
- (xviii) $P_2$ transitive ↛ $P_1$ or $P_3$ or S transitive;
- (xix) $P_3$ transitive ↛ $P_1$ or $P_2$ or S transitive.

Proofs. (i) to (ix) are easy;

(x) to (xii) result from $I=S\cap S^-$, $R = S^d\cap(S^d)^-$ and $P_1 = S\cap S^d$ and from the fact that the intersection of two transitive relations is also transitive; given two transitive relations S and T, we have

$$\begin{aligned}\mu_{S\cap T}(a,c) &= \min[\mu_S(a,c),\ \mu_T(a,c)]\\ &\geq \min[\min[\mu_S(a,b),\mu_S(b,c)],\min[\mu_T(a,b),\mu_T(b,c)]],\forall b\\ &= \min[\min[\mu_S(a,b),\mu_T(a,b)],\min[\mu_S(b,c),\mu_T(b,c)]],\forall b\\ &= \min[\mu_{S\cap T}(a,b),\mu_{S\cap T}(b,c)],\forall b.\end{aligned}$$

(xiii) see ORLOVSKY (1978)

(xiv) results from (xiii) and from the fact that for each a,b:

$$\mu_S(a,b) - \mu_S(b,a) = \mu_{P_1}(a,b) - \mu_{P_1}(b,a);$$

indeed, if $\mu_S(a,b) \leq 1-\mu_S(b,a)$, then $\mu_S(b,a) \leq 1-\mu_S(a,b)$

so that $\mu_{P_1}(a,b) = \mu_S(a,b)$ and $\mu_{P_1}(b,a) = \mu_S(b,a)$; if $\mu_S(a,b)>1-\mu_S(b,a)$, then $\mu_S(b,a)>1-\mu_S(a,b)$ so that $\mu_{P_1}(a,b) = 1-\mu_S(b,a)$ and $\mu_{P_1}(b,a) = 1-\mu_S(a,b)$

(xv) see OVCHINNIKOV (1981).

(xvi) consider $\mu_S(a,b)=.7$, $\mu_S(b,a)=.7$, $\mu_S(b,c)=.7$, $\mu_S(c,b)=.7$, $\mu_S(a,c)=1$ and $\mu_S(c,a)=.8$

(xvii) consider $\mu_S(a,b)=.8$, $\mu_S(b,a)=.5$, $\mu_S(b,c)=.8$, $\mu_S(c,b)=.3$, $u_S(a,c)=.6$ and $\mu_S(c,a)=.2$

(xviii) consider $\mu_S(a,b)=.4$, $\mu_S(b,a)=.2$, $\mu_S(b,c)=.6$, $\mu_S(c,b)=.4$, $\mu_S(a,c)=.3$ and $\mu_S(c,a)=0$

(xiv) consider $\mu_S(a,b)=.2$, $\mu_S(b,a)=0$, $\mu_S(b,c)=.7$, $\mu_S(c,b)=.5$, $\mu_S(a,c)=.5$ and $\mu_S(c,a)=.4$.

## 5. THE RANKING PROBLEM

Many decision problems in which preferences between alternatives from a given set are described by a single preference, consist in providing a ranking of the alternatives from the best to the worse.

The most usual tool used to solve the ranking problem is provided by the definition of a utility function g which is a real-valued function calculated for all the alternatives of the set A due to, e.g.,

$$g_1(a) = \underset{b \in A}{\text{MAX}}\ \mu(a,b)$$

$$g_2(a) = \sum_{b \in A} \mu(a,b)$$

$$g_3(a) = \sum_{b \in A} \left[\mu(a,b)-\mu(b,a)\right]$$

where $\mu$ is considered as $\mu_S$ or $\mu_P$.

The ranking is obviously obtained by the rule:

a is better than b (a>b) iff g(a)>g(b).

The definition of $g_1$, where $\mu(a,b) = \mu_P(a,b)$, is linked to the concept of dominance which was introduced by ZADEH (1971) in the context of fuzzy partial orders (reflexive, antisymmetric and transitive fuzzy relations) and was also studied by BLIN (1974), DUBOIS and PRADE (1980), ORLOVSKY (1978), SISKOS et al. (1984), TAKEDA and NISHIDA (1980), etc.

The <u>non-domination degree</u> $\mu_{ND}(a)$ - see ZADEH (1971) and ORLOVSKY (1978) - and the <u>non-dominance degree</u> $\mu_{Nd}(a)$ for an alternative a in A are respectively defined as:

$$\mu_{ND}(a) = 1 - \underset{b\in A}{MAX}\ \mu_{P_2}(b,a)$$

$$\mu_{Nd}(a) = 1 - \underset{b\in A}{MAX}\ \mu_{P_2}(a,b) = 1-g_1(a)$$

The real valued function g is then provided by $\mu_{ND}$ (or $1-\mu_{Nd}$) and the alternatives can be ranked according to the decreasing values of $\mu_{ND}$ or the increasing values of $\mu_{Nd}$.

One can also obtain a crisp partial preorder S on set A according to the following rules:

$a\ P^+\ b$ iff $\mu_{ND}(a) > \mu_{ND}(b)$,

$a\ I^+\ b$ iff $\mu_{ND}(a) = \mu_{ND}(b)$,

$a\ P^-\ n$ iff $\mu_{Nd}(a) < \mu_{Nd}(b)$,

$a\ I^-\ b$ iff $\mu_{Nd}(a) = \mu_{Nd}(b)$.

The preference structure (P,I,R), where $S = P \cup I$, corresponds to

$$S = (P^+ \cup I^+) \cap (P^- \cup I^-)$$

with: aPb iff $a\ P^+\ b$ and $a\ P^-\ b$,

or $a\ P^+\ b$ and $a\ I^-\ b$,

or $a\ P^-\ b$ and $a\ I^+\ b$,

aIb iff $a\ I^+\ b$ and $a\ I^-\ b$,

aRb otherwise.

S is obviously reflexive and transitive.

The function $g_2$ is used by KACPRZYK (1985) in the probabilistic situation with $P=P_2$ and is called "strength of (strict) preference".

The function $g_3$ is called the score when $\mu(a,b) = \mu_S(a,b)$ and will be used in section 6.

Another possibility is to find the ranking (complete order) which is "the nearest" to the fuzzy preference relation. This implies the choice of a distance which is as subjective as the choice of g in the previous method.

Some more sophisticated methods have been proposed like the "distillation algorithm" of ROY in ELECTRE III (1978).

In any case, there is a lack of theoretical basis allowing the comparison of the results of these different approaches and an axiomatic justification of a choice. However, let us mention some works which could be useful in this respect.

HASHIMOTO (1983) and ZADEH (1971) have extended to the fuzzy situation the well-known SZPILRAJN's theorem allowing one to complete a partial order to obtain a complete order. The resulting fuzzy relation implicitly contains a natural complete order given by

a better than b iff $\mu(a,b) > 0$.

Some results on the numerical representation of a fuzzy relation can also be useful in this context, as for example:

<u>Proposition 5</u>. The necessary and sufficient condition for the existence of a real-function g such that, $\forall a, b \in A$

$$\mu_S(a,b) > 0 \rightarrow g(a) \geqq g(b) + \mu_S(a,b)$$

is that the valued graph $(A,S^-)$ does not contain any circuit of positive value.

This proposition is a immediate consequence of theorem VIII.1 of ROY (1969).

In the probabilistic situation where $\mu_S(a,b)+\mu_S(b,a)=1$, for all $a \neq b$, and $\mu_S(a,a) = \frac{1}{2}$, for all a, FISHBURN (1973) introduced some stochastic transitivity conditions. One of these conditions is called "strong stochastic transitivity", briefly SST, and corresponds to:

$$\text{MIN}[\mu_S(a,b),\mu_S(b,c)] \geqq \frac{1}{2} \rightarrow \mu_S(a,c) \geqq \text{MAX}[\mu_S(a,b),\mu_S(b,c)], \quad \forall a,b,c \in A.$$

Due to the probabilistic situation, transitivity and negative transitivity are equivalent and

S probabilistic and transitive →

$$\text{MIN}[\mu_S(a,b),\mu_S(b,c)] \leqq \mu_S(a,c) \leqq \text{MAX}[\mu_S(a,b),\mu_S(b,c)], \forall a,b,c \in A.$$

We then have

<u>Proposition 6</u>. There holds:

(i) : S probabilistic and transitive → SST

(ii): S probabilistic and SST $\not\rightarrow$ S transitive.

<u>Proof</u>. If S is probabilistic, transitive and $\text{MIN}[\mu_S(a,b),\mu_S(b,c)] \geqq \frac{1}{2}$ : $\frac{1}{2} \leqq \mu_S(a,c) = \text{MAX}[\mu_S(a,c), \mu_S(b,c)]$

In order to prove this, suppose that $\mu_S(a,c) < \text{MAX}[\mu_S(a,b), \mu_S(b,c)]$ with $\mu_S(a,b) \geqq \mu_S(b,c)$ (the proof is still valid in the complementary situation). We thus obtain $\frac{1}{2} \leqq \mu_S(a,c) < \mu_S(a,b)$. Transitivity implies that $\mu_S(a,b) \leqq \text{MAX}[\mu_S(a,c), \mu_S(c,b)]$. $\mu_S(b,c) \geqq \frac{1}{2} \rightarrow \mu_S(c,b) < \frac{1}{2}$ and $\mu_S(a,b) \leqq \mu_S(a,c)$ which is impossible.

As pointed out by ROBERTS (1979), SST is related to some functional representations. The interested reader will find in ROUBENS and VINCKE (1984) characterizations of fuzzy relations leading to a representation by real intervals for the probabilistic case. Generalizations of these results and applications to other fields than preference modelling are presented in DOIGNON, MONJARDET, ROUBENS and VINCKE (submitted).

## 6. THE CHOICE PROBLEM

In this section we consider decision problems for which we want to rationally determine the best alternatives, i.e. those which are better than all other alternatives or the non-dominated alternatives, i.e. those alternatives for which better ones do not exist.

The class of alternatives with maximum non-domination (resp. non-dominance) degree is called the class of non-dominated (resp. non-dominating) elements.

An element is unfuzzily non-dominated (resp. unfuzzily non-dominating) iff $\mu_{ND}(a) = 1$ or equivalently $\mu_{P_2}(b,a) = 0$, all $b \in A$, (resp. $\mu_{Nd}(a) = 1$).

ORLOVSKY (1978) proved that any fuzzy preorder (reflexive and transitive relation) has unfuzzily non-dominated and non-dominating elements.

He also gave some sufficient conditions for the existence of unfuzzily non-dominated elements.

Starting with the $\mu_S$-tableau $\{\mu_S(a,b)\}$ we consider the $\mu_S$-board which is an ordered 3-tuple $(S, L_1, L_2)$ where S is a fuzzy binary relation on A and $L_1, L_2$ are two linear orders on A. The $\mu_S$-board corresponds to a representation of the fuzzy relation S by the $\mu_S$-tableau whose lines (columns) are labelled by the elements of A, ranked according to $L_1(L_2)$.

The $\mu_S$-board is monotone (MONTJARDET (1984), ROUBENS and VINCKE (1983)) iff there exists a linear order L on A such that, for all $a,b,c \in A$,

$$bLa(a<b) \rightarrow \mu_S(a,c) \geqq \mu_S(b,c) \text{ and } \mu_S(c,a) \leqq \mu_S(b,c).$$

When S presents a monotone $\mu_S$-board, it can be represented by a tableau with non-decreasing monotonicity of the elements $\mu_S$ in a line and in a column. In this particular situation, it is interesting to consider the score for each element $a \in A$:

$$s_\mu(a) = \sum_{b \in A} \{\mu_S(a,b) - \mu_S(b,a)\},$$

the crisp score relation $S_\mu$ such that

$a\ S_\mu\ b$ iff $s_\mu(a) \geqq s_\mu(b)$

and the trace which is a crisp relation $T_\mu$ such that

$a\ T_\mu\ b$ iff $\forall c : \mu_S(a,c) \geqq \mu_S(b,c)$ and $\mu_S(c,a) \leqq \mu_S(b,c)$.

In any case, the score relation is a weak order (a complete, reflexive and transitive crisp relation) and the trace is a quasi order (a reflexive and transitive crisp relation).

It can be proved that the following statements are equivalent (MONJARDET (1984)), when $\mu_S(a,a)=0$, for all a in A:

(i) S is linear,
(ii) S represents a monotone $\mu_S$-board,
(iii) Trace $T_\mu$ is a weak order,
(iv) $T_\mu = S_\mu$.

Condition (iv) is a very efficient way to recognize if a fuzzy relation is linear: for all $a,b \in A$ with $a\ S_\mu\ b$, one should test if $a\ T_\mu\ b$.

In MONJARDET (1984) some other results related to a linear fuzzy relation are reported. They generalize the results of ROUBENS and VINCKE (1983) obtained in the context of probabilistic relations.

Dealing with a fuzzy preorder S it is clear from proposition 4 that $P_2$ is irreflexive, antisymmetric and transitive. The fuzzy graph $G(A,P_2)$ - where A is the set of nodes and $P_2$ the set of arcs (a,b) with values $\mu_{P_2}(a,b)$ - is acyclic and set A can be decomposed in subsets $N_0,\ldots,N_k$ such that

$\mu_{P_2}(a,b) > 0 \rightarrow a \in N_i,\ b \in N_j,\ i<j.$

For every element $x \in N_0$, we have $\mu_S(y,x) = 0$, all $y \in A$, and for every element $y \in N_k$, $\mu_S(x,y) = 0$, all $x \in A$. $N_0$ ($N_k$) can be defined as the class of unfuzzily non-dominated (non-dominating) elements.

If S is a linear fuzzy relation, the $\mu_S$-board is monotone and

$a\ S_\mu\ b \rightarrow \mu_{P_2}(a,b) \geqq 0,\ \mu_{P_2}(b,a) = 0.$

We then obtain the following upper-triangular $\mu_{P_2}$ board

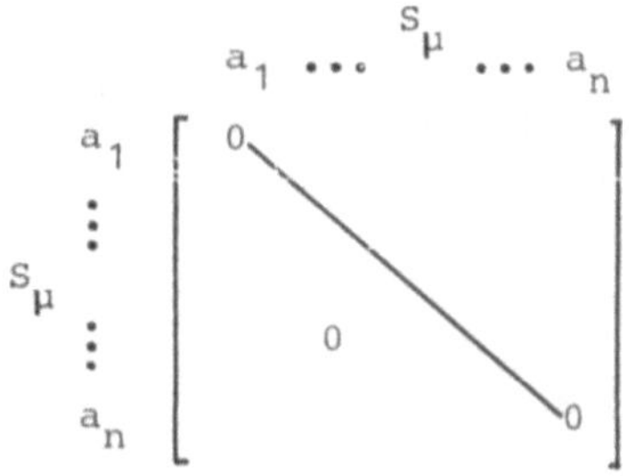


and $\mu_{P_2}(a_i,a_1) = 0$, all $i = 1,\ldots,n$, $\mu_{P_2}(a_n,a_j) = 0$, all $j = 1,\ldots,n$. Alternative $a_1$ is unfuzzily non-dominated and $a_n$ is unfuzzily non-dominating.

## 7. FUZZY OUTRANKING RELATIONS AND MULTIPLE CRITERIA DECISION MAKING (MCDM) METHODS

Let g(a) be the evaluation of an action $a \in A$ for a given criterion. Considering two thresholds PT and IT (called strict preference threshold and indifference threshold, respectively) the fuzzy outranking relation between a and b is given by the following membership function (see Fig. 1):

$$\mu_S(a,b) : \begin{cases} = 1 \text{ if } g(b) \leqq g(a)+IT \\ \text{is decreasing for } [g(b)-g(a)] \in [IT,PT] \\ = 0 \text{ if } g(b) \geqq g(a)+PT \end{cases}$$

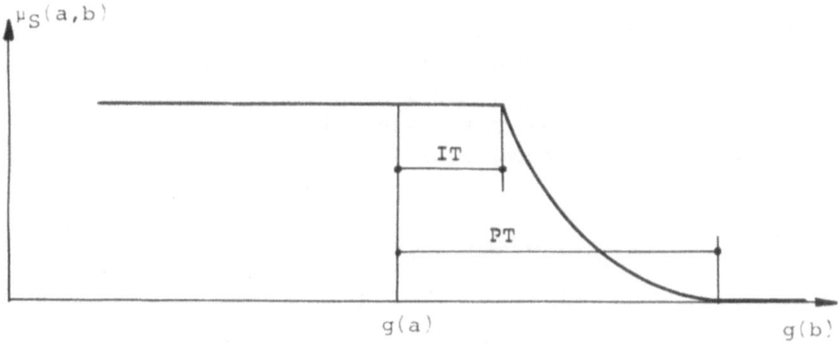


Fig. 1

$\mu_S(a,b)$ describes the degree to which the statement "a is not worse than b" is true.

It is easily seen that the dual relation $S^d$ is given by the following membership function (see Fig. 2)

$$\mu_{S^d}(a,b) : \begin{cases} =1 \text{ if } g(b) \leqq g(a)-PT \\ \text{is decreasing for } g(a)-g(b) \in [IT,PT] \\ =0 \text{ if } g(b) \geqq g(a)-IT \end{cases}$$

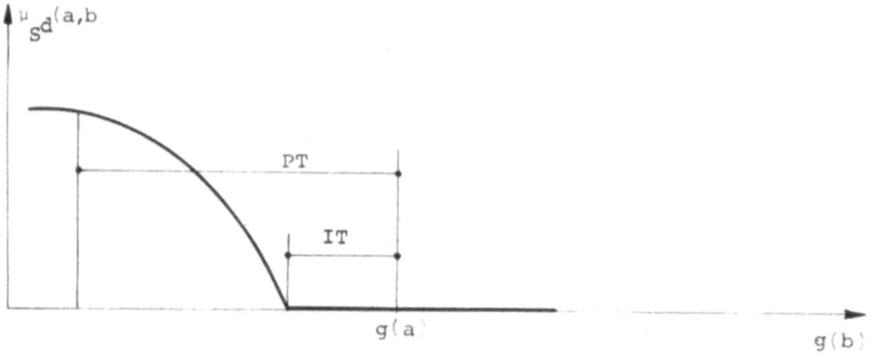


Fig. 2

It is interesting to consider a preference fuzzy relation $\mu_P$ to be understood as the degree to which the statement "a is strictly preferred to b" is true. In this particular context, it is worth while to notice that

$\mu_{S^d}(a,b) = \mu_{P_1}(a,b) = \mu_{P_2}(a,b)$ for all possible values of the pair $(g(a),g(b))$, $a,b \in A$.

Finally, we easily obtain that $\mu_I(a,b)$ looks like in Fig. 3 and corresponds to the statement "a is indifferent to b"; $\mu_R(a,b) = 0$, for all values of $(g(a),g(b))$ such that a and b are never considered incomparable; $u_S(a,b)$, with a linearly

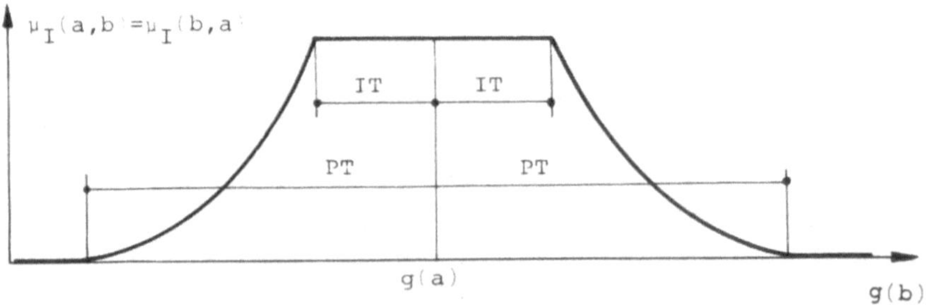


Fig. 3

decreasing shape, was considered by ROY (1978) in the ELECTRE III method and by SISKOS, LOCHARD and LOMBARD (1984). BRANS, MARESCHAL and VINCKE (1984) introduced a preference fuzzy relation when dealing with an MCDM method called PROMETHEE.

In some MCDM techniques, the fuzzy outranking relations corresponding to different criteria are aggregated using various kinds of procedures (see, e.g., ROY, 1978; SISKOS, LOCHARD and LOMBARD, 1984, ORLOVSKY, 1984) to obtain a single fuzzy outranking relation that gives a degree of overall outranking of one alternative by another. A ranking procedure is then derived using the non-dominance degree (see section 5), distillation proposed by ROY (1978) or network flows as in BRANS, MARESCHAL and VINCKE (1984).

Real problems were recently solved using these techniques: choice of metro stations to be renovated (ROY and HUGONNARD (1985) and comparison of energy alternatives (SISKOS and HUBERT (1983)).

REFERENCES

Brans, J.P., B. Mareschal, and Ph. Vincke (1984). PROMETHEE: a new family of outranking methods in Multicriteria Analysis. In J.P. Brans (ed.), Operational Research'84. North-Holland, Amsterdam. (Proc. of the Tenth IFORS Intern. Conference on Operational Research, Washington, D.C.).

Blin, J.M. (1974). Fuzzy relations in group decision theory. J. Cyber. 4, 17-22.

Doignon, J.P., B. Monjardet, M. Roubens, and Ph. Vincke (1985). Biorders families, valued relations and preference modelling. To appear.

Dubois, D., and H. Prade (1980). Fuzzy Sets and Systems: Theory and Applications. Academic Press, New-York.

Fishburn, P.C. (1973). Binary choice probabilities: on the varieties of stochastic transitivity, J. Math. Psychology 7, 327-352.

Hashimoto, H. (1983). Szpilrajn's theorem on fuzzy ordering. Fuzzy Sets and Syst. 10, 101-108.

Kacprzyk, J. (1984). Collective decision making with a fuzzy majority. Proc. 5th WOGSC Congress. AFCET, Paris.

Kacprzyk, J. (1985). Some 'commonsense' solution concepts in group decision making using fuzzy linguistic quantifiers. In J. Kacprzyk and R.R. Yager (eds.), Management Decision Support Systems Using Fuzzy Sets and Possibility Theory. Verlag TUV Rheinland, Cologne.

Monjardet, B. (1984). Probabilistic consistency, homogeneous families of relations and linear $\wedge$-relations. In E. Degreef and J. Van Buggenhaut (eds.), Trends in Mathematical Psychology. North-Holland, Amsterdam, 271-281.

Orlovsky, S.A. (1978). Decision making with a fuzzy preference relation. Fuzzy Sets and Syst. 1, 155-167.

Orlovsky, S.A. (1984). Two Approaches to Multiobjective Programming Problems with Fuzzy Parameters. Working paper, 84-37, IIASA, Laxenburg.

Ovchinnikov, S.V. (1981). Structure of fuzzy binary relations. Fuzzy Sets and Syst. 6, 169-195.

Roberts, F.S. (1979). Measurement Theory. Addison-Wesley, Reading, Mass.

Roubens, M., and Ph. Vincke (1983). Linear fuzzy graphs. Fuzzy Sets and Syst. 10, 79-86.

Roubens, M., and Ph. Vincke (1984). On families of semiorders and interval orders imbedded in a valued structure of preference: a survey. Inf. Sci. 34, 187-198.

Roubens, M., and Ph. Vincke. Preference Modelling. Springer Verlag, Berlin. To appear.

Roy, B. (1969). Algebre Moderne et Theorie des Graphes Orientées vers les Sciences Economiques et Sociales. Dunod, Paris.

Roy, B. (1978). ELECTRE III: a classification algorithm based on a fuzzy preference representation in the presence of multiple criteria (in French). Cahiers du C.E.R.O. 20, 3-24.

Roy, B., and J.C. Hugonnard (1985). A programming method for determining which Metro stations should be renovated. Eur. J. Op. Res. 22.

Siskos, J., and Ph. Hubert (1983). A survey and a new comparative approach. Eur. J. Op. Res. 13, 278-299.

Siskos, J., J. Lochard, and J. Lombard (1984). A multi-criteria Decision-making methodology under fuzziness: application to the evaluation of radiological protection in nuclear power plants. In H.J. Zimmermann, L. Zadeh and B. Gaines (eds.), Fuzzy Sets and Decision Making. North-Holland, Amsterdam.

Takeda, E., and T. Nishida (1980). Multicriteria decision problems with fuzzy domination structures. Fuzzy Sets and Syst. 3, 123-136.

Zadeh, L.A. (1971). Similarity relations and fuzzy orderings. Inf. Sci. 3, 177-200.

# PREFERENCE AND CHOICE IN A FUZZY ENVIRONMENT

Serge Ovchinnikov

Mathematics Department
San Francisco State University
San Francisco, CA 94132, USA

To Lotfi Zadeh on the occasion of the 20th anniversary of Fuzzy Sets Theory

Abstract. This paper is concerned with two models of decision making in a fuzzy environment. The first one is based on Zadeh's idea of a maximizing fuzzy set and on an approach to decisionmaking suggested by Bellman and Zadeh. The second model employs fuzzy preferences to describe the choice of "best" alternatives. Recent results in the area of representation theory for fuzzy binary relations are used to study relationships between these models.

Keywords: fuzzy choice, fuzzy decision making, fuzzy preference relation, transitivity.

## 1. INTRODUCTION

In his 1972 paper Zadeh (1972) introduces a maximizing set as follows. Let A be a set and f a real-valued positive function on A. (For our purposes it is convenient to consider A as a finite set of alternatives and f as a goal function or criterion.) Then a maximizing set $M^f$ is defined by its membership function

$$M^f(x) = \frac{f(x)}{\sup_A(f)}, \quad x \in A$$

"Intuitively, a maximizing set $M^f$... is a fuzzy subset of A such that the grade of membership of a point x in $M^f$ represents the degree to which f(x) approximates to $\sup_A(f)$..." (Zadeh, 1972). We regard a maximizing set $M^f$ as a fuzzy goal associated with a given criterion f.

If a set F of real-valued positive functions on A is given (vector criterion) then the corresponding maximizing set can be defined as

$$M^F(x) = \inf_{f \in F}\left\{\frac{f(x)}{\sup_A(f)}\right\}, \quad x \in A.$$

In this definition we follow Bellman-Zadeh's (1970) idea of a fuzzy goal in multiple-criteria decision making.

The notion of a maximizing set can be generalized as follows. Let $\varphi$ be a strictly increasing function mapping a unit interval $[0,1]$ onto itself. We define a maximizing set M by

$$M(x) = \varphi\left(\frac{f(x)}{\sup_A(f)}\right), \quad x \in A.$$

In the multiple-criteria case a maximizing set is defined as

$$M(x) = \inf_{f \in F} \varphi\left(\frac{f(x)}{\sup_A(f)}\right), \quad x \in A.$$

One can consider maximizing sets as values of a linguistic variable "goal" (see Zadeh, 1975 for definition of a linguistic variable). It follows from the results of Ovchinnikov (1981) that different generalized maximizing sets are synonyms; we will discuss in this paper how the particular choice of a maximizing set affects the decision making model. One can find it helpful to regard a maximizing set as a fuzzy dominating subset with respect to a given (vector or scalar) criterion (see Ovchinnikov, 1982 for details).

A fuzzy binary relation R on A is defined as a fuzzy subset of the direct product $A \times A$; it is completely determined by its membership function $R(x,y)$. We regard fuzzy binary relations as fuzzy preference relations considering $R(x,y)$ as a degree of preference of x over y. The fuzzy upper bound is defined in Zadeh (1971) as a fuzzy subset $\inf_{y \in A} R(x,y)$. A decision making model based on a fuzzy preference relation introduces the fuzzy upper bound as a fuzzy goal. Usually, certain restrictions are imposed on a fuzzy binary relation in order to guarantee nonvoidness of the fuzzy upper bound. As in the classical theory, these restrictions are the reflexivity and transitivity conditions. We introduce properties of fuzzy binary relations in the next section.

It is convenient to compare decision making models in the general framework of choice theory. Let us suppose that for every subset $X \subseteq A$ a fuzzy subset $C(X)$ of X is defined. Elements of $C(X)$ are considered as "best" elements in X, where, of course, "best" is a value of the fuzzy linguistic variable "best". A correspondence $X \to C(X)$ is called a fuzzy choice function. Decision making models based on maximizing sets and on fuzzy preference relations generate fuzzy choice functions which are most important in applications. We demonstrate in this paper that these two fuzzy choice mechanisms are, essentially, the same - the result well known in the classical choice theory.

An excellent review of the classical choice theory can be found in Aizerman and Malishevski (1981); fuzzy choice is discussed in Ovchinnikov (1981a, 1982, 1983).

## 2. TRANSITIVITY AND t-NORMS

A fuzzy binary relation R on a set A is said to be reflexive iff $R(x,x) = 1$ for all $x \in A$; only reflexive relations are considered in this paper. The most general definition of a fuzzy transitive relation was suggested by Zadeh (1971) : a relation R is called a fuzzy transitive relation iff

$$R(x,y)*R(y,z) \leqslant R(x,z) \quad \text{for all} \quad x,y \in A,$$

where * is a binary operation on [0,1]. In contemporary fuzzy set theory, the operation * is defined as a triangular norm (t-norm). We begin with a brief discussion of t-norms (see Alsina, 1983 and Klement, 1981 for details).

A (two-place) function T from $[0,1]\times[0,1]$ into [0,1] is a t-norm if it satisfies the following conditions for all $x,y,u,v$ in [0,1]:

(i) $T(x,1) = T(1,x) = x$,
(ii) $T(x,y) \leqslant T(u,v)$ if $x \leqslant u$, $y \leqslant v$,
(iii) $T(x,y) = T(y,x)$,
(iv) $T(x,T(y,v)) = T(T(x,y),v)$.

There are numerous examples of t-norms; we list below some of the most interesting ones:

$$T_o(x,y) = \min(x,y),$$
$$T_1(x,y) = x\cdot y,$$
$$T_\infty(x,y) = \max(x+y-1,0),$$
$$T_w(x,y) = \begin{cases} \min(x,y), & \text{if } \max(x,y) = 1, \\ 0, & \text{otherwise,} \end{cases}$$
$$T_s(x,y) = \log_s\left(1 + \frac{(s^x-1)\cdot(s^y-1)}{s-1}\right), \quad \text{if} \quad s \in (0,1).$$

A t-norm is said to be Archimedean if it is continuous and satisfies

$$T(x,x) < x, \text{ for all } 0 < x < 1$$

For example, a product (t-norm $T_1$) is an Archimedean t-norm.

Any Archimedean t-norm $T(x,y)$ can be represented as

$$T(x,y) = f(g(x) + g(y))$$

where g is a continuous strictly decreasing function from [0,1] into $R^+$ such that $g(1) = 0$, and f is a function from $R^+$ into [0,1] such that $f(0) = 1$ and $f(x) = 0$, for all $x > g(0)$, f is continuous and strictly decreasing on $[0,g(0)]$ where $f(x)=g(x)$.

The t-norms $T_w$ and $T_o$ are extreme cases of t-norms because of the following important inequalities

$$T_w \leqslant T \leqslant T_o \quad \text{for any} \quad T.$$

Now let T be a t-norm. A fuzzy binary relation R on A is said to be transitive (or, better, T-transitive) iff

$$T(R(x,y),R(y,z)) \leqslant R(x,z), \text{ for all } x,y,z \in A$$

Classical examples are the max-min transitivity

$$\min(R(x,y),R(y,z)) \leqslant R(x,z),$$

and the max-product transitivity

$$R(x,y)\cdot R(y,z) \leqslant R(x,z), \quad \text{for all} \quad x,y,z \in A.$$

We call R a fuzzy transitive relation if it is T-transitive for some t-norm T. Reflexive and transitive fuzzy binary relations are called fuzzy preferences in this paper.

The following proposition establishes simple necessary and sufficient conditions for a fuzzy binary relation to be transitive.

Proposition 1. A fuzzy binary relation R on A is transitive iff $R(a,b) = 1$ implies

$$R(x,a) \leqslant R(x,b) \quad \text{and} \quad R(a,x) \geqslant R(b,x)$$

for all $a,b,x \in A$.

This is a rather trivial but important and simple characterization of fuzzy transitive relations. One can recognize the $T_W$-transitivity in the conditions given above; the statement of Proposition 1 follows immediately from the fact that $T_W$ is the least t-norm.

A fuzzy similarity relation is defined as a reflexive, symmetric and transitive relation, i.e. S is a similarity relation iff:

(i) $S(x,x) = 1$,
(ii) $S(x,y) = S(y,x)$, and
(iii) $T(S(x,y),S(y,z)) \leqslant S(x,z)$,

for all $x,y,z \in A$ and some t-norm T.

Let us introduce a crisp (nonfuzzy) binary relation

$$\{(x,y) : S(x,y) = 1\}$$

for a given fuzzy similarity relation S. It is an equivalence relation which we denote by $\sim$. The following proposition establishes a nice characterization of fuzzy similarity relations.

Proposition 2. A reflextive symmetric fuzzy binary relation S is a similarity relation iff its corresponding crisp relation is an equivalence relation and

$$S(x,y) = S(u,v), \quad \text{whenever } x \sim u \text{ and } y \sim v.$$

These two propositions clearly illustrate the fact that transitivity is not a very demanding property and may be substituted by rather weak conditions when special classes of fuzzy binary relations like the fuzzy preference and similarity relations are considered.

Let R be a fuzzy preference relation. There are infinitely many t-norms T providing the T-transitivity of R. It is an interesting and difficult problem to describe and study the class of all these t-norms. One can always construct, say, an Archimedean t-norm providing transitivity of a given fuzzy preference relation on a finite set A.

It is convenient for our future purposes to introduce a different representation for some Archimedean t-norms. Namely, let $\varphi$ be a strictly increasing function mapping a unit interval $[0,1]$ onto itself. We define a two-place function T on $[0,1]\times[0,1]$ by

$$T(x,y) = \varphi^{-1}(\varphi(x)\cdot\varphi(y)).$$

It is easy to verify that T is, indeed, an Archimedean t-norm. Moreover, if A is a finite set, then any reflextive fuzzy binary relation on A satisfying conditions of Proposition 1 is a fuzzy preference relation for some T belonging to the class of t-norms just introduced. Let us write a T-transitivity condition for a fuzzy preference relation R

$$\varphi^{-1}(\varphi(R(x,y))\cdot\varphi(R(y,z))) \leqslant R(x,z),$$

or, equivalently,

$$\varphi(R(x,y))\cdot\varphi(R(y,z)) \leqslant \varphi(R(x,z)).$$

In summary, we have the following

Proposition 3. A fuzzy binary relation R is a preference relation iff there exists a strictly increasing mapping $\varphi$ from $[0,1]$ onto itself such that $\varphi(R)$ is a reflexive and max-product transitive fuzzy binary relation.

If a linguistic variable "preference" is introduced with values being fuzzy preferences on a finite set A, then following an approach developed in Ovchinnikov (1981b) one can say that any fuzzy preference is a synonym of a max-product transitive reflexive fuzzy binary relation.

## 3. REPRESENTATION THEOREMS

Representation theorems for transitive fuzzy binary relations were introduced in Ovchinnikov (1982) and Ovchinnikov (1984) for the max-product transitivity, and generalized in Valverde's (1982) Ph.D. thesis. We are concerned in this paper with only the max-product transitivity because of the results discussed in the previous section.

Let R and S be, respectively, a fuzzy preference and similarity relation on a finite set A, i.e.:

(i) $R(x,x) = 1$,
(ii) $R(x,y)\cdot R(y,z) \leqslant R(x,z)$,

and

(i) $S(x,x) = 1$,
(ii) $S(x,y) = S(y,x)$,

(iii) $S(x,y) \cdot S(y,z) \leqslant S(x,z)$.

The proofs of the following theorems can be found in Ovchinnikov (1984).

Theorem 1. A fuzzy binary relation R is a preference relation iff there exist a family F of positive functions on A such that

$$R(x,y) = \inf_{f \in F} \min\left\{\frac{f(x)}{f(y)}, 1\right\}.$$

Theorem 2. A fuzzy binary relation S is a similarity relation iff there exists a family F of positive functions on A such that

$$S(x,y) = \inf_{f \in F} \min\left\{\frac{f(x)}{f(y)}, \frac{f(y)}{f(x)}\right\}.$$

Note again that these results are established for the max-product transitivity only.

For completeness, we present here Valverde's general representation theorem for the similarity relations. For a given t-norm, T, a quasi-inverse $\hat{T}$ is defined by

$$\hat{T}(x|y) = \sup\left\{u \in [0,1] : T(u,x) \leqslant y\right\}.$$

Then we have the following

Theorem 3. Let S be a fuzzy binary relation on A and T a continuous t-norm. Then S is a similarity relation (assuming the T-transitivity) if and only if there exists a family of fuzzy subsets $\{h_j\}_{j \in J}$ such that

$$S(x,y) = \inf_{j \in J} \hat{T}(\min(h_j(x), h_j(y)) | \max(h_j(x), h_j(y))).$$

## 4. CHOICE IN A FUZZY ENVIRONMENT

Let A be a finite set of alternatives. We say that a fuzzy choice function C is given if, for every nonempty set $X \subseteq A$, a fuzzy subset C(X) of X is defined. Following ideas presented in the introduction, we define two choice mechanisms.

Let $F = \{f\}$ be a family of positive functions on A (a vector criterion). We define

$$C_X^{F,\varphi}(x) = \inf_{f \in F}\left\{\varphi\left(\frac{f(x)}{\sup_X (f)}\right)\right\}, \quad x \in X.$$

where $\varphi$ is an automorphism of the unit interval [0,1], i.e. a strictly increasing function mapping [0,1] onto itself.

Let now R be a fuzzy preference relation on A. We define

$$C_X^R(x) = \inf_{y \in X}\left\{R(x,y)\right\}, \quad x \in X.$$

We show below that the two classes of choice functions defined above are the same. Indeed, for a given family F a fuzzy binary relation R defined by

$$R(x,y) = \inf_{f\in F} \min\left\{\frac{f(x)}{f(y)}, 1\right\}$$

is, in accordance with Theorem 1, a reflexive max-product transitive fuzzy binary relation; therefore $\varphi(R)$ is a fuzzy preference relation. We have

$$C_X^{\varphi(R)}(x) = \inf_{y\in X}\left\{\varphi(R(x,y))\right. = C_X^{F,\varphi}(x).$$

Hence, a fuzzy choice function $C_X^{F,\varphi}$ generated by the family F and automorphism $\varphi$ coincides with the fuzzy choice function generated by the fuzzy preference relation $\varphi(R)$.

Conversely, let R be a fuzzy preference relation on A. Then there is an automorphism $\varphi$ such that $R = \varphi(R^{-})$ where $R^{-}$ is a reflexive max-product transitive fuzzy binary relation. By Theorem 1 there is a family F of positive functions on A such that

$$R^{-}(x,y) = \inf_{f\in F} \min\left\{\frac{f(x)}{f(y)}, 1\right\}.$$

Hence

$$R(x,y) = \inf_{f\in F} \min\left\{\varphi\left(\frac{f(x)}{f(y)}\right), 1\right\}.$$

Then we have

$$C_X^{R}(x) = \inf_{y\in X}\left\{R(x,y)\right\} = \inf_{f\in F} \min\left\{\varphi\left(\frac{f(x)}{\sup_X(f)}\right)\right\} = C_X^{F,\varphi}(x).$$

Therefore, a choice function generated by a fuzzy preference relation R coincides with a choice function generated by vector criterion F and automorphism $\varphi$.

We proved the following

Theorem 4. Choice mechanisms based on maximized sets and on fuzzy preference relations generate identical families of choice functions.

## 5. CONCLUSIONS

The results presented in this paper generalize some classical theorems concerning decision making models based upon multiple-criterion optimization and preference choice. It is interesting to note that in a fuzzy environment transitivity properties of preference relations do not play the same crucial role as they do in the crisp case. Transitivity can be substituted by rather weak conditions which are easily verified in applications. The ideas developed in this paper can be employed to expand the classical choice theory; some preliminary results in this direction are found in Ovchinnikov (1981, 1982).

ACKNOWLEDGEMENTS

Research sponsored by NSF Grant IST-8403431.

REFERENCES

Aizerman, M., A. Malishevski (1981). General theory of best variant choice: some aspects. IEEE Trans. AC-26, 1030-1040.
Alsina, C. (1983). A primer of t-norms. Proc. of FISAL-83. Porto Colom, Universitat de Palma de Mallorca, 27-36.
Bellman, R., L. Zadeh (1970). Decision-making in a fuzzy environment. Mang. Sci. 17, 8141-8164.
Klement, E.P. (1981). Operations on fuzzy sets and fuzzy numbers related to triangular norms. Proc. of 11th ISMVL. Oklahoma, 218-225.
Ovchinnikov, S. (1981a). Structure of fuzzy binary relations. Fuzzy Sets and Syst. 6, 169-195.
Ovchinnikov, S. (1981b). Representation of synonymy and antonymy by automorphisms. Stochastica V, 95-107.
Ovchinnikov, S. (1982). Choice theory for cardinal scales. In M.M. Gupta and E. Sanchez (eds.), Fuzzy Information and Decision Processes. North-Holland, Amsterdam, 323-338.
Ovchinnikov, S. (1983). Fuzzy choice functions. Proc. of the IFAC Symp. on Fuzzy Inform., Knowledge Representation and Decision Analysis. Marceille, 359-365.
Ovchinnikov, S. (1982). On fuzzy relational systems. Proc. of the 2nd World Conference on Mathematics at the Service of Man. Las Palmas, Canaries, 566-569.
Ovchinnikov, S. (1984). Representations of transitive fuzzy relations. In H. Skala, S. Termini, E. Trillas (eds.), Aspects of Vaguness. D. Reidel, Dordrecht, 105-118.
Valverde, L. (1982). Contributions to Studying Mathematical Models for Multivalued Logics (in Catalonian). Ph.D. Thesis, Barcelona.
Zadeh, L. (1971). Similarity relations and fuzzy orderings. Inf. Sci. 3, 177-200.
Zadeh, L. (1972). On Fuzzy Algorithms. Memo UCB/ERL M325, UC Berkeley.
Zadeh, L. (1975). The concept of a linguistic variable and its application to approximate reasoning - Part I. Inf. Sci. 8, 199-249.

# FUZZY CHOICE

V.B. Kuz'min and S.I. Travkin

Institute for Systems Studies
Moscow, USSR

Abstract. The operations of fuzzy maximum, $\widetilde{\text{Max}}$, and minimum, $\widetilde{\text{Min}}$, defined on a crisp set, and their multivariable analogon, i.e. the concept of a fuzzy Pareto optimum, are introduced. A model of fuzzy choice based on the concept of fuzzy concentration and on rules of fuzzy logic is presented. The rules of fuzzy choices are classified by a number of options chosen by infinitely concentrated, therefore crisp, rules. The correctness of the given definitions is shown by applying the extension principle. The probability distribution of the height of a fuzzy Pareto optimum, and an asymptotic result for the expected value of the height are calculated. The rate at which the concentrated fuzzy set converges to the limit crisp set is estimated. Some properties of fuzzy choice classes are discussed.

Keywords: fuzzy choice, generated fuzzy $\widetilde{\text{Max}}$ and $\widetilde{\text{Min}}$, limiting concentration, fuzzy Pareto optimum, fuzzy choice rule.

## 1. INTRODUCTION

Choice theory supplies methods to cope with imprecise optimization problems when their traditional statements are impeded by conflicting adjectives. This replacement of an optimization problem by a choice problem is only one, not necessarily the best, way to take into account the underlying uncertainty. The problem of choice is dealt with in a number of studies conducted within the framework of fuzzy sets theory applied to multicriteria programming (e.g., Sakawa 1983, Takatsu, 1984, etc.). Zadeh (1976) proposed a fuzzy version of a trade-off procedure to limit the number of Pareto optimal options.

In this paper a somewhat different approach is adopted. We are taking only one standard fuzzy set to generate a rather wide system of fuzzy choice rules.

The operations of fuzzy $\widetilde{\text{Max}}$ and $\widetilde{\text{Min}}$ defined on a crisp set, and their multivariable analogon, i.e. the concept of the fuzzy Pareto optimum are introduced. A model of fuzzy choice, based on the concept of fuzzy concentration and on rules of fuzzy logic, is presented. The rules of fuzzy choices are classified by a number of options chosen by infinitely concentrated, therefore crisp rules. The correctness of the given definitions is shown by applying the extension principle.

The probability distribution of the height of a fuzzy Pareto optimum, and an asymptotic result for the expected value of the height are calculated. The rate at which the concentrated fuzzy set converges to the limit crisp set is estimated. Some properties of fuzzy choice classes are discussed. In particular, it is shown that the rule $R \subset R(I, \neg, \wedge)$ selects no more than two options and a rich choice is possible by the rules of higher order classes.

## 2. FUZZY SETS AS SOME RESULTS OF FUZZY OPTIMIZATION

Let $\Omega$ be a countable, linearly ordered set and $B = \left\{X \subset \Omega : |X| < \infty\right\}$ be a family of finite subsets of the set $\Omega$. To each set $X = \left\{x_i\right\}_{i=1}^{n} \subset \Omega$ we assign a vector $\bar{x} = (x_1, \ldots, x_n)$ and define two functions Min: $X^n \to X$ and Max: $X^n \to X$ as

$$\text{Min } \bar{x} = \min\left\{x; x \in X\right\},$$

$$\text{Max } \bar{x} = \max\left\{x; x \in X\right\}.$$

Suppose we have a fuzzy set $A = \int_0^1 x/x$ which is defined on [0,1] and whose membership function is $\mu(x) = x$. Then a function $\varphi : [0,1] \to \left\{\frac{i}{n-1}\right\}_{i=0}^{n-1}$ maps the interval [0,1] into n equidistant points and generates a fuzzy set $\tilde{Q}_n = \sum_0^1 \varphi(x)/x$ according to the extension principle. If members of a set $X \in B$ are enumerated according to their increasing values, then by application of the extension principle to the function $\Theta(\frac{i}{n-1}) = x_{i+1}$ we will produce the fuzzy set $\tilde{X} = \Sigma \frac{i}{n-1}/\Theta(\frac{i}{n-1})$ with the membership function $\mu_n(x_i) = \frac{i}{n-1}$. Denote the set of all such fuzzy sets by $\tilde{B}$.

<u>Definition 1</u>. The operation which assigns to each set $X \in B$ a fuzzy set $\tilde{X}$ is called fuzzy maximization, written $\tilde{X} = \widetilde{\text{Max}}\ X$.

Let us show that the given definition agrees with the definition of $\widetilde{\text{Max}}$ introduced in Dubois and Prade (1978). For this we have to prove that for any subset $X_m = \left\{x_i\right\}_{i=1}^{m}$ of a set $X_n \subset B$, there holds the equality $\mu_{\widetilde{\text{Max}}}(x) = \mu_m(x)$ where $\mu_{\widetilde{\text{Max}}}(x)$ is defined due to the extension principle as

$$\mu_{\widetilde{\text{Max}}}(z) = \sup_{\bar{x}: z = \text{Max } \bar{x}} \min\left\{\mu(x_i)\right\},\ i \in \overline{1,m}.$$

But, due to monotonicity of $\mu(x)$, we have

$$\mu_{\widetilde{\text{Max}}}(z) = \sup_{\bar{x}: z = \text{Max } \bar{x}} \mu(\text{Min } \bar{x}) = \mu(z),$$

because the maximum value of $\mu(\text{Min}\ \bar{x})$ among the vectors of a fixed maximum of coordinates is reached on the vector $z \cdot 1 = (z, z, \ldots, z)$. A similar consideration with the use of the fuzzy set $\int_0^1 (1-x)/x$ leads to the definition of $\widetilde{\text{Min}}$.

Note that $\widetilde{\text{Max}} = \neg\, \widetilde{\text{Min}}$, in the sense that $\mu_{\widetilde{\text{Max}}}(\cdot) = 1-\mu_{\widetilde{\text{Min}}}(\cdot)$, and that instead of standard membership functions $\mu(x) = x$ (or $\mu(x) = 1-x$) one may choose any strictly increasing (or decreasing) function, for instance $\mu(x) = 1-e^x/(1-e^{-1})$ and its complement.

## 3. THE FUZZY PARETO CHOICE RULE

The introduced operations of fuzzy optimization map every crisp set into its fuzzy subset. By comparison with the known definition of a choice function as the decreasing mapping of a boolean B, we may classify $\widetilde{\text{Max}}$ and $\widetilde{\text{Min}}$ as fuzzy choice functions.

This enables us to give the following definition.

Definition 2. The fuzzy Pareto choice rule is a mapping of the finite set $X_n = \{\bar{x}^i\}_{i=1}^n$, $\bar{x}^i \in R^m$, of m-dimensional vectors into the family of fuzzy Pareto subsets, such that

$$\tilde{\Pi}_j = \int_{\Pi_j} \frac{\min\ \text{rank}_i \bar{x} - 1}{n-1} / \bar{x} ,$$

where: $\Pi_1 = \pi X_n = \{\bar{x} : x \in X_n,\ (\bar{y} \in X_n)\ \&\ (\bar{y} > \bar{x}) \Rightarrow \bar{y} = \bar{x}\}$,

$\Pi_j = \pi(X_n \setminus \bigcup_{k=1}^{j-1} \Pi_k)$,

$\text{rank}_i \bar{x}$ is the rank of vector $\bar{x}$ in $X_n$ ordered by the decreasing values of coordinates of its members.

Denote $\text{Rank}\ \bar{x} = (\text{rank}_1 \bar{x}, \text{rank}_2 \bar{x}, \ldots, \text{rank}_m \bar{x})$. To clarify the underlining assumptions let us try to arrive at this definition by the use of the extension principle starting with the standard fuzzy set $\int_0^1 x/x$; thus

$$v(y) = \sup_{\bar{x} : y \in \varphi(x)} \mu(x).$$

Construct the Cartesian product $(\int_0^1 x/x)^n$ and a step funcction $\varphi_n : [0,1]^m \to \overline{0,n}^m$, such that

$$\varphi_n(x) = \begin{cases} 0 & \text{if } \bar{x} = 0 \\ \bar{i} & \text{if } \bar{i}'' x \le i \end{cases}$$

where the j-th coordinate $i_j''$ of vector-index $\bar{i}''$ is equal to $i_j - 1$ if the j-th coordinate $i_j$ of vector-index $\bar{i}$ is not zero, and $i_j'' = 0$ in the opposite case. The extension principle gives us the fuzzy lattice

$$\tilde{L} = \int_{[0,1]^m} \operatorname{Min} \bar{x}/\varphi_n(\bar{x}) = \Sigma \frac{\operatorname{Min} i}{n-1} \Big/ \frac{i}{n-1} .$$

Till now we were able to transfer fuzziness of the standard set $\int_0^1 x/x$ on the lattice $L^n = \{ i \}$ by using a simple function $\varphi_n$. But now we have to pass from $L^n$ to $X^n$ and there is no obvious functional mapping and therefore we are forced to use a more general extension principle. It gives the freedom to choose any sort of a mapping $L^n \to X^n$ in which some points $\bar{i}$ may have no images whatsoever while several images can be assigned to the other. A straightforward way to define the needed mapping is to retract the fuzzy set $\tilde{L}^n$ on

$$\Pi_j = \pi_j(\{\text{Rank } x^k\}_{k=1}^{n})$$

and then a family of functions $\Theta_j : \Pi_j \to \pi_j(X^n)$ will bring us to the desired fuzzy sets

$$\tilde{\Pi}_j = \sum_{\bar{x} \in X^n} \frac{\min_i \text{rank}_i \bar{x} - 1}{n-1} / \bar{x} .$$

Perhaps a more promising approach which is not persued here consists in mapping each point $\bar{i}$ into a part of the Pareto set $\Pi_j$ whose elements dominate $\bar{i}$. The difference between these two definitions becomes meaningful if we have in mind the possibility of a more complicated standard fuzzy set than $\int_0^1 x/x$.

## 4. THE HEIGHT OF THE PARETO SET

In any case, the fuzzy Pareto optimization on a sample $X^n$ leads to a subnormal fuzzy set. The height of it varies considerably with the variation of elements in $X^n$.

Let us look at $X^n$ as on n independent realizations of a random vector $\xi = (\xi_1,\ldots,\xi_n)$. Suppose that the coordinates of $\xi$ are continuously distributed independent random variables. Thus the height

$$H = \sup \mu_{P-\widetilde{\max}}(x)$$

becomes a discrete random variable over $L = \left\{\frac{i}{n-1}\right\}_{i=0}^{n-1}$. It is important that its distribution is free from the distributions of the vector $\xi$. The exact result is that

$$P(H \leq h-1) = \left(\frac{h!}{n!}\right)^m \prod_{i=h+1}^{n} (i^m-(n-h)^m).$$

The proof is based on the consideration of random ranks $\left\{\text{Rank } \xi^j\right\}_{j=1}^{n}$ and calculation of the number of assignments of points to the nodes of lattice $L^n$ which fulfill the following requirements:

1) if a point is assigned to the node $\bar{i} = (i_1,\ldots,i_m)$, then no other points may be assigned to the node $\bar{i}' = (i_1',\ldots,i_m')$, such that for some $j \in \overline{1,m}$, $i_j = i_j'$ holds;

2) no points are assigned to the nodes $\bar{i} : \bar{i} \geqslant (h,h,\ldots,h)$.

The expected height H of a random fuzzy Pareto set tends to 1 with the increase of the size of a sample. An asymptotic estimation has the form

$$1 - \xi H \approx \frac{1}{m} B\left(\frac{1}{m},n+1\right) \approx n^{-\frac{1}{m}} \Gamma\left(1 + \frac{1}{m}\right), \; n \to \infty .$$

## 5. THE GENERALIZED CONCENTRATION

To design a general system of a choice rule we have to quarantee that a system of rules will include the crisp rules of choice at least as extreme cases. For this purpose we utilize the concept of concentration in a somewhat generalized form.

Let X be a fuzzy set which later will be interpreted as a result of a fuzzy choice with the membership function $\mu(x)$. Let $\gamma$ and $\Delta$ be two real numbers subjected to the restrictions: $\gamma > 1$ and $0 \leqslant \gamma - \Delta < 1$.

The operation of concentration $I = I(\gamma,\Delta)$ transforms the fuzzy set $\widetilde{X}$ into the fuzzy set $I\widetilde{X}$ with the membership function

$$\mu_1(x) = (\sup_x \mu(x))^{-\Delta}\mu^{\gamma}(x)$$

Note that if $\tilde{X}$ is a normalized set, then $I(\gamma,\Delta)$ equals $I(\gamma)$, the concentration operator by Zadeh (1975) for $\gamma = 2$. Operator $I(1,1)$ is the normalizer, i.e.

$$\mu_1(x) = \mu(x)/\sup \mu(x).$$

Dilation is defined as $I = I^{-1}$ such that

$$\mu_{-1}(x) = (\sup \mu(x))^{\frac{1}{\gamma-\Delta} - \frac{1}{\gamma}} (\mu(x))^{\frac{1}{\gamma}}$$

Repeated application of the introduced operators generates a compound modificator. It is easily seen that the superposition of operators I and $I^{-1}$ gives the identity operator, i.e. $II^{-1}=E$. The set of modificators is a free group with one generator.

By means of several concentrations we will arrive at higher fuzzy sets. To find how close we are to the crisp set after a number of concentrations, let us derive the operator of n-tuple concentration.

Denote $h_o = \sup \mu(x)$ and let $\mu_n$ be the membership function of the fuzzy set $I^n\tilde{X}$, and $h_{n-1} = \sup \mu_n$.

As a result of the second concentration we get $I^2X = I(IX)$ with

$$\mu_2 = \frac{(\mu_1)^{\gamma}}{h_1^{\Delta}} \quad ,$$

where

$$h_1 = \frac{(\sup_{x \in X} \mu)^{\gamma}}{h^{\Delta}} = h^{\gamma-\Delta}$$

Then

$$\mu_2 = \frac{\mu^{\gamma^2}}{h^{\Delta\gamma + (\gamma-\Delta)\Delta}} \quad \underline{\underline{def}} \quad \frac{\mu^{\gamma_2}}{h^{\Delta_2}}$$

Suppose that after n-concentrations we get a fuzzy set with the membership function

$$\mu_n = \frac{\mu^{\gamma_n}}{h^{\Delta_n}}$$

Then $h_n = \dfrac{h^{\gamma_n}}{h^{\Delta_n}} = h^{\gamma_n-\Delta_n}$ and

$$\mu_{n+1} = \frac{(\mu^{\gamma_n})^{\gamma}}{h^{\Delta_n\gamma + (\gamma_n - \Delta_n)n}} = \frac{\mu^{\gamma_{n+1}}}{h^{\Delta_{n+1}}}$$

Hence $\gamma_{n+1} = \gamma.\gamma_n$ and $\Delta_{n+1} = \Delta_n\gamma + (\gamma_n - \Delta_n)\Delta$. But $\mu_o = \mu$ and hence $\gamma_o = 1$ and $\Delta_o = 0$. Therefore $\gamma_n = \gamma^n$ and

$$\Delta_{n+1} = \Delta_n \ \gamma + (\gamma^n - \Delta_n)\Delta$$

or

$$\Delta_{n+1} = (\gamma - \Delta)\Delta_n + \gamma^n \ . \ \Delta \ .$$

Denote $\tau_n = \Delta_n/(\gamma-\Delta)^n$, then $\tau_{n+1} = \tau_n + (\frac{\gamma}{\gamma-\Delta})^n \frac{\Delta}{\gamma-\Delta}$ .

From this

$$\tau_n = \frac{\Delta}{\gamma-\Delta} \sum_{k=1}^{n} (\frac{\gamma}{\gamma-\Delta})^k = (\frac{\gamma}{\gamma-\Delta})^n - 1 .$$

Returning to the sought variables $\Delta_n$, we get

$$\Delta_n = \gamma^n - (\gamma - \Delta)^n .$$

Finally the process of n-concentration bring us to the fuzzy set

$$I^h\tilde{X} = \int h^{(\gamma-\Delta)^n} \ (\frac{\mu(x)}{h})^{\gamma^n}/x .$$

The first factor $h^{(\gamma-\Delta)^n}$ is the height of this fuzzy set.

When $\gamma=\Delta$, then $I(\gamma,\gamma)$ is the operation of concentration for the normalized fuzzy set and that leaves only the second factor $((\mu/h)^{\gamma})^n$.

The rate at which the height $h(\tilde{X})$ converges to 1 is exponential, i.e.

$$\max_x \mu_n(x) = h_n = 1 + (\gamma - \Delta)^n \ln h(1+o(1)) .$$

The other values of the membership function converge to zero at a more than exponential rate, i.e. $\mu_n(x) = (\mu/h)^{\gamma^n} + o(1)$.

## 6. A MODEL OF FUZZY CHOICE

The techniques of concentration allow one to express the fuzzy optimization problem in the form $I^n\widetilde{Max}$, $n \in N = \{0, \pm 1, \ldots\}$. An obvious next step is to use the statements of fuzzy logic and to construct a whole system of fuzzy choice rules. But the operations of negation and concentration do not commute, for instance, $1-x^{\gamma^n} \neq (1-x)^{\gamma^n}$, and thus $\neg I^n \widetilde{Max}\ X \neq I^n \neg \widetilde{Max}\ X$. To present the class $R(I,\neg)$ of rules generated by these two operations I and $\neg$, let us introduce the set of finite sequences $N = \{\bar{n}:\bar{n}=(\bar{n}_1,\ldots,\bar{n}_k),\ k\in N^+=\{1,2,\ldots\}\}$ whose elements $n_j \in N$, but none of the sequences $\bar{n}$ except $\bar{n} = (0,0)$ has two consecutive zero elements, i.e. $n_j = 0 \Rightarrow (n_{j-1} \neq 0\ \&\ (n_{j+1} \neq 0)$. Now define

$$R(I,\ ) = \{R(\bar{n}):R(\bar{n}) = I^{n_1}\neg I^{n_2}\neg\ldots\neg I^{n_k}P\text{-}\widetilde{Max},\ \bar{n}\in N\}.$$

It can be shown that $\bar{n} \neq \bar{n}' \Rightarrow R(\bar{n}) \neq R(\bar{n}')$.

The conjunction of two rules from R will give us another rule $T = R(\bar{n})\wedge R(\bar{n}')$ which need not belong to $R(I,\neg)$. This will certainly be the case if $|\dim \bar{n} - \dim \bar{n}'| = 2k+1$ because the membership function $\mu_T(x)$ is not monotonic. Thus

$$R(I,\ ,\wedge) = \{R:R = \bigwedge_{j=1}^{k} R(\bar{n}^j),\bar{n}^j\in N\} \supset R(I,\neg).$$

Concentration of the rules from $R(I,\neg,\wedge)$ leads to an element of the class $R(I,\neg,\wedge,I)$. This class may be treated in the same way as the class $R(I) = \{I^n\widetilde{Max},n\in N\}$ to generate a class $R_2(I,\neg) = R(I,\neg,\wedge,I,\neg)$, and, following the same procedure, to generate classes $R_3(I,\neg)$, $R_4(I,\neg)$,...

Note that De Morgan's formulae imply $R_2(I,\ ) \supset R(I,\vee)$ and that the order of operations in a class is important. For instance, $R(I,\wedge,\neg) \neq R(I,\neg,\wedge)$ because

$$R(I,\wedge) = \{I^{n_1}\ \widetilde{Max} \wedge I^{n_2}\ \widetilde{Max},\ n_i\in N\} =$$
$$= \{I^{\max(n_1,n_2)}\ \widetilde{Max}\} = R(I).$$

7. THE LIMIT CRISP CHOICE

How many options will be chosen if we apply a given choice rule, is a natural question in the crisp theory of choices. But, as soon as fuzzy choice leads to an answer in terms of fuzzy subsets, we may then say that all the elements are chosen but with a different membership degree. An alternative to this classical interpretation is based on the following definition.

Definition 3. A fuzzy rule of choice R(.) selects as many options as can be chosen by its crisp counterpart, that is the infinitely concentrated rule $\ddot{R}(.)$.

By means of infinite concentration we may arrive at the same crisp choice rule starting from different fuzzy choice rules. The only requirement to be made is that the fuzzy rules reach their heights on the same set of points. From now on we will restrict ourselves to the one-dimensional case where we can utilize the equation $\neg\widetilde{\text{Max}} = \widetilde{\text{Min}}$. It is quite obvious that any rule from the class R(I) has only one limit crisp rule which chooses max $X_n$. The rules from $R(I,\neg)$ are of two types, i.e. with increasing and decreasing membership functions. If an element of $R(I,\neg)$ is presented in the form $I^{n_k}\neg I^{n_{k-1}}\neg\ldots\neg I^{n_1}$, then $\widetilde{\text{Max}} = R(n_1,\ldots,n_k) = R(n)$, and if it is such that dim $\bar{n}$ is even, then $\mu_{R(n)}(x)$ increases, and if dim $\bar{n}$ is odd, then $\mu_{R(n)}(x)$ decreases.

Hence for the rules of the first type the limit crisp rule is max $X_n$, and for the second type of fuzzy rules, its limit rule is min $X_n$.

Proposition 1. Any fuzzy choice from the class $R(I,\neg,\wedge,I)$ selects no more the two options from $X_n$.

First note that instead of $R(I, ,\wedge,I)$ it is sufficient to deal with the class $R(I,\neg,\wedge)$, and that any element $R\in R(I,\neg,\wedge)$ may be represented as

$$R = (\bigwedge_{i=1}^{k} R(\bar{n}_i)) \wedge (\bigwedge_{j=1}^{l} R(\bar{m}_j))$$

where dim $\bar{n}_i$, $i\in\overline{1,k}$ is even for all $i\in\overline{1,k}$, and dim $\bar{m}_j$ is odd for all $j\in\overline{1,l}$.

If we find $(i^o,j^o)$ such that

$$H(R_o) = \min_{i,j} H(R(\bar{n}_i) \wedge R(\bar{n}_j))$$

where

$$R_o = R(\bar{n}_{i^o}) \wedge R(\bar{m}_{j^o})$$

then we assert that the choice by the limit crisp rule will indicate an option $x\in X_n$ for which $\mu_{R_o}(x) = H(R_o)$. To demonstrate that there are no more than two options, it suffices to consider the above construction applied to a system of fuzzy choices not on $X_n$ but on [0,1]. In this case for each rule R we have instead of $RX_n$ a fuzzy set R[0,1] which is concave in the fuzzy sense and unimodal. Implementation of the extension principle as shown above leads to the retraction of the membership function $\mu_R(x)$ on $X_n$. So the retracted membership function cannot have more then two maximum points.

To gain insight into the structure of the rules from $R(I,\neg,\wedge)$ let us consider three examples.

Example 1. The limit choice rule for the rule from $R(I,\neg,\wedge)$ which has the form $R = I^m\widetilde{Max} \wedge I^n \widetilde{Min}$ is the same for the rule $I^{m-n}\widetilde{Max} \wedge \widetilde{Min}$. Therefore, even for the problems of choice from the whole interval [0,1] of options, there is no more than a countable set of limit choices which are given by the roots of the equation $x^{\gamma^k} = 1-x$. For the rule $R$ the limit rule selects arg max $H(R)$ which is a root of $x^{\gamma^m} = (1-x)^{\gamma^k}$. Hence no concentration of the rule of this form may result in the selection of an option from the interval $[1/2, x_o]$ where $x_o$ is a root of $x^\gamma = 1-x$.

Note that fixing n but varying the parameter $\gamma$ may make $x_o$ to be any given number from (0,1).

Example 2. Let a fuzzy choice rule be $R = \neg I^n\widetilde{Max}\ 1 \wedge \wedge\ I^{-m}\ \widetilde{Max}(n,m>0)$, then, no limit crisp choice rule is in the interval $(1/2, x_o)$ where $x_o$ is the same as above.

Let us prove this statement by contradiction. Suppose that for some arbitrary but fixed $n$ we may find a value of $m$ such that $y = \max_x \mu(Y,x) \in (1/2,x_o)$. Then $y$ is a single root of $1-y^{\gamma^n} = y^{1/\gamma m}$ in (0,1). By taking the logarithm, we get $\gamma^m \ln \frac{1}{1-y^{\gamma n}} = \ln y$ and doing this once more we get

$$m \ln \gamma + \ln \ln \frac{1}{1-y^{\gamma n}} = \ln \ln \frac{1}{y} .$$

Hence, for $m$ we have

$$m = \frac{1}{\ln\gamma} (\ln \frac{\ln y}{\ln(1-y)^{\gamma^n}}) .$$

But $\gamma > 1$ so $1/\ln\gamma > 0$. For $m$ to be more than zero, it is necessary (and sufficient) that $\ln y/\ln(1-y^{\gamma n}) > 1$, i.e. $y^{\gamma n} < 1-y$. The root of $1-y^{\gamma n} = y^{1/\gamma m}$ which is in the interval (0,1) is an increasing function of $n$. (Assume for a moment that n is a real number and take the differential of y given as an implicit function of n by the equation $1-y^{\gamma n} = y^{\gamma m}$. Then it will be seen that $y'(n) > 0$ follows from y (0,1).) This means that $y > \min y(\gamma^n) = y(\gamma) = x_o$ which contradicts the assumption.

Example 3. In this example we illustrate the case of limit (crisp) choice from [0,1] by the rule resulting in the selection of two options (see Fig. 1)

$(I \neg \widetilde{Max} \wedge Max) \vee (I\ \widetilde{Max} \wedge \neg \widetilde{Max}) = (I\ \widetilde{Min} \wedge \widetilde{Max}) \vee (I\ \widetilde{Max} \wedge \widetilde{Min}) \in$
$\in R_3(I, \neg)$.

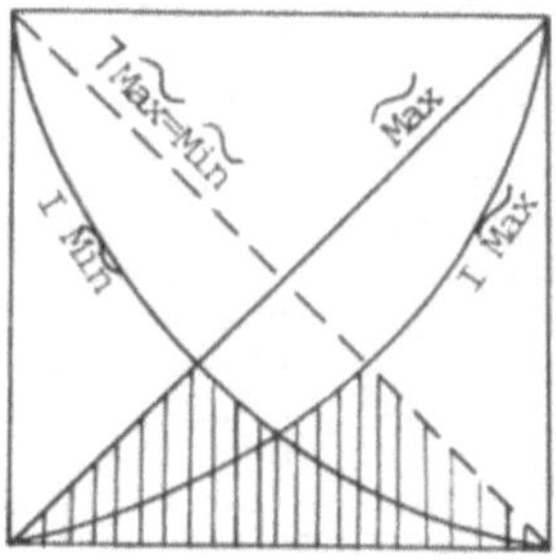


Fig. 1

REFERENCES


Dubois, D., and H. Prade (1978). Operations on fuzzy numbers. Int. J. Syst. Sci. 9, 613-626.

Sakawa, M. (1983). Interactive computer programs for fuzzy linear programming with multiple objectives. Int. J. Man-Machine Stud. 18, 489-503.

Takatsu, S. (1984). Multiple-objective satisficing decision problems. Kybernetes 13, 21-26.

Zadeh, L.A. (1975). The concept of a linguistic variable and its application to approximate reasoning. Part 1, 2, and 3. Inf. Sci. 8, 199-249; 8, 301-357; 9, 43-80.

Zadeh, L.A. (1976). The linguistic approach and its application to decision analysis. In Y.C. Ho, and S.K. Mitter (eds.), Directions in Large-Scale Systems, Many-Person-Optimization and Decentralized Control. Plenum Press, New York.

# PREFERENCES DEDUCED FROM FUZZY QUESTIONS

Bernadette Bouchon

CNRS - Laboratoire C.F. Picard
Université Paris VI - Tour 45
4, place Jussieu
75230 Paris Cédex 05 - France

Abstract. Preferences can be deduced by questioning a given population, which may be either reduced to a group of experts or considered as a set of inquired persons. Vagueness frequently appears in the asked questions or proposed criteria, and in the answers or characteristics given by the individuals. Fuzzy preferences are then necessary and, in some cases, they must lead to crisp conclusion . We study several means of evaluating the preferences in the case of fuzzy or imprecise answers to crisp questions (or criteria) or in the case of deliberately vague or subjective questions (or criteria). We deal with the problem of coming to the "best conclusion" with respect to the obtained results and we propose to use fuzzy relations and a method based on a measure of crispness of the classes of fuzzy opinions they determine.

Keywords: preference modelling, questionnaires, vague answers, fuzzy preferences, measure of crispness, fuzzy partitions.

## 1. INTRODUCTION

Decision making processes can be based on the opinion of members of a population with regard to a given problem; they are asked questions which often deal with fuzzy concepts such as a qualitative characterization - "useful", "important" -, a subjective appreciation - "beautiful", "good"-, imprecise measurements - "approximately one meter", "about 20 years old"-, incomplete information - "no opinion" -, modifiers of qualitative concepts - "very high", "almost achieved" -, imprecise evaluation - "most customers"-, or fuzzy probabilities - "likely", "unlikely"-. The fuzziness is involved either in the questions themselves, in order to allow the inquired persons to express their opinion in a flexible way, or in the answers because of non-understanding, uncertainty, unreliability, difficulty of these persons to express their feelings.

To deal with this vagueness before making a decision, we

propose to use concepts of fuzzy classes of a given universe, fuzzy partitions, measure of proximity and crispness.

More precisely, let $X = \{x_1, \dots, x_n\}$ be a finite universe of discourse which is linked with a set of decisions $D=\{d_1, \dots, d_r\}$, $r \leq n$, by a mapping from X onto D. A decision tree is built by using sequences of questions in order to obtain more and more precise classes of X until each element of X is identified and can be associated with a decision of D. When questions or answers involve some fuzziness, an inquired person cannot correspond to a unique path of the decision tree as in the non-fuzzy (or crisp) case and it is impossible to deduce a decision from its preferences without any treatment allowing to cope with the fuzziness and to determine crisp preferences not too far from the vague ones.

The problem of determining the choice of a population with regard to a decision when the preferences are not certain has been studied from several points of view. A fuzzy majority may be expressed as an impresice evaluation of the part of the population in favor of the decision, Kacprzyk (1984). The problem of defining a consensus when preferences are defined by means of fuzzy relations is studied for instance in Bezdek, Spillman and Spillman (1979). The existence of a socially satisfactory group function using fuzzy information about the opinions of the individuals is based on the definition of elementary requirements generalizing classical ones, Dimitrov (1983). The utilization of questions asked to the population to determine their opinions has been introduced in Bouchon (1982), and the construction of questionnnaires as sequences of such questions is presented in Bouchon (1985).

## 2. CREDIBILITY OF ANSWERS

### 2.1. Description of the problem

The set X corresponds to the description of the opinions of the population which would enable us to make a decision if questions and answers were not fuzzy. In this crisp case, every asked question q would produce a partition of X, finer and finer as we progress from the first question to the last ones. Let $X(q) = \{x^1, \dots, x^m\}$ be the set of crisp classes of X determined by q, and let us note $M = \{1, \dots, m\}$. In the fuzzy case, questions produce fuzzy classes, i.e. fuzzy subsets $\{\tilde{c}_1, \dots, \tilde{c}_m\}$ of X(q), defined by means of membership functions $f_1 : X(q) \to ]0, 1]$, $i \in M$, such that

$$\sum_{x \in X(q)} f_i(x) \leq 1, \quad \forall i \in M \tag{1}$$

Their values may be represented by a binary relation R(q) defined by the expert analyzing the answers and described by a matrix with elements $R_{ij} = f_i(c_j)$, $i \in M$, $j \in M$.

Example 1. Consider an inquiry about a new product to be put on the market, with $X = \{x_1, \dots, x_5\}$ and $D = \{d_1, d_2, d_3\}$ as portrayed in Fig. 1.

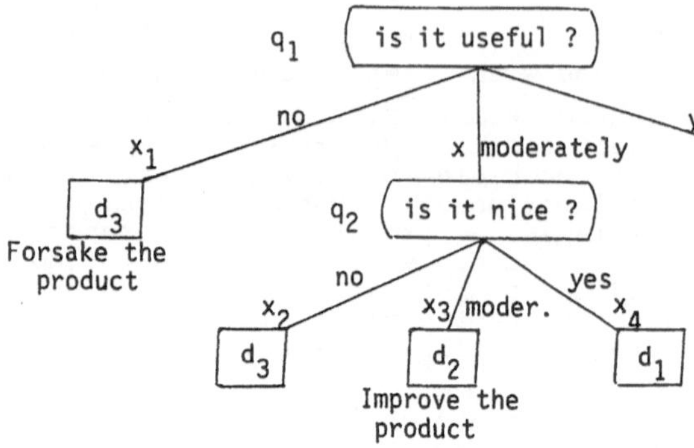


Fig. 1. Example 1

Example 2. The opinion of inquired persons is not completely reliable and the expert processing this inquiry weights the obtained results with coefficients of credibility (or certainty), yielding fuzzy subsets of X, as exhibited in Fig. 2.

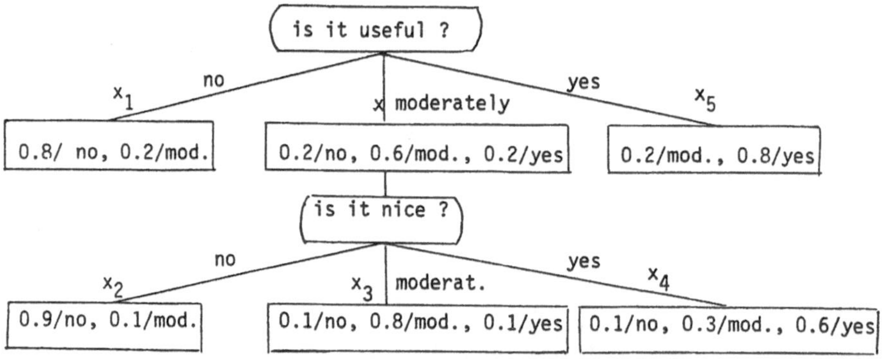


Fig. 2. Example 2

We get $X(q_1) = \{c_1, c_2, c_3\}$ in example 1 and $f_1(c_1) = 0.8$, $f_1(c_2) = 0.2$, $f_1(c_3) = 0$, $f_2(c_1) = 0.2, \ldots$ in example 2.

A fuzzy partition $E(q) = \{E_1, \ldots, E_m\}$ of M is then deduced, with each class $E_j$ defined by $\sum_{i \in M} R_{ij}/c_i$. In order to make a crisp decision, it is necessary to determine a crisp partition of M as close to E(q) as possible, which means that we decide to which element of X(q) every fuzzy answer $c_i$ must be associated.

In the previous examples, where $M = \{1, 2, 3\}$, it is obvious that the partition $\{\{1\}, \{2\}, \{3\}\}$ is close to E(q), but it can be difficult to produce such a partition in more complicated cases and a criterion is necessary to measure the proximity between fuzzy and crisp partitions.

We further suppose that a probability distribution p is defined on X(q), P(X(q)), which can be defined either from the probability of making the decision associated with every $c_i$ or from the frequencies of answers obtained in every $\tilde{c}_i$, or in such a way that $p(c_i) = 1/m$ for every $i \in M$. We denote $p_i = p(c_i)$.

## 1.2. Determination of crisp partitions

A threshold s is chosen in (0, 1] and we consider, for every $c_j \in X(q)$, the elements i of M such that $R_{ij} \geqslant s$ to construct the class $\bar{E}_j$ of a crisp partition e of X(q). We can restrict ourselves to the values of s in the interval $[s_1, s_2]$, with $s_1 = \min_j \min_i R_{ij}$, $s_2 = \min_j \max_i R_{ij}$, to be able to put every element of X(q) in a class of a crisp partition.

It is obvious that this process defines either a partition or a covering of X(q). As we need a unique decision for every fuzzy answer, we only take an interest in crisp partitions e of X(q) defined for a given s. Let $E^s(q)$ be their family; we look for a tool describing their proximity from E(q).

For any $e = \{\bar{E}_1, \ldots, \bar{E}_m\} \in E^s(q)$, we define the <u>relative s-probability</u> of $E_j$, $j \in M$, as

$$p_{s,e}(E_j) = \sum_{i \in \bar{E}_j} R_{ij} p_i \tag{2}$$

which satisfies

$$p_{s,e}(E_j) \leqslant p(\bar{E}_j) \tag{3}$$

for every $e \in E^s(q)$, with $p(\bar{E}_j) = \sum_{c_i \in \bar{E}_j} p_i$.

If $E_j$ is a crisp subset of M, then the choice of $E_j = \bar{E}_j$ implies the equality in (3). If $p_1$ is the frequency of answer $\tilde{c}_i$, then $p_{s,e}(E_j)$ represents the average credibility in $c_i$ when using e as a reference. If $p_i$ is equal to 1/m, then $p_{s,e}(E_j)$ is proportional to the fuzzy cardinality of the s-level set of $E_j$ (i.e. the subset of X(q) containing the elements $c_i$ such that $R_{ij} \geqslant s$).

The problem is to choose one crisp partition in $E^s(q)$ which is the closest to E(q) in order to use the greatest amount of information included in E(q) when restricting ourselves to crisp situations instead of directly utilizing the fuzzy opinions we get.

We propose to measure the proximity between $e \in E^s(q)$ and E(q) by means of the following quantity

$$F_s(e,E(q)) = \sum_i h(p(\bar{E}_i), P_{s,e}(E_i)) \tag{4}$$

for a given real-valued function h, positive and continuous, defined on [0, 1] x [0, 1], such that h(x,y) = 0 if and only if x = y, and nonincreasing in y.

A particular case of such a quantity is given by the gain of information defined for incomplete probability distributions ( $\sum_j P_{s,e}(E_j) \leqslant 1$, $\sum_j p(\bar{E}_j) = 1$ because of condition (1), (see Renyi, 1970))

$$h(x,y) = x \log (x/y) \tag{5}$$

1.3. s-crispness of fuzzy partitions

Let us consider a partial order defined on the family of fuzzy partitions of X(q). We study the variations of $F_s$ with respect to this order, to prove that $F_s$ may be regarded as a measure of the fuzziness of E(q), allowing to evaluate the closeness of their resemblance to crisp partitions.

Let E(q) and E'(q) be two fuzzy partitions of X(q) respectively defined by values $R_{ij}$ and $R'_{ij}$, $\forall i \in M$, $\forall j \in M$, as indicated previously. We say that E'(q) is s-sharper than E(q) if, for every $j \in M$, we have

$$\begin{cases} R'_{ij} \geqslant R_{ij}, & \forall i \in M \text{ such that } R_{ij} \geqslant s \\ R'_{ij} \leqslant R_{ij}, & \forall i \in M \text{ such that } R_{ij} < s. \end{cases}$$

We write $E(q) \prec_s E'(q)$.
This means that E'(q) is closer to a crisp situation than E(q) and this implies that they admit the same family of crisp partitions for the threshold s, $E^s(q) = E'^s(q)$. We get the following:

Property 1: If $E(q) \prec_s E'(q)$, then $F_s(e,E(q)) \geqslant F_s(e,E'(q))$, for every $e \in E^s(q)$.

Proof. For every $e \in E^s(q)$, we get $P_{s,e}(E_i) \leqslant P_{s,e}(E'_i)$, for each $i \in M$. Thus, the monotony of h entails the property.

Example 3: Let us consider a question $q_1$ and another question $q_2$, for instance "Is it really useful: yes? no? can be discussed?".

We consider the following coefficients:

$R(q_1)$

| i \ j | 1 | 2 | 3 |
|---|---|---|---|
| 1 | 0.8 | 0.2 | 0 |
| 2 | 0.2 | 0.6 | 0.2 |
| 3 | 0 | 0.2 | 0.8 |

$R(q_2)$

| i \ j | 1 | 2 | 3 |
|---|---|---|---|
| 1 | 0.9 | 0.1 | 0 |
| 2 | 0.2 | 0.7 | 0.1 |
| 3 | 0 | 0.1 | 0.9 |

Then, $E(q_2)$ is 0.5-sharper than $E(q_1)$ and we get, for the crisp partition e defined by $\bar{E}_i = \{c_i\}$, i = 1, 2, 3, and function h introduced in (5), with $p_i = 1/3$, i = 1, 2, 3:

$$F_{0.5}(e,E(q_1)) = 0.46 \quad \text{and} \quad F_{0.5}(e,E(q_2)) = 0.27$$

which proves that $F_{0.5}(e,E(q_1)) \geqslant F_{0.5}(e,E(q_2))$.

We call <u>s-crispness</u> of E(q) the quantity

$$F_s(E(q)) = \min_{e \in E^s(q)} F_s(e,E(q)) \tag{6}$$

It is null if and only if $P_{s,e}(E_j) = p(\bar{E}_j)$ which means that $R_{ij} = 1$ for every $c_1 \in \bar{E}_j$ and $j \in M$. It is particularly the case if E(q) is crisp and it is the only possibility if $\sum_i R_{ij} = 1 \ \forall j \in M$.

<u>Property 2</u>: If $E(q) \prec s\ E'(q)$, then $F_s(E(q)) \geqslant F_s(E'(q))$, for every $e \in E^s(q)$.

Proof. We have

$$F_s(E(q)) = \min_{e \in E^s(q)} F_s(e,E(q)) = F_s(e_o,E(q)) \geqslant$$

$$\geqslant F_s(e_o,E'(q)) \geqslant \min_{e \in E^s(q))} F_s(e_o,E'(q)) = F_s(E'(q))$$

This property means that if the fuzzy partition E(q) is s-sharper than another one, then it is also closer to its nearest crisp partition.

<u>Property 3</u>. For two thresholds s and t, such that $s \leqslant t$, it holds

$$F_t(E(q)) \geqslant F_s(E(q))$$

<u>Proof</u>. For $s \leqslant t$, every $e \in E^t(q)$ belongs to $E^s(q)$. Then

$$F_t(E(q)) = F_t(e_o,E(q)) = F_s(e_o,E(q)) \geqslant$$

$$\min_{e \in E^s(q)} F_s(e,E(q)) = F_s(E(q)).$$

Consequently, the s-crispness of E(q) is non-decreasing with respect to s. We preserve more information from the fuzzy classes when using a small threshold. We can state the following:

Proposition 1. For a given threshold s, the s-crispness of a fuzzy partition E is a measure of its closeness to the nearest crisp partition $\bar{E}$ and of the loss of information deduced from the replacement of E by $\bar{E}$. A means of determining $\bar{E}$ is to choose s in $[s_1, s_2]$, and to find a crisp partition $\bar{E}$ such that $F_s(\bar{E},E) = F_s(E)$.

Example 4. If we consider question $q_1$, the crisp partition e exhibited previously is such that $F_{0.5}(e,E(q_1)) = F_{0.5}(E(q_1))$. But, for s = 0.2, e would not be the only crisp partition corresponding to $E(q_1)$. For instance, e' defined by $\bar{E}'_1 = \{c_1, c_2\}$, $\bar{E}'_2 = \emptyset$, $\bar{E}'_3 = \{c_3\}$ would give $F_{0.2}(e',E(q_1)) = 0.44 = F_{0.2}(E(q_1))$ and $F_{0.2}(e,E(q_1)) = 0.46$.

1.4. Choice between questions

Let us suppose that two different questions q and q' may be asked by the expert to the inquired persons to get their opinion about the same criterion, in such a way that the results $\{c_1,\ldots,c_m\}$ are obtained in both cases when no fuzziness is taken inco account. When coefficients of credibility are added by expert, fuzzy partitions E(q) and E'(q) of X(q) are deduced.

To choose which question is more interesting to ask, we calculate their s-crispness, for a as low as possible threshold s, and the smallest obtained value corresponds to a question loosing little information and which we recommend to use.

For instance, if we substitute $q'_1$ to $q_1$ in example 1, we obtain the fuzzy results defined in example 5.

Example 5. Let the situation be as in Fig. 3, and

$R(q_1)$

| i \ j | 1 | 2 | 3 |
|---|---|---|---|
| 1 | 0.8 | 0.2 | 0 |
| 2 | 0.2 | 0.6 | 0.2 |
| 3 | 0 | 0.2 | 0.8 |

$R(q'_1)$

| i \ j | 1 | 2 | 3 |
|---|---|---|---|
| 1 | 0.9 | 0.1 | 0 |
| 2 | 0.2 | 0.6 | 0.2 |
| 3 | 0 | 0.5 | 0.5 |

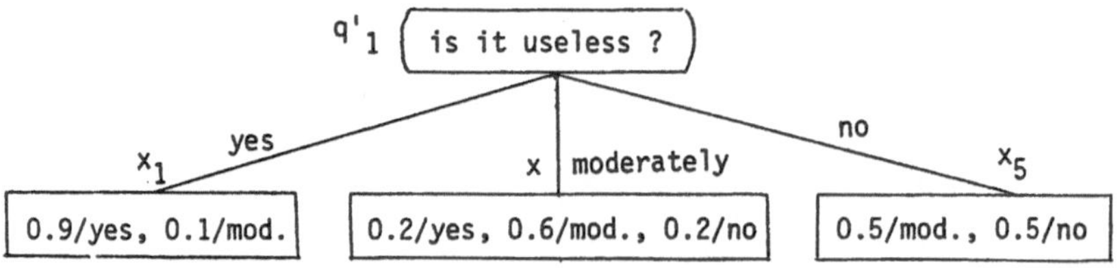


Fig. 3. Example 5

We suppose once more that $p_i = 1/m$, $\forall i \in M$. we get:
$F_{0.5}(E(q_1)) = 0.46$, corresponding to e introduced previously, and $F_{0.5}(E(q'_1)) = 0.63$, corresponding to $e'' = \{\bar{E}''_1, \bar{E}''_2\}$, with $\bar{E}''_1 = \{c_1\}$ and $\bar{E}''_2 = \{c_2, c_3\}$.

Thus, individuals answering "no" to question $q'_1$ are treated jointly with those answering "moderately" and question $q_2$ may further be asked of all of them, in the case of $q'_1$. Nevertheless, $q_1$ is more interesting than $q'_1$ because its 0.5-crispness is smaller than that of $q'_1$. The latter is compatible with an intuitive point of view, since it gives less precise information about the opinions, due to the given coefficients of credibility. We also verify that the crisp partition e" intuitively close to $E(q_1)$ admits a provimity of $E(q_1)$ equal to its 0.5-crispness; then, decision $d_3$ corresponds to individuals answering "no" to $q_1$, decision $d_1$ to those answering "yes", and question $q_2$ is asked to those answering "moderately". We are then facing a classical decision tree.

### 1.5. How to trade on questions with coefficients of credibility

As a conclusion of this section, we suggest the two following propositions.

Proposition 2. When coefficients of credibility are assigned to answers of a question, we make a crisp decision in partitioning the population giving fuzzy responses which split it in a fuzzy partition E into crisp classes of a partition e yielding $F_s(e,E)$ as the s-crispness of E. The question is acceptable if its s-crispness is not greater than a given level v.

If the s-crispness of the question has a high value, we can conclude that the question is not well-defined (or well-formulated) and it is not interesting to be utilized.

Proposition 3. When several questions exist giving information about the same criterion but corresponding to different values of the coefficients of credibility, we choose the one giving the smallest s-crispness.

## 2. VAGUE QUESTIONS

A different point of view consists in the determination of preference coefficients by the questioned person himself, when answering a question q with possible issues $c_1, \ldots, c_m$.

Consider question $q_3$ in example 6, concerning preferences of a population about a product to be put on the market.

Example 6. Let

$q_3$: Indicate your preference order with regard to the color of the product, by assigning a number between 0 and 1 to the possible issues $c_1$ = "red", $c_2$ = "blue", $c_3$ = "yellow".

Let $R_{ij}$ be the value assigned by the individual number i to $c_j$. We get a fuzzy partition $E(q) = \{E_1, \ldots, E_m\}$ of the given finite population Y into m classes.

A crisp determination of the preferences would correspond to values of $R_{ij}$ equal to 0 or 1. The proximity between this crisp situation e and the fuzzy one is measured by $F_s(e,E(q))$. We take $p_i = 1/k$, with $k = |Y|$, $\forall i \in Y$. For a given threshold s and a crisp partition $e \in E^s(q)$, we obtain $P_{s,e}(E_j)$, $j \in M$, as the average preference coefficient assigned by the persons corresponding to the s-level set of $E_j$, that is to say giving a coefficient at least equal to s for the issue $c_j$. Let $d_{u(j)}$ denote the decision corresponding to every $c_j$ in the decision tree, which can be either an element of D or the decision to ask an additional question. We can also interpret $P_{s,e}(E_j)$ as the index of satisfaction of the population if the decision $d_{u(j)}$ is made.

Then, $F_s(e,E(q))$ represents the lack of satisfactoriness in making decisions $d_{u(1)}, \ldots, d_{u(m)}$ corresponding to $c_1, \ldots, c_m$. If in the case of crisp preferences, there is no ambiguity in the determination of the decision, then $e = E(q)$ and $F_s(e(q))=0$. In the case of complete indifference of the population ($R_{ij} = 1/m$, $\forall i \in \{1, \ldots, k\}$, $\forall j \in M$), to make any decision corresponding to one of the $c_j$, i.e. a crisp partition with one class $e = \{Y\}$, would give

$$F_s(e,E(q)) = k.\ h(1,\ 1/m)$$

and for instance with h defined in (5)

$$F_s(e,E(q)) = k.\ \log m$$

which has a high value.

Suppose that, in a second interview, every individual increases all the coefficients at least equal to the s he has given in the first interview. This means that if his indifference to the proposed issues is diminishing, then property 1 entails that $F_s(e,E(q))$ decreases towards zero.

Proposition 4. If several persons are asked a vague question, their preferences, described by a fuzzy partition E, can be aggregated by splitting them into a crisp partition e such that $F_s(e,E)$ is as small as possible for some threshold s. The value of $F_s(e,E)$ indicates the non-satisfactoriness of this crisp partition, which can be regarded as acceptable if $F_s(e,E)$ is not greater than a given level v.

We are then reducing our study to a crisp classical problem. If the number of persons is small, $F_s(E)$ can easily be reached, for a crisp partition e of Y. Otherwise, the combinatorial description of all the crisp partitions compatible with the threshold s and with the fuzzy partition E is too long, and it is sufficient to choose one of them, say e, such that $F_s(e,E)$ is smaller than a given level of acceptability v.

Another possibility consists in preserving the fuzziness of the answers and defining the following fuzzy decision

$$\sum_{j\in M} P_{s,e}(E_i)/d_{u(i)} \tag{7}$$

Example 7. Coming back to question $q_3$ indicated in example 6, we consider the following answers, for a population of 4 persons (k = 4):

$R(q_3)$

| i \ j | 1 | 2 | 3 |
|---|---|---|---|
| 1 | 0.6 | 0.2 | 0.2 |
| 2 | 0.5 | 0.5 | 0 |
| 3 | 0 | 0.1 | 0.9 |
| 4 | 0 | 0.7 | 0.3 |

We get $F_{0.5}(E(q_3)) = 0.178$, corresponding to $e = \{\bar{E}_1, \bar{E}_2, \bar{E}_3\}$, with $\bar{E}_1 = \{1\}$, $\bar{E}_2 = \{2, 4\}$, $\bar{E}_3 = \{3\}$.

Then, individual number 1 is considered as in favor of decision $d_{u(1)}$, individuals 2 and 4 are in favor of $d_{u(2)}$ and individual 3 in favor of $d_{u(3)}$. A fuzzy decision could be the following, as indicated in (7)

$$0.15 / d_{u(1)},\ 0.3 / d_{u(2)},\ 0.225 / d_{u(3)}.$$

## 3. CONCLUSION

The model proposed here can be used in several situations: if a group of experts is asked to give coefficients of credibility, or certainty, to several evidences and their opinions must be concentrated in a single one, if a set of persons is questioned before a decision is made which must satisfy them as far as possible, if an inquiry is hold to verify a supposed hypothesis, in social sciences for instance, or for market research.

What we have called a question may also be a physical, mechanical or medical test, the results of which are not completely reliable because of difficulties in measurement, observation or utilization of subjective criteria such as the strength of suffering, the color of a chemical product, etc.

The measure of proximity $F_s$ and the s-crispness are interesting quantities because of their isotony with regard to the s-sharpness of partitions. They could be utilized in other types of problems than those studied here, such as identifications, classifications or pattern recognition, for example. They permit the measurement of the closeness of a crisp partition to a fuzzy one, to give an idea of the accuracy of fuzzy classes of a universe, of the fuzziness of imprecise partitions with overlapping classes. They indicate the "quality" of a fuzzy partitioning which can be reasonably precise or so vague that no information can be deduced from its study. They also express the fact that a crisp partition is adequate for making a decision or that it is so far from the original fuzzy situation that it has no sense.

In conclusion, we have dealt with the problem of determining crisp decisions from answers to questions assigned with coefficients of credibility, certainty, reliability, and with the problem of utilizing answers to questions or results of tests expressed by means of grades of preference, either to reduce the situation to a near classical one, or to define fuzzy decisions. Numerous other types of problems could be regarded from an analogous point of view and treated with the studied tools or with related ones.

## REFERENCES

Bezdek, J., B. Spillman, and R. Spillman (1979). Fuzzy relation spaces for group decision theory: an application. Fuzzy Sets and Syst. 2, 5-14.

Bouchon, B. (1982). Decisions based on fuzzy information (in French). Proc. 19th Meeting on Multiple Criteria Decision Aid, Liege, 1982 (Public. European Institute for Advanced Studies in Management).

Bouchon, B. (1985). Questionnaires in a fuzzy setting. In J. Kacprzyk and R.R. Yager (eds.), Management Decision Support Systems Using Fuzzy Sets and Possibility Theory. Verlag TUV Rheinland, 189-197.

Dimitrov, V. (1983). Group choice under fuzzy information. Fuzzy Sets and Syst. 6, 25-39.

Kacprzyk, J. (1984). Collective decision-making with fuzzy majority rule. Proc. of 6th International Congress of Cybernetics and Systems. Paris, 153-159.

Renyi, A. (1970). Probability Theory. North Holland, Amsterdam.

## II.2. Aspects of Fuzzy Decision Making and Optimization

# OPTIMAL ALTERNATIVE SELECTION IN THE FACE OF EVIDENTIAL KNOWLEDGE

Ronald R. Yager

Machine Intelligence Institute
Iona College
New Rochelle, NY 10801, USA

Abstract. We discuss the problem of selecting an optimal alternative in the face of uncertain knowledge about the state of nature. We then introduce a general framework for representing the knowledge about the state of nature. This general framework is based upon the use of fuzzy sets, possibility theory and the Dempster-Shafer mathematical theory of evidence. We then suggest an approach to the selection of an optimal alternative in this more general framework.

Keywords: Decision Theory, Fuzzy Sets, Possibility Theory, Evidence Theory.

## 1. INTRODUCTION

A classic decision making paradigm involves the selection of a best alternative in which the payoffs are a function of the state of nature which is in some sense unknown. The classic solutions to this problem concern themselves with the situations in which the knowledge about the state of nature is either purely probabilistic or purely possibilistic.

In many cases the information available about the state of nature is obtained from some type of auxiliary evidence and is much subtler than that which can be represented by a purely probabilistic or complete ignorance characterization. In particular our knowledge about the state of nature may be better represented by a possibility distribution or more generally a Dempster-Shafer type evidential belief structure (D-S granule) (Zadeh, 1978, 1979; Shafer, 1976; Yager, 1985a, 1985b). These D-S granules provide a framework in which the uncertainty associated with a state variable can manifest possibilistic, fuzzy as well as probabilistic uncertainty components.

The main concern here is to suggest a methodology for solving the alternative selection problem in the situation in which the state of nature is described by a D-S granule. As we may see, the classic characterizations become special cases of this more general framework.

Special attention is given to the fact that in the face of possibilistic uncertainty one must choose a "decision attitude".

The classic decision attitudes of pessimism, optimism and least regret are studied in this new framework.

## 2. A DECISION MAKING PARADIGM

A problem of considerable interest to decision theorists can be captured in terms of the following matrix,

$$\begin{array}{c} \\ A_1 \\ \cdots \\ A_i \\ \cdots \\ A_q \end{array} \begin{array}{c} S_1 \ldots S_j \ldots S_n \\ \begin{bmatrix} C_{11} & & \\ & & \\ & C_{ij} & \\ & & \\ & & C_{qn} \end{bmatrix} \end{array}$$

In the above matrix $A_1$ corresponds to a particular action alternative available to the decision maker. $S_j$ sorresponds to a variable called the "state of nature". $C_{ij}$ corresponds to the payoff to be received by the decision maker when selecting action $A_i$ when the state of nature is $S_j$. The problem faced by the decision maker is to select the action which provides him with the "best" payoff. In its simplest form, if the decision maker knows that the state is $S_k$, his problem is simply reduced to that of finding the $A^*$ such that $C_{ik}$ is maximized, that is find $i^*$ such that

$$C^* = \underset{i}{\text{Max}}\ C_{ik}$$

This is called decision making under certainty.

In most realistic problems the decision maker does not know the value of the state and is thus faced with the problem of making a decision in the face of uncertainty. Two situations in regards to the uncertainty have been considerably studied in the literature. In the first case the decision maker assumes he has knowledge as to the probabilities of the occurrence of each state of nature. This is called decision making in the face of probability. In this case we let $p_j$ equal the probability that the value of the state is $S_j$. The accepted procedure for solving the problem in this situation is first to find the expected payoffs for each $A_i$, which we denote

$$E(A_i/P) = \sum_{j=1}^{n} (C_{ij} \times p_j)$$

and then to select the alternative $A^*$ which has the maximum expected payoff. That is, select as the action alternative $A^*$ such that

$$E(A^*/P) = \max_i E(A_i/P)$$

In the second situation the decision maker assumes no knowledge of the state of nature other than the list of possibilistic states. This is sometimes called decision making under ignorance. In this situation the decision maker must introduce some additional criteria called the decision attitude to help in making the decision. A number of decision attitudes have been suggested. Four of these shall be considered here.

The maximin strategy is essentially a pessimistic attitude. In this strategy the decision maker first finds the worst possible payoff for each alternative and then selects the alternative with the best worst outcome.

The maximax strategy is essentially an optimistic attitude. In this strategy the decision maker first finds the best possible payoff for each alternative and then selects the alternative with the maximum best outcome.

The so-called Hurwicz strategy is a combination of these two. One first finds the worst possible outcome for each alternative, call this $W_1$ and then finds the best outcome for each alternative, call this $B_1$. Next one selects a coefficient of pessimism $h$ and calculates $H_i = h*W_i + (1-h)*B_i$. Then the decision maker selects the alternative with the highest Hurwicz value $H_i$. We note that if h=1, we get the maximin strategy, and if h=0, we get the maximax strategy.

The fourth approach is called the least regret strategy. In this approach we create a new matrix called the regret matrix denoted R. To calculate the regret matrix, each element in this new matrix, $r_{ij}$, is obtained as

$$r_{ij} = C_{ij} - T_j$$

where $T_j$ is the maximal value obtainable under the occurence of state $S_j$. Then for each alternative we calculate minimum value of $r_{ij}$ in that row, the max regret. Finally, we select the alternative with the largest value, the least regret. All these approaches can be formalized under the following unifying framework.

Let S be the state of nature. Let $E(A_i/S)$ be the evaluation of the alternative $A_i$ under this state of nature where

$$E(A_i/S) = f/C,i,S)$$

We then select as best alternative the alternative $A^*$ such that

$$E(A^*/S) = \max_i f(C,i,S)$$

We denote $E(A^*/S)$ as $E(S)$.

In decision making under certainty $S = S_a$ and therefore

$$E(A_i/S) = f(C,i,S_a) = C_{ia}$$

In decision making under ignorance we know that the value of the state of nature S is a member of the set $(S_1,\ldots,S_n)$ but we do not know which element it is. In this case the evaluation function is dependent upon the decision attitude we choose.

(1) For the pessimistic attitude

$$E(A_i/S) = \underset{j}{\text{Min}}\ C_{ij}$$

(2) For the optimistic attitude

$$E(A_i/S) = \underset{j}{\text{Max}}\ C_{ij}$$

(3) For the Hurwicz attitude with coefficient of pessimism h

$$E(A_i/S) = h * \underset{j}{\text{Min}}\ C_{ij} + (1-h) * \underset{j}{\text{Max}}\ C_{ij}$$

(4) For least regret attitude

$$E(A_i/S) = \underset{j}{\text{Min}}\ r_{ij}$$

where $r_{ij} = C_{ij} - T_j$ in which $T_j = \underset{i}{\text{Max}}\ C_{ij}$.

In all cases the selected alternative is the one which maximizes the evaluation function $E(A_i/S)$ over all i.

The above approaches are limited in that they are valid for distinct formulations of our knowledge about the state, certainty, randomness or ignorance. In the following we present a formalism which captures all of these types of uncertainty in one general structure, D-S granule.

## 3. REPRESENTATION OF STATE KNOWLEDGE

In this section we shall present a more general framework in which to represent the knowledge about the state of nature. This approach is based upon the theory of possibility developed by Zadeh (1978) and the theory of mathematical evidence developed by Shafer (1976).

Assume V is a variable which can assume its value in the set X. Let A be a subset of X. The knowledge that

V is in A

can be seen to induce a possibility distribution $\pi_V$ on X such that

$$\pi_V(x) = \begin{cases} 1 & \text{if } x \in A \\ 0 & \text{if } x \notin A \end{cases}$$

The function $\Pi_V(x)$ is interpreted as being the <u>possibility</u> that V is x. Thus if all we know is that V is some element in the set A, we get a possibility distribution as its representation.

More generally, if A is a fuzzy subset of X, the knowledge that

$$V \text{ is } A$$

induces a possibility distribution $\Pi_V$ on X such that

$$\Pi_V(x) = A(x)$$

where A(x) is the membership grade of x in the fuzzy subset A of X, Zadeh (1965).

The use of fuzzy subsets, which parenthetically include crisp sets as a special case, becomes particularly useful for the representation of knowledge that is vague as well as possibilistic. For example if V is a variable corresponding to the interest rates, then the knowledge that the interest rates are "high" can be represented as

$$V \text{ is } H$$

where H is a fuzzy subset representing the concept "high interest rates".

An even more general framework can be obtained by the introduction of basic probability assignment functions which will allow for probabilistic as well as possibilistic and fuzzy uncertainty.

Assume m is a mapping from the fuzzy subsets of X into the unit interval

$$m: I^X \rightarrow [0,1]$$

Let $A_i$, $i=1,\ldots,p$ be the collection of fuzzy subsets for which $m(A_i) = a_i \neq 0$. We call these $A_i$'s the focal elements of m. If

(1) $\sum_{i=1}^{p} m(A_i) = 1$

(2) $m(\emptyset) = 0$

we shall call m a basic probability assignment (bpa) function.

Let V be a variable which takes its values in the set X and let m be a bpa. We shall call a statement of the form

$$V \text{ is } m$$

a D-S (Dempster-Shafer) granule.

The statement "V is m" is an appropriate way to model our knowledge about the variable V in the following situation.

Consider a random experiment in which the outcome space is the set

$$Y = \{y_1, \ldots, y_p\}$$

in which $a_i$ equals the probability of $y_i$ occurring. Furthermore, assume that if $y_i$ occurs as a result of the performance of the experiment, then we know that V is in the set $A_i$, $A_i \neq \emptyset$, but do not know which element in $A_i$.

We feel that these D-S granules can provide a more general structure in which to represent the knowledge about the state associated with the decision problem previously introduced.

Let V be a variable indicating the state of nature. Let $S = \{S_1, \ldots, S_q\}$ be the set of possible states of nature. The situation in which our knowledge about the variable V is complete ignorance can be represented by the statement

$$V \text{ is } m$$

in which $m(S) = 1$.

This is equivalent to the information "V is in S".

The situation which we previously called decision making under uncertainty, that is the one in which we have probability $p_i$ associated with each state $S_i$ in S, can be represented as

$$V \text{ is } m$$

in which

$$m(A_i) = p_i, \quad i=1,\ldots,q$$

where $$A_i = \{S_i\}$$

The situation in which we have complete knowledge of the state of nature, V is $S^*$, can be represented as

$$m(\{S^*\}) = 1, \quad \text{where } S^* \in S$$

This is what we call decision making under certainty.

In addition to being able to represent these three notable types of knowledge in a uniform framework, the use of D-S granules allows us to easily represent more complex pieces of knowledge about the state of nature in this uniform format.

Let A be a subset of S. If our knowledge about V is such that we know that the probability that V is in A is at least α, we can represent this as

$$V \text{ is } m$$

where m is such that

$$m(A) = \alpha \quad \text{and} \quad m(S) = 1-\alpha$$

In Yager (1985b) the representation of various forms of knowledge in this framework is discussed.

4. DECISION MAKING WITH D-S GRANULES

In this section we shall extend the decision making paradigm to the case in which our knowledge about the statement of nature V is formulated in terms of a D-S granule, "V is m".

Again consider a decision situation as captured by the matrix

$$\begin{array}{c} \\ A_1 \\ \cdots \\ A_i \\ \cdots \\ A_q \end{array} \begin{array}{c} S_1 \;\cdots\; S_j \;\cdots\; S_n \\ \left[\begin{array}{ccc} & & \\ & & \\ & C_{ij} & \\ & & \\ & & \end{array}\right] \end{array}$$

Thus in this case V is a variable taking its values in the set

$$S = \{ S_1, \ldots, S_n \}$$

Let our knowledge about V be represented by

$$V \text{ is } m$$

We shall let $E(A_i/m)$ indicate the value of selecting alternative $A_i$ in the case of our knowledge, V is m. As before the optimal alternative $A^*$ is selected as

$$E(A^*/m) = \underset{i=1,\ldots,q}{\text{Max}} \; E(A_i/m)$$

Our first goal is to be able to evaluate $E(A_i/m)$. We shall initially restrict ourselves to the situation in which m is of the form $m(B) = 1$, B is some crisp subset of S. This corresponds to the case in which all we know about V, the state of nature, is that it lies in the crisp subset B of S. For ease of notation we shall indicate the valuation in this situation as $E(A_i/B)$. Note that the decision under ignorance is a special case of this, $E(A_i/S)$.

Let $C_i(B)$ indicate the set of possible payoffs in the situation in which we know the state of nature is B and the selected alternative is $A_i$, therefore

$$C_i(B) = \{ C_{ij} / \text{over all } j \text{ such that } S_j \in B \}.$$

Our evaluation $E(A_i/B)$ then becomes dependent upon $C_i(B)$. At this point we must select a <u>decision attitude</u> as we have previously had to do. Four possible attitudes are:

(1) Pessimistic
(2) Optimistic
(3) Regret
(4) Hurwicz

We shall indicate the evaluation function under these situations as $E_P(A_i/B)$, $E_0(A_i/B)$, $E_R(A_i/B)$, and $E_H(A_i/B)$. In the pessimistic attitude we are in the mind of maximizing the worst possibilities. In the optimistic we are interested in maximizing our best possibilities and in the regret we are interested in optimizing our least regret. In the Hurwicz case we use some balance between the pessimistic and optimistic.

Thus in the pessimistic case

$$E_P(A_i/B) = \underset{i}{\text{Min}}\ C_i(B) = \underset{x \in C_i(B)}{\text{Min}}\ x$$

thus $E_P(A_i/B)$ is the minimum payoff that occurs in the subset of payoffs under $A_i$ from the subset $B$. This is exactly the procedure originally suggested.

In the optimistic case

$$E_0(A_i/B) = \underset{i}{\text{Max}}\ C_i(B) = \underset{x \in C_i(B)}{\text{Max}}\ x$$

In the regret situation let $R_i(B)$ be the matrix obtained as follows

$$R_i(B) = \left\{ C_{ij} - T_j /\ j \text{ such that } S_j \in B \right\}$$

where $T_j = \underset{i}{\text{Max}}\ C_{ij}$, then

$$E_R(A_i/B) = \text{Min}\ R_i(B) = \underset{x \in R_i(B)}{\text{Min}}\ x$$

In the Hurwicz case

$$E_H(A_i,B) = h * E_P(A_i/B) + (1-h) * E_0(A_i/B)$$

It can easily be seen that these formulations are the same as in the classic approach. We shall now extend the analysis to the situation in which again our knowledge about the state is still

$$m(B) = 1$$

but in this case we shall allow B to be a fuzzy subset of S.

In this case the set $C_i(B)$, the possible payoffs under alternative $A_i$, becomes a fuzzy subset of payoffs; in particular

$$C_i(B) = \bigcup_{j=1}^{n} \left\{ B(S_j)/C_{ij} \right\} = \left\{ B(S_1)/C_{i1}, B(S_2)/C_{i2}, \ldots, B(S_n)/C_{in} \right\},$$

in which $B(S_j)$ is the membership grade of state of nature $S_j$ in the set B.

In order to calculate $E_p(A_i/B)$ we must be able to find $\underset{i}{\text{Min}}\ C_i(B)$. However, in this case $C_i(B)$ is a fuzzy subset of numbers and therefore the calculation of $E_p(A_i/B) = \underset{i}{\text{Min}}\ C_i(B)$ requires us to find the minimum element of a fuzzy subset. In order to calculate this value we shall draw upon some ideas suggested by Yager (1981).

Let D be a fuzzy subset of real numbers. We define the α-level set of D as

$$D^{(\alpha)} = \{ y: D(y) \geq \alpha) \}$$

Thus $D^{(\alpha)}$ is the crisp subset of numbers which have the membership grade of at least α. Let us furthermore define Min $D^{(\alpha)}$ to be the smallest element in D. We note that Min $D^{(\alpha)}$ is defineable since $D^{(\alpha)}$ is crisp. We then define

$$\text{Min } D = \int_0^1 \text{Min } D^{(\alpha)}\, d\alpha$$

Thus Min $D^{\alpha}$ is an effective minimum element in D. It can be easily shown that if D is crisp, then the above procedure leads to the usual smallest element in D. More generally, if Sup D = $= \{ x/D(x) > 0 \}$, i.e. the set of elements with non-zero membership grade, and if the Min(Sup $D^{(\alpha)}$), i.e. the smallest element in Sup $D^{(\alpha)}$, has membership grade one, then Min $D^{(\alpha)}$ equals this element.

Using the above methodology in the case where B and therefore $C_1(B)$ are fuzzy subsets, we obtain

$$E_p(A_1/B) = \int_0^1 \text{Min } C_1^{(\alpha)}(B) \quad d\alpha$$

which is an ordinary number. Therefore we can easily obtain $A^*$ as

$$E_p(A^*/B) = \underset{i=1,\ldots,q}{\text{Max}} \quad E_p(A_i/B)$$

The optimistic situation can easily be obtained by replacing the min by the max. Thus

$$E_o(A_1/B) = \int_0^1 \text{Max } C_1^{(\alpha)}(B) \quad d\alpha$$

In the case of minimum regret we replace $C_1(B)$ by $R_1(B)$ in which

$$R_1(B) = \bigcup_{i=1}^{n} \{ B(S_j)/C_{ij} - T_j \}$$

where $T_j = \underset{i}{\text{Max}}\ C_{ij}$, and then

$$E_R(A_1/B) = \int_0^1 \text{Min}\ R_1^{(\alpha)}(B) \quad d\alpha$$

An example at this point will help clarify the methodology.

Example. Consider the matrix

$$\begin{array}{c} \\ A_1 \\ A_2 \\ A_3 \end{array} \begin{array}{cccc} S_1 & S_2 & S_3 & S_4 \\ \end{array}$$

$$\begin{matrix} & S_1 & S_2 & S_3 & S_4 \\ A_1 & 10 & -15 & 5 & 20 \\ A_2 & 5 & 5 & 10 & 10 \\ A_3 & 0 & 10 & 15 & 0 \end{matrix}$$

Assume what we know about the state is "V is m" where m(B) = 1 and

$$B = \{1/S_1,\ 1/S_2,\ .5/S_3,\ .2/S_4\}$$

Then it follows that

$$C_1(B) = \{1/10,\ 1/-15,\ .5/5,\ .2/20\}$$
$$C_2(B) = \{1/5,\ 1/5,\ .5/10,\ .2/10\}$$
$$C_3(B) = \{1/0,\ 1/10,\ .5/15,\ .2/0\}$$

The associated level sets are:

| $C_1(N)$ | for | Max | Min |
|---|---|---|---|
| $\{-15,5,10,20\}$ | $0 \leq \alpha \leq .2$ | 20 | -15 |
| $\{-15,5,10\}$ | $2 < \alpha \leq .5$ | 10 | -15 |
| $\{-15,10\}$ | $\alpha > .5$ | 10 | -15 |

| $C_2(B)$ | for | Max | Min |
|---|---|---|---|
| $\{5,10\}$ | $0 \leq \alpha \leq .5$ | 10 | 5 |
| $\{5\}$ | $\alpha > .5$ | 5 | 5 |

| $C_3(B)$ | for | Max | Min |
|---|---|---|---|
| $\{0,10,15\}$ | $0 \leq \alpha \leq .5$ | 15 | 0 |
| $\{0,10\}$ | $\alpha > .5$ | 10 | 0 |

Therefore

$$E_p(A_1/B) = \int_0^{.2}(-15)\,d\alpha + \int_{.2}^{.5}(-15)\,d\alpha + \int_{.5}^{1}(-15)\,d\alpha = -15$$

$$E_p(A_2/B) = \int_0^{.5} 5\,d\alpha + \int_{.5}^{1} d\alpha = 5$$

$$E_p(A_3/B) = \int_0^{.5} 0\,d\alpha + \int_{.5}^{1} 0\,d\alpha = 0$$

Hence the maximum $E_p$ comes with the selection of $A_2$.
To calculate $E_o$ we proceed as follows

$$E_o(A_2/B) = \int_0^{.2} 20\, d\alpha + \int_{.2}^{.5} 10\, d\alpha + \int_{.5}^{1} 10\, d\alpha =$$

$$= (.2)(20) + (.3)(10) + (-5)(10) = 12$$

$$E_o(A_2/B) = \int_0^{.5} 10\, d\alpha + \int_{.5}^{1} 5\, d\alpha = 5 + 1.5 = 7.5$$

$$E_o(A_3/B) = \int_0^{.5} 15\, d\alpha + \int_{.5}^{1} 10\, d\alpha = 7.5 + 5 = 12.5$$

Hence the maximum $E_o$ is under the selection of $A_3$. To calculate the regret approach our regret matrix is

$$\begin{array}{c} \\ A_1 \\ A_2 \\ A_3 \end{array} \begin{array}{c} \begin{array}{cccc} S_1 & S_2 & S_3 & S_4 \end{array} \\ \begin{bmatrix} 0 & -25 & -10 & 0 \\ -5 & -5 & -5 & -10 \\ -10 & 0 & 0 & -20 \end{bmatrix} \end{array}$$

Then

$$R_1(B) = \{1/0,\ 1/-25,\ .5/-10,\ .2/0\}$$
$$R_2(B) = \{1/-5,\ 1/-5,\ .5/-5,\ .1/-10\}$$
$$R_3(B) = \{1/-10,\ 1/0,\ .5/0,\ .2/.20\}$$

The associated level sets are

| $R_1(B)$ | | Min |
|---|---|---|
| $\{0, -25, -10\}$ | $0 \leqslant \alpha \leqslant .5$ | -25 |
| $\{0, -25\}$ | $\alpha > .5$ | -25 |

| $R_2(B)$ | | |
|---|---|---|
| $\{-5, -10\}$ | $0 \leqslant \alpha \leqslant .2$ | -10 |
| $\{-5\}$ | $\alpha > .2$ | -5 |

| $R_3(B)$ | | |
|---|---|---|
| $\{-10, 0, -20\}$ | $0 \leqslant \alpha \leqslant .2$ | -20 |
| $\{-10, 0\}$ | $\alpha > .2$ | -10 |

Therefore

$$E_R(A_1/B) = -25$$

$$E_R(A_2/B) = \int_0^{.2} -10d\alpha + \int_{.2}^{1} -5\ d\alpha = -2 + (-4) = -6$$

$$E_R(A_3/B) = \int_0^{.2} -20d\alpha + \int_{.2}^{1} -10\ d\alpha = -4 - 8 = -12$$

Therefore under the condition of selecting using least regret we would select $A_2$.

We shall now consider the most general case in which our knowledge about the state variable V is a D-S granule of the form "V is m" where m is bpa which satisfies

$$m(B_k) = a_k, \ k = 1,2,\ldots,n$$

where the $B_k$ are fuzzy subsets of S. Without loss of generality we shall assume our decision maker has a pessimistic attitude. Thus in this case we shall let $E_p(A_i/m)$ indicate the evaluation of the choice of alternative $A_i$ in the face of the evidence "V is m".

We note that if as a result of the "underlying experiment" the outcome $y_k$ occurred resulting in the fact that "V is $B_k$", then the evaluation could be calculated as $E_p(A_i/B_k)$ by the procedure we have already discussed. Since the probability of any $B_k$ occurring is $a_k$, we can calculate $E_p(A_i/m)$ as the expected value of these $E_p(A_i/B_k)$'s; thus

$$E_p(A_i/m) = \sum_{k=1}^{p} (E_p(A_i/B_k) * a_k)$$

In the case of an optimistic decision attitude we get

$$E_o(A_i/m) = \sum_{k=1}^{p} (E_o(A_i/B_k) * a_k)$$

In the case of the Hurwicz attitude with coefficient of pessimism h we obtain

$$E_H(A_i/m) = h * E_p(A_i/m) + (1-h) * E_o(A_i/m).$$

Finally the case of a least regret attitude yields

$$E_R(A_i/m) = \sum_{i=1}^{p} (E_R(A_i/B_k) * a_k)$$

Again in all cases the best alternative is selected by finding the $A^*$ which maximizes the appropriate evaluation conditioned upon m.

<u>Example</u>. Consider the situation in which the payoff matrix is the same as in the first problem. However, now assume that the

knowledge about the value of the state variable V is that we believe that there is at least a .7 probability that V lies in the set $D = \{S_1, S_2\}$.

In this case we represent our knowledge by

$$V \text{ is } m$$

where m is a bpa such that

$$m(B_2) = .7$$

$$m(B_2) = .3$$

where: $B_1 = D = \{S_1, S_3\}$ and $B_2 = S = \{S_1, S_2, S_3, S_4\}$.

In this case

| | Max | Min |
|---|---|---|
| $C_1(B_1) = \{10, 5\}$ | 10 | 5 |
| $C_2(B_1) = \{5, 10\}$ | 10 | 5 |
| $C_3(B_1) = \{0, 15\}$ | 15 | 0 |

therefore

$$E_p(A_1/B_1) = 5 \qquad E_o(A_1/B_1) = 10$$

$$E_p(A_2/B_1) = 5 \qquad E_o(A_2/B_1) = 10$$

$$E_p(A_3/B_1) = 0 \qquad E_o(A_1/B_1) = 15$$

Furthermore

| | Max | Min |
|---|---|---|
| $C_1(B_2) = \{10, -15, 5, 20\}$ | 20 | -15 |
| $C_2(B_2) = \{5, 10\}$ | 10 | 5 |
| $C_3(B_2) = \{0, 15, 10\}$ | 15 | 0 |

therefore

$$E_p(A_1/B_2) = -15 \qquad E_o = (A_1/B_2) = 20$$

$$E_p(A_2/B_2) = 5 \qquad E_o = (A_2/B_2) = 10$$

$$E_p(A_3/B_2) = 0 \qquad E_o = (A_3/B_2) = 15$$

Since

$$E_p(A_i/m) = \sum_{k=1}^{p} E_p(A_i/B_k) * a_k$$

we get

$$E_p(A_1/m) = (5)(.7) + (-15).3 = -1$$

$$E_p(A_2/m) = (5)(.7) + (5)(.3) = 5$$

$$E_p(A_3/m) = (0)(.7) + (0)(.3) = 0$$

Thus $A_2$ is the best alternative under the pessimistic attitude.

Since in the optimistic case

$$E_o(A_i/m) = \sum_{k=1}^{p} E_o(A_i/B_k) * a_k$$

we get $E_o(A_1/m) = (10)(.7) + (20)(.3) = 13$

$$E_o(A_2/m) = (10)(.7) + (10)(.3) = 10$$

$$E_o(A_3/m) = (15)(.7) + (15)(.3) = 15$$

Thus $A_3$ is the best alternative under the optimistic attitude. We shall not do the least regret and Hurwicz cases in this example.

While in general the four different attitudes may lead to different selections of best alternative there are two special cases of knowledge about the state in which all four attitudes always lead to the same selection of best alternative.

The first is the case of decision making under certainty. In this case the knowledge about the variable is

V is m

where

$$m(B) = 1$$

and $B = \{ s^{\perp} \}$ where $s^{\perp}$ is some element in S.
In this case

$C_i(B) = \{ c_i^{\perp} \}$, where $c_i^{\perp}$ is the unique payoff under the selection of $A_i$ with the value of nature S . In this case

$$\text{Min } C_i(B) = \max C_i(B) = c_i^{\perp}$$

hence

$$E_p(A_i/B) = E_o(A_i/B) = c_i^{\perp}$$

Furthermore since $m(B) = 1$ then

$$E_p(A_i/m) = E_o(A_i/m) = c_i^{\perp}$$

In addition, since

$$E_H(A_i/m) = h(E_p(A_i/m) + (1-h) * E_o(A_i/m)$$

then

$$E_H(A_i/m) = E_p(A_i/m) = E_o(A_i/m) = C_i$$

Therefore the alternative $A^*$ which has the maximum payoff for any alternative with S fixed at $S^\perp$ is the same for all three cases. In the case of the least regret criterion

$R_i(B) = \{C_i^\perp - T^\perp\}$, where $C_i^\perp$ is as above and $T^\perp$ is the biggest payoff under $S^\perp$. Therefore

$$E_R(A_i/m) = C_i^\perp - T^\perp$$

and again $A^*$ is the alternative with the largest payoff under S fixed at $S^\perp$.

The second situation in which the selection of optimal alternatives is indifferent to the choice of decision attitude is the purely probabilistic case.

<u>Theorem</u>. Assume that our knowledge about the state of nature is purely probabilistic. That is "V is m" where

$$m(B_k) = a_k \qquad k = 1,\ldots,n$$

$$B_k = \{S_k\}$$

Then the four attitudes toward decision making lead to the same selection of optimal alternative, the $A^*$, such that $\sum_{k=1}^{n} C_{ik} * a_k$ is maximum.

<u>Proof</u>. If $B_k = \{S_k\}$ then $C_i(B_k) = \{C_{ik}\}$ and therefore

$$\text{Max } C_i(B_k) = \text{Min } C_i(B_k) = C_{ik}$$

Hence $E_p(A_i/B_k) = E_o(A_i/B_k) = C_{ik}$ and therefore

$$E_p(A_i/m) = E_o(A_i/m) = \sum_{k=1}^{n} C_{ik} * a_{ik}$$

$$E_H(A_i/m) = h\, E_p(A_i/m) + (1-h) \cdot E_o(A_i/m) = \sum_{k=1}^{n} C_{ik} \cdot a_{ik}$$

In the least regret approach, the regret matrix R is such that $R_{ij} = C_{ij} - T_j$, where $T_j$ is maximum of $C_{ij}$. Therefore $R_i(B_k) = \{r_{ik}\} = \{C_{ik} - T_k\}$ and hence

$$E_R(A_i/B_k) = \text{Min } R_i(B_k) = C_{ik} - T_k$$

hence

$$E_R(A_i/m) = \sum_{k=1}^{n} E_R(A_i/B_k) = \sum_{k=1}^{n} (C_{ik} * A_k) - \sum_{k=1}^{n} (T_k * a_k)$$

letting $\sum_{k=1}^{n} T_k * a_k = T$, therefore

$$E_R(A_i/m) = E_p(A_i/m) - T$$

and thus the ordering is the same.

In general we see that if the focal elements are specific, i.e. consist of one and only one element and therefore introduce no possibilistic uncertainty; the selection of best alternative is indifferent to the decision attitude.

5. UNCERTAINTY IN THE PAYOFFS

In this section we shall further generalize the situation by allowing for some uncertainty in the payoff values, the $C_{ij}$'s.

Under this situation assume we have the knowledge that the state of nature is B, that "V is B" where B is a fuzzy set of S. In this case

$$C_i(B) = \bigcup_{j=1}^{n} \{(B(S_j)/C_{ij})\}$$

In this case we are now allowing the $C_{ij}$'s to be uncertain. Initially we shall assume that $C_{ij}$ is simply possibilistic, that this our knowledge of $C_{ij}$ is that

$$C_{ij} \text{ is } F_j$$

in which $F_j$ is a fuzzy subset of the reals indicating the value of $C_{ij}$, for example, F could be "about 50", "high", etc. Thus in this case

$$C_i(B) = \bigcup_{j=1}^{n} \{(B(S_j)/F_j)\}$$

Thus $C_i(B)$ is a fuzzy subset over the set F where F is a set of fuzzy subsets of real numbers. In particular we have a fuzzy subset over fuzzy subsets. We shall use an alternative representation developed by Yager (1983) for these types of situations. Let R be the set of real numbers, let $x \in R$. In Yager (1983), it is suggested that a fuzzy subset of the type mentioned above can be converted to a fuzzy subset over the base set of the $F_j$'s, the real numbers, as follows. For $x \in R$ the membership grade of $x$ in $C_i(B)$ is

$$C_i(B)(x) = \max_{j=1,\ldots,n} (B(S_j) \wedge F_j(x))$$

Thus $C_i(B)$ is now simply a fuzzy subset over the set of real numbers. We can now use our previously defined methodology to obtain the max and min of $C_i(B)$ necessary for using the optimistic or pessimistic attitude approach.

Let us now consider the situation in which our knowledge about the value of the payoff involves independent D-S granules. In particular assume that for each i,j

$$C_{ij} \text{ is } m_{ij}$$

where $m_{ij}$ is a bpa in which the focal elements are $F_{ijk}$, such that

$$m(F_{ijk}) = a_{ijk}$$

where $k=1,\ldots,n(ij)$.

We further assume independence of the $m_{ij}$'s. To calculate $C_i(B)$ we proceed as follows. Let $G_{ijk} = B(S_j)/F_{ijk})$, then using the previous methodology we can represent $G_{ijk}$ as a fuzzy subset of the real line where for each $x \in R$

$$G_{ijk}(x) = B(S_j) \wedge F_{ijk}(x)$$

Then we let

$$G_{ij} = \sum_{k=1}^{n(i,j)} (G_{ijk} * a_{ijk})$$

Therefore $G_{ij}$ becomes a fuzzy subset of the real numbers such that

$$G_{ij} = \bigcup_{\text{all } x} \left\{ \frac{\underset{k}{\text{Min}}\ G_{ijk}(x_k)}{\sum_k (x_k * a_{ijk})} \right\}$$

that is, for any $r \in R$

$$G_{ij}(r) = \underset{\substack{\text{all} \\ x_1,\ldots,x_{n}(ij) \\ \text{such that} \\ \sum_k (x_k * a_{ijk}) = r}}{\text{Max}} (\text{Min}\ G_{ijk}(x_k))$$

Then

$$C_i(B) = \bigcup_{j=1}^{n} G_{ij}$$

We note in this case $C_i(B)$ is again simply a fuzzy subset of the real line and thus all the previously developed mechanisms are available to us.

## 6. CONCLUSION

We have provided a general framework for the selection of best alternatives in the face of state knowledge obtained by evidential reasoning in the form of D-S granules.

## REFERENCES


Richmond, S.B. (1968). Operations Research for Management Decisions. Ronald Press, New York.

Shafer, G. (1976). A Mathematical Theory of Evidence. Princeton University Press, Princeton.

Yager, R.R. (1981). A procedure for ordering fuzzy subsets of the unit interval. Inf. Sci. 24, 143-161.

Yager, R.R. (1983). Membership in a compound fuzzy subset. Syst. and Cyber. 14, 173-184.

Yager, R.R. (1985a). Toward a General Theory of Reasoning with Uncertainty. Part I: Non-Specificity and Fuzziness. Tech. Report MII-509, Machine Intelligence Institute, Iona College.

Yager, R.R. (1985b). Toward a General Theory of Reasoning with Uncertainty.Part II: Probability. Tech. Report MII-510, Machine Intelligence Institute, Iona College.

Zadeh, L.A. (1965). Fuzzy sets. Inf. and Cont. 8, 338-353.

Zadeh, L.A. (1978). Fuzzy sets as a basis for a theory of possibility. Fuzzy Sets and Syst. 1, 3-28.

Zadeh, L.A. (1979). Fuzzy sets and information granularity. In M.M. Gupta, R.K. Ragade and R.R. Yager (eds.), Advances in Fuzzy Set Theory and Applications. North-Holland, Amsterdam, 3-18.

# ANALYSIS OF FUZZY EVIDENCE IN DECISION MAKING MODELS

V.I. Glushkov and A.N. Borisov

Department of Automatized Control Systems
Riga Polytechnic Institute
Riga, Latvian SSR, USSR

Abstract. Analysed are contradictions, incompleteness and redundance of information in the sets of subjective conditional statements of the form "IF L THEN C" where L is a logical expression including values of parameters that are used for the description of alternatives and C contains an assertion about the value of some criterion that is used for the choice of alternatives. The coordination of information in the parametric description of alternatives and in the sets of such conditional statements is tested. A method for choosing the best alternatives is presented, based on computation of the degree of fulfilling demands for the values of criteria.

Keywords: information analysis, fuzzy evidence, decision making in uncertain environment, compositional rule of inference.

## 1. INTRODUCTION

The interest in expert systems (see, e.g., Alekseeva and Stefanyuk, 1984) and logical-linguistic models (e.g., Pospelov, 1982; or Eshkova and Pospelov, 1978) has lately been growing. The knowledge bases of such systems contain sets of heuristic rules of the type "IF L THEN C". These rules are constructed on the basis of subjective opinions and may contain contradictory, incomplete or redundant information.

If parameters in the statements are probabilistic, and probabilities are fuzzy or linguistic, then description of alternatives may also be contradictory, incomplete or redundant. Contradictions in the description of alternatives become evident when a subjective probability distribution has no objective support, while incompleteness and redundancy - when the probabilities for some possible values of a parameter are not defined or are defined repeatedly.

## 2. THE USE OF CONDITIONAL STATEMENTS IN DECISION MAKING

Let us consider the set of alternatives $A = \{a_d : d = 1, \ldots, D\}$. Each alternative is described by the set of parameters $Y = \{Y_j : j=1,\ldots,J\}$ and is evaluated on the set of criteria $X = \{X_i : i=1,\ldots,I\}$ dependent on the parameters. Suppose that dependence of each criterion on the set of parameters can be expressed only approximately, by means of a set of statements "IF L THEN C", where L is a logical expression containing an assertion about the values of parameters, C is an assertion about the value of some criterion.

If the values of L and C may be not only numerical but also fuzzy or linguistic, then the dependence of a criterion on the subset of parameters is expressed by the set of fuzzy conditional statements (Zadeh, 1979)

$$g_t \ : \ \text{IF } L_t \text{ THEN } (X_i = H_t), \tag{1}$$

where $L_t$ is a logical expression of the form

$$L_t \ : \ (Y_{j1} = G_{j_1 t}) \wedge \ldots \wedge (Y_{jm} = G_{j_m t}),$$

$$\{Y_{j_1}, \ldots, Y_{j_m}\} \subseteq Y.$$

More complicated statements may consist of several simple conditional statements joined by the connective "ELSE"

$$\begin{cases} \text{IF } L_1 \text{ THEN } (X_i = H_1) \text{ ELSE} \\ \ldots \ldots\ldots\ldots \ldots\ldots\ldots\ldots \\ \text{IF } L_{n-1} \text{ THEN } (X_i = H_{n-1}) \text{ ELSE } (X_i = H_n) \end{cases} \tag{2}$$

Each conditional rule may have a degree of confidence $\alpha \in [0, 1]$.

For choosing the best alternative one must first of all compute the values of criteria using information about their dependence on parameters.

Alternatives may be probabilistic, i.e., parameters in their description may take values $G_{j_k}$ with probabilities $\lambda_{j_k}$, where $\lambda_{j_k}$ can be numerical, fuzzy or linguistic.

## 3. COMPUTATION OF FUZZY RELATIONS AND THEIR GRAPHIC REPRESENTATION

Let us denote $L_t = \bar{L}_1^* \wedge \bar{L}_2^* \wedge \ldots \wedge \bar{L}_{t-1}^* \wedge \bar{L}_t^*$. Then, (2) may be written as

$$\begin{cases} \text{IF } L_1 \text{ THEN } (X_i = H_1) \\ \ldots\ldots\ldots\ldots\ldots\ldots\ldots\ldots \\ \text{IF } L_n \text{ THEN } (X_i = H_n) \end{cases} \tag{3}$$

Each string in (3) is a simple conditional granule but together they are evidence (Zadeh, 1979)

$$E = \{g_1, g_2, \ldots, g_n\} \tag{4}$$

In the simplest case an evidence may consist of a single granule $E = \{g\}$. Taking into account the information contained in the evidence (4), one can construct a fuzzy relation $R$ between a criterion $X_i$ and parameters $Y_j \in Y$ which are used in evidence E.

Let $Y^E$ and $Y^t$ be the sets of parameters which are used in the evidence $E$ and the granule $g_t$, respectively. The power $|Y^E|$ of the set $Y^E$ is equal to $m_E$, $|Y^t| = m_t$, $Y^t = \{Y^t_{j1}, \dots, Y^t_{j_{mt}}\}$, $Y^E = \{Y^E_{j1}, \dots, Y^E_{jmE}\}$, $Y^E = \bigcup_{t=1}^{n} Y^t$.

Let us designate by $R_t$ a relation between the parameters and criterion, induced by a granule $g_t$. $R_t$ is a fuzzy set with the membership function (Dubois and Prade, 1980)

$\mu_{Rt}(y^t_j, \dots, y^t_{j_{mt}}, x) = 1 - \mu_{L_t}(y^t_j, \dots, y^t_{j_{mt}}) + \mu_{H_t}(x)$, where

$y^t_{j_k} \in Y^t_{j_k}$, $x \in X$; $Y^t_{j_k}$, $X$ are the sets of nonfuzzy basic meanings for $Y^t_{j_k}$ and X, and $\mu_{L_t}(.)$ is a membership function for the fuzzy set defined by expression $L_t$.

The cylindrical extension of the relation $R_t$ to $Y^E = Y^E_{j_1} \times Y^E_{j_2} \times \dots \times Y^E_{j_{mE}}$ can be obtained as follows (Zadeh, 1975)

$$c(R_t) = \int_{Y^E} \mu_{R_t}(y^t) \, y^E$$

And, finally, the relation $R$ induced by evidence $E$ is (Dubois and Prade, 1980)

$$\mu_R(y^E, x) = \min_t \mu_{c(R_t)}(y^t, x)$$

If several evidences $\{E_h\}$ exist, then each of them induces a fuzzy relation $R^h$.

Knowing the parameter valuations for the alternatives and using the compositional rule of inference (Mizumoto, 1981), we can define the fuzzy value $H$ of criterion $X$ as follows

$$\mu_H(x) = \max_{y^E} \odot \left(\mu_{c(G^E_o)}(y^E, x),\ \mu_R(y^E, x)\right) \tag{5}$$

where $G^E_o = G_{j_1o} \wedge \dots \wedge G_{j_{mE}o}$, $c(G^E_o)$ is the cylindrical extension of $G^E_o$ to $Y^E_{j1} \times \dots \times Y^E_{j_{mE}} \times X$, and the operation of "bounded-product" is defined by the following expression

$$\mu_F(u) \odot \mu_G(u) = \max(0, \mu_F(u) + \mu_G(u) - 1)$$

The fuzzy value $H$ can also be obtained using fuzzy linear

interpolation methods (Alekseev, 1982) but in a multidimensional case they demand more time-consuming computations.

Each evidence E can be depicted by a sector of a semantic network (Fig. 1), where vertices correspond to criteria or parameters, and vectors of parameters or criteria; the non-marked arcs show the direction in which the values of parameters and criteria are being passed on; the marked arcs define transformation in accordance with a fuzzy relation R; a dot near a vertex denotes synopsis and shows that for the excitation of a vertex it is necessary to have information from all vertices joined by this synopsis.

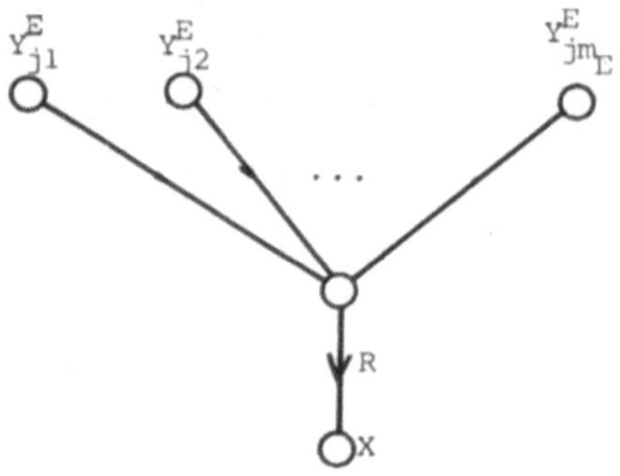


Fig. 1. Representation of the statement by the sector of the network

If the value of X can be derived on the basis of evidence E, then we write

$$X = E(Y^E) \tag{6}$$

Generally speaking, $Y^E$ may include not only parameters of an alternative's description but also other variables, the values of which must be computed using other evidences, for example

$$Z = E_1(Y_1, Y_2), \quad X = E_2(Z, Y_3) \tag{7}$$

The network corresponding to the set of evidences (7) is depicted in Fig. 2.

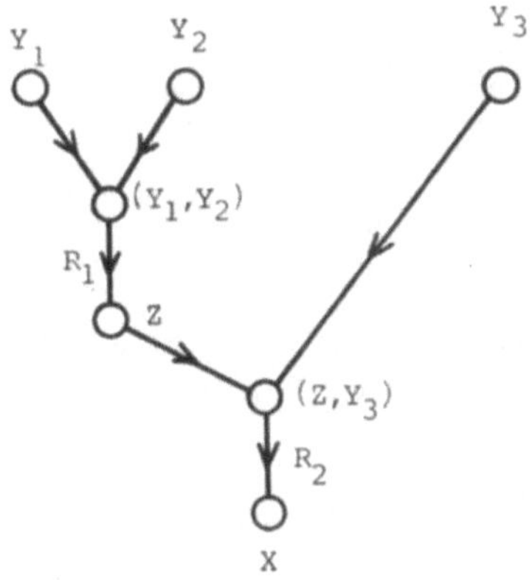


Fig. 2. The network for representation of the set of evidences (7)

## 4. REVEALING OF CONTRADICTIONS, INCOMPLETENESS AND REDUNDANCE OF A SET OF STATEMENTS

Let us divide the whole set of statements into subsets such that all the statements of one subset have common parameters and the same criterion. Let us consider one of such subsets. Suppose that it includes statements $g_1, g_2, \ldots, g_{n_o}$ . This subset can be considered as an evidence

$$E = \{ g_1, g_2, \ldots, g_{n_o} \} \tag{8}$$

Assume for simplicity that evidence (8) describes a dependence of criterion $X_i$ on a parameter $Y_{j_o}$ .

The subjectivity of statements, and perhaps their different origins, can lead to contradictions. Generally speaking, the dependence $X_i = E(Y^E)$ may be of various kinds, therefore the only manifestation of contradictions in $E$ is the existence of such $g_{t'}$ and $g_{t''}$ that $G_{j_o t'} \subseteq G_{j_o t''}$ and $H_{t'} \not\subseteq H_{t''}$, where $G_1 \subseteq G_2 \Leftrightarrow \mu_{G_1}(y) \leqslant \mu_{G_2}(y)$.

The origin of statements (1) and (2) can be not available when the alternatives are analysed, that is why it is necessary to get more information about the dependencies of criteria on parameters while the statements are collected.

The degree of incompleteness of evidence can be expressed as

$$\mathrm{ine} = \max_{y_j \in Y_j} \max(0,\ 1 - \sum_{t=1}^{n_o} \mu_{G_{jt}}(y_j)) \tag{9}$$

Some statements can express values of the criterion for the same or close values of the parameters. The degree of redundance of evidence is equal to

$$\mathrm{rede} = \max_{y_j \in Y_j} \max\ (0,\ \sum_{t=1}^{n_o} \mu_{G_{jt}}(y_j) - 1) \tag{10}$$

Similar evidences may be obtained for an evidence in which $Y^E$ consists of more than one parameter. We must only replace $\mu_{G_{jt}}(y_j)$ in formulas (9) and (10) by $\mu_{G_t^E}(y^E)$, where

$$\mu_{G_t^E}(y^E) = \min_{j: Y_j \in Y^E} \mu_{G_{jt}}(y_j)$$

## 5. DEGREES OF CONFIDENCE IN EVIDENCE

Conditional statements "IF ... THEN ..." are commonly gathered from different experts. If the number of statements is rather high , for example, in the MYCIN expert system - about 500, then some of them can be mutually conflicting. That is why after dividing the set of statements into evidences (8) it is expedient with the help of experts to assign the degree of confidence $\alpha \in [0,1]$ to each. These $\alpha$'s will "escort" each expression of the type (6) as follows

$$X = E(Y^E), \alpha$$

If the criterion value is computed on the basis of $m$ conflicting evidences, then the vertex, corresponding to this criterion, must be supplied with a special kind of synopsis (Fig. 3).

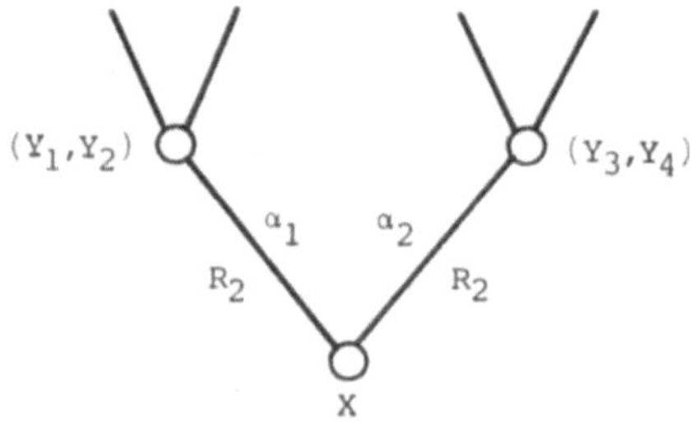


Fig. 3. The network for the representation of two conflicting evidences

The degrees of confidence are assigned directly in a heuristic manner or are defined by using statistical methods and training algorithms (Lesmo, Saitta and Torasso, 1983). We do not consider here the methods of obtaining $\alpha$'s assuming they are known.

Let the evidences $E_1$ and $E_2$ be given, i.e.
$X = E_1(Y_1, Y_2), \alpha_1$; $X = E_2(Y_3, Y_4), \alpha_2$.

If we compute the value of criterion $X$ using each of these evidences, we shall obtain a different $H_1$ and $H_2$. Methods of combining evidences are discussed in, e.g., Borisov and Glushkov(1983), Dempster (1961), Shafer (1976), Yager (1981), Zadeh (1979). In Zadeh, (1979) the resulting value of $H$ is suggested to be

$$\mu_H(x) = \min(\mu_{H_1}(x), \mu_{H_2}(x)), \tag{11}$$

or, taking into consideration the degrees of confidence (Yager, 1981),

$$\mu_H(x) = \min(\mu_{H_1}^{\alpha_1}(x), \mu_{H_2}^{\alpha_2}(x)) \tag{12}$$

If the "product" - operation is used instead of "min", then (Smets, 1983)

$$\mu_H(x) = \frac{\mu_{H_1}(x)\ \mu_{H_2}(x)}{\max_x(\mu_{H_1}(x)\ \mu_{H_2}(x))} \tag{13}$$

Formulas (11), (12) and (13) were derived from the results of Dempster (1967) and Shafer (1976) on the basis of fuzzy sets theory. But in Shafer (1976) it is pointed out that if the mutual contradiction of evidences is rather great, then the adding of "belief masses" leads to more natural results. For fuzzy evidences it corresponds to the computation of $H$ by the formula (Glushkov and Derkach, 1985):

$$\mu_H(x) = \frac{\alpha_1}{\alpha_1 + \alpha_2} \mu_{H_1}(x) - \frac{\alpha_2}{\alpha_1 + \alpha_2} \mu_{H_2}(x)$$

Analysis of the merits and shortcomings of different methods of combination (Glushkov and Derkach, 1985) allows a generalized formula to be suggested, which, depending on the degree of mutual conflict and the degrees of confidence in separate evidences, gives a result close to the one mentioned earlier.

Let $M$ be the set of evidences which enables the computation of the value of criterion $X$ and for which all values of parameters are known. Suppose $M_1$ is the set of all nonempty subsets of $M$. Then, the membership function of $H$ is defined by

$$\mu_H(x) = \bigoplus_{m \in M_1} ((1 - \max_{i \in M-m} \alpha_i) \min(\min_{i \in m} \mu_{H_i}(x), \min_{j \in M-m} (1-\mu_{H_j}(x))) \quad (14)$$

where $\bigoplus(a_1, a_2, \ldots, a_n) = \min(1, a_1 + a_2 + \ldots + a_n)$.

## 6. COMPUTATION OF THE VALUES OF CRITERIA FOR THE VALUES OF ALTERNATIVES

The value of a criterion $X_i$ may be computed using the so-called algorithm of the consecutive excitement of vertices. Values of the parameters must be known. Vertices corresponding to these parameters are considered to be excited. The algorithm consists of consecutive excitements of the vertices that have at least one excited synopsis. A synopsis is excited if all its incoming arcs come out of the excited vertices.

The algorithm of the consecutive excitements of vertices is:

Step 1. The set $M_1$ of all vertices corresponding to the criteria of alternatives is determined.

Step 2. The set $M_2$ of all vertices of the network is determined.

Step 3. A vertex $m \in M_2$ is sought from which it is possible to immediately reach any vertex $s \in M_1$. If such $m$ does not exist, go to Step 5.

Step 4. The vertex $m$ is included into $M_1; M_1 := M_1 \cup m$. The vertex $m$ is excluded from $M_2; M_2 := M_2 - m$. Go to Step 3.

Step 5. The set $S_1$ of excited vertices $s$, such that $s \in M_1$, is determined.

Step 6. The set $S_2 = M_1 - S_1$ is determined.

Step 7. Each $s \in S_2$ is tested to determine whether one of its synopses can be excited if all the vertices in $S_1$ are excited. If a synopsis can be excited, then the corresponding value of the criterion or other variable (see Fig. 2) is computed using the relations, marked on incoming arcs, and the vertex $s$ is included in $S_1$; $S_1 := S_1 \cup s$; $S_2 := S_2 - s$.

Step 8. If in Step 7 at least one vertex is included in $S_1$, then repeat Step 7. If $S_2 = \emptyset$, then the algorithm is successfully finished, otherwise it is unsuccessfully finished.

## 7. NONDETERMINISTIC VALUES OF PARAMETERS

Let it be known that for some alternative parameters $Y_j \in Y^E$ for evidence E take values $G_{jk_j}$ with probabilities $P_{jk_j}$;

$k_j = 1,\ldots,n_j$; $j \in M^E$; $M^E$ is the set of indices of parameters of evidence E. The algorithm for computing values of X remains almost the same, only the exciting of the vertex is accompanied by assigning to a vertex not one value H but a set of values $\{H_k : k=1,\ldots,m\}$. Each of them has the probability $p_k$, $k=1,\ldots,n$, $n=n_1 * n_2 * \ldots * n_{mE}$. The method of computing $p_k$ s is determined by the existence of a dependence between parameters $Y_j$. If parameters are mutually independent, then

$$p_k = \prod_{j \in M^E} p_{jk_j} \tag{15}$$

If the degree of mutual dependence is unknown, then $p_k$ can be estimated by the interval

$$p_k = [0, \min_{j \in M^E} p_{jk_j}] \tag{16}$$

If the probabilities of parameter values are fuzzy and equal to $\lambda_{jk_j}$, then for the independent parameters

$$\lambda_k = \prod_{j \in M^E} \lambda_{jk_j} \tag{17}$$

but for the parameters with an unknown degree of mutual dependence

$$\lambda_k = \widetilde{\min}(\delta, \widetilde{\min}\, \lambda_{jk_j}) \tag{18}$$

where $\widetilde{\min}$ is an operation of extended minimum (Dubois and Prade, 1980) and $\delta$ is a fuzzy number with the membership function

$$\mu_\delta(\eta) = \begin{cases} 1 & 0 \leqslant \eta \leqslant 1 \\ 0 & \eta < 0 \text{ or } \eta > 1. \end{cases}$$

8. CONTRADICTIONS, INCOMPLETENESS AND REDUNDANCE IN THE DESCRIPTION OF ALTERNATIVES

The description of an alternative is contradictory if the subjective probability distribution for at least one parameter has no objective support, i.e. no probability distribution $f(y_j)$ satisfies the conditions

$$\begin{cases} \int_{y_j \in Y_j} f(y_j)\, dy_j = 1 \\ p_{jk_j} \supseteq \int_{y_j \in Y_j} \mu_{G_{jkj}}(y_j) f(y_j)\, dy_j, \quad k_j \in \{1,\ldots,n_j\} \end{cases} \tag{19}$$

where $P_{kj_j}$ is the nonfuzzy probability of $G_{jk_j}$ or the 1-level subset of $\lambda_{jk_j}$.

The degree of contradictoriness is expressed as follows

$$\text{contra} = \min_{f \in F} \left(1 - \int_{y_j \in Y_j} f(y_j)\,dy_j\right)$$

with the restrictions (19) being fulfilled, where $F$ is the set of all possible probability distributions. The degree of redundance of an alternative's description can be computed by the formula

$$\text{ina} = \min\left(0, 1 - \max_{p \in \operatorname{supp} \Lambda^1} p\right)$$

where $\Lambda^1$ is the $\alpha$-level subset of $\Lambda = \sum_{k_j=1}^{n_j} \lambda_{jkj}$ for $\alpha=1$. The degree of redundance in an alternative's description is equal to

$$\text{reda} = \min\left(0, \max_{p \in \operatorname{supp} \Lambda^1} p - 1\right)$$

It is expedient to eliminate undesired characteristics of information while statements defining the dependences and descriptions of alternatives are being collected. It is not difficult for the decision maker (DM) to correct his set of statements (to add, delete or change some of them) if the origin of inaccuracy is pointed out.

Correction of an alternative's description is more difficult so that if it cannot be eliminated in dialogue with DM, then a heuristic method of normalization of probabilities (Slyadz and Borisov, 1982) should be used.

## 9. THE CHOICE OF ALTERNATIVES

The choice is performed on the basis of computed degrees of fulfilling the DM requirements for each alternative. These requirements must be given by an unconditional statement about ideal (hypothetical) values of criteria for the best alternative.

We shall determine the degree of fulfilling the simplest unconditional statement $(X_i = Q)$ where $Q$ is some fuzzy value of $X_i$ by two values: the expected certainty $EC(Q)$ (Borisov and Glushkov, 1983)

$$EC(Q) = \sum_{k=I}^{n} \lambda_k (\sup H_k - \sup(H_k \cap \bar{Q})) \qquad (20)$$

and expected possibility (Zadeh, 1979)

$$E\pi(Q) = \sum_{k=1}^{n} \lambda_k \sup(H_k \cap Q) \tag{21}$$

where $\sup A = \max_x \mu_A(x)$, $\mu_{\bar{Q}}(x) = 1 - \mu_Q(x)$, and $\lambda_k$ are computed by formulas (15) - (18). If the probabilities are non-fuzzy, then EC(Q) and Eπ(Q) may be directly computed by the formulas (20) and (21). If the probabilities $\lambda_k$ are fuzzy or linguistic, then their computation comes to solving of the set of linear programming problems (Borisov and Glushkov, 1983). For a more complicated hypothetic statement $(X_{i_1} = Q_{i_1}) \wedge \ldots \vee \ldots \ldots \wedge (X_{i_m} = Q_{i_m})$ evaluation of its fulfillment can be done in the same manner. If, for example, the statement is

$$X_1 = Q_1 \wedge (X_2 = Q_2) \vee (X_3 = Q_3)$$

then $H_k$ and $Q$ in formulas (15) and (16) must be substituted by $(H_1 \wedge H_2)_k$ and $(Q_1 \wedge Q_2 \vee Q_3)$ which have membership functions

$$\mu_{(H_1 \wedge H_2)_k}(x_1, x_2) = \min(\mu_{H_{1k}}(x_1), \mu_{H_{2k}}(x_2))$$

$$\mu_{Q_1 \wedge Q_2 \vee Q_3}(x_1, x_2, x_3) = \max(\min(\mu_{Q_1}(x_1), \mu_{Q_2}(x_2)), \mu_{Q_3}(x_3))$$

It has already been shown that it is not necessary to know the values of all parameters for some criterion value computation. The set of parameters that are used for computation of any criterion value has been defined in Steps 1-4 of the algorithm given in Section 5. On the other hand, if any vertex corresponding to the primary parameter is included in the list of vertices to be excited, but no value is assigned to that parameter, then the procedure of inference must have a means to elicit the missing information from the data base or from the decision maker in an interactive manner.

## 10. REVEALING THE SET OF BEST ALTERNATIVES

If the set of alternatives $\{a_d: d=1,\ldots,D\}$ is given and valuations $EC_d(Q)$ and $E\pi_d(Q)$ are obtained, then we can determine the set of nondominated alternatives $A_{ND}$. Alternative $a_{d_1}$ belongs to $A_{ND}$ if there is no alternative $a_{d_2} \in A$ such that:

$$EC_2(Q) \geqslant EC_1(Q) \quad \text{and} \quad E\pi_2(Q) > E\pi_1(Q)$$

or

$$EC_2(Q) > EC_1(Q) \quad \text{and} \quad E\pi_2(Q) \geqslant E\pi_1(Q)$$

Some other methods of revealing the best alternatives are

described in Borisov and Glushkov (1983).

## 11. CONCLUDING REMARKS

The proposed method of choosing alternatives enables us to update information about nondeterministic fuzzy values of parameters, to compute the fuzzy values of criteria and, at the same time, to take into consideration the degrees of confidence in statements defining the dependence between the parameters of descriptions of alternatives and criteria of their evaluation. In future it is necessary to develop a more effective procedure of inference and ways to accelerate the computations for large networks of statements.

## REFERENCES


Alekseev, A.V. (1982). Using fuzzy algorithms for control in fuzzy environment (in Russian). - In Decision-making in non-statistical uncertainty environment. Riga Polytechnical Institute, Riga.

Alekseeva, E.F., and V.L. Stefanyuk (1984). Expert Systems - the state and perspectives (in Russian). - News of Academy of Sciences of USSR, Technical Cybernetics, 5.

Borisov, A.N., and V.I. Glushkov (1983). Using the concept of information granularity in decision-making problems. - In Proceedings of IFAC/IFORS International Symposium on Large Scale Systems (Warsaw), Pergamon Press, Oxford.

Dempster, A.P. (1967). Upper and lower probabilities induced by a multivalued mapping. - Annals of Mathematical Statistics, 38.

Dubois, D., and H. Prade (1980). Fuzzy Sets and Systems: Theory and Applications. Academic Press, New York.

Eshkova, I.V., D.A. Pospelov (1978). Decision-making on fuzzy foundation. II. Scheme of inference (in Russian). News of Academy of Sciences of USSR, Technical Cybernetics, 2.

Glushkov, V.I., and O.I. Derkach (1985). The choice of alternatives with fuzzily described consequences (in Russian). In Methods and Systems of Decision Making. Automatized Decision Support Systems in Management and Designing. Riga

Lesmo, L., L. Saitta and P. Torasso (1983). Fuzzy Production Rules: a Learning Methodology. - In P.P. Wang and S.K. Chang (Eds.) Fuzzy Sets and Possibility Theory with Applications, Plenum, New York.

Mizumoto, M. (1981). Note on the arithmetic rule by Zadeh for fuzzy conditional inference. Cybernetics and Systems: An Int. J. 12.

Pospelov, D.A. (1981). Logical-linguistic Models in Management Systems (in Russian). Energoizdat, Moscow.

Shafer, G. (1976). A Mathematical Theory of Evidence. Princeton University Press, Princeton.

Slyadz, N., and A.N. Borisov (1982). Analysis of fuzzy initial information in decision making models. - In R. Trappl (ed.), Cybernetics and System Research. North-Holland, Amsterdam.

Smets, Ph. (1983). Information contents of an evidence. Int. J. Man-Machine Studies, 19, 33-43.

Yager, R.R. (1981). Application of information granularity to political and other decisions. Policy and Information, 5.
Zadeh, L.A. (1975). Calculus of fuzzy restrictions. - In L.A. Zadeh et al. (eds.), Fuzzy Sets and their Applications to Cognitive and Decision Processes. Academic Press, New York.
Zadeh, L.A. (1979). Fuzzy Sets and Information Granularity. In M.M. Gupta, R.K. Ragade and R.R. Yager (eds.), Advances in Fuzzy Set Theory and Applications. North-Holland, Amsterdam.

# FUZZY INCLUSIONS AND FUZZY DICHOTOMOUS DECISION PROCEDURES

L.M. Kitainik
MLTI, Moscow, USSR

Abstract. An approach to decision making with an arbitrary antireflexive fuzzy binary relation (FR) is developed, generalizing some well-known ordinary constructions: undomination, intrinsic and external stability, kernels of a graph, etc.

The main concept of the method is a fuzzy dichotomous decision procedure (FDDP). An optimal decision is defined due to the general maximal decision principle. Basically, it gives a "soft" estimate of the applicability of a decision procedure and of its quality. Subprocedures of coordinating the decision maker's a priori preferences and the results of FDDP application are presented as a basis for examining the decision maker's qualities (competence, resoluteness), and also for improving the final choice. The conventional multiattribute majority approach to formation of binary relations is axiomatically extended to FR construction. A practical example of application of the method is given.

Keywords: decision making, choice, fuzzy preference, fuzzy set inclusion, undominance.

## 1. INTRODUCTION

A fuzzy binary relation (FR) is an acknowledged object in decision making. It is known that "ordinary fuzzy" and L-fuzzy FR's describe some well-defined group preferences (e.g., Kuzmin, 1982; Danilov, 1984). The nonfuzzy multiattribute "majority approach" to decision making may also be extended to formation of FR's, carrying more knowledge of the decision maker's (DM) preferences (see Section 4).

A general "maximal decision" principle in fuzzy decision making (Bellman and Zadeh, 1970) is to be adapted to decision making problems involving FR analysis procedures. Consider first ordinary analogues. Any ordinary binary relation (OR) $G$ is related to a graph (all graphs in the paper are supposed to be oriented), also called $G$ (Hasse diagram), and most admissible decision making structures are associated with the way this graph affects subsets $Z \subseteq X$, $X$ being the set of initial alternatives. Such are the subsets of all undominated nodes $M = \mathrm{Max}(G)$, the sets of all intrinsically stable - $I(G)$, externally stable - $\mathcal{E}(G)$, G-invariant - $S(G)$ subsets of $X$, and the set of all kernels $\mathcal{K}(G) = I(G) \cap \mathcal{E}(G)$ [not by chance this concept is central in Roy, 1972 - see Section 4]. Note that any of the concepts can be expressed in terms of inclusions of a subset $Z \subseteq X$, the supplement $\bar{Z}$, and G - image GZ, that is, by means of a "dichotomy" on X being defined by G. More precisely: $Z \subseteq M \Leftrightarrow GZ \subseteq \bar{Z}$ & $G\bar{Z} \subseteq \bar{Z}$; $Z \in J(G) \Leftrightarrow GZ \subseteq \bar{Z}$; $Z \in \mathcal{E}(G) \Leftrightarrow GZ \supseteq Z$; $Z \in S(G) \Leftrightarrow \bar{G}Z \subseteq Z$; $Z \in \mathcal{K}(G) \Leftrightarrow GZ \subseteq \bar{Z}$ & $\bar{G}Z \supseteq \bar{Z}$. To generalize such dichotomies, consider two concepts: "$\bar{G}$-independence" and "G-domination". For $V, W \subseteq X$, we say that V is G-independent of W iff $GW \subseteq \bar{V}$ ($GW \cap V = \emptyset$), and VG-dominates W iff $GV \supseteq W$. Replacing V,W by $Z, \bar{Z}$ in the previous definitions, one can easily obtain all the foregoing structures. To be meaningful, these structures call for antireflexivity of G, and this is the only constraint on FR's assumed below.

Fuzzy decision procedures can be designed in the same way. The notion $\bar{a}$ for a fuzzy subset (f.s.) a is traditional; ga (g being a FR) may be any v-o composition (we use the most "economical" $\vee$-$\wedge$ composition). However, the choice of a fuzzy inclusion is not so evident and needs further discussion (see Section 2).

The subsequent analysis of the nonfuzzy case shows that the "optimal decision" is unquestionably unique with a transitive G; under this condition, $\mathcal{K}(G)$ reduces to M = Max(G) [this concept is basic in the classical fuzzy decision making - see Orlovski, 1978], and M is beyond comparison, since Max generates the well-defined "graphodominant choice function" (Berezovski, Borzenko and Kempner, 1981). But with intransitive G, M may be empty (a nonempty M may be inadmissible - see Section 3), and all of the remaining above-mentioned optimal decisions are generally not unique. This allows us (or, maybe, requires) to use less formalized structures for the final choice, that is, to enlist <u>descriptive</u> decision theory problems. To mention only two of them (Larichev, 1980):

1) different estimates by the DM of those "entire" alternatives and the criterial evaluations; in particular, the DM may have "a priori" preferences often varying from his criterial choice;

2) intransitivity of DM preferences leading specifically to the above intransitivity of binary relations. Therefore, we consider subprocedures of coordination between optimal decisions and a priori preferences as an integral part of any <u>normative</u> decision making procedure, the subprocedures to be applied, when possible, to the exploration of both DM qualities and decision making psychology.

Quality measuring for optimal decisions is a more specific fuzzy problem. While in an ordinary case, the preferences of the chosen alternatives to the remaining ones are 1 : 0, the fuzzy optimal decision $k \in \tilde{\Phi}(X)$ with $k|_Z \equiv 1$, $k|_Z \equiv 0.999$ can be hardly considered as a consistent one; furthermore, if any optimal decision is like this, then either the initial FR or the decision making procedures are unfit.

The following problems will be dealt with here: fuzzy inclusions study (Section 2), construction and research of fuzzy dichotomous decision procedures (Section 3); a few aspects of coordinating "a priori" and "a posteriori" preferences as a basis for learning certain qualities of the DM (Section 4).

Common notation: $I = [0,1]$; $X$ - a finite set containing $n$ ordinary initial alternatives; $S_n$ - symmetric group of order $n$; $Y$ - an arbitrary set; $\tilde{\mathcal{P}}(Y) \equiv \tilde{\mathcal{P}}_1(Y)$ - the set of all f.s. of $Y$; $\mathcal{P}(Y) \equiv \mathcal{P}_1(Y)$ - the set of all ordinary subsets of $Y$; $Y^m$ - the m-th Cartesian power; $\tilde{\mathcal{P}}_m(Y) = \tilde{\mathcal{P}}(\tilde{\mathcal{P}}_{m-1}(Y))$ - the set of all level m f.s. of $Y$ (usually denoted by $\tilde{\mathcal{P}}^m(Y)$; however, we reserve superscripts for Cartesian powers); $\tilde{\tilde{\mathcal{P}}}_m(Y) = \mathcal{P}(\mathcal{P}_{m-1}(Y))$; $\mu_a(y)$ - membership function for $a \in \mathcal{P}(Y)$; nonfuzzy subsets of $Y$ are denoted either by $A, B, C, \ldots$, and $\{y_i\}$ (with usual operations $\cup, \cap, -, \subseteq$) or, when associated with elements of $\mathcal{P}(Y) \subset \tilde{\mathcal{P}}(Y)$, by characteristic functions, i.e. $a \in \mathcal{P}(Y) \Leftrightarrow \mu_a = \chi_A$ with $A = \text{supp}(a) = \mu_a^{-1}(1)$; zero, unit elements, and the "median" in $\tilde{\mathcal{P}}(Y)$ are denoted by $0$, $\mathbb{1}$, $\mathbb{1}/\mathbb{2}\,(0_Y, \mathbb{1}_Y, \mathbb{1}/\mathbb{2}_Y)$. For brevity, $\vee$ is max, sup, f.s. union, and disjunction, $\wedge$ - min, inf, f.s. intersection, and conjunction; $\overset{\circ}{\vee} a = \bigvee_{y \in Y} \mu_a(y)$; $\overset{\circ}{\wedge} a = \bigwedge_{y \in Y} \mu_a(y)$. Other notations are introduced when necessary.

## 2. AXIOMATICS AND PROPERTIES OF FUZZY INCLUSIONS

There are several reasons for the subsequent discussion on fuzzy inclusions. We share the opinion that the conventional "$\subseteq$" is very strict (Dubois and Prade, 1980). For instance, let $Y = \{y_1, y_2, y_3\}$; $a = 0.5/y_1 + 0.51/y_2 + 0.001/y_3$; $b = 1/y_1 + 0.5/y_2 + 0.5/y_3$; $c = 1/y_1 + 1/y_2 + 0/y_3$. The statement "$a \subseteq c$" is obviously much more reliable than "$b \subseteq c$"; however, not a single pair in $\{a; b; c\}$ satisfies "$\subseteq$". Turning to different fuzzy inclusions, we note the $\varepsilon$- inclusions for $I_5(a,b) = \overset{\circ}{\wedge}(a \vee b)$, $I_4(a,b) = \overset{\circ}{\wedge}(a \cup b)$, and also the family of weak inclusions $\prec_\alpha$ (Dubois and Prade, 1980). One can easily find discouraging "strict examples" for any of these inclusions. It is not a matter of specific record form, but of ordinarity (any of the pointed out inclusions is an OR on $\tilde{\mathcal{P}}(Y)$), and hence, discontinuity. We see no reason why <u>fuzzy</u> inclusion must be an <u>ordinary</u> concept; below, this constraint is rejected. Moreover, the diversity of the available fuzzy inclusions calls for an axiomatic study.

<u>Definition 1</u>. A Fuzzy Inclusion (FI) is a FR inc on $\tilde{\mathcal{P}}(X)$ (we take for simplicity a finite $X$ so as not to deal with topological details), inc $\tilde{\mathcal{P}}(\tilde{\mathcal{P}}^2(X))$ satisfying four axioms given in Table 1. The set of all FIs is denoted by Inc.

Axioms 1 and 2 link FI with the algebraic structure on $\tilde{\mathcal{P}}(X)$; Axiom 3 sets the symmetry of alternatives ($\mu_{inc}(a,b)$ as dependent only on the relative position of $\mu_a, \mu_b$); Axiom 4 requires inc to be an extension of the usual $\subseteq$. First of all, we state four simple properties of any FI.

Table 1: FI Axioms

| Ordinary prototype: $\subseteq$ <br> $(A,\ B,\ C \subseteq X;\quad \sigma \in S_n)$ | Fuzzy generalization: inc <br> $(a,b,c \in \tilde{\mathcal{P}}(X);\quad \sigma \in S_n)$ |
| --- | --- |
| 1. $A \subseteq B \Leftrightarrow \bar{B} \subseteq \bar{A}$ | 1. $\mu_{inc}(a,b) = \mu_{inc}(\bar{b},\bar{a})$ |
| 2. $A \cup B \subset C \Leftrightarrow A \subseteq C \ \&\ B \subseteq C$ | 2. $\mu_{inc}(a \vee b,c) = \mu_{inc}(a,c) \wedge \mu_{inc}(b,c)$ |
| 3. $A \subseteq B \Leftrightarrow \sigma A \subseteq \sigma B$ | 3. $\mu_{inc}(a,b) = \mu_{inc}(\sigma a,\sigma b)$ |
| | 4. $inc\|_{\mathcal{P}^2(X)} = \subseteq$ $(\mu_{inc}(\chi_A,\chi_B) = \begin{cases} 1, & A \subseteq B \\ 0, & \text{otherwise} \end{cases}$ |

1. $\mu_{inc}(a,b\wedge c) = \mu_{inc}(a,b) \wedge \mu_{inc}(a,c)$.

2. $\mu_{inc}(a,b) \searrow_a \nearrow_b$

Property 1 directly follows from Axioms 1, 2; 2 - from Property 1, Axiom 2, and also the equivalences $(a_1 \subsetneqq a_2) \Longleftrightarrow (a_1 \vee a_2 = a_2)$, $(b_2 \subsetneqq b_1) \Longleftrightarrow (b_1 \wedge b_2 = b_2)$

3. For any $\alpha, \beta \in I$, $x, z \in X$, $x \neq z$ implies $\mu_{inc}(\alpha\chi_{\{x\}}, \overline{\beta\chi_{\{z\}}}) = 1$.

Proof. $\mu_{inc}(\alpha\chi_{\{x\}}, \overline{\beta\chi_{\{z\}}}) \geqslant \mu_{inc}(\chi_{\{x\}}, \overline{\beta\chi_{\{z\}}}) \geqslant \mu_{inc}(\chi_{\{x\}}, \chi_{\overline{\{z\}}}) = 1$; both inequalities are due to Property 2, the equality is from Axiom 4, since $x \neq z \Longrightarrow \{x\} \subseteq \overline{\{z\}}$.

4. $\mu_{inc}(a,b) = \bigwedge\limits_{x \in X} \mu_{inc}(\mu_a(x)\chi_{\{x\}}, \overline{\overline{\mu_b(x)\chi_{\{x\}}}})$.

Proof. The evident identities $a = \bigvee\limits_{x \in X} \mu_a(x)\chi_{\{x\}}$, $b = \bigwedge\limits_{z \in X} \overline{\overline{\mu_b(z)\chi_{\{z\}}}}$ result, because of Axiom 2 and Property 1, in the equality $\mu_{inc}(a,b) = \bigwedge\limits_{x,z \in X} \mu_{inc}(\mu_a(x)\chi_{\{x\}}, \overline{\overline{\mu_b(z)\chi_{\{z\}}}})$. Next, Property 3 makes any term with $x \neq z$ a unit; hence, a minimum is achieved with $x = z$. Q.E.D.

The main result of this section is as follows:

Theorem 1. (Inc realization). Inc is in a one-to-one correspondence with the set $\Phi$ of functions defined on the triangle $T = \{(\alpha,\beta) \in I^2 \mid \alpha \geqslant \beta\}$. $\Phi = \{\varphi : T \to I \mid \varphi(\alpha,\beta) \searrow_{\alpha,\beta};\ \varphi(0,0) = \varphi(1,0) = 1;$

$\varphi(1,1) = 0\}$. (throughout in the paper, both increase ↗ and decrease ↘ are not strict, and the ordering of $\mathfrak{P}(X)$ is due to $\subseteq$).

Proof. Let $\mu_{inc}(\alpha,\beta) = \mu_{inc}(\alpha\chi_{\{x\}}, \overline{\beta\chi_{\{x\}}})$. We prove that a mapping $\nu$ : Inc $\to \varphi$, $\nu(inc) = \varphi_{inc}$ is a bijection. $\varphi_{inc}$ correctness is due to Axiom 3; the belonging of $\varphi$ to $\Phi$ follows from Property 2 and Axioms 1 and 4; thus, $\varphi_{inc}(0,0) = \mu_{inc}(\emptyset, \mathbb{1}) = 1$, $\mu_{inc}(\mathbb{1},\mathbb{1}) = \mu_{inc}(\mathbb{1},\emptyset) = 0$. Injectivity of $\nu$ is implied by Property 4, since inc is uniquely defined by the set $\{\mu_{inc}(\alpha\chi_{\{x\}}, \overline{\beta\chi_{\{x\}}})\}_{(\alpha,\beta)\in I^2} = \{\mu_{inc}(\alpha\chi_{\{x\}}, \overline{\beta\chi_{\{x\}}})\}_{(\alpha,\beta)\in T} = \{\varphi_{inc}(\alpha,\beta)\}$. To prove surjectivity, we define, with $\varphi\in\Phi$, an inc $\varphi \in \tilde{\mathfrak{P}}(\tilde{\mathfrak{P}}^2(X))$:

$$\mu_{inc}\ (a,b) = \bigwedge_{z\in X} \varphi(\mu_a(z) \vee \overline{\mu_b(z)},\ \mu_a(z)\wedge \overline{\mu_b(z)}).$$

First, we establish inc$\varphi \in$ Inc. Axioms 1 and 3 are evidently valid. Axiom 4 is easily verified due to $\varphi$ values in the apexes of T; e.g., $(\neg(A\subseteq B)) \Longleftrightarrow (\exists y\in A\cap\bar{B}) \Longrightarrow (\chi_A(y) = \bar{\chi}_B(y) = 1) \Longrightarrow (\mu_{inc\varphi}(\chi_A,\chi_B) = \varphi(1,1) = 0)$. Next, we note that monotonicity of $\varphi$ implies the equality $(\forall\alpha,\beta,\gamma \in I)\ \varphi(\alpha\vee\beta\vee\gamma, (\alpha\vee\beta)\wedge\gamma) = \varphi(\alpha\vee\gamma,\alpha\wedge\gamma) \wedge \varphi(\beta\vee\gamma,\beta\wedge\gamma)$, yielding Axiom 2 for inc$\varphi$ with $\alpha = \mu_a(z)$, $\beta = \mu_b(z)$, $\gamma = \overline{\mu_c(z)}$. Furthermore, with $(\alpha,\beta) \in$ T, we have $\varphi_{inc}\ (\alpha,\beta) = \mu_{inc}\ (\alpha\chi_{\{x\}}, \overline{\beta\chi_{\{x\}}}) = \bigwedge_{z\in X}\varphi(\alpha\chi_{\{x\}}(z) \vee \overline{\beta\chi_{\{x\}}(z)}, \alpha\chi_{\{x\}}(z)\wedge\overline{\beta\chi_{\{x\}}(z)}) = (\bigwedge_{z\neq x}\varphi(0,0))\wedge\varphi(\alpha,\beta) = \varphi(\alpha,\beta)$; hence, $\varphi_{inc\varphi} = \varphi$, that is, $\varphi \to$ inc$\varphi$ is $\nu^{-1}$. Q.E.D.

Corollary 1. The only linear function in $\Phi$ is $\varphi_*(\alpha,\beta) = \bar{\beta} = 1 - \beta = \nu(inc_*)$ with $\mu_{inc_*}(a,b) = I_5(a,b) = \mathring{\wedge}(\bar{a}\vee b)$. In turn, $inc_*$ is the unique FI representable as $\mathring{\wedge}\,p(a,b)$, p being a polynomial form of two fuzzy variables.

Proof. The statements concerning $\varphi_*$ are evident. The fact $inc_* \in$ Inc is tested directly. Next, $(\alpha,\beta) \in$ T implies: $\varphi_{inc_*}(\alpha,\beta) = \mu_{inc_*}(\alpha\chi_{\{x\}}, \overline{\beta\chi_{\{x\}}}) = \mathring{\wedge}(\overline{\alpha\chi_{\{x\}}} \vee \overline{\beta\chi_{\{x\}}}) = 1 \wedge (\bar{\alpha}\vee\bar{\beta}) = \bar{\beta} = \varphi_*(\alpha,\beta)$, so that $\nu(inc_*) = \varphi_*$. Let now $\mathring{\wedge}p(a,b)\in$ Inc; write p in a disjunctive form. Axiom 4 implies $\mathring{\wedge}p(\chi_A,\chi_B) \equiv_{A,B} \mathring{\wedge}(\bar{\chi}_A \vee \chi_B)$. It follows from the usual properties of ordinary Boolean functions that any minterm in p, except for $\bar{a}$ and b, is a trivial Boolean conjunction, that is, contains either $\bar{a}\wedge a$ or $\bar{b}\wedge b$; however, any such min-term is absorbed by $\bar{a}\vee b$, and the reduced form of $p$ is $\bar{a}\vee b$; hence, $\overset{\circ}{\wedge} p = inc_*$. Q.E.D.

This is a quite natural tie, due to Corollary 1, of the FI (an object of fuzzy set theory) to the implication (a function of fuzzy logic).

Theorem 1 yields a convenient device for an FI analysis according to the $\varphi_{inc}$ behavior, as shown in

Corollary 2. Let $T' = \{(\alpha,\beta) \in T \mid \alpha \leq \bar{\beta}\}$, $T'' = T\setminus T'$, $T''_1 = \{(\alpha,\beta) \in T'' \mid \beta < \bar{\beta}\}$, $T''_2 = T''\setminus T''_1$, $n = \mathrm{card}(X) \geq 2$.

(1) An FI inc is reflexive iff $\varphi_{inc}|_{T'} \equiv 1$, that is, inc is consistent with $\subseteqq$ ($a \subseteqq b \Rightarrow \mu_{inc}(a,b) = 1$).

(2) inc is [perfectly] antisymmetric iff $\varphi_{inc}|_{T''} \equiv 0$ (clearly, only $\subseteqq$ is both reflexive and [perfectly] antisymmetric).

(3) The three simple one-sided conditions for inc transitivity are as follows ($\vee$-$\wedge$ transitivity criterion for FI's is a rather complicated one): necessary - "$\varphi_{inc}|_{T''_2}$ depends only on $\alpha$"; sufficient - (i) "$\varphi_{inc}|_{T''}$ depends only on $\alpha$"; (ii) "$\varphi_{inc}|_{T''_2} \equiv 0$ & ($\varphi_{inc}|_{T''_1}$ depends only on $\beta$)".

(4) No FI is weakly linear (see Dubois and Prade, 1980; Kuzmin, 1982; Orlovski, 1978). The proofs are evident, except, maybe, for (3).

Corollary 2 shows that any perfectly antisymmetric FI is transitive. To review weak inclusions in "$\varphi$-terms", note that $\prec_\gamma$ is nothing but a $\gamma$-cut of $inc_*$, and $\varphi_{\prec_\gamma} = \chi_{\{(\alpha,\beta)\in T \mid \beta \leq \bar{\gamma}\}}$. Therefore, Corollary 2 yields all $\prec_\gamma$ properties; reflexivity with $\gamma \leq 1/2$, transitivity with $\gamma > 1/2$, etc. (see Dubois and Prade, 1980). Furthermore, some of the "inclusion grades measures" are not FI's: Axioms 1 and 2 fail for $I_1(a,b) = \|a\wedge b\|/\|a\|$; Axiom 2 - for $I_2(a,b) = \|\bar{a}\cup b\|$ and $I_3(a,b) = \|\bar{a}\vee b\|$. A FI $inc_{\subseteqq *} = I_4$ is a "quasilinear" one; that is, $\varphi_{\subseteqq *}(\alpha,\beta) = \bar{\beta}\cup\bar{\alpha} = 1\wedge(\bar{\beta}+\bar{\alpha})$ is the only element in $\Phi$ being both consistent with $\subseteqq$ and linear on $T''$; hence, $inc_{\subseteqq *}$ is, unlike $inc_*$, reflexive; however, both are intransitive. As to the other "polynomially represented" FI's, with degree 2, the result is: $\varphi(\alpha,\beta) = \bar{\beta}\hat{+}(t\bar{\alpha}+u\bar{\beta})$ $(u\in[-1;1],\ t\in[0;1-u]$, $\hat{+}$ is a probabilistic sum), the only $\{\vee;\wedge;^{-};\hat{+};\cdot\}$ algebraic preimages being $inc_*\cdot inc_* = \nu^{-1}(\bar{\beta}^2)$, $inc_*\hat{+}inc_* = \nu^{-1}(\bar{\beta}^2)$ and $inc_{\hat{+}} = \nu^{-1}(\overline{\alpha\beta})$ with $\mu_{inc_{\hat{+}}}(a,b) = \mathring{\wedge}(\bar{a}+b)$.

Finally, we describe global algebraic structures of Inc and $\Phi$. Assume $\circ$ to be a binary commutative operation on I, extended to any $\tilde{\mathcal{P}}(Y)$ as $\mu_{a\circ b} = \mu_a \circ \mu_b$. The result is:

(i) Inc is closed under $\circ$ iff $\circ$ is a "generalized conjunction", i.e. $\alpha\circ\beta \equiv f_o(\alpha\wedge\beta)$, with $f_o: I\nearrow I$, $f_o(0) = 0$, $f_o(1)=1$;

(ii) $\Phi$ is closed under $\circ$ $(\Phi\subset\tilde{\mathcal{P}}(T))$ iff $\alpha\circ\beta\nearrow\alpha,\beta$, $0\circ 0 = 0$, $1\circ 1 = 1$.

Hence, $\Phi$ is much more tolerant, being closed under $\wedge,\vee,\hat{+},\cdot,\uplus$, $\Cap$, etc., whereas Inc – only under $\wedge$. However, $\nu$ is evidently a "local homomorphism", that is, $(inc_1, inc_2, inc_1\ inc_2 \in Inc) \Longrightarrow (\nu(inc_1\circ inc_2) = \nu(inc_1)\circ\nu(inc_2))$.

## 3. CONSTRUCTION AND ANALYSIS OF FUZZY DICHOTOMOUS DECISION PROCEDURES

In this section, we try to answer the following questions:

(?) how can one define procedures for FR analysis, an optimal decision and quality of a decision?

(??) what properties of FR´s and the procedures do guarantee high quality decisions?

An answer to (?) comes from what is stated in Section 1, also involving the concept of FI. Below $\tilde{\mathcal{P}}_o(X^2)$ denotes the set of all antireflexive FR´s on X; we also set $Y_o^2 = Y^2\setminus\{(y,y)\}$.

Definition 2. Let $g\in\tilde{\mathcal{P}}_o(X^2)$, $inc\in Inc$.

(i) fuzzy (g,inc)-independence and fuzzy (g,inc)-domination are FR´s indp, domn on $\tilde{\mathcal{P}}(X)$ defined as:

$\mu_{indp}(a,b) = \mu_{inc}(gb,\bar{a})$ (we omit for brevity some parameters; formally, $\mu_{indp}(a,b;g,inc)$) "a is independent of b";

$\mu_{domn}(a,b) = \mu_{inc}(b,ga)$ "a dominates b".

(ii) Basic (g,inc) dichotomies are the level 2 f.s. $\Delta_1,\Delta_2,\Delta_3\in\tilde{\mathcal{P}}_2(X)$: $\mu_{\Delta_1}(a) = \mu_{indp}(a,\bar{a})$; $\mu_{\Delta_2}(a) = \mu_{indp}(a,a)$; $\mu_{\Delta_3}(a) = \mu_{domn}(a,\bar{a})$ (a is independent of $\bar{a}$ $(\Delta_1)$ and of a $(\Delta_2)$, a dominates $\bar{a}$ $(\Delta_3)$)

(iii) A fuzzy dichotomous decision procedure (FDDP) is a monotone nondecreasing polynomial form $p(\Delta_1,\Delta_2,\Delta_3)$, $p\in\tilde{\mathcal{P}}_2(X)$.

(iv) An optimal decision when applying a FDDP p to a FR g is an ordinary set of f.s. $\mathfrak{D}_p(g) = \mu_p^{-1}(\mu_p^*)\subset\tilde{\mathcal{P}}(X)$, with

$\mu_p^* = \bigvee_{a\in\tilde{\mathcal{P}}(X)} \mu_p(a)$. The last item is nothing but a "maximal decision principle" for p.

In an ordinary case, ($g = G$, inc $= \subseteq$), any FDDP coincides with a certain notion of Section 1 ($\mathcal{D}_{\Delta_1\wedge\Delta_2}(G) = \mathcal{P}(M)$, $\mathcal{D}_{\Delta_2}(G) = I(G)$, etc.).

An approach to quality measuring for optimal decisions stems from a general analysis of f.s. "resolution".

Definition 3. (i) The dichotomousness of an f.s. $a \in \tilde{\mathcal{P}}(Y)$ is

$$\delta(a) = \bigvee_{Z\in Y} \bigwedge_{u\in Z, v\in \bar{Z}} (0 \vee (\mu_a(u) - \mu_a(v))).$$

(ii) The dichotomousness of an ordinary set $\mathfrak{X}$ of f.s. ($\mathfrak{X} \subseteq \tilde{\mathcal{P}}(Y)$) is a value $\delta(\mathfrak{X}) = \bigwedge_{a\in\mathfrak{X}} \delta(a)$.

In other words, $\delta(a)$ is the "maximal gap" in the $\mu_a(y)$ graph, and $\delta(\mathfrak{X})$ is a gap of that kind guaranteed for any $a\in\mathfrak{X}$.

Examples. (1) Y is a connective Hausdorff space, $\mu_a(y)$ is continuous; $\delta(a) = 0$. (2) X is finite, as defined is Section 1, $a \in \tilde{\mathcal{P}}(X)$; $\delta(a) \geqslant \frac{\vee a - \wedge a}{n}$. (3) $\delta(\mathcal{P}(Y)\setminus\{\mathbf{0},\mathbf{1}\}) = 1$. (4) $\alpha\cdot\mathbf{1} \in \mathfrak{X}(\alpha\in I)$. $\delta(\mathfrak{X}) = 0$.

Definition 4. Let p be an FDDP, g an FR, $g\in\tilde{\mathcal{P}}_0(X^2)$. Set $\delta_p(g) = \delta(\mathcal{D}_p(g))$. A pair (g,p) is called dichotomously contensive (DC) iff $\delta_p(g) > 0$, and dichotomously trivial (DT), otherwise. A FDDP (FR) is called DC/DT iff there exists (does not exist) a DC pair containing this FDDP (FR).

To explain the meaning of $\delta_p(g)$ as an evaluation for the quality of an optimal decision, we note that any element of $\mathcal{D}_p(g)$ is an f.s. of X, yielding a certain variant of "a posteriori" preferences, being consistent with (g,p,inc), i.e. resulting from the analysis of an FR g by means of an FDDP p. In general, we cannot expect, due to the lack of transitivity and/or antisymmetry of g, that $\mathcal{D}_p(g)$ should be "coherent", thus ordering X or at least any subclasses of alternatives (e.g., equivalence classes). Moreover, nonfuzzy optimal choice itself is often ambiguous (see Section 1). However, the ultimate end of decision making is an ordinary choice; therefore, one would like to have a guaranteed preference of arbitrarily selected "optimal alternatives" over the remaining ones. Just such a guarantee is proposed by $\delta_p(g)$. For instance, if $\mathcal{D}_p(g)$ contains a constant $\alpha\cdot\mathbf{1}$, then no positive partition of X, based on (g,p), can be made, and at least the applicability of p to g is dubious. So we believe that $\delta_p(g)$ characterizes the two decision making aspects: applicability of FDDP to FR analysis, and a certain superiority of the alternatives to be chosen over all the other ones.

To answer (??), we describe all DC FDDP's and propose a method for $\mathcal{D}_p(g)$ design and $\delta_p(g)$ computing in the case inc=inc$_*$(that is, with a linear FI model).

First, we modify the $\alpha$-cut definition. Consider the set $\underline{\ \ }=\{\geqslant;>;\leqslant;<;=\}$, owing an unary idempotent operation $\bar{\ }$: $\overline{<}=>$, $\overline{<}=>$, $\overline{(=)}=(=)$. We assume the $\alpha$-cut to be a mapping $\lambda_{\tau\alpha}: \tilde{\mathcal{P}}(X)\longrightarrow\mathcal{P}(X)$ $(\tau\in\mathcal{R}, \alpha\in I)$, $\lambda_{\tau\alpha}(a) = a_{\tau\alpha} = \chi_{A_{\tau\alpha}}$, $A_{\tau\alpha}=\{x\in X\mid \mu_a(x)\,\tau\,\alpha\}$. The needed algebraic properties of the "cut method" are presented in

Lemma 1. (i) With any $\tau\in\{>;>;=\}$, $\alpha\in I(a,b\in\tilde{\mathcal{P}}(X), g\in\tilde{\mathcal{P}}(X^2))$, $\lambda_{\tau\alpha}$ is a homomorphism of a lattice $\tilde{\mathcal{P}}(X)$ with operators in $\tilde{\mathcal{P}}(X^2)$ (the action being a $\vee-\wedge$ composition) onto a lattice $\mathcal{P}(X)$ with operators in $\mathcal{P}(X^2)$, that is, $(a\vee b)_{\tau\alpha}=a_{\tau\alpha}\vee b_{\tau\alpha}$; $(a\wedge b)_{\tau\alpha}=a_{\tau\alpha}\wedge b_{\tau\alpha}$; $(ga)_{\tau\alpha}=g_{\tau\alpha}a_{\tau\alpha}$.

(ii) With any $\tau\in\mathcal{R}$, $\alpha\in I$, $a\in\tilde{\mathcal{P}}(X)$, $(\bar{a})_{\tau\alpha}=a_{\bar{\tau}\bar{\alpha}}$

The proof easily follows from the definitions (see, e.g., Dubois and Prade, 1980).

We begin with examining min-terms $\Delta_\xi=\bigwedge_{i\in\xi}\Delta_i$ $(\xi\in\mathcal{P}(\{1;2;3\}))$ that will do for all FDDP´s. For brevity, we will write in the subscripts $\xi$ for $\Delta_\xi$ $(\delta_{123}(g), \mathfrak{D}_{23}(g)$, etc.).

Let $\Psi_\xi$ be the following functions on $\overline{\tilde{\mathcal{P}}_o(X^2)\times\mathcal{P}(X)}$:

$\Psi_1(g,Z) = \bigvee_{x\in\bar{Z},y\in Z}\mu_g(x,y)$; $\Psi_2(g,Z) = \overline{\bigvee_{x,y\in Z}\mu_g(x,y)}$; $\Psi_3(g,Z) = \bigwedge_{y\in\bar{Z}}\bigvee_{x\in Z}\mu_g(x,y)$ $(\Psi_1(g,X) = \Psi_3(g,X)=1)$; $\Psi_\xi(g,Z) = \bigwedge_{i\in\xi}\Psi_i(g,Z)$.

For $\alpha\in I_{1/2}=]1/2;1]$, $Z\subseteq X$, let $L(\alpha,Z)=[\alpha\chi_{\bar{Z}};\chi_Z\vee\bar{\alpha}\chi_{\bar{Z}}]$ be an interval in $\tilde{\mathcal{P}}(X)$ $([a;b]=\{c\in\tilde{\mathcal{P}}(X)\mid a\underset{\approx}{\subseteq}c\underset{\approx}{\subseteq}b\})$. The complete description of FDDP´s is due to connections between the sets $I(G_{>\alpha})$, $\mathcal{E}(G_{>\alpha})$, $S(G_{>\alpha})$, $\mathcal{K}(G_{>\alpha})$, $Max(G_{>\alpha})$ (see Section 1), the functions $\Psi_\xi(g,A_{\geqslant\alpha})$, and the intervals $L(\alpha,A_{\geqslant\alpha})$, $a\in\mathfrak{D}_p(g)$.

Below in the paper, $\mathcal{K}(G_{>1/2})$ is denoted by $\mathcal{K}$, and $Max(G_{>1/2})$ - by M.

Lemma 2. (For (1), (2), $g\in\tilde{\mathcal{P}}_o(X^2)$, $a\in\tilde{\mathcal{P}}(X)$, $\alpha,\beta\in I_{1/2}$, $Z\subseteq X$):

(1) (i) $\mu_{23}(a)=\alpha$ implies $A_{>\bar{\alpha}}=A_{\geqslant\alpha}\in\mathcal{K}$ & $\alpha\leqslant\Psi_{23}(g,A_{\geqslant\alpha})$.

(ii) $\Psi_{23}(g,Z)=\beta$ & $b\in L(\beta,Z)$ implies $Z\in\mathcal{K}$ & $\beta\leqslant\mu_{23}(b)$.

(2)(i) $\mu_{123}(a)=\alpha$ implies $A_{>\bar{\alpha}}=A_{\geqslant\alpha}=M\neq\emptyset$ & $\mathcal{K}=\{M\}$ & $\alpha\leqslant\Psi_{123}(g,A_{\geqslant\alpha})$.

(ii) $\Psi_{123}(g,Z)=\beta$ & $b\in L(\beta,Z)$ implies $Z=M$ & $\mathcal{K}=\{Z\}$ & $\beta\leqslant\mu_{123}(b)$.

(3) For any FDDP p, $\mu_p(1/2)\geqslant 1/2$.

Proof. (1)(i) $\mu_{23}(a)=\alpha$ implies $\underline{\mu_2(a)\geqslant\alpha}$; hence, $\mu_2(a) = \mu_{indp}(a,a) = \mu_{inc_*}(ga,\bar{a}) = \mathring{\wedge}(\overline{ga}\vee\bar{a}) = \mathring{\vee}\, ga\wedge a\geqslant\alpha$, and

$\overset{\circ}{\vee} ga\wedge a \leqslant \bar\alpha$. Application of the homomorphism $\lambda_{>\bar\alpha}$ to $ga\wedge a$ yields, due to Lemma 1, (i), $G_{>\bar\alpha} A_{>\bar\alpha} \cap A_{>\bar\alpha} = \emptyset$; it follows that $A_{>\bar\alpha} \in J(G_{>\bar\alpha})$. Since $\gamma > \bar\alpha$ implies $G_{>\gamma} \subseteq G_{>\bar\alpha}$, we also have $A_{>\bar\alpha} \in J(G_{>\gamma})$; hence, $A_{>\bar\alpha} \in J(G_{>1/2})$. In the same way, using $\mu_3(a) = \mu_{domn}(a,\bar a) = \mu_{inc_*}(\bar a, ga) = \overset{\circ}{\wedge}(ga\vee a) \geqslant \alpha$, and applying $\lambda_{\geqslant\alpha}$ to $ga\vee a$, we obtain that $A_{\geqslant\alpha} \in \mathcal{E}(G_{\geqslant\gamma})$; owing to $(\gamma<\alpha \Longrightarrow G_{>\gamma} \supseteq G_{\geqslant\alpha})$, $A_{\geqslant\alpha}\in\mathcal{E}(G_{>\gamma})$ also holds, leading to $A_{\geqslant\alpha} \in \mathcal{E}(G_{>1/2})$. Moreover, $\bar\alpha<\alpha$ implies $A_{>\bar\alpha} \supseteq A_{\geqslant\alpha}$, so that $J(G_{>1/2}) \ni A_{>\bar\alpha} \supseteq A_{\geqslant\alpha} \in \mathcal{E}(G_{>1/2})$. It follows easily that both $A_{>\bar\alpha}$ and $A_{\geqslant\alpha}$ belong to $J(G_{>1/2}) \cap \mathcal{E}(G_{>1/2}) = \mathcal{K}$. however, the proper inclusion of kernels of an ordinary graph is evidently impossible; therefore, $A_{>\bar\alpha} = A_{\geqslant\alpha} \in \mathcal{K}$. Furthermore, $A_{>\bar\alpha}\in J(G_{>\bar\alpha})$ is equivalent to $(\forall x,y \in A_{>\bar\alpha})(y \notin G_{>\bar\alpha}\{x\})$; hence, $\bigvee_{x,y\in A_{>\bar\alpha}} \mu_g(x,y) \leqslant \bar\alpha$, and $\alpha \leqslant \overline{\bigvee_{x,y\in A_{>\alpha}} \mu_g(x,y)} = \Psi_2(g, A_{>\alpha})$ Similarly, $A_{\geqslant\alpha}\in\mathcal{E}(G_{\geqslant\alpha})$ leads to $\Psi_3(g,A_{\geqslant\alpha}) \geqslant\alpha$. Finally,

$$\alpha \leqslant \Psi_2(g,A_{>\bar\alpha})\wedge\Psi_3(g,A_{\geqslant\alpha}) = \Psi_2(g,A_{\geqslant\alpha})\wedge\Psi_3(g,A_{\geqslant\alpha}) = \Psi_{23}(g,A_{\geqslant\alpha}).$$

(ii) is inverse to (i). Indeed, $\Psi_{23}(g,Z)=\beta$ implies $\bigvee_{x,y\in Z}\mu_g(x,y) < \bar\beta$ & $\bigwedge_{y\in\bar Z}\bigvee_{x\in \bar Z}\mu_g(x,y)\geqslant\beta$; arguing exactly as in (1) (i), we obtain that $Z\in J(G_{>\bar\beta})$ & $Z\in\mathcal{K}$. Next, $b\in L(\beta,Z)$ is the same with $B_{>\bar\beta} = B_{\geqslant\beta} = Z$; therefore, $\overline{(gb\wedge b)}_{>\bar\beta} = G_{>\bar\beta}Z\cap Z = \emptyset$; hence, $\overset{\circ}{\vee}\, gb\wedge b \leqslant\bar\beta$, and $\mu_2(b) = \overline{\overset{\circ}{\vee}\, gb\wedge b} \geqslant\beta$. Similarly, $(gb\vee b)_{\geqslant\beta} = G_{\geqslant\beta}Z\cup Z = X$, and $\mu_3(b) = \overset{\circ}{\wedge}(gb\vee b)\geqslant\beta$. Finally, $\mu_{23}(b) = \mu_2(b)\wedge\mu_3(b)\geqslant\beta$.

(2) (i) Using (1)(i) (the initial $\mu_2(a)\geqslant\alpha$ & $\mu_3(a)\geqslant\alpha$ is valid), we come up to $A_{>\bar\alpha} = A_{\geqslant\alpha}\in\mathcal{K}$. Calculation of $\mu_1(a)$ results in $\mu_1(a) = \overset{\circ}{\vee}\, g\bar a\wedge a\geqslant\alpha$, so that $\overset{\circ}{\vee}\, g\bar a\wedge a\leqslant\bar\alpha$; hence, $G_{>\bar\alpha}(\bar A)_{>\bar\alpha}\cap A_{>\bar\alpha} = \emptyset$. Since $(\bar A)_{>\bar\alpha} = A_{<\alpha}$ (see Lemma 1, (ii), we obtain $G_{>\bar\alpha}A_{<\alpha}\cap A_{>\bar\alpha} = \emptyset$, that is, $G_{>\bar\alpha}A_{<\alpha}\subseteq\overline{(A_{>\bar\alpha})} = A_{\leqslant\bar\alpha}$. The already proved $A_{>\bar\alpha} = A_{\geqslant\alpha}$ easily results in $G_{>\bar\alpha}A_{\leqslant\bar\alpha}\subseteq A_{\leqslant\bar\alpha}$, and $\overline{(A_{\geqslant\alpha})} = A_{\leqslant\bar\alpha}\in S(G_{>1/2})$. It follows that $A_{\geqslant\alpha}$ is a kernel of $G_{>1/2}$ with $G_{>1/2}$-invariant supplement, thus coinciding with M and being the unique kernel. Next, just as for $\Psi_2$ in (1)(i), we obtain that $\alpha\leqslant\Psi_1(g,A_{\geqslant\alpha})$; together with (1) (i), this yields $\alpha\leqslant\Psi_{123}(g,A_{\geqslant\alpha})$.

(ii) is proved by inverting of (2)(i) exactly as (1)(ii).

(3) Clearly, $g\overline{\mathbf{1/2}} = g\mathbf{1/2}\subsetneqq\mathbf{1/2}$. Using the formulas for $\mu_i(a)$ from (1)(i) and (2)(i), one can easily obtain that

$\mu_1(\mathbb{1}/\mathbb{2}) = \mu_2(\mathbb{1}/\mathbb{2}) \geq 1/2$ and $\mu_3(\mathbb{1}/\mathbb{2}) = 1/2$. For any different FDDP p, the statement is implied by the monotonicity of p with respect to $\Delta_1, \Delta_2$ and $\Delta_3$. Q.E.D.

A complete analysis of DCity is now given by the following

Theorem 2. (1) Let $P_t = \{\Delta_1; \Delta_2; \Delta_3, \Delta_1 \wedge \Delta_2; \Delta_1 \wedge \Delta_3\}$. Any element in $P_t$ is both a normal f.s. of $\mathfrak{F}(X)$ and a DT FDDP.

(2) $\mathcal{K} = \emptyset$ implies both $\mu^*_{23} = 1/2$ and $\delta_{23}(g) = 0$. For $\mathcal{K} \neq \emptyset$, set $\Psi^*_{23} = \bigvee_{K \in \mathcal{K}} \Psi_{23}(g,K)$ and $\mathcal{K}^* = \{K \in \mathcal{K} \mid \Psi_{23}(g,K) = \Psi^*_{23}\}$. The following formulas are valid: $\mu^*_{23} = \Psi^*_{23}$; $\mathfrak{D}_{23}(g) = \bigcup_{K \in \mathcal{K}^*} L(\Psi^*_{23}, K)$; $\delta_{23}(g) = \begin{cases} 2\Psi^*_{23} - 1, & \mathcal{K}^* \neq \{X\} \\ 0, & \mathcal{K}^* = \{X\} \end{cases}$.

A pair $(g, \Delta_2 \wedge \Delta_3)$ is DC iff $\mathcal{K} \neq \emptyset$ & $\mathcal{K}^* \neq \{X\}$ & $\Psi^*_{23} > 1/2$.

(3) $\mathcal{K} \neq M$ implies $\mu^*_{123} = 1/2$ and $\delta_{123}(g) = 0$. $\mathcal{K} = \{M\} \neq \emptyset$ leads to:

$$\mu^*_{123} = \Psi^*_{123} = \Psi_{123}(g,M); \ \mathfrak{D}_{123}(g) = L(\Psi^*_{123}, M); \ \delta_{123}(g) = \begin{cases} 2\Psi^*_{123} - 1, & M \neq X \\ 0, & M = X \end{cases}$$

A pair $(g, \Delta_1 \wedge \Delta_2 \wedge \Delta_3)$ is DC iff $M \neq \emptyset$ & $\mathcal{K} = \{M\} \neq \{X\}$ & $\Psi^*_{123} > 1/2$.

(4) Any DC case is either (2) or (3).

Proof. (1) Normality immediately follows from the equalities $\mu_1(\mathbb{0}) = \mu_1(\mathbb{1}) = \mu_2(\mathbb{0}) = \mu_3(\mathbb{1}) = \mu_{12}(\mathbb{0}) = \mu_{13}(\mathbb{1}) = 1$; hence, for any $p \in P_t$ either $\mathbb{0}$ or $\mathbb{1}$ is in $\mathfrak{D}_p(g)$, and p is DT (see Example 4).

(2) Let $\alpha = \mu^*_{23}$, $a \in \mathfrak{D}_{23}(g)$. With respect to Lemma 2, (3), $\alpha = 1/2$ implies $\mathbb{1}/\mathbb{2} \in \mathfrak{D}_{23}(g)$ and $\delta_{23}(g) = 0$, so that $(g, \Delta_2 \wedge \Delta_3)$ is DT. For $\alpha > 1/2$, Lemma 2, (1)(i) results in $A_{>\bar{\alpha}} = A_{\geq\alpha} \in \mathcal{K} \neq \emptyset$; hence, $\mathcal{K}^*$ is also nonempty; moreover, $\alpha = \mu^*_{23} \leq \Psi_{23}(g, A_{\geq\alpha}) \leq \Psi^*_{23}$. Let now $K \in \mathcal{K}^*$, $b \in L(\Psi^*_{23}, K)$. Lemma 2, (1)(ii) leads to $\Psi^*_{23} = \Psi_{23}(g,K) \leq \mu_{23}(b) \leq \mu^*_{23}$. The comparison of the two inequalities yields both $\mu^*_{23} = \Psi^*_{23} = \Psi_{23}(g, A_{\geq\alpha})$ (hence, $A_{>\bar{\alpha}} = A_{\geq\alpha} \in \mathcal{K}^*$, and $a \in L(\Psi^*_{23}, A_{\geq\alpha})$, and $\mu_{23}(b) = \mu^*_{23}$ (so that $b \in \mathfrak{D}_{23}(g)$). Therefore, $\mathfrak{D}_{23}(g) = \bigcup_{K \in \mathcal{K}^*} L(\Psi^*_{23}, K)$. Next, with $\mathcal{K}^* \neq \{X\}$ we obtain from $L(\alpha, Z)$ the definition that $\delta_{23}(g) = \Psi^*_{23} - \overline{\Psi^*_{23}} = 2\Psi^*_{23} - 1$, whereas $\mathcal{K}^* = \{X\}$ leads to $\mathbb{1} \in \mathfrak{D}_{23}(g)$ and $\delta_{23}(g) = 0$. More results on $\Delta_2 \wedge \Delta_3$ are given in (Kitainik, 1981).

(3) The proof is a repetition of the previous one, involv-

ing Lemma 2, (2).

(4) Let p be represented in a disjunctive form, containing none of $\Delta_1, \Delta_2, \Delta_3$ (the representation possibility is equivalent to the monotonicity of p), $p = \bigvee_i \pi_i$, $\pi_i$ being min-terms. If, for some i, $\pi_i \in P_t$, then (1) makes p both normal and DT. Hence, only the two FDDP's, $\Delta_2 \wedge \Delta_3$ and $\Delta_1 \wedge \Delta_2 \wedge \Delta_3$, are the DC ones, the disjunction being $\Delta_2 \wedge \Delta_3$. Q.E.D.

<u>N.B.</u> $G_{>1/2}$ is nothing but the nearest OR to a FR g (Kaufmann, 1977).

We discuss first the structural intention of the results.

A. Both a definitive and qualitative choice based on a FR g analysis is due to the case when $\mathcal{K}^* = \{K_*\}$ contains exactly one of the $G_{>1/2}$ kernels. $\Delta_1 \wedge \Delta_2 \wedge \Delta_3$ suits $K_* = M$, and DCity is estimated as $\delta_{123}(g)$; otherwise $(K_* \neq M)$, $\Delta_2 \wedge \Delta_3$ and $\delta_{23}(g)$ are to be used. A sufficient but by far not necessary condition for the A-case is the $\vee$-$\wedge$ transitivity of g.

To classify other situations, we set $K_\cap = \bigcap_{K \in \mathcal{K}^*} K$.

B. $K_\cap \neq \emptyset$. In this case $K_\cap$ is contained in any DC choice $K \in \mathcal{K}^*$, being amplified by some "peripheral elements" $K \setminus K_\cap$ (a proper choice of $K_\cap$ is DT). For $M \subseteq K_\cap$, evidently, any undominated node in $G_{>1/2}$ is always chosen.

C. $K_\cap = \emptyset$, $\mathcal{K}^* \neq \emptyset$. The choice is most ambiguous. It may well occur that $\mathcal{K}^*$ defines a partition of X, that is, any alternative is optimal when considered in a certain neighborhood (see Example 8 below). Clearly, M is empty, as well as $K_\cap$.

D. $\mathcal{K}^* = \emptyset$. $G_{>1/2}$ has no kernels (though M may be non-empty); any choice is inadmissible since the dichotomousness estimate of any $(Z, \bar{Z})$, due to arbitrary $\mathfrak{D}_p(g)$, is 0.

In any of the cases A-C, the common choice formula is in effect: take any $K \in \mathcal{K}^*$ and reject $\bar{K}$; the preference of K to $\bar{K}$ is at least $\mu^*_p : \overline{\mu^*_p}$, p being in $\{\Delta_2 \wedge \Delta_3; \Delta_1 \wedge \Delta_2 \wedge \Delta_3\}$; the difference $\mu^*_p - \mu_p$ is $\delta_p(g)$.

<u>Note</u>. A "consistency condition" is $\delta_p(g) >> \frac{\mathring{v}_g}{n^2}$ (see Example 2).

To improve the argumentation of Theorem 2, (1) concerning DTity, let us consider in detail the FDDP $\Delta_1 \wedge \Delta_2$. The technique of Lemma 2 and Theorem 2 easily results in the equality $\mathfrak{D}_{12}(g) = \{a \in \mathfrak{P}(X) \mid A_{>0} \subseteq M_0 = \mathrm{Max}(G_{>0})\}$; hence, any alternative being dominated in $G_{>0}$ is to be rejected; however, there exists no superiority estimate for $M_0$, since $\forall x \in M_o$, $\mu_a(x)$ may be anywhere in I. This is another reason to avoid graphodominant choice with intransitive g. As to the transitive g (A-case), the usual nonfuzzy approach is supplied by the choice efficiency evaluation $\delta_{123}(g)$ (not $\delta_{12}(g)$!).

Let us examine several special cases.

<u>Examples</u>.

(5) $g \subsetneqq \mathbb{1}/2_{X^2}$. $G_{>1/2}$ is a trivial graph, $\mathcal{K} = \{X\}$, the dichotomy is also trivial - a "pure uncertainty" case.

(6) $g|_{X_o^2} \supsetneqq \mathbb{1}/2_{X_o^2}$. $G_{>1/2}$ is complete, $\mathcal{K} = \{\{x\}\}_{x \in X}$, the choice is due to the usual max-min procedure assuming $\mu_g(x,y)$ to be a payoff.

(8) With a reciprocal g (Bezdek, 1978), the following result is easy: either g is DT or the A-case with $\Delta_1 \wedge \Delta_2 \wedge \Delta_3$ and a one-point M is valid.

Multistage FDDP's may be constructed by means of subsequent analysis of $g|_{K^2} \in \tilde{\mathcal{P}}_0(K^2)$; note that, due to Theorem 2 and Example 5, any $K \in \mathcal{K}^*$ is unimprovable, since $g|_{K^2} \subsetneqq \mu_p \cdot \mathbb{1}_{K^2} \lesssim \mathbb{1}/2_{K^2}$.

The decision making process algorithmization, based on Theorem 2, is evident, excluding, maybe, the search for $\mathcal{K}$; to this end, one can use "kernels ascent" from $G_{>1/2}/(C)$, (C) being the retraction of cycles.

## 4. COORDINATION OF PREFERENCES

Let $apr \in \tilde{\mathcal{P}}(X)$ be an a priori estimate by the DM of a fuzzy notion of an "optimal alternative", i.e. DM preferences (see Section 1). With $g \in \tilde{\mathcal{P}}_o(X^2)$, p being a FDDP, we define "concordance of decisions" as a fuzzy concept con, $\mu_{con}(apr,g,p) = \bigvee_{a \in \mathcal{D}_p(g)} \mu_{eq}(apr,a)$, eq being the fuzzy equivalence, $\mu_{eq}(a,b) = \mu_{inc}(a,b) \wedge \mu_{inc}(b,a)$ ($inc \in Inc$). Here, the concordance may be interpreted as a degree of "anticipation" by the DM of the "objective analysis result" due to (g,p), thus yielding an evaluation of DM competence. For the $inc_*$-

based con, we have $\mu_{con}(apr,g,p) = \overline{\mu_p^*} \vee (\bigvee_{K \in \mathcal{K}^*} (\mathring{\wedge}\, \overline{apr|_K}) \wedge \wedge(\mathring{\wedge}\, apr|_K))$.

One can also discover an analogy "dichotomousness-resoluteness": the more polarized are the DM preferences the more resolute his position seems to be. Most promising is a joint analysis of con and res, provided that an "ideal DM" achieves a maximum values of both $\mu_{con}$ and $\mu_{res}$ in any well-defined decision making problem. Let us consider a special case with an ideterminate initial FR, $g \in \Gamma \subseteq \tilde{\mathcal{P}}_o(X^2)$. By setting $\mu^*_{con} = \bigvee_{g \in \Gamma} \mu_{con}(apr,g,p)$, $\Gamma^* = \{g \in \Gamma \mid \mu_{con}(apr,g,p) = \mu^*_{con}\}$, one can examine the values $\bigwedge_{g \in \Gamma^*} \delta_p(g)$, $\bigvee_{g \in \Gamma^*} \delta_p(g)$ as an interval, implicit estimation of $\mu_{res}$, being compatible with DM competence (dually, $\mu^*_{con}$ may be used for the final choice determination). This approach is most attractive in multiattribute decision making problems when using preferences of the same DM for establishing "attribute significance". We dwell on this situation from the viewpoint of FR formation.

Let $C(x) = \{c_i(x)\}_{i \in Q, x \in X}$ be a multiattribute decision making problem. For the majority approach, the sets $I^+(x,y) = \{i \in Q \mid x \succ_{c_i} y\}$ are conventional (Roy, 1978; Beresovsky, Borzenko and Kempner, 1981; etc.), and a OR G is constructed as ($G(x,y) = 1$ iff $\sum_{I^+(x,y)} \alpha_i > \sum_{I^+(y,x)} \alpha_i$), with "attribute additive weights" $\alpha_i$. Fuzzy refinement may be made immediately; e.g., set $\mu_g(x,y) = \sum_{I^+(x,y)} \alpha_i / \bigvee_{(x,y) \in X_o^2} \sum_{I^+(x,y)} \alpha_i$. However, one can use a more axiomatic method. Let $\alpha_i = \mu_{sgca}(c_i)$, sgca being a fuzzy concept of "significant attribute". Clearly, the elements of g must be in accord with an extended Pareto ordering, say, Par.s (for $\alpha = \{\alpha_i\}_{i \in F}$, $\beta = \{\beta_i\}_{i \in H}$, we say that $\alpha$ Par.s $\beta$ iff there exists an injection $\varrho : H \to F$, yielding $\beta_i < \alpha_{\varrho(i)}$, for any $i \in H$) of the sets $\varphi^+(x,y) = \{\alpha_i \mid i \in I^+(x,y)\}$ (note that Par.s turns $X_o^2$ into a transitive graph $\Pi$). So we propose the following "fuzzy majority axioms".

FM1. $(\forall (x,y),(u,v) \in X_o^2)(\varphi^+(x,y) \text{ Par.s } \varphi^+(u,v) \Rightarrow \mu_g(x,y) \geq \mu_g(u,v))$.

FM2. $(\forall (x,y) \in X_o^2)(\mu_g(x,y) \in [\wedge \varphi^+(x,y); \vee \varphi^+(x,y)])$ (that is, $\mu_g(x,y)$ must belong to a corresponding interval of $\mu_{sgca}$ values).

The set $\Gamma$ of all FR's satisfying (FM1, FM2) for a given decision making problem is a nonempty convex subset of $[g_\wedge ; g_\vee]$; $g_\vee = \bigvee_\Gamma g$ has $\mu_{g_\vee}(x,y) = \bigvee_{I^+(x,y)} \alpha_i$; $g_\wedge = \bigwedge_\Gamma g$ is easily restored by means of the transitive skeleton of $\Pi$.

To illustrate the developed approach, we suggest a practical example (for details, see Alyabyev, Kitainik and Perelmuter, 1982).

Example 8. Table 2 contains the estimates of six systems of machines for wood-road engineering based on seven attributes, and also the DM's subjective evaluations of apr and sgca. $g_\wedge : g_\vee$ is given in Table 3. For $g_\vee$, it is the A-case ($\mathcal{K}^* = \{\{x_6\}\}$), whereas $g_\wedge$ is due to the C-case, and $\mathcal{K}^* = \{\{x_1;x_2\};\{x_3;x_4;x_5\};\{x_6\}\}$ defines the partition of X. Moreover, $\bigcap_{g\in\Gamma} = \{\{x_1;x_2\};\{x_6\}\}$; hence, no $g \in \Gamma$ is transitive. Using the above formula for $\mu_{con}$, we obtain that $\mu^*_{con} = 0.4$ is achieved with $K = \{x_3;x_4\}$, yielding a pointwise value $\mu_{res} = 0.2$ and a dichotomy $0.6|_K : 0.4|_{\bar{K}}$. However, both the resoluteness and the competence of the DM are questionable since apr $= \chi_{\{x_6\}}$ (most stable optimal choice for the given sgca) yields $\mu_{con} = 1$ and $\mu_{res} = 0.6$, and a dichotomy $0.8|_{\{x_6\}} : 0.2|_{\overline{\{x_6\}}}$. Therefore, one can hardly regard the initial (apr; sgca) to be consistent. The DM may be advised either to revise apr or to lower $\alpha_1$ and $\alpha_5$ with a simultaneous more resolute choice of K (thus, with $\alpha_1 = \alpha_5 = 0.2$, and apr $= \chi_K$ the choice of the DM is "ideal").

ACKNOWLEDGEMENTS

I wish to express my gratitude to V.B. Kuzmin, S.I. Travkin, and A.V. Yarkho for their valuable advice and their interest in my work.

REFERENCES

Alyabyev,V.I., Kitainik L.M. and Perelmuter Yu.N. (1982). Multiattribute Production Control Problem Solving Using a priori Leader's Preferences, Scientific Works of MLTI, Issue 142, 5-15 (in Russian).

Bellman,R.E. and Zadeh L.A. (1970). Decision-Making in a Fuzzy Environment. Management Science 17, 141-164.

Beresovsky,B.A., Borzenko V.I. and Kempner L.M. (1981). Binary Relations in Multicriterial Optimization. Nauka, Moscow 1981 (in Russian).

Table 2. Alternatives and Attributes (DM problem data)

| Attributes ($c_i$) | | Alternatives ($x_j$) | | | | | |
|---|---|---|---|---|---|---|---|
| denomination | Significance $\mu_{sgca}$ ($\alpha_i$) | $x_1$: 2B+T+ +M+R | $x_2$: 3B+T+ +R | $x_3$: 2B+E +T+M | $x_4$: 2B+E +T+R | $x_5$: 3B +2R | $x_6$: 3B +R [1] |
| I. Quantitative (calculated relative[2] values) | | | | | | | |
| $C_1$: specific layout | 0.3 | 6 | 0 | 64 | 46 | 100 | 50 |
| $C_2$: cost price | 0.8 | 0 | 7 | 89 | 94 | 100 | 62 |
| $C_3$: labour productivity | 0.8 | 0 | 0 | 100 | 96 | 98 | 75 |
| $C_4$: labour mechanization level | 0.7 | 0 | 0 | 100 | 97 | 99 | 71 |
| $C_5$: specific metal content | 0.4 | 0 | 15 | 33 | 48 | 100 | 65 |
| $C_6$: specific energy capacity | 0.6 | 1 | 0 | 95 | 100 | 57 | 26 |
| II. Qualitative (estimated by DM) | | | | | | | |
| $C_7$: equipment unification | 0.4 | sf. | gd | isf. | isf. | gd | ex[3] |
| DM a priori preferences $\mu_{apr}$ | | 0.6 | 0.5 | 0.7 | 0.8 | 0.3 | 0.5 |

1) B - a bulldozer, T - a tip-lorry, M-a motor-grader, R- a roller, E - an excavating machine

2) The scale for attributes: $100\times(c_i(x) - \bigwedge_x c_i(x))/(\bigvee_x c_i(x) - \bigwedge_x c_i(x))$

3) sf. = sufficient, gd = good, isf.-insufficient, ex.-excellent

Calculation accuracy: $c_1$-$c_3$ - to 10%, $c_4$-$c_6$ - to 4%

(this, $x_1$ and $x_2$ are indistinguishable due to $c_1$, $c_2$ and $c_6$)

Table 3. Membership function values for $g_\wedge : g_\vee$

| Alternatives | Alternatives | | | | | |
|---|---|---|---|---|---|---|
| | $x_1$ | $x_2$ | $x_3$ | $x_4$ | $x_5$ | $x_6$ |
| $x_1$ | 0 | 0.4:0.4 | 0.6:0.8 | 0.6:0.8 | 0.6:0.8 | 0.6:0.8 |
| $x_2$ | 0.4:0.4 | 0 | 0.6:0.8 | 0.6:0.8 | 0.6:0.8 | 0.6:0.8 |
| $x_3$ | 0.7:0.8 | 0.7:0.8 | 00.4 | 0.4:0.6 | 0.4:0.8 | 0.7:0.7 |
| $x_4$ | 0.7:0.8 | 0.7:0.8 | 0.3:0.3 | 0 | 0.4:0.4 | 0.7:0.8 |
| $x_5$ | 0.7:0.8 | 0.7:0.8 | 0.4:0.6 | 0.4:0.6 | 0 | 0.7:0.8 |
| $x_6$ | 0.7:0.8 | 0.7:0.8 | 0.6:0.8 | 0.6:0.8 | 0.6:0.8 | 0 |

Bezdek, J. et al. (1978). A fuzzy relation space for group decision theory. Fuzzy Sets and Systems 1, 255-268.

Danilov, V.I. (1984). Structure of binary rules for preferences aggregation. Econ. and Math. Methods XX, 882-893 (in Russian).

Dubois, D. and Prade H. (1980). Fuzzy Sets and Systems: Theory and Spplications. Academic Press, New York.

Kaufmann, A. (1977). Introduction a la Théorie des Sous-ensembles Flous. Masson, Paris.

Kitainik, L.M. (1981). On a certain method of technico-economical estimate of alternatives involving DM preferences. In: "Problems of Improving Efficiency and Quality in Forest Industry". Moscow, VNIPIEILESPROM, 124-132 (in Russian).

Kuzmin, V.B. (1982). Construction of Group Choice in Spaces of Ordinary and Fuzzy Binary Relations. Nauka, Moscow (in Russian).

Larichev, O.I. (1980). Tracing of estimation, comparison and choice with multiattribute alternatives in decision problems. Collected Papers of the Institute for System Studies, Moscow, Issue 9, 26-35 (in Russian).

Podinovsky, V.V. (1978). Relative importance of criteria in multiobjective decision problems. In: Multiobjective Decision Problems , Mashinostroyenye, Moscow, 48-92 (in Russian).

Orlovski, S.A. (1978). Decision-making with a fuzzy preference relation. Fuzzy Sets and Systems 1, 155-168.

Roy, B. (1972). Decisions avec criteres multiples. Problemes et methodes. METRA International 11, 121-151.

# COMBINATORIAL SEARCH WITH FUZZY ESTIMATES

Didier Dubois, Henri Farreny and Henri Prade

Langages et Systemes Informatiques
Université Paul Sabatier
118, route de Narbonne
31062 Toulouse, France

Abstract. Combinatorial search methods known as graph-search techniques aim at finding optimal or feasible solutions to a problem, among a usually finite but possibly very large set of alternatives, while obviating brute-force enumeration. The search process is guided by evaluations of how close is a solution. Usually evaluation functions are supposed to provide precise estimates. Here we discuss two kinds of situations where such estimates are naturally represented as fuzzy intervals: first, the case when several evaluation functions are available; a fuzzy interval can synthetize this information. The second case is when data defining the problem are themselves imprecise; the traveling salesman problem is used to illustrate the latter point.

Keywords: combinatorial optimization, combinatorial search, graph search, traveling salesman problem

## 1. INTRODUCTION

Most combinatorial problems can be solved by graph search procedures (e.g., Nilsson, 1980). Such procedures consider the solution of a combinatorial problem as one of building a search graph, whose initial node represents the initial state of the solution, until a goal node providing a feasible or optimal solution is reached. The efficiency of these procedures heavily relies upon the search strategy, i.e. the branching principle by which new nodes are created. Usually the combinatorial problem involves some cost to be minimized, and the search strategy uses estimates of the cost of solutions which can be reached from each pending node. The branching principle then says that one should start from the most promising node, in terms of the estimate at this node, to build a path to the goal.

In this paper we discuss some issues pertaining to the use of fuzzy intervals (Dubois and Prade, 1980, 1985c) in graph search procedures. First, the branching principle usually assumes that only one cost estimate is used in the search strategy. It is indicated that a multiple criteria strategy can be contemplated when bounds on the real cost, together with several cost estimates, are available. A fuzzy interval can be built to synthetize the heuristic information; the branching is carried out by comparing fuzzy intervals. Another situation is when the input data are imprecise. Data are then naturally represented by fuzzy intervals, and induce fuzzy estimates. In both situations one is led to compare fuzzy intervals in order to make a branching decision. This problem has received attention by the authors (Dubois and Prade, 1983b) previously, and finds here some application to graph search strategy. To our knowledge, no work using fuzzy intervals in combinatorial search has been published except an early attempt by Farreny and Prade (1982). However, combinatorial problems with fuzzy constraint sets or fuzzy goals have been solved in the past, especially via dynamic programming (see Esogbue and Bellman, 1984, or Kacprzyk, 1982) and more recently integer programming problems (Fabian and Stoica, 1984; Zimmermann and Pollatschek, 1984). In these works, following Bellman and Zadeh (1970), a membership function is built by aggregation of fuzzy constraints and goal sets, and the question is then to find a solution with the greatest membership grade. Formally, this is but a classical optimization problem which our concern clearly departs from.

The first two sections are brief refreshers on graph search and fuzzy intervals. Then the questions of multiple criteria branching and combinatorial search with fuzzy data are successively addressed, and illustrated on the traveling salesman problem.

## 2. GRAPH SEARCH WITH HEURISTIC INFORMATION

Solving a combinatorial problem can be viewed as finding a path in a directed graph, called the search graph, from a node r called the root to a node t called the goal node, belonging to a prescribed subset of nodes. The root is the representation of the initial state of a system, or of the solution to a problem (in the latter case it may be a possibly infeasible solution or a set of unassigned decision variables). A goal node embodies an admissible final state of a system, or a feasible, or even optimal solution to a problem. A path from the root to a goal node is viewed as the set of elementary transformations modelling the trajectory of the system, or the process of building a feasible solution to the problem at hand.

The search graph, which can be interpreted as a state space, is only potentially defined, in the sense that only the root and the goals are explicitly available, together with a set of rules which specify how to build the successors of a current node. Sometimes goal nodes are described by means of constraints.

Using applicable rules on some node ("the father") creates new nodes ("the sons"), together with an arc from the father to each son. This is called node expansion. Each arc is valuated by a cost which is supposedly a positive number. The cost of a path is the sum of the costs of its arcs. An optimal solution corresponds to a minimal cost path from the root to a goal node. The graph search methodology consists in applying rules, where possible, to nodes until a goal node is reached. As long as a goal node is not reached, the main problem is to select the proper node to be expanded.

This choice can be driven by the knowledge of a so-called evaluation function f, so that its value, f(n), at any node n estimates the sum of the minimal cost from the root to node n (denoted g(n)) and the minimal cost from node n to a goal state (denoted h(n)). g(n) is called the <u>backward evaluation</u>, and h(n) the <u>forward estimate.</u> Let $g^*(n)$ (respectively, $h^*(n)$) be the cost of an optimal path from r to n (respectively: n to t). Generally, an upper bound $\bar{g}(n)$ of $g^*(n)$ is available; it is the cost of the best path from r to n which has been actually built by the procedure. If the search graph is a tree, then $\bar{g}(n) = g^*(n) = g(n)$. Upper and lower bounds of $f^*(n)$ are denoted $\underline{f}(n)$ and $\bar{f}(n)$, respectively. An upper bound of $f^*(n)$ is available as soon as a feasible solution to the problem is available. $\bar{f}(n)$ is the cost of the corresponding path and can be used to reduce the width of the search graph by pruning pending (= not expanded yet) nodes n' such that $\bar{f}(n) \leqslant \underline{f}(n')$. An estimate f(n) which is not established as an upper or lower bound is called <u>heuristic</u>.

A graph search procedure can then be defined as follows, when the search graph is a tree (otherwise, see Nilsson, 1980):

1 - Start with the root r. Put it on a list PENDING.
2 - Create a list EXPANDED which is empty.
3 - If PENDING is empty, exit with failure.
4 - Select the first node n in PENDING, put it on EXPANDED.
5 - If n is a goal node, exit successfully with the path from r to n.
6 - Expand node n, generating a list M of nodes (They are new ones due to the tree assumption); add members of M to PENDING.
7 - Reorder the list PENDING according to f(n).
8 - Go to 3.

If $h(n) \leqslant h^*(n)\ \forall\ n$, then such an algorithm is called A* by Nilsson (1980) and has interesting properties: it terminates in a finite number of steps (if the cost of each arc is bounded from below by a positive number); when it terminates, either it provides an optimal path (exit 5) or such a path does not exist (exit 3). The algorithm is then said to be <u>admissible</u>. When h(n) is only heuristic, then the optimality of a solution discovered by the algorithm is no longer guaranteed. However, the use of a heuristic evaluation may enable such a solution to be discovered faster than using a bad lower bound $\underline{h}(n)$ (for instance $\underline{h}(n) = 0$ only produces a uniform-cost algorithm). Besides, if for each node n, a dedicated upper bound $\bar{h}(n)$ is available together with a lower bound $\underline{h}(n)$, then it is possible to cancel nodes in PENDING in step 7 by pruning. Pruning is only heuristic if only a heuristic estimate is available.

In the following, we shall try to synthetize all the available information concerning the value of $h^*(n)$, in order to allow for a more elaborate reordering of PENDING in step 7 i.e. reordering in terms of the whole information, not only a single heuristic term. For this purpose we need the notion of a fuzzy interval.

N.B. A* algorithms are very close to branch and bound algorithms, long since used in operations research (see Geoffrion and Marsten, 1972 for an old but yet excellent introduction).

## 3. FUZZY INTERVALS

A fuzzy interval is a fuzzy set (Zadeh, 1965) of real numbers, denoted M, with a membership function $\mu_M$ which is unimodal and upper semi-continuous, that is $\forall\alpha\in ]0,1]$, $M_\alpha = \{ r \mid \mu_M(r) \geq \alpha \}$ (the α-cut of M) is a closed interval (cf. Fig. 1).

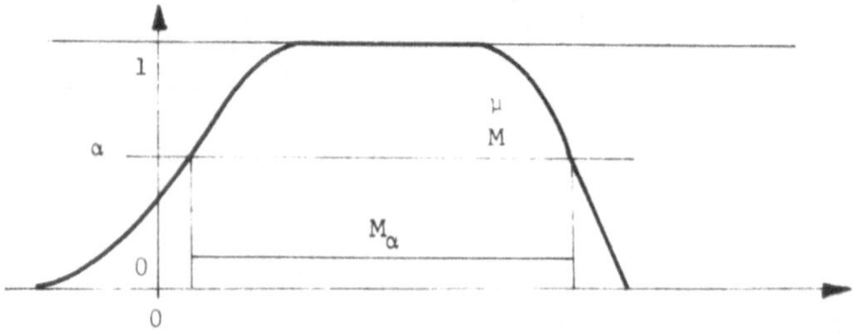


Fig. 1.

A fuzzy interval generalizes the concept of a closed interval, including that of a real number. It may model the range of some variable x with more sophistication than a usual interval. Namely, the support $S(M) = \{ r \mid \mu_M(r) > 0 \}$ is the widest range for x (x cannot take values outside S(M)), while the peak $\mathring{M} = \{ r \mid \mu_M(r) = 1 \}$ is the set of most plausible values of x, called modal values. Interval analysis (Moore, 1966) generalizes to fuzzy interval analysis by Zadeh's extension principle (e.g., Dubois and Prade, 1980, 1985c): Let * be an operation between real numbers, M and N two fuzzy intervals associated to variables x and y respectively, the fuzzy range of x*y, denoted by M*N, is obtained by

$$\mu_{M*N}(w) = \sup \{ \min(\mu_M(u), \mu_N(v)) \mid u*v = w \} \quad (1)$$

This formula can be derived by applying the rules of computation of possibility theory (Dubois and Prade, 1980; Zadeh, 1978), and assumes that x and y are not linked (non-interactive). As soon as the supremum is attained in (1), we have

$$(M*N)_\alpha = M_\alpha * N_\alpha = \{ u*v \mid u \in M_\alpha, v \in N_\alpha \} \quad (2)$$

which is interval analysis.

The operations used in classical graph search algorithms

are addition and comparison. We shall assume that the involved fuzzy numbers are all of the same type, i.e. there are shape function L, R, modal values $\underline{m}$, $\bar{m} \in R$, spreads $\alpha, \beta \geqslant 0$ such that

$$u \leqslant \underline{m} : \mu_M(u) = L\left(\frac{\underline{m} - u}{\alpha}\right)$$

$$\underline{m} \leqslant u \leqslant \bar{m} : \mu_M(u) = 1$$

$$u \geqslant \bar{m} : \mu_M(u) = R\left(\frac{u - \bar{m}}{\beta}\right)$$

where L (or R) is an upper semi-continuous monotonic function with $L(0) = 1$, $1 > L(x) > 0 \ \forall x \in (0,1)$, $L(x) = 0 \ \forall x > 1$. M is said to be of type L-R and denoted $(\bar{m}, \underline{m}, \alpha, \beta)_{LR}$. Closed intervals are modelled by $\alpha = \beta = 0$. If moreover $N = (\bar{n}, \underline{n}, \gamma, \beta)_{LR}$, we have (Dubois and Prade, 1980)

$$M+N = (\underline{m}+\underline{n},\ \bar{m}+\bar{n}, \alpha+\gamma, \beta+\delta)_{LR} \tag{3}$$

when * is the addition in (1), i.e. M+N is still of the L-R type.

The comparison of fuzzy intervals can be carried out consistently with interval overlapping analysis, in the setting of possibility theory (Dubois and Prade, 1983b). Let M and N be usual intervals $[\underline{m}, \bar{m}]$ and $[\underline{n}, \bar{n}]$. We consider four respective locations of M and N:

i) $\forall x \in M, \forall \ y \in N, x > y$ (equivalent to $\underline{m} > \bar{n}$)

ii) $\forall x \in M, \exists \ y \in N, x \geqslant y$ (equivalent to $\underline{m} \geqslant \underline{n}$)

iii) $\exists x \in M, \forall \ y \in N, x > y$ (equivalent to $\bar{m} > \bar{n}$)

iv) $\exists (x,y) \in M \times N, x \geqslant y$ (equivalent to $\bar{m} \geqslant \underline{n}$)

These are the only four possible statements expressing that "M is greater than N". i) is the strongest one and means that M is on the right of N and does not overlap it. iv) is the weakest statement, and means that finding x at least as great as y is possible (simple overlapping). Statements ii) and iii) are implied by i) and each implies iv); they, respectively, refer to least and greatest values of x and y .

If M is a fuzzy interval describing the range of a variable x, then $[M, +\infty)$ and $]M, +\infty)$ denote the fuzzy sets of numbers, respectively, greater or equal to x, and strictly greater than x. They are defined by their membership functions:

$$\mu_{[M,+\infty)}(u) = \sup\{\mu_M(x) \mid u \geqslant x\} \tag{4}$$

$$\mu_{]M,+\infty)}(u) = \inf\{1-\mu_M(x) \mid u \leqslant x\} \tag{5}$$

$$= 1-\mu_{(-\infty,M]}(u)$$

where $1-\mu_F$ defines the membership function of the complement $\bar{F}$

of F. If M is an interval, $[M,+\infty) = [\underline{m},+\infty)$, $]M,+\infty) = (\bar{m},+\infty)$.

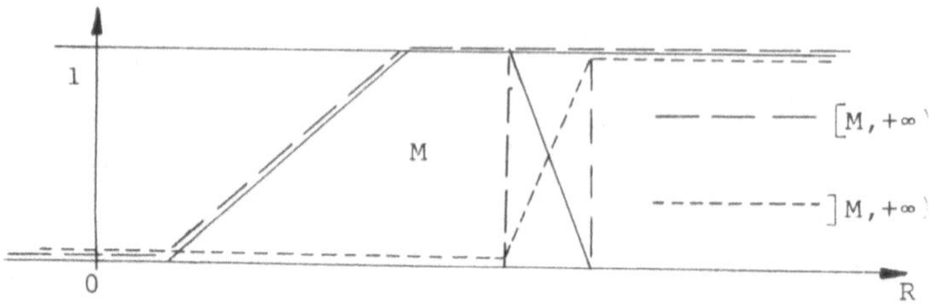


Fig. 2.

The extent to which M is greater than N can thus be discussed in terms of overlapping or inclusion of M in $[N,+\infty)$ or $[N,+\infty)$. The extent to which an overlapping between fuzzy sets F and G exists is assessed by the following index (Zadeh, 1978)

$$\pi(F;G) = \sup_u \min(\mu_F(u), \mu_G(u)) = \pi(G;F) \tag{6}$$

called possibility of the fuzzy event F (given G) or G (given F). The quantity $N(F;G) = 1-\pi(\bar{F};G)$ can be used as a grade of inclusion of G in F, and is called the necessity of the fuzzy event F (given G) (Dubois and Prade, 1985a). Namely

$$\exists u,\ \mu_F(u) = \mu_G(u) = 1 \Rightarrow \pi(F;G) = 1$$

$$F \cap G = \emptyset \Leftrightarrow \pi(F;G) = 0$$

$$\mu_G \leqslant \mu_F \text{ (Zadeh fuzzy set inclusion)} \Rightarrow N(F;G) \geqslant 0.5$$

$$S(G) \subseteq \overset{\circ}{F} \Leftrightarrow N(F;G) = 1.$$

The four following indices were suggested in Dubois and Prade (1983b) to assess the extent to which M is greater than N: $Nec(\underline{x} > \bar{y}) = N(]N,+\infty);M)$; $Nec(\underline{x} \geqslant \underline{y}) = N(]N,+\infty);M)$; $Pos(\bar{x}>\bar{y}) = \pi(]N,+\infty);M)$; $Pos(\bar{x} \geqslant \underline{y}) = \pi(]N,+\infty);M)$. $\underline{x}$ (resp. $\bar{x}$) is short for the least (resp.: greatest) values of x. Nec and Pos are short for necessity and possibility. These four indices are fuzzy counterparts of statements (i-iv) since:

i) is true $\Leftrightarrow Nec(\underline{x} > \bar{y}) = 1$

ii) is true $\Leftrightarrow Nec(\underline{x} \geqslant \underline{y}) = 1$

iii) is true $\Leftrightarrow Pos(\bar{x} > \bar{y}) = 1$

iv) is true $\Leftrightarrow Pos(\bar{x} \geqslant \underline{y}) = 1$

The links between properties relating (i - iv) are reflected in the following inequalities:

$$\mathrm{Pos}(\bar{x} \succcurlyeq \underline{y}) \geq \max(\mathrm{Pos}(\bar{x} > \bar{y}), \mathrm{Nec}(\underline{x} \succcurlyeq \underline{y})) \tag{7}$$

$$\mathrm{Nec}(\underline{x} > \bar{y}) \leq \min(\mathrm{Pos}(\bar{x} > \bar{y}), \mathrm{Nec}(\underline{x} \succcurlyeq \underline{y})) \tag{8}$$

Moreover, the following dependencies can be observed when M and N are exchanged in the indices:

$$\mathrm{Pos}(\bar{x} \succcurlyeq \underline{y}) = 1 - \mathrm{Nec}(\underline{y} > \bar{x}) \tag{9}$$

$$\max(\mathrm{Pos}(\bar{x} \succcurlyeq \underline{y}), \mathrm{Pos}(\bar{y} \succcurlyeq \underline{x})) = 1 \tag{10}$$

$$\mathrm{Pos}(\bar{x} > \bar{y}) + \mathrm{Pos}(\bar{y} > \bar{x}) = 1 \tag{11}$$

$$\mathrm{Nec}(\underline{x} \succcurlyeq \underline{y}) + \mathrm{Nec}(\underline{y} \succcurlyeq \underline{x}) = 1 \tag{12}$$

(11) and (12) may fail to hold in pathological configurations (e.g. M and N are not fuzzy, $\underline{m} = \underline{n}$, and then $\mathrm{Nec}(\underline{x} > \underline{y}) = \mathrm{Nec}(\underline{y} > \underline{x}) = 1$). From a computational point of view, obtaining the index values is a simple matter of intersecting membership functions $\mu_M$, $\mu_N$, $1-\mu_M$, $1-\mu_N$. The soundness and intuitive appeal of the approach to the ranking of fuzzy intervals, based on the four indices has been experimentally checked by Degani and Bortolan (1985).

## 4. FUZZY BRANCHING IN TREE SEARCH

A drawback of search algorithms based on a single evaluation function is that one uses only part of the available information in many cases. When both a lower bound $\underline{h}$ and a heuristic estimate h are at hand, it may not be interesting to ask whether $\underline{h}$ or h must be chosen to guide the search for a new node to expand. A more relevant question may be how to use <u>both</u> items of information. Similarly the knowledge of an upper bound $\bar{h}$ on the cost of an optimal solution can be useful not only for pruning purposes, but can also be involved in the branching decision.

The idea of using an upper and a lower bound to guide the search is not new. A similar concern appears in Berliner (1979), in the framework of 2-person searches, where one player tries to maximize a given function while the other tries to minimizes it. For tree search algorithms considered in this paper, the following facts are easy to establish:

- If the upper bound $\bar{h}(n)$ of a node is available, then the upper bound associated to any ancestor n′ of n can be updated as $\min(\bar{h}(n'), \bar{h}(n))$
- If the lower bounds $\underline{h}(n')$ of all sons n′ of n are available, the lower bound of n can be updated as $\max(\min_{n'} \underline{h}(n'), \underline{h}(n))$

Hence, as the search tree develops, it is possible to shrink the intervals $[\underline{h}(n), \bar{h}(n)]$ each time newly expanded nodes are evaluated. If, eventually, there is a son n of the root such that for any other son n′ of the root $\bar{h}(n) \leq \min_{n'} \underline{h}(n')$, then the optimality of arc (r,n) is proved and all n′ can be pruned. This remark applies also to all subtrees of the search tree, and is the basis of Berliner's (1979) B* algorithm.

Knowing $\underline{h}(n)$ and $\bar{h}(n)$, any estimate h(n) of the form $\alpha\underline{h}(n) + (1-\alpha)\bar{h}(n)$ is a heuristic forward evaluation ($\alpha\in[0,1]$). The value of α may reflect the tightness of the lower bound with respect to the tightness of the upper bound. An idea to preserve the possible branching efficiency of the heuristic term, while keeping the optimality checking capability of the B* algorithms is to synthetize the available information regarding node n under the form of a fuzzy interval F(n) restricting the estimate variable f(n), and defined as follows:

- Its support is $g^*(n) + H(n)$ where $H(n) = [\underline{h}(n), \bar{h}(n)]$,
- Its peak is $g^*(n) + h(n)$ ; if several heuristic estimates are available, the peak is the interval bounded from below (respectively: above) by the least (respectively: greatest) estimate value,
- The membership function is triangular or trapezoid-shaped, accordingly.

The problem of ranking pending nodes for further expansion becomes less trivial. It requires a ranking of fuzzy estimates of the cost to be performed. Using the results of section 3, we can proceed in the following way:

- For each node n belonging to the set P of pending nodes, define four dominance indices expressing the extent to which f(n) is smaller than other f(n′), $n'\in P - \{n\}$; let $k(n) = \min_{n'\neq n} f(n)$; the four indices are:

  $I_j(n), j = \overline{1,4}$ respectively defined by:

  $I_1(n) = \mathrm{Nec}(\underline{k(n)} > \overline{f(n)})$; $I_2(n) = \mathrm{Nec}(\underline{k(n)} \geq \underline{f(n)})$

  $I_3(n) = \mathrm{Pos}(\overline{k(n)} > \overline{f(n)})$; $I_4(n) = \mathrm{Pos}(\overline{k(n)} \geq \underline{f(n)})$
- These four indices equip P with a partial ordering structure, which is that of the set of vectors $I(n) = \{I_1(n), I_2(n), I_3(n), I_4(n) \mid n \in P\}$. Find the set M of maximal elements of P (in the sense of Pareto-optimization). M contains the set of nodes to be further expanded.

The fuzzy interval K(n) associated with variable $\widetilde{\min}_{n'\neq n} f(n)$ is defined by the extension principle, substituting 'minimum' to * in (1). In order to account for the convention of linear membership function, a reasonable approximation is to define K(n) as a trapezoid-shaped fuzzy interval such that (Dubois and Prade, 1985a)

$$\mathring{K}(n) = \widetilde{\min}_{n'\neq n} \mathring{F}(n') \; ; \qquad S(K(n)) = \widetilde{\min}_{n'\neq n} S(F(n'))$$

and

$$\min([a,b],[c,d]) = [\min(a,c),\min(b,d)] \tag{13}$$

The advantages of this tree search methodology seem to be the following:

- *reduction of the horizon effect*: it is well known that the ordering of node evaluations at a given level of depth l in the search tree may be misleading in the sense that the ordering of nodes at level l+1 can be completely different. Hence best-first strategies selecting only one expandable node may lead to explore non-optimal paths too far ahead. The selection in terms of fuzzy estimates is a "limited-breadth" strategy which only leaves behind pending nodes which are dominated in terms of the 4 ranking indices. Hence the horizon effect is coped with. It is also possible to discriminate further in the set M of non-dominated pending nodes, if we use, as done by Pearl and Kim (1982), an estimate of the computational effort required for completing the search. For instance, if n, $n' \in M$ and the depth level of n is significantly greater than that of n', it may be better to expand n first and get a solution earlier.

- *bring together the merits of admissible algorithms and the efficiency of heuristic search*. Indeed, because the peak of F(n) reflects the heuristic evaluation, the value of the ranking indices is more influenced by this evaluation than by the upper and lower bounds. The latter introduce a perturbation in the ranking only if they significantly contradict the ordering along the heuristic estimate. Hence, if a good heuristic estimate is at hand, a feasible solution will be found fast. On the other hand, the support of the fuzzy estimate is useful for optimality-proving by updating the fuzzy estimate supports of ancestors of pending nodes as done in the B* algorithm (Berliner, 1979). Optimality-proving is achieved by considering index $I_1$ since it checks for disjointness of fuzzy intervals.

## 5. TREE SEARCH WITH FUZZY DATA

In this section, we assume that due to incomplete knowledge regarding the data of the considered problem, the elementary costs involved in the evaluation function f are available under the form of error intervals. Evaluation estimates are then obtained by means of the rules of interval analysis (Moore, 1966)(see (2)). In order to adapt tree search algorithms to interval-valued estimates, three questions must be answered.

- What can be the branching strategy?
- What is a reasonable stopping rule for the algorithms?
- What are the properties of the obtained results?

The choice of a node to expand is a matter of comparing the respective location of intervals of the form $\tilde{F}(n) = [f(n), f(n)]$ for $n \in P$. But the *meaning* of this interval is quite different from the meaning of interval $[\underline{f}(n), \overline{f}(n)]$ used in the previous section. In the case of an A* algorithm, F(n) is the possible range of the *lower bound* of the cost of the best solution of the subtree rooted in node n; this range reflects the imprecision of the data and is *not* the range of possible costs of the best solution rooted in node n. Particularly, F(n) does not allow for optimality-proving as in the B* algorithm.

In order to rank the set $\{\widetilde{F}(n) \mid n \in P\}$, we consider the four points of view expressed in statements (i) - (iv) of section 3, applied to $\widetilde{F}(n)$ and $\widetilde{\min}\ \widetilde{F}(n')$, $\forall n \in P$, i.e. the four selection criteria can be used for each $n' \neq n$,

$n \in P$:

C1 - select n such that $\min_{n \neq n'} \underset{\vee}{f}(n') > \hat{f}(n)$,

C2 - select n which minimizes $\underset{\vee}{f}(n)$,

C3 - select n which minimizes $\hat{f}(n)$,

C4 - select n such that $\min_{n \neq n'} \hat{f}(n') \geqslant \underset{\vee}{f}(n)$.

If C1 is verified for some $n \in P$, the evaluation $\widetilde{F}(n)$ is certainly better than the evaluation of other nodes, in spite of the imprecision. However, if n satisfies only C4, $\widetilde{F}(n)$ is only possibly better than other evaluations. Let $M_i$ be the subset of pending nodes satisfying $C_i$, then $M_1 \subseteq M_2 \cap M_3$, $M_2 \cup M_3 \subseteq M_4$. $M_1$ contains at most one node and can be empty, while $M_4$ is never empty but can be P itself. When $n \in M_2 \cap M_3$, one may write, introducing a new selection criterion

$$\text{C23}: \quad \widetilde{\min_{n' \in P}}\ \widetilde{F}(n') = \widetilde{F}(n)$$

where $\widetilde{\min}$ is defined by (13). This criterion is less drastic than C1 (intervals may overlap) but intuitively satisfactory. C2 and C3 rank intervals according to their lower and upper bounds, respectively, and cannot be compared, while C1, C23 and C4 are decreasingly demanding in this order. It is natural to select nodes to be expanded on the basis of the most demanding criterion which is able to discriminate among pending nodes. If C23 does not discriminate, C2 will be given priority over C3 if there is some evidence that the lowest values in $\widetilde{F}(n)$ are more plausible than the greatest values. Hence the selection procedure can be summarized as follows:

- Find $M_1$ ; if $M_1 \neq \emptyset$, $M = M_1$
- Find $M_2$, $M_3$; if $M_2 \cap M_3 \neq \emptyset$, $M = M_2 \cap M_3$
- If C2 has the highest priority $M = M_2$ otherwise $M = M_3$
- If no priority has been assigned, $M = M_2 \cup M_3$.

Note that $M_2 \neq \emptyset$, $M_3 \neq \emptyset$ anyway , so that criterion C4 is not useful for the branching strategy. A further refinement in the selection of the nodes to be expanded can be introduced on the basis of other evaluations (for instance, the proximity to a solution).

When a terminal node is reached, thus yielding a solution, one may wish to keep on searching when the stopping rule used too weak a criterion. Indeed, each of the above introduced criteria can be used as a stopping rule, once a node has been acknowledged as meeting the constraints which define feasibility. However, because $M_1$ and $M_2 \cap M_3$ can be empty, there may not exist any optimal terminal node in the sense of criteria C1 or C23. In that case the algorithm may never stop, or stops proving no optimal solution exists (if the search tree is finite). Hence when an admissible algorithm (such as A*) is used with fuzzy data, optimality becomes uncertain, depending upon the stopping rule criterion; if C4 is adopted, then the algorithm stops earlier but the obtained solution is only possibly optimal. If C1 is adopted, then if the algorithm comes up with a solution it is optimal with certainty; but such a solution may not exist. Intermediary criteria C2 and C3 provide sure optimality only with respect to lower or upper estimates of the data.

Now the data can be available under the form of fuzzy intervals as well, and the ranking indices introduced in section 2 can be used to deal with the admissible algorithm in a fuzzy environment. Once again the interpretation of the fuzzy estimate $\tilde{F}(n)$ is quite different from section 4; $\tilde{F}(n)$ only reflects the imprecision of the data and its support is not the range of costs of solutions rooted in n. As a branching rule, Farreny and Prade (1982) suggested the selection of pending node n such that:

$$\tilde{F}(n) = \widetilde{\min_{n' \in P}}\ \tilde{F}(n') \tag{14}$$

where $\widetilde{\min}$ is the minimum operation extended to fuzzy arguments by (1). However, such a node may not exist, because the fuzzy interval $\widetilde{\min}(M,N)$ may be neither M or N (see Dubois and Prade, 1980); in that case any pending node may be expanded! Moreover (14) is a very strong selection criterion because it is equivalent to the simultaneous application of criterion C23 to all $\alpha$-cuts of the fuzzy evaluations.

Another idea is to select a node to expand by means of the four ranking indices, as done in section 4. Note that although the fuzzy estimates of the cost do not have the same interpretation, the same branching methodology applies, i.e. to build the set M of non-dominated elements in P. Note that this strategy is consistent with the selection procedure used in the case of usual intervals since:

- if $I_1(n) = 1$, then from (7) and (8), $I(n) = (1, 1, 1, 1)$ and $n \in M$. Moreover, $\forall\, n' \neq n$, $I_1(n') = 0$ so that $M = M_1 = \{n\}$.
- if $\forall\, n, I_1(n) = 0$, but $\exists\, n: I_2(n) = I_3(n) = 1$, then from (8) $n \in M$ and $M = M_2 \cap M_3$ is easy to conclude with.
- If $\forall\, n$, $I_1(n) = 0 = \min(I_2(n), I_3(n))$, then if $I_2(n) = 1$, node n is not dominated ($I(n) = (0, 1, 0, 1)$) and if $I_3(n) = 1$, node n is not dominated either ($I(n) =$

(0, 0, 1, 1)); but if I(n) = (0, 0, 0, 1) n is dominated; thus $M = M_2 \cup M_3$.

i.e. choosing M as the set of candidate pending nodes for expansion reduces to the selection procedure of nodes with interval-valued estimates when no priority has been assigned to C2 and C3. Such a priority assignment can reduce the set of candidates (to $M_i = \{n \mid \forall n' \; I_i(n) \geqslant I_i(n')\}$, i = 2 or 3) in the fuzzy case too.

As previously, a stopping rule can be defined by choosing one of the ranking indices $I_i$, and searching for a terminal node n such that $I_i(n)$ is greater than some threshold $\theta > 0$. The counterpart of criterion C23 is the rule $I_{23}(n) = \min(I_2(n), I_3(n)) > \theta$. Once again with $I_1$ or $I_{23}$ the stopping rule may never apply to any terminal node. $\theta$ can be interpreted as a grade of certainty of optimality (if $I_1$ is chosen) or possibility of optimality (if $I_4$ is chosen); etc...

Example. Consider a traveling salesman problem with a non-directed graph G defined in Fig. 3.a, together with edge costs $C_{ij}$ of edges $(S_i, S_j)$:

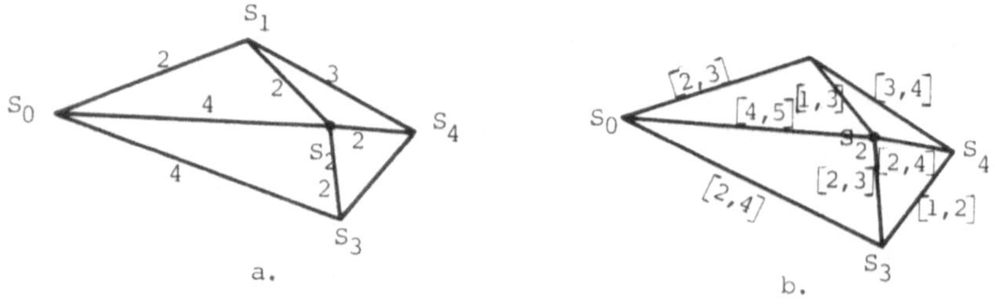


Fig. 3

Using any A* algorithm, it is easy to obtain $(S_0S_1S_2S_4S_3S_0)$ as an optimal Hamiltonian circuit, with cost 11. In Fig. 3.b, the data, i.e. the edge costs, become fuzzy, here interval-valued. We consider an extension of an A* algorithm defined as follows:

- Any non-terminal node n of the search tree is a path in G of the form $S_{i0}S_{i1}\ldots S_{ik}$ containing no circuit. The root is $S_{i0} = S_0$. A terminal node is a length-N(=5)-Hamiltonian circuit, i.e. k = 6, $S_{i6} = S_0$.
- Node expansion consists in adding one edge at the end of the path corresponding to the node to be expanded.

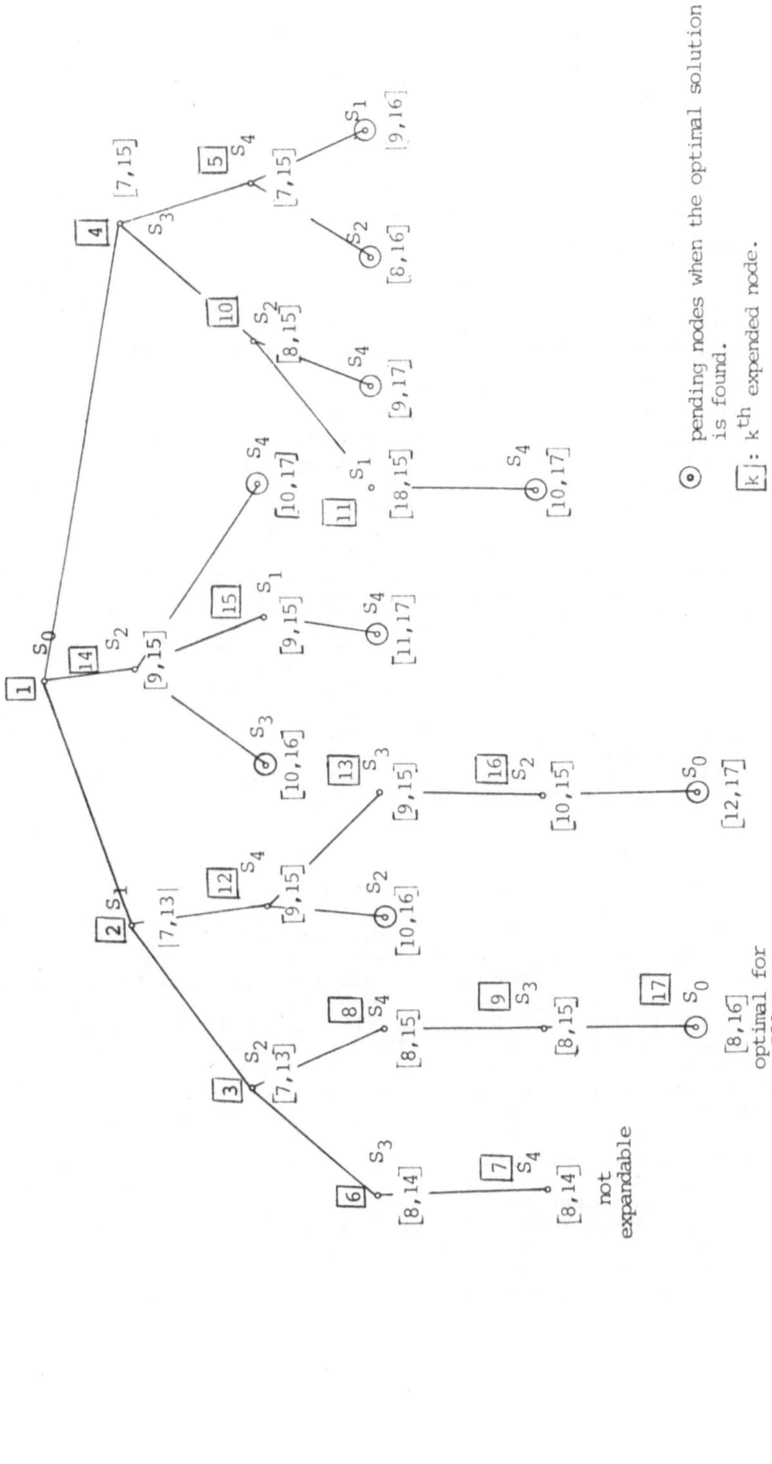

not expandable
optimal for C23
pending nodes when the optimal solution is found.
k : k th expended node.


Fig.4

. A lower bound $\underline{f}(n)$ is obtained for $n = S_{i0}\, S_{i1} \ldots S_{ik}$ as (Pearl and Kim, 1982):

$$\underline{f}(n) = \sum_{j=1}^{k} C_{i_{j-1} i_j} + \sum_{j \notin \{1,2,\ldots,k\}} \min_{l} C_{i_j l} \qquad (15)$$

in the case of precise data. With interval-valued costs $C_{ij}$, the rules of interval analysis are applied to (15), and yield $F(n)$.

- The branching rule is to select pending nodes according to criteria C1, C23, C2, C3, C4 in this order. When several nodes are selected, the left-most one only is expanded if they are of the same depth level.
- The stopping rule is based on criterion C23.

The obtained tree is given in Fig. 4 Note that the same Hamiltonian circuit is obtained as in the case of precise data (note that G in Fig. 3.a is a special instance of the interval-valued graph) in spite of imprecision. Criterion C23 is indeed satisfied for node $(S_0 S_1 S_2 S_4 S_3 S_0)$, against the 10 other pending nodes at the end of the search. Note that if only C2 had been used as a stopping rule, only 9 nodes would have been expanded. Besides it is clear that using a branching strategy based solely on C2 (respectively: C3) (as well as the stopping rule), yields the regular A* algorithm with edge costs equal to the lower (respectively: upper) bounds of the cost intervals, i.e. we are back to the precise data case. More particularly if the optimal solutions obtained using C2, then C3, correspond to one Hamiltonian circuit (at least), this circuit is optimal for C23 and is obtained by the extended A* algorithm with C23 as a stopping rule. This is what happens here. Note that an optimal solution in the sense of C1 does not exist since the sons of 9 and 16 are not comparable with respect to C1.

## 6. CONCLUSIONS

In this paper, we have discussed the possible use of fuzzy intervals in graph search methods. Not only can these methods be extended to deal with fuzzy data, but the concept of a fuzzy interval might be useful to synthetize the available branching information and guide the search strategy in an attempt to improve upon previous approaches. The types of extension of graph search methods proposed here clearly differ from search procedures based on non-additive evaluation functions as studied by Pearl(1984) or Yager (1986). The preliminary nature of this paper is not hidden: computational experiments with fuzzy intervals should be run in the immediate future in order to verify conjectures regarding efficiency of the fuzzy branching strategy. Both extensions, presented in sections 4 and 5, respectively, do not compete with each other. One may be interested in applying a fuzzy branching strategy to a problem with fuzzy data. In that case intervals containing the optimal solution costs would be intervals with fuzzy bounds. Processing this type of information requires the concept of a twofold fuzzy set, Dubois, Prade (1983, 1985b).

## REFERENCES


Bellman, R,E., L.A. Zadeh (1970). Decision-making in a fuzzy environment. Mang. Sci. 17, B141-B164.

Berliner, H. (1979). The B* tree search algorithm: a best-first proof procedure. Art. Intel. 12, 23-40.

Degani, R., G. Bortolan (1985). A review of some methods for ranking fuzzy subsets. Fuzzy Sets and Syst. 15, 1-19.

Dubois, D., H. Prade (1980). Fuzzy Sets and Systems: Theory and Applications. Academic Press, New-York.

Dubois, D., H. Prade (1983a). Twofold fuzzy sets. Fuzzy Math. (China) 3, 53-76.

Dubois, D., H. Prade (1983b). Ranking fuzzy numbers in the setting of possibility theory. Inf. Sci. 30, 183-224.

Dubois, D., H. Prade (1985a). Theorie des Possibilités: Applications a la Représentation des Connaissances en Informatique, Masson, Paris, 1985.

Dubois, D., H. Prade (1985b). Twofold fuzzy sets and rough sets: some issues in knowledge representation. Fuzzy Sets and Syst. To appear.

Dubois, D., H. Prade (1985c). Fuzzy numbers: An overview. In J.C. Bezdek (ed.), The Analysis of Fuzzy Information. Vol. 1, CRC Press, Boca Raton, Fl.

Esogbue, A.O., R.E. Bellman, (1984). Fuzzy dynamic programming and its extensions. In Zimmermann, Zadeh, and Gaines (1984), 147-169.

Fabian, C., M. Stoica (1984). Fuzzy integer programming. In Zimmermann, Zadeh, and Gaines (1984), 123-132.

Farreny, H., H. Prade (1982). Search methods with imprecise estimates. Proc. 26th Int. Symp. on Gen. Syst. Meth. 442-446.

Geoffrion, A.M., R.E. Marsten (1972). Integer programming algorithms: a framework and state of the art survey. Mang. Sci. 18.

Kacprzyk, J. (1982). Multistage decision-processes in a fuzzy environment: a survey. In M.M. Gupta and E. Sanchez (eds.), Fuzzy Information and Decision Processes, North-Holland, Amsterdam, 251-266.

Moore, R.A. (1966). Interval Analysis. Prentice Hall, Englewood Cliffs, N.J.

Nilsson, N. (1980). Principles of Artificial Intelligence. Tioga Pub. Co., Palo Alto, Ca.

Pearl, J. (1984) Heuristics. Intelligent Search Strategies for Computer Problem Solving. Addison-Wesley, Reading, Ma.

Pearl, J., J.H. Kim (1982). Studies in semi-admissible heuristics. IEEE Trans. Pattern Anal. and Machine Intell. 6, 392-399.

Yager, R.R. (1986). Paths of least resistance in possibilistic production systems. Fuzzy Sets and Syst. 19, 121-132.

Zadeh, L.A. (1965). Fuzzy sets. Inf. and Control 8, 338-353.

Zadeh, L.A. (1978). Fuzzy sets as a basis for a theory of possibility. Fuzzy Sets and Syst. 1, 3-28.

Zimmermann, H.J., M.A. Pollatschek (1984). Fuzzy 0-1 linear programs. In Zimmermann, Zadeh, and Gaines (1984).

Zimmermann, H.J., L.A. Zadeh, B.R. Gaines (eds.)(1984). Fuzzy Sets and Decision Analysis. North-Holland, Amsterdam.

# LINEAR REGRESSION ANALYSIS BY POSSIBILISTIC MODELS

H. Tanaka*, J. Watada** and K. Asai*

* Department of Industrial Engineering
College of Engineering
University of Osaka Prefecture
591 Osaka, Sakai, Mozu-Umemachi 4-8-4, Japan

** Faculty of Business Administration
Ryukoku University
612 Kyoto, Fushimi, Fukakusa, Japan

Abstract. Fuzziness must be considered in systems where human estimation is influential. Since Zadeh has introduced the concept of possibility, fuzziness of system equations has been grasped by possibility distributions. Possibilistic linear systems have been studied as fuzzy arithmetic operations of fuzzy numbers by the extension principle. In this paper, possibilistic linear systems are applied to the linear regression analysis which is called possibilistic linear regression. In the background of usual regression models, deviations values are supposed to be due to observation errors. Here, on the contrary, it is assumed that these deviations depend on the possibility of parameters in systems structure. More specifically, linear systems with parameters of fuzzy numbers are considered as possibilistic linear models. The estimated values obtained from the possibilistic linear model are fuzzy numbers which represent the possibility of the system structure, while the conventional confidence interval is related to the observation errors. This possibilistic linear regression analysis might be useful for finding a fuzzy structure in a fuzzy environment.

Keywords: linear regression analysis, possibilistic linear regression, possibility theory, possibilistic linear models.

## 1. INTRODUCTION

Linear regression analysis is an important and general method to analyze situations in which one observed variable is assumed to be a linear function of other variables. Statistical regression models are constructed in the framework in which the difference between observed data and estimated values comes from the statistical observation error.

Tanaka, Kejima and Asai (1982) have proposed fuzzy regression analysis based on fuzzy set theory. In that fuzzy regression model the difference between observed and estimated values is not considered as a statistical error, but is assumed to result from the fuzziness of a system structure itself. Regarding the deviation between the data as the reflection of fuzziness on system parameters, the fuzzy structure is represented as a linear function with fuzzy parameters, which is called a fuzzy linear function.

In this paper we give a possibilistic interpretation to a fuzzy regression model. The possibilistic concept in fuzzy sets theory has been discussed in many articles since its proposal by Zadeh (1978). In the possibilistic regression model, fuzzy parameters expressed as fuzzy numbers represent possibility distributions of parameters. Furthermore, the possibility of the value of a fuzzy linear function can be calculated from the viewpoint of possibility measure (Zadeh, 1978). The possibilistic linear regression analysis is formulated by a fuzzy linear function as a model of the possibilistic structure of systems. The fuzzy linear function is calculated from possibility distributions of parameters by the extension principle (Dubois and Prade, 1980; Negoita and Ralescu, 1975; Zadeh, 1975) which can also be explained in terms of possibility measure.

A decision problem concerning the number of local public service workers is considered as an application. Generally, the number of the staff is determined through the analysis of the staff's individual duties, but here we have analyzed the number of the staff of local governments in the Osaka Prefecture by a possibilistic linear model. This model tells us whether the number of public service workers of some city is determined on the same basis as that of similar cities. It is understood in this example that real data are interpreted through a fuzzy linear function taken as the model.

## 2. POSSIBILITY DISTRIBUTION AND POSSIBILITY MEASURE

Zadeh (1978) has proposed possibilistic interpretation of a fuzzy set wherein a possibility distribution is defined by a membership function of a fuzzy set.

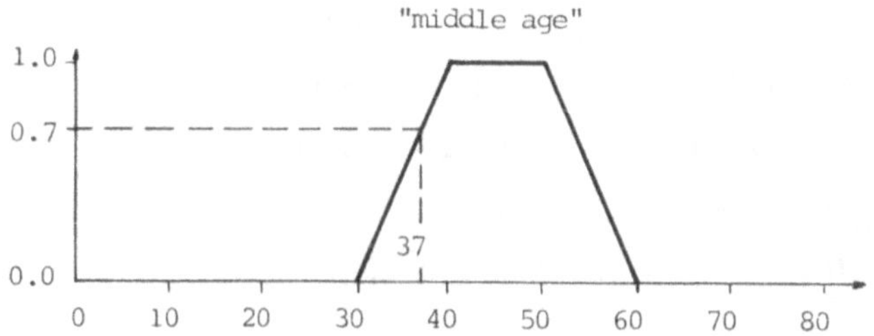


Fig. 1. Fuzzy set "middle age"

Let us consider the proposition "John is middle aged", where "middle age" is defined as a fuzzy set in R characterized by a trapezoidal membership function as shown in Fig. 1. In Fig. 1, consider a numerical age, say 37, whose grade of membership in the fuzzy set "middle age" is 0.7. This grade 0.7 can be interpreted as the degree of possibility that John is 37, given that proposition "John is middle aged".

Let F be a fuzzy subset of a universe X whose membership function is denoted by $\mu_F(x)$. Given a proposition "u is F", the degree of possibility that u=x is denoted by $\Pi(x)$ and is defined to be numerically equal to the membership of x in F, i.e.

$$\Pi(x) = \mu_F(x) \tag{1}$$

where u is a variable on the universe X. Given a possibility distribution $\Pi(x)$, a possibility measure of a crisp set E is defined as

$$\Pi(E) = \sup_{x \in E} \Pi(x) \tag{2}$$

In a similar way, the possibility measure of a fuzzy set A is defined as

$$\Pi(A) = \sup_{x \in X}(\mu_A(x) \wedge \Pi(x)) \tag{3}$$

Figure 2 illustrates a possibility measure.

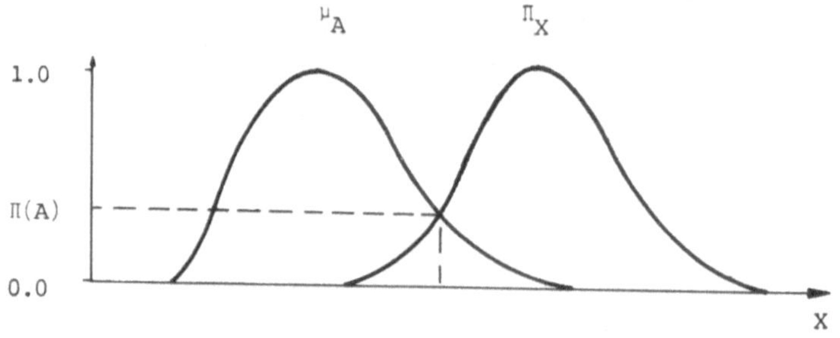


Fig. 2. Possibility measure of a fuzzy set

A possibility measure has the following properties:

(i) $\Pi_X(\emptyset) = 0, \quad \Pi_X(X) = 1$ (4)

(ii) $\Pi_X(A_1 \cup A_2) = \Pi_X(A_1) \vee \Pi_X(A_2)$ (5)

(iii) $A_1 \subset A_2 \Rightarrow \Pi(A_1) \leqslant \Pi(A_2)$ (6)

(iii) derived from (ii) shows that $\Pi_X$ is monotonous. Let us denote a possibility space (Nahmias, 1978; Sugeno, 1983) as $(X, \mathfrak{F}(X), \Pi_X(.))$, where $\mathfrak{F}(X)$ is a set of all fuzzy subsets of $X$. Given two universal sets $X$ and $Y$ and a function $f: X \to Y$, a possibility space $(Y, \mathfrak{F}(Y), \Pi_Y(.))$ can be induced from the given possibility space as follows:

Denoting set $E$ as $E = \{(x \mid y = f(x)\}$, a possibility distribution of $y$ is induced from $\Pi_X(.)$ as

$$\Pi_Y(y) = \Pi_X(E) \tag{7}$$

which can be rewritten by the definition of possibility measure as

$$\Pi_Y(y) = \sup_{\{x \mid y = f(x)\}} \Pi_X(x) \tag{8}$$

Hence, we have for $B \in \mathfrak{F}(y)$

$$\Pi_Y(B) = \sup_y (\mu_B(y) \wedge \Pi_Y(y)) \tag{9}$$

which leads to a possibility space $(Y, \mathfrak{F}(Y), \Pi_Y(.))$. (8) brings about the relations:

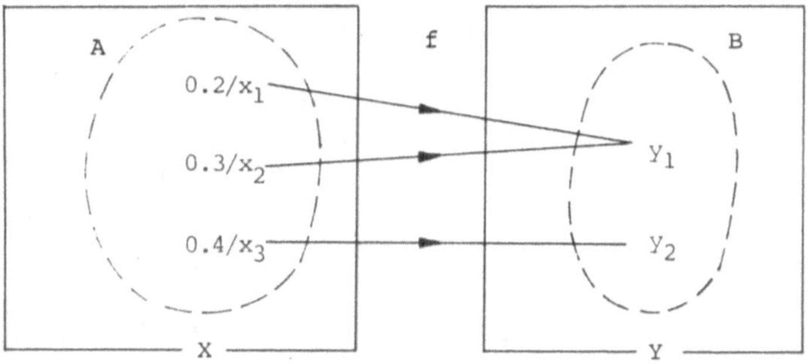


$$A = 0.2/x_1 + 0.3/x_2 + 0.4/x_3 \xrightarrow{f} f(A) = 0.3/y_1 + 0.4/y_2$$

$$f^{-1}(B) = 0.3/x_1 + 0.3/x_2 + 0.4/x_3 \xrightarrow{f} B$$

Fig. 3. Mapping of a fuzzy set

$$f(A) = B : \mu_B(y) = \begin{cases} \sup_{\{x \mid y = f(x)\}} \mu_A(x) & : \{x \mid y = f(x)\} \neq \emptyset \\ 0 & : \text{otherwise} \end{cases} \tag{10}$$

$$f^{-1}(B) = A : \mu_A(x) = \mu_B(f(x)) \tag{11}$$

Figure 3 explains briefly the above-mentioned relations.

Relating to the function $f: \mathfrak{X}(X) \to \mathfrak{X}(Y)$, we have the properties:

(i) Given a fuzzy set A in X, for the fuzzy set B in Y such that $f(A) = B$, we have $\Pi_X(A) \leqslant \Pi_Y(B)$.

(ii) Given a fuzzy set B in Y, for the fuzzy set A in X such that $f^{-1}(B)=A$, we have $\Pi_X(A) = \Pi_Y(B)$.

These are proved as follows:

(i)
$$\begin{aligned}
\Pi_Y(B) &= \sup_y \left\{ \mu_B(y) \wedge \Pi_Y(y) \right\} \\
&= \sup_y \left\{ \mu_{f(A)}(y) \wedge \Pi_Y(y) \right\} \\
&= \sup_y \left\{ \sup_{y=f(x)} \mu_A(x) \wedge \sup_{y=f(x)} \Pi_X(x) \right\} \\
&\geqslant \sup_y \sup_{y=f(x)} \left\{ \mu_A(x) \wedge \Pi_X(x) \right\} \\
&= \sup_x \mu_A(x) \wedge \Pi_X(x) \\
&= \Pi_X(A)
\end{aligned}$$

(ii)
$$\begin{aligned}
\Pi_Y(B) &= \sup_y \left\{ \mu_B(y) \wedge \Pi_Y(x) \right\} \\
&= \sup_y \left\{ \mu_B(y) \wedge \sup_{y=f(x)} \Pi_X(x) \right\} \\
&= \sup_x \left\{ \mu_A(f(x) \wedge \Pi_X(x) \right\} \\
&= \sup \left\{ \mu_{f^{-1}(A)}(x) \wedge \Pi_X(x) \right\} = \Pi_X(A)
\end{aligned}$$

A fuzzy function whose parameters are fuzzy numbers is denoted by $f(y,A)$. By the above-mentioned relation (10), the fuzzy number of $Y=f(y,A)$ can be calculated as

$$\mu_Y(y) = \sup_{y=f(x,a)} \mu_A(a) \qquad (12)$$

where $\mu_Y(y)=0$ for the case of $\left\{ a \mid y=f(x,a) \right\}=\emptyset$.

## 3. EXTENSION PRINCIPLE

The extension principle introduced by Zadeh(1975) plays a central role in dealing with a possibilistic model. It is a method for extending nonfuzzy mathematical concepts to fuzzy quantities. Let us explain the extension principle for handling possibilistic linear functions.

<u>Definition 1</u>. Let  f  be a mapping from the Cartesian product of universes  $X_1$  and  $X_2$, $X_1 \times X_2$  to a universe  Y  such that  $y = f(x_1, x_2)$. The fuzzy function of fuzzy numbers  $A_1 \in X_1$  and  $A_2 \in X_2$  is defined as

$$Y = f(A_1, A_2) \; ; \; \mu_Y(y) = \sup_{f(a_1, a_2) = y} (\mu_{A_1}(a_1) \wedge \mu_{A_2}(a_2)) \quad (13)$$

In the case that f is a binary operation *, we have

$$Y = A_1 * A_2 \quad ; \; \mu_Y(y) = \sup_{a_1 * a_2 = y} (\mu_{A_1}(a_1) \wedge \mu_{A_2}(a_2)) \quad (14)$$

A function whose parameters are fuzzy numbers  $A_j$  is defined as

$$Y = f(x, A_1, \ldots, A_n) ; \; \mu_Y(y) = \sup_{y = f(x, a)} (\mu_{A_1}(a_1) \wedge \ldots \wedge \mu_{A_n}(a_n)) \quad (15)$$

Now, let us define the form of a symmetric fuzzy number considered in this paper.

<u>Definition 2</u>. A symmetric fuzzy number  $A_j$  is denoted by  $(\alpha_j, c_j)$  and defined as

$$A_j = (\alpha_j, c_j)_L : \mu_{A_j}(a_j) = L(\frac{a_j - \alpha_j}{c_j}) \quad (16)$$

where  $L(x)$  is a reference function satisfying:

(i)      $L(x) = L(-x)$

(ii)     $L(0) = 1$

(iii)    L is a monotone function which strictly decreases in $[0, +\infty)$.

Example 1. Let us give  some examples of the reference function:

$$L_1(x) = \max(0, 1 - |x|^P)$$
$$L_2(x) = e^{-|x|^P}$$
$$L_3(x) = 1/(1 + |x|^P)$$

where  p > 0. In the case when p=1, $L_1(x) = 1 - |x|$, the membership function of fuzzy parameter  $A_j$  is written as

$$\mu_{A_j}(a_j) = L(|a_j - \alpha_j|/c_j)$$
$$= 1 - |a_j - \alpha_j|/c_j \quad (17)$$

Fuzzy parameters are defined by such fuzzy sets as illustrated in Fig. 4. In Fig. 4, the fuzzy parameter $A_j$ is

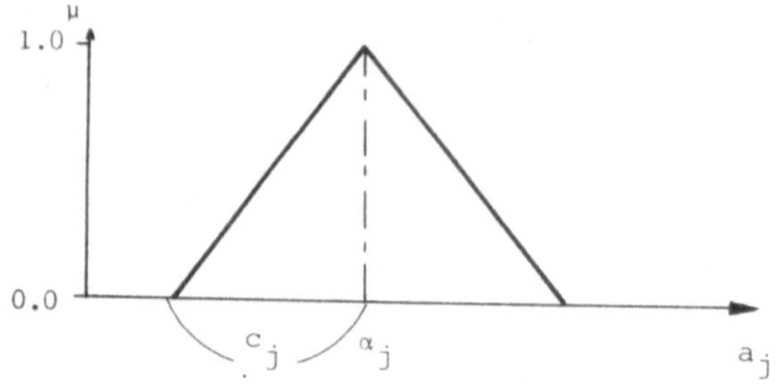


Fig. 4. Fuzzy set of parameter $A_j$ : $A_j \subset$ "approximately $\alpha_j$"

denoted as $(\alpha_j, c_j)$; where $\alpha_j$ and $c_j$ stand for the center and width of fuzzy set $A_j$, respectively.

Theorem 1. Given a fuzzy parameter $A = (\alpha, c)$, $j = 1, \ldots, n$, the fuzzy linear function

$$Y = A_1 x_1 + \ldots + A_n x_n = Ax \tag{18}$$

is obtained as the following membership function (Tanaka, Uejima and Asai, 1982)

$$u(y) = \begin{cases} ((y - x\alpha)/c'|x|)_L & : x \neq 0 \\ 1 & : x = 0,\ y = 0 \\ 0 & : x = 0,\ y \neq 0 \end{cases} \tag{19}$$

where $|x| = (|x_1|, \ldots, |x_n|)$ and $\mu_Y(y) = 0$, when $c'|x| < |y - x'|$, and $c'$, $x'$ denote the transpose.

This type of fuzzy linear function can be symbolically rewritten as

$$(\alpha_1, c_1)_L x_1 + \ldots + (\alpha_n, c_n)_L x_n = (\alpha' x, c'|x|)_L$$

In contrast to the above, the linear function whose parameters are Gaussian random variables can be symbolically written as

$$N(\mu_1, \sigma_1^2) x_1 + \ldots + N(\mu_n, \sigma_n^2) x_n = N(\mu' x, (\sigma^2) x) \tag{20}$$

where $\sigma^2 = (\sigma_1^2, \ldots, \sigma_n^2)$.

The above calculations are similar in form, but their meanings are different as reflected by the terms "Possibility" and "Probability", respectively. Hence, the fuzzy output Y is calculated in view of the possibility of parameter A.

## 3. POSSIBILISTIC LINEAR REGRESSION MODEL

In ordinary linear regression models, data such as shown in Table 1 are given and the coefficients of a linear function are determined so as to coincide well with the data. Let a linear regression model be $y=a'x$. In this case the difference between the estimated values and the data

$$y_i - a'x_i = e_i \;;\; i=1,\dots,N \tag{21}$$

is considered to be the observation error, where $x_i=(x_{i1},\dots,x_{in})$. In the possibilistic linear regression model of this paper, deviation of data from a linear model is supposed to be due to the possibility of the parameters of the linear function. The problem here is to choose a fuzzy linear function

$$Y = A_1x_1 + \dots + A_nx_n \tag{22}$$

as a model and to determine the fuzzy parameters A so that the output $y_i$ may be contained in the estimated fuzzy set $Y_i = A\,x_i$ to more than to a certain degree.

We will obtain the fuzzy number A of a parameter vector in a linear model which fits well with the ordinary data as given in Table 1. The following are supposed to obtain a possibilistic linear regression model:

Table 1. Input-output data

| Sample Number | Output $y_i$ | Input $x_{i1}, x_{i2}, \dots, x_{in}$ |
|---|---|---|
| 1 | $y_1$ | $x_{11}, x_{12}, \dots, x_{1n}$ |
| 2 | $y_2$ | $x_{21}, x_{22}, \dots, x_{2n}$ |
| ⋮ | ⋮ | ⋮ |
| N | $y_N$ | $x_{N1}, x_{N2}, \dots, x_{Nn}$ |

(ii) The type of fuzzy parameter $A_j$ denoted as $(\alpha_j, c_j)$ is given by $L_1(x)$ with p=1.

(iii) Given input-output relations $(x_i, y_i)$ i=1,...,N and a threshold h, it must hold that

$$\mu_{Y_i^*}(y_i) \geq h \qquad i = 1,\dots,N \tag{23}$$

This means that the estimated fuzzy set $Y_i^*$ contains the observed values $y_i$ at least to the degree of more than h.

(iv) The index of fuzziness of the possibilistic linear model is

$$s = k_1c_1 + \ldots + k_nc_n \tag{24}$$

where $k_j$ is a weight parameter.

Given the condition that all the data are contained in the possibilistic interval with more than h degree, (i)-(iii) and (19) lead to the following inequality

$$|y_i - x' \,| \leqslant (1-h) \sum_{j=1}^{n} c_j \,|x_{ij}| \qquad i=1,\ldots,N \tag{25}$$

The possibilistic linear regression model is obtained by fuzzy parameters A = (α, C) that minimize s in (iv) subject to constraints (25). Given a threshold h, this problem leads to the three linear programming problems:

$$\min_{\alpha,c}\left\{s = k_1c_1 + \ldots + k_nc_n\right\} \tag{26}$$

subject to:

$$\begin{aligned} &(1-h) \sum_{j=1}^{n} c_j |x_{ij}| + x_i'\alpha \geqslant y_i \\ & \qquad\qquad\qquad\qquad\qquad\qquad i=1,\ldots,n \\ &(1-h) \sum_{j=1}^{n} c_j \,|x_{ij}| - x\,\alpha \geqslant y_i \end{aligned} \tag{27}$$

The possibilistic parameters $A_j = (\alpha_j, c_j)$, $j=1,\ldots,n$, can easily be obtained by solving the above linear programming problem. As for threshold h, it should be noted that if sufficient sample numbers are given, we can set h=0 in terms of possibility. In this case, only the possibility given by the data is considered; if the given sample number is half as compared to the ideal one, it is suggested from theorem 5 that h=0.5. This means that the Y is obtained at h=0.

Another type of a possibilistic regression model (Tanaka, Shimomura, Watada and Asai, 1984) has been formulated by minimizing the maximum fuzziness among all parameters.

## 4. APPLICATION

Let us consider a problem of determining the number of the staff of self-governing bodies in the Osaka Prefecture by means of possibilistic linear regression analysis. Thirteen cities which are considered to be small or middle sized were chosen from the self-governing bodies in the Osaka Prefecture, as shown in Table 2 (see Annual Reports..., 1979a, 1979b). The sample numbers 1-13 in Table 2 are used for constructing a possibilistic linear regression model, and cities 14-24 are taken as new samples. As explanation variables, total revenue

Table 2. The data analyzed by the possibilistic linear regression model

| | City | Number of Staff | Total Revenue | Population | State of Utilization of Computer | State of Commition to Outside |
|---|---|---|---|---|---|---|
| Sample for Modeling | 1:Izumi | 1581 | 19550392 | 121729 | 13 | 5 |
| | 2:Izumiotsu | 863 | 104440666 | 68148 | 9 | 5 |
| | 3:Kaizuka | 913 | 10224683 | 80972 | 18 | 4 |
| | 4:Kashihara | 642 | 8809703 | 67349 | 18 | 6 |
| | 5:Katano | 537 | 7584870 | 59213 | 10 | 6 |
| | 6:Kadoma | 1407 | 17849205 | 142226 | 12 | 7 |
| | 7:Kawachina-gano | 536 | 10849969 | 74398 | 15 | 7 |
| | 8:Kishiwada | 1914 | 26567217 | 176575 | 8 | 7 |
| | 9:Shijonawate | 490 | 8321935 | 52289 | 8 | 2 |
| | 10:Takaishi | 607 | 9521894 | 66917 | 21 | 6 |
| | 11:Habikino | 723 | 11976781 | 101538 | 12 | 4 |
| | 12:Fujiidera | 535 | 8226626 | 62614 | 9 | 7 |
| | 13:Matsubara | 1114 | 17003076 | 135929 | 9 | 6 |
| New Sample | 14:Ikeda | 1347 | 16613856 | 102115 | 8 | 1 |
| | 15:Izumisano | 1134 | 11650192 | 89317 | 7 | 4 |
| | 16:Ibaraki | 1802 | 31837478 | 224155 | 23 | 8 |
| | 17:Setsu | 778 | 11554942 | 80780 | 16 | 10 |
| | 18:Sennan | 473 | 8168829 | 50682 | 8 | 11 |
| | 19:Daito | 1245 | 23442218 | 114364 | 24 | 2 |
| | 20:Tondabayashi | 710 | 16552059 | 95711 | 8 | 7 |
| | 21:Neyagawa | 2375 | 35010176 | 259364 | 9 | 9 |
| | 22:Mino | 963 | 22354793 | 91932 | 13 | 4 |
| | 23:Moriguchi | 1985 | 24375945 | 168125 | 20 | 3 |
| | 24:Yao | 2516 | 38416142 | 271491 | 10 | 0 |

$x_1$, population $x_2$, state of utilization of computers $x_3$, and state of commission to the outside $x_4$ are taken. Thereby a possibilistic linear model which explains the real number of the staff $y$ is obtained. Here, the state of utilization means the number of computerized duties from 51 duties performed by the city, such as personal management, management of various sorts of city rates and national pensions, management of water supply and drainage. The state of commission to the outside means the number of duties committed to outside traders from 18 duties, such as cleaning, levy of rental fee for public housing and management of public facilities. The data in Table 2 are standardized in obtaining a possibilistic linear regression model in practice.

In a fuzzy parameter $A_j = (\alpha_j, c_j)$ for each explanation variable $x_j$, $\alpha_j \geqslant 0$ is taken if $y$ increases as the value of $x_j$ increases, and $\alpha_j \leqslant 0$ is taken if $y$ decreases as the value of $x_j$ increases. That is, $\alpha_j \geqslant 0$ or $\alpha_j \leqslant 0$ is added to the constraint condition (23). In this example, as an increase in the values of $x_1$ and $x_2$ causes $y$ to increase, and an increase in the values $x_3$ and $x_4$ causes $y$ to decrease, the following constraint is added

$$\alpha_1 \geqslant 0, \quad \alpha_2 \geqslant 0, \quad \alpha_3 \leqslant 0, \quad \alpha_4 \leqslant 0$$

In conventional regression models, if some explanation variables are correlated with other variables, the sign of the obtained parameter cannot be explained well. On the contrary, in the possibilistic model, when independency among explanation variables is broken, $\alpha_j$ for a certain variable becomes zero. This means that the obtained parameter $A_j$ of a possibilistic linear model reflects the input-output relation in the above sense.

Let us denote the possibilistic linear regression model as

$$Y = A_o + A_1x_1 + \ldots + A_4x_4$$

The optimal fuzzy parameters obtained from samples 1-13

Table 3. The optimal fuzzy parameters

| Constant $A_0$ | Total Revenue $A_1$ | Population $A_2$ | State of Utilization of Computer $A_3$ | State of Commission to the Outside $A_4$ |
|---|---|---|---|---|
| (903.199) | (466, 0) | (0. 0) | (0. 0) | (-51. 72) |

are shown in Table 3, where $h = 0.4$ and $k_o = k_1 = \ldots = k_4 = 1$ in (24). Since there is no knowledge about the weights, identical ones are chosen.

Lack of conformity in the data of the 13 cities used for modeling is represented by fuzzy parameters. Therefore, in this model, dispersion of data is represented by giving fuzziness to the parameters corresponding to the constant and to the number of commissions to the outside. Figure 5 shows possibilistic estimates of cities only in odd numbers, using optimal parameters. In Figure 5 a possibilistic estimate of the i-th city is denoted as $Y_i$ where $i=1,3,\ldots,13$, and also a symbol shows the real number of the staff $y_i$ of the i-th city and the degree to which $y_i$ is contained in its possibilistic estimate $Y_i^*$. This degree, which is called the grade of fitness of $y_i$ to $Y_i^*$, is shown in Table 4. Since the data 1-13 are used

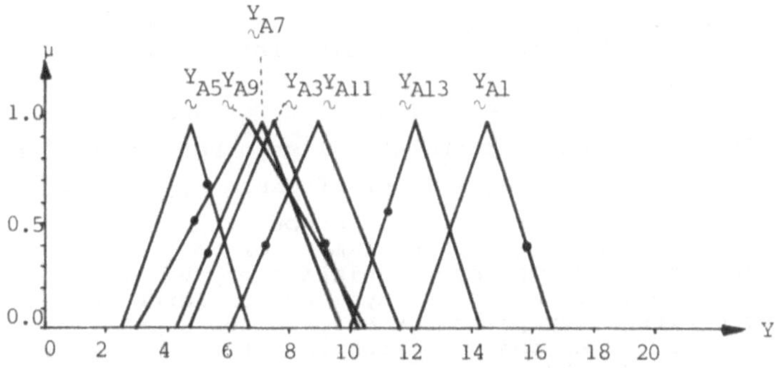


Fig. 5. The estimated fuzzy sets Y

for modeling with $h = 0.4$, it is assured for $i=1,\ldots,13$ that $\mu_{Y_i^*}(y_i) > 0.4$. From the grade of fitness of the samples in Table 4, cities which are close to the model of the training group can be found. If the city with a higher grade of fitness than 0.4 is considered to be the same as that of the modeling group, cities 14, 17, 18, 21 and 23 belong to the modeling group and cities 15, 16, 19, 20, 22 and 24 do not belong to this group. As for the cities not belonging to the model group, the number of the staff of city 15 is rather large, and that of the cities 19, 20 and 22 is extremely small. These cities are considered to have another structure than that of the modeling group.

Table 4. The grade of fitness of $y_i$ to $Y_i^*$, $(\mu_{Y_i^*}(y_i))$

| Sample for Modeling | | New Sample | |
|---|---|---|---|
| City | Y | City | Y |
| 1:Izumi | 0.40 | 14:Ikeda | 0.99 |
| 2:Izumiotsu | 0.42 | 15:Izumisano | 0.00 |
| 3:Kaizuka | 0.40 | 16:Ibaraki | * |
| 4:Kashihara | 0.68 | 17:Setsu | 0.69 |
| 5:Ktano | 0.72 | 18:Sennan | 0.74 |
| 6:Kadoma | 0.40 | 19:Daito | * |
| 7:Kawachinagano | 0.40 | 20:Tondabayashi | * |
| 8:Kishiwada | 0.95 | 21:Neyagawa | 0.60 |
| 9:Shijonawate | 0.52 | 22:Mino | * |
| 10:Takaishi | 0.91 | 23:Moriguchi | 0.71 |
| 11:Habikino | 0.40 | 24:Yao | * |
| 12:Fujiidera | 0.84 | | |
| 13:Matsubara | 0.55 | | |

From the results of the application mentioned above the following can be asserted about possibilistic linear regression models.

When the center of the optimal fuzzy parameter $\alpha_1$ is 0, the i-th item is strongly correlated with other items. Since we have $A_2^* = (0.0)$ and $A_3^* = (0.0)$ of Table 3, the population and the state of the utilization of computers can be explained through the total revenue and the number of commissions to the outside, respectively. In the formulation of this possibilistic linear regression, if there are any items of large correlation, the center of its fuzzy parameter becomes 0 automatically. Therefore, there is no contradiction in interpreting a fuzzy linear function $Y = Ax$ (see Table 3).

In this model, the whole fuzziness of the parameters is defined additively, and the optimal fuzzy parameter which minimizes this fuzziness occurs only in the items of $x_2$ and $x_3$.

Since an estimate of a new sample by a possibilistic model is obtained as a fuzzy set, it can be interpreted as the possibility of an estimate. The maximum possibility of the estimate $Y_i^* = A^* x_i$ of input data $x_i$ is the closure of $\{ y \mid \mu_{Y_i^*}(y) > 0 \}$ which can be considered to be an interval estimation (see Fig. 5).

As described above, a possibilistic linear regression model is constructed from a new point of view, and is convenient for a model that analyzes and explains possibilistic phenomena in which human recognition is involved.

## 5. CONCLUDING REMARKS

Possibilistic linear regression models are formulated and applied to a model of the number of the staff of local self-governing bodies. The emphasis in this paper is placed on the possibility by which the fuzziness of data is interpreted in terms of possibility measure. Thus, our linear regression analysis is based on a possibilistic model. This formulation is a new trial, and will be effective for modeling vague phenomena in management systems and social systems as shown in this paper.

## REFERENCES


Dubois, D., and H. Prade (1980). Fuzzy Sets and Systems: Theory and Application. Academic Press, New York.

Nahmias, S. (1978). Fuzzy variables. Fuzzy Sets and Syst. 1, 97-111.

Negoita, C.V., and D.A. Ralescu (1975). Application of Fuzzy Sets to Systems Analysis. Birkhauser Verlag, Basel.

Sugeno, M. (1983). Fuzzy theory IV (lecture note). (in Japanese), J. SICE 22, 554-559.

Tanaka, H., T. Shimomura, J. Watada and K. Asai (1984). Fuzzy linear regeression analysis of the number of state in local government. FIP-84 at Kauai, Hawaii.

Tanaka, H., S. Uejima and K. Asai (1982). Linear regression analysis with fuzzy model. IEEE Trans. on SMC 12, 903-907.

The Annual Report of Statistic Data in Osaka Prefecture.(1979a). Department of Statistic in Osaka Prefecture.

The Annual Report of Utilization of Computers in Osaka Prefecture. (1979b). Department of Information in Osaka Prefecture.

Watada, J., H. Tanaka, and K. Asai (1984). Fuzzy quantification theory type I. The Jap. J. Behaviormetr. 11.

Zadeh, L.A. (1975). The concept of a linguistic variable and its application to approximate reasoning - Part I. Inf. Sci. 8, 199-249.

Zadeh. L.A. (1978). Fuzzy sets as a basis for a theory of possibility. Int. J. Fuzzy Sets and Syst. 1, 3-28.

## II.3. Fuzzy Multicriteria Decision Making, Optimization, and Mathematical Programming: Analysis, Solution Procedures, and Interactive Approaches

# A FUZZY MULTICRITERIA DECISION MAKING MODEL

Vladimir E. Zhukovin

Institute of Cybernetics
Academy of Sciences of the Georgian SSR
S. Euli 5, Tbilisi, Georgian SSR, USSR

Abstract. A multicriteria decision making model with a collection of fuzzy preference relations is presented. A concept of the nonfuzzy set of nondominated alternatives is formulated for this case. It is analogous to the Pareto set in the multicriteria decision making theory. The possibility of representation of that collection by one fuzzy preference relation, called the convolution, is studied. An interconnection of the multicriteria and fuzzy decision making models is examined.

Keywords: fuzzy preference relation, multicriteria decision making, nondominated alternatives, Pareto set, effective convolution.

## 1. INTRODUCTION

With the growing number of applications of fuzzy sets in decision making, one is confronted with the multicriteria decision making problem. Decision making is the most important and popular aspect of application of mathematical methods in various fields of human activity and many spheres of research include it. Two of them are important for this paper. These are multicriteria decision making and decision making on the basis of fuzzy information. The terms "multicriteria" and "fuzzy" are now being used often in many decision making publications. But the theory of fuzzy multicriteria decision making is not developed yet. Usually the decision making problem is formulated as a fuzzy multicriteria one. But in the first stage of research, a collection of fuzzy preference relations is substituted for one fuzzy preference relation by a certain method, and then a scalar fuzzy decision making problem is studied. This method is often intuitive and its effectiveness cannot be estimated since the Pareto set for fuzzy decision making is not determined. In this paper an essential interconnection of the multicriteria and fuzzy decision making models will be examined. We shall combine them and develop a fuzzy multicriteria decision making theory on this basis.

First we present some definitions and results which shall be needed in the sequel. The decision making situation is formulated as the pair <X,R>, where $X$ is a set of competitive alternatives and $R$ is a binary preference relation defined in $X$. For any preference relation $R$, the relations $R^{-1}$, $R^e = R \cap R^{-1}$ and $R^s = R/R^{-1}$ can be determined. Also we maximal (effective) alternatives set, which we call the "Pareto set" and denote by $X_\Pi(R)$, corresponds to any preference relation $R$ defined in $X$. It is the core of the relation $R$.

Definition 1.1. For the subset $X_o$ of the set $X$, the property "external stability" holds if for any $x \in X - X_o$ there exists a $y \in X_o$ such that $(y,x) \in R^s$.

Theorem 1.1. If the set $X$ is finite and $R$ is transitive, then the Pareto set $X\ (R) \neq \emptyset$ and the property "external stability" holds for it.

The proof of this theorem is not presented here since it is known.

Theorem 1.2. Given is a finite set $X$ of the alternatives with two preference relations, $R_1$ and $R_2$, defined in this set. If $R_1^s \subseteq R_2^s$, then $X\ (R_2) \subseteq X\ (R_1)$.

Proof. Let us examine the sets $A_1 = X - X\ (R_1)$ and $A_2 = X - X_\Pi(R_2)$. Let $x$ be in $A_1$. Then there exists an alternative $y \in X$ such that $(y,x) \in R_1^s$ and hence $(y,x) \in R_2^s$. This means that $x \in A_2$. Thus $A_1 \subseteq A_2$ and hence $X\ (R_2) \subseteq X\ (R_1)$. Q.E.D.

If a preference relation can be presented as $R = \{R_1, R_2, \ldots, R_m\}$, where any $R_j$, $j=\overline{1,m}$, is a binary preference relation defined in $X$, then we have a multicriteria decision making situation. In this case we denote $R$ as the Vector Preference Relation (VPR). For it, the Pareto set and Pareto Domination Relation (PDR) are introduced as $X_\Pi(R_k) = X_\Pi^k$ and $R_k = \bigcap_{j=1}^{m} R_j$, respectively.

Theorem 1.3. For a given VPR, if all $R_j$, $j=\overline{1,m}$, are linear (connected) quasi-orders, then $R_k$ is a quasi-order which is not necessarily linear.

The proof of this theorem is known and is not presented here.

For a given vector criterion $K(x) = \{K_1(x), K_2(x), \ldots, K_n(x)\}$, where $K_j(x)$, $j=1,m$, is a scalar function, defined in $X$, the preference relation $R_j$ can be presented as $R_j = \{(x,y) : K_j(x) \geqslant K_j(y)\}$, $(x,y) \in E$ and $E = X \times X$.

Definition 1.2. Any binary relation $\bar{R} = f(R_1, R_2, \ldots, R_n)$ is a convolution of the initial VPR if $\bar{R} \subseteq \bigcup_{j=1}^{m} R$ and $X_\Pi(\bar{R}) \neq \emptyset$.

Definition 1.3. A convolution is effective if $R_k^s \subseteq \bar{R}^s$ when $X_\Pi^k \neq \emptyset$.

The last fact is important because the decision making model with the empty Pareto set can be studied. But we do not consider this case in this paper. From Theorem 1.2 we conclude that $X_\Pi(\bar{R}) \subseteq X_\Pi^k$ for an effective convolution R.

## 2. FORMULATION OF THE PROBLEM

A fuzzy preference relation (FPR) has been introduced in Zadeh (1965) and is denoted as $P = [E, \mu(x,y)]$, where $(x,y) \in E = X \times X$ and $\mu(x,y)$ is the membership function (MF) of the corresponding FPR. Let $\mu(x,y) \in [0,1]$. A fuzzy decision making situation is formulated as the pair $\langle X,P \rangle$. If the FPR can be presented as $P = \{P_1, P_2, \ldots, P_m\}$, where $P_j$, $j=\overline{1,m}$ is a partial FPR, defined in X, then we have a fuzzy multicriteria decision making problem. In this case we call P the Vector Fuzzy Preference Relation (VFPR). The principles of effective choice must be studied for this problem. The methods and results of the multicriteria decision making theory which have been developed well enough will be taken as a point of departure for this study.

## 3. SOME NECESSARY INFORMATION ON FPR

Some results on FPR are presented in this section. We need the following properties of it: reflexivity, symmetry, asymmetry, transitivity and connectivity. If an FPR is given, we can determine the corresponding binary fuzzy relations: $P^{-1} = [E, \mu^{-1}(x,y)]$, $P^e = [E, \mu^e(x,y)]$ and $P^s = [E, \mu^s(x,y)]$, where $\mu^{-1}(x,y) = \mu(y,x)$, $\mu^e(x,y) = \min[\mu(x,y), \mu^{-1}(x,y)]$ and

$$(x,y) = \begin{cases} \Delta(x,y) = \mu(x,y) - \mu(y,x), & \text{if } \Delta(x,y) > 0, \\ 0, & \text{if } \Delta(x,y) \leqslant 0. \end{cases} \tag{3.1}$$

On the basis of the last binary relation we shall present the nonfuzzy set of nondominated alternatives which has been introduced by Orlovski (1978).

The membership function of the fuzzy set of nondominated alternatives (decisions) is

$$\mu^{ND}(x) = 1 - \max_{y \in X} \mu^s(y,x) \tag{3.2}$$

and then

$$X^{UND}(\mu) = \{ x : u^{ND}(x) = 1 \} \tag{3.3}$$

is the nonfuzzy set of nondominated alternatives which is analogous to the Pareto set in the multicriteria decision making problem for the case of one FPR.

Definition 3.1. $x_o \in X$ is the maximal alternative corresponding to the FPR $P = [E, \mu(x,y)]$ if there does not exist

$y \in X$ such that $(y,x_o) \in P^s$ (this means that $\mu^s(y,x_o) = 0$ for all $y \in X$).

The minimal, best and worst alternatives in X corresponding to an FPR can be defined analogously, but we do not present them here. The set of all maximal alternatives, corresponding to a FPR P, is the core of the FPR or the Pareto set in the multicriteria decision making terminology and denoted as $X_\Pi(P)$.

Theorem 3.1. $X^{UND}(\mu) = X_\Pi(P)$.

Proof. Let x be in the set $X_\Pi(P)$. Due to Definition 3.1, this means that $\mu^s(y,x) = 0$ for all $y \in X$, including x. Thus $\mu^{ND}(x) = 1$ and hence $x \in X^{UND}(\mu)$. Now let x be in the set $X^{UND}(\mu)$. This means that $\mu^{ND}(x) = 1$ and hence $\mu^s(y,x) = 0$ for all $y \in X$, including x. Thus $x \in X\ (P)$. Q.E.D.

It has been proved by Orlovski (1978) that the set $X^{UND}(\mu)$ is not empty if set X is finite. We add that the property of "external stability" (Definition 1.1) holds for it too if the FPR P is transitive. This fact can be formulated and proved as a theorem analogous to Theorem 1.1 but using fuzzy sets terminology.

If $X^{UND}(\mu) \neq \emptyset$, any effective choice procedure or rule must produce an alternative (or some equivalent alternatives) from this set.

Theorem 3.2. (analogous to Theorem 1.2). Given is a finite set X of competitive alternatives with two FPR, $P_1$ and $P_2$, defined in this set, such that $P_1^s \subseteq P_2^s$. Then $X^{UND}(\mu_2) \subseteq X^{UND}(\mu_1)$.

Proof. Let us introduce two sets: $A_1 = X/X^{UND}(\mu_1)$ and $A_2 = X - X^{UND}(\mu_2)$, where $\mu_1(x,y)$ and $\mu_2(x,y)$ are the membership functions corresponding to the FPR's $P_1$ and $P_2$. Let x be in set $A_1$. Then an alternative $y \in X$ exists such that $\mu_1^s(y,x) > 0$. Because $P_1^s \subseteq P_2^s$, this means that $\mu_1^s(x,y) \leqslant \mu_2^s(x,y)$ for any pair $(x,y) \in E$, the inequality $\mu_2^s(y,x) > 0$ holds, and hence $x \in A_2$ which means that $A_1 \subseteq A_2$ and hence $X^{UND}(\mu_2) \subseteq X^{UND}(\mu_1)$. Q.E.D.

We want to extend the concept of the nonfuzzy set of dominated alternatives for the VFPR case. Unfortunately, formula (3.3) cannot be used with this aim in view. And now we present some results which we need later on for this problem.

Definition 3.2. Two preference relations, crisp - R and a fuzzy - P, are consentient if the following conditions hold:

$$(x,y) \in R^s \iff \mu^s(x,y) > 0, \qquad (3.4)$$
$$(x,y) \notin R^s \iff \mu^s(x,y) = 0.$$

Theorem 3.3. The following equality is implemented for the consentient crisp and fuzzy preference relations, R and P:

$$X_{\Pi}(R) = X\ (P) = X^{UND}(\mu).$$

Proof. Let $x$ be in the set $X_{\Pi}(R)$. This means that there does not exist a y from X such that $(y,x) \in R^S$. Thus $(y,x) \bar{\in} R^S$ for all $y \in X$, including $x$. By Definition 3.2, $\mu^S(y,x) = 0$ for all $y \in X$ and hence $\mu^{ND}(x) = 1$ and $x \in X^{UND}(\ )$. Now let $x$ be in the set $X^{UND}(\mu)$. This means that $\mu^{ND}(x) = 1$ and $\mu^S(y,x) = 0$ for all $y$ from $X$, including $x$. By Definition 3.2 this means that $(y,x) \bar{\in} R^S$ for all $y$ from $X$ and hence $x \in X_{\Pi}(R)$. Q.E.D.

Definition 3.3. Let us introduce a crisp preference relation $F$ corresponding to FPR as follows

$$F = \left\{ (x,y)\ :\ \Delta(x,y) \geqslant 0 \right\}.$$

For it we can form the relations $F^{-1}$, $F^e$, $F^s$ and the Pareto set $X_{\Pi}(F)$.

Theorem 3.4. The preference relation $F$ is consentient with the initial FPR and hence $X_{\Pi}(F) = X^{UND}(\mu)$.

## 4. FUZZY MULTICRITERIA DECISION MAKING

In this case the choice situation is presented as the pair <X,P>, where $P$ is VFPR or the collection of the FPRs. The crisp relation $F_j$ corresponds to the fuzzy relation $P_j$, $j=\overline{1,m}$, (see Definition 3.3), with the membership function $\mu_j(x,y)$. Let us introduce the Pareto domination relation for $P$ as $F_P = \bigcap_{j=1}^{m} F_j$, and the Pareto set as $X_{\Pi}(F_P)$.

Definition 4.1. The Pareto set $X\ (F_P)$ is the set of nonfuzzy nondominated alternatives for the VFPR case. We denote it as $X_P^{UND}$, so that $X_P^{UND} = X_{\Pi}(F_P)$.

The introduced set is analogous to the Pareto set $X^k$ in the multicriteria decision making problem and will be used for the estimation of the effectiveness of choice procedures and rules.

Let all the components $P_j$, $j=\overline{1,m}$, of the VFPR be transitive. By Definition 3.3 and Theorems 3.3 and 3.4, the corresponding to them crisp preference relations $F_j$ are also transitive, and hence the relation $F_P$ is a quasi-order by Theorem 1.3. Since the set $X$ is finite, we can conclude on the basis of Theorem 1.1 that $X_P^{UND}$ is not empty and the property of "external stability" holds for it.

Definition 4.2. The choice procedure or rule is effective if it produces an alternative (or some equivalent alternatives) from the set $X_P^{UND}$.

Thus, for the one FPR case the choice rule based on formula (3.3) is effective. The proof of this result is not presented here.

Now we shall study the problem of representation of the VFPR by one FPR. Usually the set $X_P^{UND}$ contains many alternatives and is too large for the decision maker (DM). Then DM uses some representation of the VFPR by one FPR or the convolution of the VFPR for his choice.

Definition 4.3. The FPR $\bar{P} = [E, \bar{\mu}(x,y)]$, where the membership function is determined as $\bar{\mu}(x,y) = f[\mu_1(x,y), \mu_2(x,y), \ldots, \mu_m(x,y)]$ will be called the convolution of the VFPR if $X^{UND}(\bar{\mu}) \neq \emptyset$.

Let us present some examples of convolutions:

1. $$\mu_V(x,y) = \min_{j=\overline{1,m}} \mu_j(x,y),$$
$$\mu_Z(x,y) = \max_{j=\overline{1,m}} \mu_j(x,y); \tag{4.1}$$

2. $$\mu_L(x,y) = \sum_{j=1}^{m} \lambda_j \mu_j(x,y), \tag{4.2}$$

where $\lambda = \{\lambda_1, \lambda_2, \ldots, \lambda_m\}$, $\lambda \in \Lambda$ and

$$\Lambda = \left\{\lambda : \lambda_j \geq 0; \sum_{j=1}^{m} \lambda_j = 1\right\}$$

It is clear that the sets $X^{UND}(\mu_V)$, $X^{UND}(\mu_Z)$ and $X^{UND}(\mu_L; \lambda)$ are determined for these convolutions (formula (3.3)).

Definition 4.4. The convolution of the VFPR is effective if $X^{UND}(\bar{\mu}) \subseteq X_P^{UND}$.

Theorem 4.1. The convolution of the VFPR, represented by formula (4.2), is effective for all $\lambda \in \Lambda$ if the FPRs $P_j$, $j=\overline{1,m}$ are transitive.

Proof. The crisp relation $F(\mu_L; \lambda)$ corresponds to the convolution $\mu_L(x,y)$ - see Definition 3.3. This relation can be represented as

$$F(\mu_L; \lambda) = \left\{ (x,y) : \sum_{j=1}^{m} \lambda_j \Delta_j(x,y) \geq 0 \right\}, \tag{4.4}$$

where $\lambda \in \Lambda$ (formula (4.3)). It can be easily proved that $F_P^S = F^S(\mu_L; \lambda)$ if $\lambda \in \Lambda$. Then, on the basis of Theorem 1.2 the following result holds:

$X_\Pi[F(\mu_L;\lambda)] \subseteq X_\Pi(F_P)$, and hence on the basis of Definition 4.1 and Theorem 3.2, $X^{UND}(\mu_L;\lambda) \subseteq X_P^{UND}$ for all $\lambda \in \Lambda$ (formula (4.3)). Since all the FPRs included in the VFPR P are transitive, the set $X^{UND}(\mu_L;\lambda)$ is not empty for all $\lambda \in \Lambda$.

This means that the convolution given by formula (4.2) is effective (Definition 4.4) Q.E.D.

Unfortunately, we cannot say anything about the effectiveness of the convolutions of VFPR given by formula (4.1). Let us note that they are the convolutions introduced by Zadeh (1965) for the intersection and union of the fuzzy sets and relations, respectively.

Let all components of the VFPR be determined in the linear space. Then for the convolution of VFPR given by formula (4.2), an interesting result holds. An analogous result in the multicriteria decision making case is called Karlin's Lemma (Karlin, 1959).

<u>Theorem 4.2 (Karlin's Lemma)</u>. Given is a VFPR, i.e. the collection of FPRs P. Let all membership functions $\mu_j(x,y)$, $j=\overline{1,m}$, be concave in X. Then for any $x_o \in X_P^{UND}$ there is such a $\lambda^o \in \Lambda$ that $x_o \in X^{UND}(\mu_L;\lambda)$.

<u>Proof</u>. If $x_o \in X_P^{UND}$, then the separating plane exists for the set of points

$$\Delta(y,x_o) = \{\Delta_1(y,x), \Delta_2(y,x), \ldots, \Delta_m(y,x)\},$$

where $y \in X$ since all the membership functions are concave in X. This plane can be represented by the following inequality

$$\sum_{j=1}^{m} \lambda_j \Delta_j(y,x) \leqslant 0$$

for all $y \in X$ and $\lambda^o = \{\lambda_1^o, \lambda_2^o, \ldots, \lambda_m^o\}$ from $\Lambda$. This means that $x_o \in X^{UND}(\mu_L;\lambda^o)$. Q.E.D.

This fact is very important, because any alternative (decision) from the set $X_P^{UND}$ is reachable for the DM if he uses a choice procedure based on the convolution (4.2).

Now we shall consider another decision making model using fuzzy preference relations. It is introduced for the case when the partial FPRs in VFPR are not the same in view of their importance for the decision maker and so are ordered lexicographically.

## 5. A LEXICOGRAPHIC DECISION MAKING MODEL

Let us consider a collection of fuzzy preference relations, where the relations are indexed so that the second relation is more important than the first one, the third relation is more important than the second one, etc.

Definition 5.1. The decisions $x$ and $y$ are equivalent with respect to a fuzzy preference relation (FPR), if the following condition holds

$$\mu^s(x,y) = \mu^s(y,x) = 0.$$

This means there exists no decision which strictly dominates another one.

Definition 5.2. The strict fuzzy lexicographic preference relation $\mu^s_{lex}(x,y)$ holds for $(x,y)$ from $E$ if one of the following relations holds:

1. $\mu^s_m(x,y) > 0$
2. $\mu^s_m(x,y) = \mu^s_m(y,x) = 0$ and $\mu^s_{m-1}(x,y) > 0$

. . . . . . . . . . . . . . . . . . (5.1)

m. $\mu^s_r(x,y) = \mu^s_r(x,y) = 0$ for all $r=2,m$, and $\mu^s_1(x,y) > 0$.

Here $\mu^s_{lex}(x,y) = \mu^s_{m-j+1}(x,y)$, where $j$ is the number of the string in which the comparison of $x$ and $y$ was ended.

If $\mu^s_r(x,y) = \mu^s_r(y,x) = 0$ for all $r=\overline{1,m}$, then the decisions $x$ and $y$ are lexicographically equivalent.

Let us introduce the following convolution

$$\bar{\mu}_{lex}(x,y) = \sum_{j=1}^{m} \lambda^{lex}_j \mu_j(x,y) \tag{5.2}$$

where $\lambda^{lex}_j$, $j=\overline{1,m}$, are the lexicographic coefficients, which are calculated by the formulae

$$\lambda^{lex}_s = \prod_{j=1}^{s-1} (1 + B_j/a),$$

where $B_j = \max_{x,y\in X} \mu^s_j(x,y)$, $a = \min A_j$,

$$A_j = \min_{x,y\in X} \mu^s_j(x,y) > 0. \qquad j=\overline{1,m}$$

The sets of nonfuzzy nondominated decisions $X^{UND}(\bar{\mu}_{lex})$ for $\bar{\mu}_{lex}(x,y)$, and $X^{UND}(\mu_{lex})$ for $\mu_{lex}(x,y)$ are defined with respect to formula (3.3).

Theorem 5.1. $X^{UND}(\bar{\mu}_{lex}) = X^{UND}(\mu_{lex})$.

Proof. Let us remember that

$$\bar{\mu}^s_{lex}(x,y) = \begin{cases} \Delta(x,y), & \text{if} \quad \Delta(x,y) \geq 0, \\ 0, & \text{if} \quad \Delta(x,y) < 0, \end{cases}$$

where

$$\Delta(x,y) = \sum_{j=1}^{m} \lambda_j^{lex} \quad \Delta_j(x,y).$$

For any element $x \in X^{UND}(\mu_{lex})$ we have $\mu_{lex}(y,x) = 0$ for all $y$ or $\mu^s_{lex}(x,y) \geqslant 0$. First we shall consider the $y$ for which the equality takes place. This means that if $\mu^s_{lex}(x,y)=0$ and $\mu^s_{lex}(y,x) = 0$, then $\mu^s_j(x,y) = \mu^s_j(y,x) = 0$. As a result we obtain that

$$\bar{\mu}^s_{lex}(x,y) = \bar{\mu}^s_{lex}(y,x) = 0 \quad \text{for all } y \text{ from } X.$$

Now we shall consider such $y$ for which $\mu^s_{lex}(x,y) > 0$. From this we have that starting from $j=m$, the first membership function which is not equal to $0$ is positive, i.e. for some $1$ $\mu^s_1(x,y) > 0$. The coefficients $\lambda_j^{lex}$ are combined so that $\bar{\mu}^s_{lex}(x,y)$ is necessarily positive and hence $\bar{\mu}^s_{lex}(y,x) = 0$. Thus we have $\mu^s_{lex}(y,x) = 0$ for any $y$ from $X$. We thus have obtained that $x \in X^{UND}(\mu_{lex})$.

Conversely, let $x$ be from $X^{UND}(\bar{\mu}_{lex})$. This means that $\bar{\mu}^s_{lex}(y,x)=0$ for all $y$ from $X$ or $\bar{\mu}^s_{lex}(x,y) \geqslant 0$. Let us consider such a y that the equality takes place, i.e. the following two equations are true at the same time: $\bar{\mu}^s_{lex}(y,x) = 0$ and $\bar{\mu}^s_{lex}(x,y) = 0$. As a consequence we obtain $\Delta(x,y) = 0$. The coefficients are selected so that $\mu_j(x,y) - \mu_j(y,x) = 0$ for all $j=\overline{1,m}$ and $\mu^s_{lex}(y,x) = 0$, $\mu^s_{lex}(x,y) = 0$. Let us consider such a y for which $\bar{\mu}^s_{lex}(x,y) > 0$. We obtain $\Delta(x,y) > 0$. This means that, in sum, components which are not equal to 0 exist. Also, the first component which is not 0, say when $j=1$, starting from $j=m$ in the converse order is necessarily positive. This also follows from the properties of the coefficients $\lambda_j^{lex}$, i.e. we have $\mu^s_1(x,y) > 0$, and hence, $\mu^s_{lex}(y,x) = 0$. Thus, $x \in X^{UND}(\mu_{lex})$. Q.E.D.

It is not difficult to prove that all the decisions in $X^{UND}(\mu_{lex})$ are lexicographically equivalent in terms of Definition 5.1.

## 6. INTERCONNECTION OF THE MULTICRITERIA AND FUZZY REPRESENTATIONS OF THE DECISION MAKING PROBLEMS

Intuitively, it is clear that a fuzzy representation of

the decision making problem must include its crisp counterpart as a special case. For example, it is known that the fuzzy set includes the ordinary set. On this basis we shall formulate the principle of the interconnection of these two representations of the decision making problem. Let us suppose that both representations are defined in the set X, and are given as the pairs <X,R> and <X,P>. For them the Pareto domination relations are given as $R_k$ and $F_P$, respectively.

Definition 6.1. Two representations of a decision making problem, multicriteria and fuzzy, are concordant (noncontradictory) if $R_k^s \subseteq F_P^s$. Let us notice that the numbers of criteria and fuzzy preference relations are not necessarily equal.

For the concordant representations of the decision making problem on the basis of Theorem 1.2, the inclusion $X_P^{UND} \subseteq X_\Pi^k$ holds. This result is important and published in Zhukovin (1983, 1984). It is known that usually the initial Pareto set $X_\Pi^k$ of a crisp multicriteria decision making problem contains too many alternatives and is hardly visible for the DM. By introducing fuzziness in it, i.e. using VFPR instead of a vector criterion, we can diminish the number of alternatives in the Pareto set. Fuzziness in the multicriteria decision making problem can be introduced in the following way. Let us consider the equality

$$\mu_j(x,y) = [K_j(x) - K_j(y)]/2d_j + 1/2, \tag{6.1}$$

where $d_j = \max[K_j(x) - K_j(y)]$ and is the $K_j$ - criterion's scale range. The set of maximal alternatives of criterion $K_j$ is determined as

$$K_j^* = \left\{x : \max_{x \in X} K_j(x)\right\}. \tag{6.2}$$

Theorem 6.1. For this case, the equality $X_j^* = S^{UND}(\mu_j)$ holds.

Proof. The next strict fuzzy relation corresponds to the initial FPR

$$\mu_j^s(x,y) = \begin{cases} \Delta K_j = [K_j(x) - K_j(y)]/d_j, & \text{if } \Delta K_j \geqslant 0, \\ 0, & \text{if } \Delta K_j < 0. \end{cases} \tag{6.3}$$

From the condition $\mu^{ND}(x) = 1$ (formula (3.3)) we conclude that $\max_{y \in X} \mu_j^s(x,y) = 0$ and hence $\Delta K_j \geqslant 0$ for all $y \in X$. Then we can determine the nonfuzzy set of nondominated alternatives for this case as

$$X^{UND}(\mu_j) = \left\{x : \max_{x \in X} ([K_j(x) - d_j]/d_j)\right\}, \tag{6.4}$$

where $d_j$ = const. This means that the equality $X_j^* = X^{UND}(\mu_j)$ holds. Q.E.D.

## 7. AN l-LEVEL FUZZY DECISION MAKING MODEL

Such models have been discussed for the first time by Orlovsky (1978). Using $\Delta(x,y)$ let us introduce the following l-level preference relations

$$F(l)=\{(x,y) : \Delta(x,y) \geqslant l\}, \quad (7.1)$$

where $0 \leqslant l \leqslant 1$. They are crisp unconnected strict (asymmetric) relations when $l \neq 0$. The Pareto sets $X_\Pi(l)$ correspond to them. On the basis of Theorem 1.2, $X_\Pi(l_1) \subseteq X_\Pi(l_2)$ holds if $l_1 < l_2$. If $l = 0$, then the equality $F(0) = F$ holds where $F(0)$ is a connected reflexive relation. In the general case these relations are not transitive. Let us now study the interconnection of the introduced relations with the nonfuzzy set of nondominated alternatives $X^{UND}(\mu)$. Let us introduce the concept of the set of r-nondominated alternatives

$$X^{ND}(r) = \{x : \mu^{ND}(x) \geqslant r\}, \quad (7.2)$$

where $0 \leqslant r \leqslant 1$.

It is clear that $X^{ND}(1) = X^{UND}(\mu)$.

Theorem 7.1. $X^{ND}(r) = X_\Pi(l)$ if $r = 1 - l$.

Proof. Let $l \neq 0$. Let also $x$ be from the set $X(l)$. It means that there does not exist a $y \in X$ such that $(y,x) \in F(l)$ and hence $\Delta(y,x) < l$ for all $y \in X$ including $x$. Let us remember that $l > 0$. Then the inequality $\max_{y \in X} \mu^S(y,x) < l$ holds and hence $\mu^{ND}(x) > 1 - l = r \neq 1$. Thus $x \in X^{ND}(r)$. Now let x be in the set $X^{ND}(r)$. This means that $\mu^{ND}(x) > r$ and $\max_{y \in X} \mu^S(y,x) < 1-r = l$. It follows that $\Delta(y,x) < l$ for all $y \in X$ and $x \in X_\Pi(l)$. Q.E.D.

An analogous result for $l = 0$ and $r = 1$ has been proved earlier.

Theorem 7.2. If $r_1 > r_2$, then $X^{ND}(r_1) \subseteq X^{ND}(r_2)$.

Proof. By Theorem 7.1, $X^{ND}(r_1) = X(l_1)$ and $X^{ND}(r_2) = X(l_2)$, where $l_1 = 1 - r_1$ and $l_2 = 1 - r_2$ and hence $l_2 > l_1$. By Theorem 1.2, $X_\Pi(l_1) \subseteq X_\Pi(l_2)$, which means that $X^{ND}(r_1) \subseteq X^{ND}(r_2)$. Q.E.D.

Thus changing the value of the level l, the decision maker has an opportunity to obtain the set of effective alternatives of different cardinality from which the final choice would be made. The effectiveness of the final choice is quaranteed. This fact is very important for an elaboration of the man-machine decision making procedures.

An analogous model can be considered for the VFPR.

## 8. APPLICATIONS

The methods and results of the fuzzy multicriteria decision making theory can be used in many practical problems.

The following points are essential while dealing with somewhat complex decision making problems:

1. The estimation and choice is carried out by a number of criteria.
2. All the criteria are qualitative.
3. All the information about comparison of alternatives on the given criteria is obtained from experts.
4. The experts' data are fuzzy.

Therefore the methods and results of this paper seem to be useful in practice.

Here are some examples of their use:

1. Yearly plan formation for a manufacturing branch.
2. Project competition.
3. Election of a candidate for a vacancy.
4. Distribution of homogeneous supply (cement, fuel, water, etc.).

The initial information for the computer is obtained from experts by an interactive mode. Through comparison of alternatives in pairs as follows: the one out of the two compared alternatives which is preferable, according to the criterion and the degree of the expert's certainty is made known. The latter represents a strict fuzzy preference relation (formula (3.1)).

The initial information is then processed in accordance with the given approach dependent on the problem (also in an interactive regime).

The fuzzy multicriteria decision making procedure is used together with some other auxiliary procedures such as:

1. Revalation and formation of criteria on the basis of a system approach.
2. Formation of a set of competitive alternatives in case when the latter is not given initially.
3. Decrease of material processed by experts.
4. The possibility for experts of returning to the previous steps (feedback) on the decision maker's demand.

## 9. SUMMARY

It is known that the multicriteria decision making theory is developed well enough. Many interesting results are proved and methods are studied in it. Using them we have developed in this paper a basis for the fuzzy multicriteria decision making theory. We have introduced the concepts of the VFPR instead of the vector criterion and the nonfuzzy set of nondominated alternatives for this case which is analogous to the Pareto set in the multicriteria decision making problem. For one FPR it agrees with the concept introduced by Orlovski (1978). We have considered some convolutions of the VFPR, defined the concept of an effective convolution of the VFPR, and proved the effectiveness of the linear convolution. We have proved Karlin's lemma for it too. The lexicographic aspect of the fuzzy multicriteria decision making problem has been studied. Interconnections of the two representations of the decision making problem, multicriteria and fuzzy, have been examined. The given approach and results and methods related to it have been realized as a man-machine decision making procedure and used in some practical problems.

REFERENCES

Karlin, S. (1959). Mathematical Methods and Theory in Games, Programming and Economics, Pergamon Press, London.
Orlovski, S.A. (1978). Decision making with a fuzzy preference relation. Int. J. Fuzzy Sets and Syst. 1, 155-167.
Zadeh, L.A. (1965). Fuzzy Sets. Inform. and Contr. 8, 338-353.
Zhukovin, V.E. (1983). Multicriteria Decision Making Models with Uncertainty (in Russian). Metsniereba, Tbilisi.
Zhukovin, V.E. (1984). The multicriteria decision making with vector fuzzy preference relation. In R. Trappl (ed.), Cybernetics and Systems Research. Amsterdam, 179-181.
Zhukovin, V.E., and F.V. Burshtein (1981). The Dialogue Procedure of the Choice of the Best Decision with Respect to Multiple Objectives and Non-Complete Information. (in Russian). Report Tbilisi University, 19-32.
Zhukovin, V.E., F.V. Burshtein, and A.B. Zaslavski (1980). Multicriterial dialogue procedure of the best decision choice under uncertainty (in Russian). Proc. KPI, Kutaisi, 66-69.

# FUZZY PROGRAMMING - A NEW MODEL OF OPTIMIZATION

Feng Yingjun

Harbin Institute of Technology
Harbin, People's Republic of China

Abstract. In this paper, limitations of the classical model of optimization are indicated and a more general model is given in terms of fuzzy subsets theory. The concept of a fuzzy value set of a function is defined and other important concepts involved in the model are established, based on the fuzzy value set. In this model an objective function and a constraint are symmetric. The former is only a special case of the latter. Moreover, the multiple objective problem and the single objective problem are also unified. Optimization problems often emerge in a fuzzy environment. A nonfuzzy environment is only its special case, so that this model is of great value in solving realistic problems of optimization in a large spectrum of fields. In this paper, a relationship between an optimal solution of fuzzy programming and important concepts of multiobjective programming, the efficient and weak-efficient solutions, is shown.

Keywords: fuzzy optimization, fuzzy mathematical programming, efficient solution, weak-efficient solution.

## 1. INTRODUCTION

Modelling is of prime concern in solving real world problems by using mathematical tools. As research goes deeper it is necessary to find some new mathematical tools, so that a mathematical model to better reflect the real situation can be established.

The classical mathematical models of optimization with a single objective, i.e.

$$f(x) \to \max_{x \in R} \quad , \quad R=\left\{x:g(x) \geqslant 0;\ i=1,2,\ldots,m\right\}$$

have some limitations.

First, the constraint conditions of real problems often occur in a fuzzy environment. If we solve it neglecting fuzziness, a more proper solution may sometimes be lost.

Second, there is no unbridgeable gap between multiple objective optimization and single objective optimization and also between the objective function and the constraint. Some problems are originally of the multiobjective type, however, in order to simplify the solution process, the function value of some objective functions is artificially restricted and the objective functions are turned into constraints, hence multiple objective optimization becomes single objective optimization. However, this artificial simplification may sometimes be inadequate. This is one of the reasons that more and more interest is recently being shown in the research on multiobjective optimization.

Generally, the problem of multiobjective optimization is stated as

$$(VP) \quad \begin{cases} F(x) \to V - \max \\ \text{Subject to: } g(x) \geqslant 0 \end{cases}$$

where $F(x) = (f_1(x), f_2(x), \ldots, f_p(x))^T$, $g(x) = (g_1(x), g_2(x), \ldots, g_m(x))^T$, $x \in E_n$.

This is only a notation, not a mathematical model, because it cannot indicate what is meant by a result.

On multiobjective optimization, Wierzbicki (1979) stated: "... its various applications still result in vexing methodological and theoretical questions. The resulting tools are often applied because of certain traditions rather than their suitability for solving a given problem. The most important questions in multiobjective optimization are how and in what form the additional information which comes from decision-maker can be obtained". Basically, this argument states that multiobjective optimization needs a suitable mathematical model.

In order to solve the above-mentioned problem, in this paper a more general model of optimization is given in terms of fuzzy subsets theory. The classical mathematical model of optimization is only a special case of this model which can contain more realistic cases that cannot be included in the classical model. In this paper, the concepts of a fuzzy feasible set, fuzzy optimal point set and fuzzy programming are defined. These concepts are of great value in solving real world problems of optimization in a variety of fields. The solution effectiveness of this model is proved. Therefore, the theoretical suitability of this solution is ensured. A method using fuzzy subset theory to solve the multiobjective problem is given in Feng (1981, 1983) and Feng and Wei (1982). This paper extends those results.

## 2. FUZZY CONSTRAINT, FUZZY OPTIMUM POINT SET AND FUZZY PROGRAMMING

In optimization problems, values of the objective function or the constraint functions need to be considered. In some practical problems it is required to take function values as fuzzy subsets. Thus we give the following definition of a fuzzy value set of a function. Other important optimization-related concepts will be established on its basis.

Definition 2.1. Let $f(x)$ be a function defined on a subset $D \subset E_n$. $\underset{\sim}{B}$ is called a fuzzy value set of the function $f(x)$

if $\underset{\sim}{B}$ is a fuzzy subset on the value set E of f(x).

Definition 2.2. Let $\underset{\sim}{A}_i$ be a fuzzy subset on $D\subset E_n$ and $\underset{\sim}{B}_i$ be a fuzzy value set of $g_i(x)$, $\underset{\sim}{A}_i$ is called a fuzzy constraint of $g_i(x)$ with respect to $\underset{\sim}{B}_i$ if $\underset{\sim}{A}_i(x) = \underset{\sim}{B}_i(g_i(x))$.

Obviously, the classical constraint is equivalent to a special case of the fuzzy constraint. If $\underset{\sim}{A}_i(x) = \underset{\sim}{B}_i(g_i(x)) = 0$ for $g_i(x) < 0$ and $\underset{\sim}{A}_i(x) = \underset{\sim}{B}_i(g_i(x)) = 1$ for $g(x) \geqslant 0$, then the fuzzy constraint with respect to $\underset{\sim}{B}_i$ is equivalent to the classical inequality constraint $g_i(x) \geqslant 0$. However, a fuzzy constraint can provide more information than a classical equality or inequality constraint. Similarly, the following fuzzy feasible set is also an extension of a classical feasible set.

Definition 2.3. Let $\underset{\sim}{R}$ be a fuzzy subset on $D\subset E_n$ and $\underset{\sim}{A}_i$ be a fuzzy constraint of $g_i(x)$ with respect to $\underset{\sim}{B}_i$, i=1,2,...,m. $\underset{\sim}{R}$ is called a fuzzy feasible set with respect to $\underset{\sim}{B}_i$ if

$$\underset{\sim}{R}(x) = \bigwedge_{1\leqslant i\leqslant m} \underset{\sim}{A}_i(x) = \min_{1\leqslant i\leqslant m} \underset{\sim}{A}_i(x)$$

Definition 2.4. Let $\underset{\sim}{B}_i^*$ be a fuzzy value set of $f_i(x)$ on $(-\infty, M_i]$ where $M_i = \sup_{x\in D} f_i(x)$. $\underset{\sim}{B}_i^*$ is called a fuzzy optimum set of component $f_i(x)$ of $F(x)$ if $\underset{\sim}{B}_i^*(y)$ is a strictly monotone increasing function on $[m_i, M_i]$ and $\underset{\sim}{B}_i^*(y) = 0$ for $f_i(x) < m_i$, where $m_i \geqslant \inf_{x\in D} f_i(x)$.

Definition 2.5. Let $\underset{\sim}{A}^*$ be a fuzzy subset on $D\subset E_n$. $\underset{\sim}{A}^*$ is called a fuzzy optimum point set of F(x) if

$$\underset{\sim}{A}^*(x) = \bigwedge_{1\leqslant i\leqslant p} \underset{\sim}{A}_i^*(x) \triangleq \bigwedge_{1\leqslant i\leqslant p} \underset{\sim}{B}_i^*(f_i(x))$$

Definition 2.6. Let $\underset{\sim}{H}^*$ be a fuzzy subset on $D\subset E_n$. $\underset{\sim}{H}^*$ is called a fuzzy optimum point set of F(x) on a fuzzy feasible set $\underset{\sim}{R}$ if $\underset{\sim}{A}^*$ is a fuzzy optimum point set of F(x) and

$$\underset{\sim}{H}^*(x) = \underset{\sim}{A}^*(x) \wedge \underset{\sim}{R}(x) \triangleq \min(\underset{\sim}{A}^*(x), \underset{\sim}{R}(x))$$

Then

$$(\underset{\sim}{P}) \quad \underset{\sim}{H}^*(x) \to \max_{x\in D}$$

is called a fuzzy programming problem with respect to $\underset{\sim}{H}^*$.

Definition 2.7. $\bar{x}$ is called on optimal solution of fuzzy programming problem $(\underset{\sim}{P})$ if

$$\underset{\sim}{H}^*(\bar{x}) = \max_{x\in D} \underset{\sim}{H}^*(x) > 0$$

The relationship between a fuzzy feasible set $\underset{\sim}{R}$ and a fuzzy optimum point set $\underset{\sim}{A}^*$ is symmetric. The unique difference is that the membership functions $\underset{\sim}{B}_i^*(y)$ of the fuzzy optimum set $\underset{\sim}{B}_i^*$, $i=1,2,\ldots,p$, whose intersection forms the fuzzy optimum point set $\underset{\sim}{A}^*$ of $F(x)$, are strictly monotone increasing functions on $[m_i, M_i]$. The fuzzy optimum set $\underset{\sim}{B}_i^*$ is only a special case of the fuzzy constraint. In this way, both the concepts of a multiobjective problem and single objective problem and the concepts of objective function and constraint are unified.

In solving multiobjective programming (VP) by classical mathematical methods, a number of difficulties are encountered. For example, the finding of an optimal solution of (VP) by adding weighted individual objective functions requires the decision maker to use many simplifying assumptions about the underlying value structure. It is easier to represent the decision maker's requirements by membership functions $\underset{\sim}{B}_i^*(y)$ of the fuzzy optimum set $\underset{\sim}{B}_i^*$. The decision maker can express different requirements for the individual objective functions by using different membership functions $\underset{\sim}{B}_i^*(y)$.

For instance, we may take a convex function as the membership function $\underset{\sim}{B}_i^*(y)$ of the fuzzy optimum set of component $f_i(x)$ if there is a desire to minimize trade-offs between it and other objectives. That is, if the value of the component $f_i(x)$ slightly decreases, its degree of membership will greatly decrease. Analogously, we may take a concave function as the membership function $\underset{\sim}{B}_i^*(y)$ of the fuzzy optimum set of some competent $f_i(x)$ if there is a willingness to exchange its performance for that of another objective. That is, if the value of the component $f_i(x)$ decreases a little, its degree of membership will decrease only a little too. In this way, various points of the feasible set can be given various "weights".

In definition 2.4 it is realistic to require the membership function $\underset{\sim}{B}_i^*(y)$ of the fuzzy optimum set $\underset{\sim}{B}_i^*$ to be strictly monotone increasing on $[m_i, M_i]$. If this were not fulfiled, the corresponding optimal solution of fuzzy programming (P) would not always be an efficient or weak-efficient solution. Hence, this solution would not be reasonable . In addition, the requirement $\underset{\sim}{A}_i^*(x) = \underset{\sim}{B}_i^*(f_i(x)) = 0$ for $f_i(x) < m_i$ means that the solution of $f_i(x) < m_i$ is rejected by the decision maker.

The determination of an optimal solution to fuzzy programming $(\underset{\sim}{P})$ is equivalent to finding a solution to a nondifferentiable optimization problem, as, e.g.,

$$\begin{cases} \lambda \to \max \\ \text{subject to: } \underset{\sim}{A}_i^*(x) = \underset{\sim}{B}_i^*(f_i(x)) \geq \lambda \qquad i=1,2,\ldots,p, \\ \qquad\qquad\quad \underset{\sim}{A}_j(x) = \underset{\sim}{B}_j(g_j(x)) \geq \lambda \qquad j=1,2,,\ldots,m. \end{cases}$$

Obviously, $0 \leqslant \lambda \leqslant 1$. This transformation leads both to a method for solving fuzzy programming problems and yields a "grade of membership" (i.e. $\lambda$ ) of a solution. If $\lambda = 0$, then there exists no solution fulfiling the decision maker's requirement.

Example. The following problem illustrates how the above approach can be used to solve optimization problems via fuzzy programming. This problem comes from Zimmermann (1978).

A company manufactures two products 1 and 2 under given capacities. Product 1 yields a profit of 2 units per piece and product 2 of 1 unit per piece. Product 2 can be exported with a profit of 2 units per piece. Product 1 needs imported raw material of 1 unit per piece. The two goals are an optimal balance of trade and a maximum profit while the capacity constraints are, e.g.:

$$-x_1+3x_2 \leqslant 21 \qquad x_1+3x_2 \leqslant 27$$
$$4x_1+3x_2 \leqslant 45 \qquad 3x_1+x_2 \leqslant 30$$
$$x_1, x_2 \geqslant 0$$

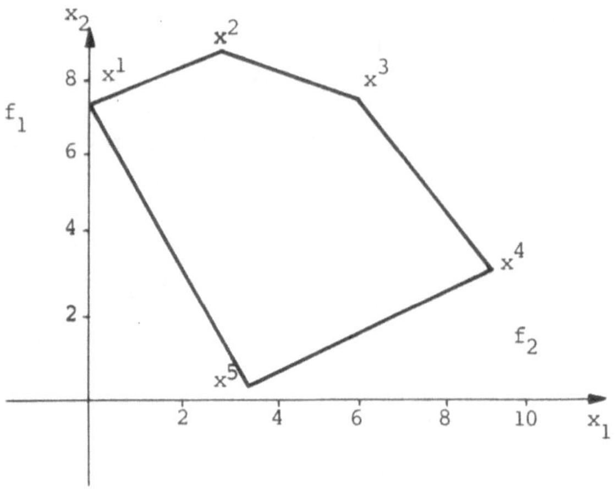


Fig. 1

Figure 1 shows the feasible set of the problem. The objective functions are $f_1(x) = -x_1+2x_2$, $f_2(x) = 2x_1+x_2$. Clearly, $x^1$ is optimal with respect to the first objective ( $f_1(x^1) = 14$, the maximum net export) while $x^4$ is optimal with respect to the profit objective ($f_2(x^4) = 21$, the maximum profit). Solution $x^5$ gives the worst result, namely $f_1(x^5) = -3$ (3 units imported) and $f_2(x^5) = 7$ (7 units of profit).

Let us apply the fuzzy programming approach to this problem. We may take $m_1 = -3$, $m_2=7$ and $M_1=14$, $M_2=21$. The decision maker can choose various membership functions $\underset{\sim}{B}_1^*(y)$ and $\underset{\sim}{B}_2^*(y)$. The choice assumed there is to use linear functions on $[m_i, M_i]$ In this way, we can obtain the membership functions of the fuzzy optimum set of $f_1(x)$ and $f_2(x)$ as

$$\underset{\sim}{B}_1^*(y) = \begin{cases} 0 & \text{for } y < -3 \\ \frac{y+3}{17} & \text{for } -3 \leqslant y \leqslant 14 \end{cases}$$

$$\underset{\sim}{B}_2^*(y) = \begin{cases} 0 & \text{for } y < 7 \\ \frac{y-7}{14} & \text{for } 7 \leqslant y \leqslant 21 \end{cases}$$

The corresponding membership functions of the fuzzy optimum point set of $F(x)$ become:

$$\underset{\sim}{A}^*(x) = \bigwedge_{1 \leqslant i \leqslant p} \underset{\sim}{B}_i^*(f_i(x)) = \min(\tfrac{1}{17}(-x_1+2x_2+3), \tfrac{1}{14}(2x_1+x_2-7))$$

The membership function of the fuzzy feasible set can be taken as

$$\underset{\sim}{R}(x) = \begin{cases} 0 & \text{for } -x_1+3x_2>21,\ x_1+3x_2>27, \\ & \quad 4x_1+3x_2>45,\ 3x_1+x_2>30, x_1, x_2<0 \\ 1 & \text{elsewhere} \end{cases}$$

The fuzzy programming becomes therefore

$$\min(\underset{\sim}{A}^*(x), \underset{\sim}{R}(x)) \to \max_x$$

As is well known, it is equivalent to solving the following linear programming problem:

$$\begin{cases} & \lambda \to \max & \\ \text{subject to:} & \frac{1}{17}(-x_1+2x_2+3) \geqslant \lambda & \frac{1}{14}/2x_1+x_2-7) \geqslant \lambda \\ & -x_1+3x_2 \leqslant 21 & x_1+3x_2 \leqslant 27 \\ & 4x_1+3x_2 \leqslant 45 & 3x_1+x_2 \leqslant 30 \\ & x_1, x_2 \geqslant 0 & 0 \leqslant \lambda \leqslant 1 \end{cases}$$

The optimal solution of the fuzzy programming is $\bar{x} = (5.03, 7.32)^T$ yielding an export of $f_1(\bar{x}) = 4.58$ and a profit of $f_2(\bar{x}) = 17.38$.

## 3. SEVERAL CONCEPTS OF AN OPTIMAL SOLUTION IN FUZZY PROGRAMMING

As fuzzy programming is involved in multiobjective optimization, it is required to show a relation between an optimal solution of fuzzy programming and important concepts of multiobjective optimization, the efficient and weak-efficient solutions, so that the theoretical suitability of that solution be ensured.

Definition 3.1. Let $\bar{x} \in R$. $\bar{x}$ is called an efficient solution of $F(x)$ on $R$ if there is no $x \in R$, such that

$$f_i(x) \geq f_i(\bar{x}), \quad i \in \{1,2,\ldots,p\}$$

and

$$f_i(x) > f_i(\bar{x}) \quad \text{for at least one } i \in \{1,2,\ldots,p\}$$

$\bar{x}$ is called a weak-efficient solution of $F(x)$ on $R$ if there is no $x \in R$, such that

$$f_i(x) > f_i(\bar{x}), \qquad i = 1,2,\ldots,p$$

Clearly, if $\bar{x}$ is an efficient solution on $R$, then $\bar{x}$ is a weak-efficient solution on $R$. The converse is true only under some restrictions. Let $F(x)$ be a strictly concave vector function and $g(x)$ - a concave vector function. If $\bar{x}$ is a weak-efficient solution on $R$, then $\bar{x}$ is a efficient solution on $R$ (Kiyotaka, 1976).

A new concept is introduced first.

Definition 3.2. Let $\bar{x}$ be an optimal solution of fuzzy programming $(\underset{\sim}{P})$. A fuzzy feasible set $R$ is not active if $\underset{\sim}{A}^*(\bar{x}) \leq \underset{\sim}{R}(\bar{x})$.

Theorem 3.1. If $\bar{x}$ is an optimal solution of fuzzy programming $(\underset{\sim}{P})$ and the fuzzy feasible set $\underset{\sim}{R}$ is not active, then $\bar{x}$ is a weak-efficient solution on $\Gamma = \{x : \underset{\sim}{A}^*(x) \leq \underset{\sim}{R}(x)\}$.

Proof. Since the fuzzy feasible set $\underset{\sim}{R}$ is not active, we can get $\bar{x} \in \Gamma$ and

$$\underset{\sim}{H}^*(\bar{x}) = \underset{\sim}{A}^*(\bar{x}) \tag{1}$$

Suppose that $\bar{x}$ is not a weak-efficient solution on $\Gamma$. Then, there is an $\bar{\bar{x}} \in \Gamma$, such that

$$f_i(\bar{\bar{x}}) > f_i(\bar{x}), \quad i = 1,2,\ldots,p \tag{2}$$

First, we prove that $\bar{\bar{x}} \in \bar{R} = \bigcap_{i=1}^{p} \{x : f_i(x) \geq m_i\}$

If $\bar{\bar{x}} \notin \bar{R}$, then there is $i_0$, such that $f_{i_0}(\bar{x}) < f_{i_0}(\bar{\bar{x}}) < m_{i_0}$,

hence $\underset{\sim}{A}^*_{i_0}(\bar{x}) = \underset{\sim}{B}^*_{i_0}(f_{i_0}(\bar{x})) = 0$

and

$$\underset{\sim}{H}^*(\bar{x}) = \underset{\sim}{A}^*(\bar{x}) = 0,$$

which constradicts $\underset{\sim}{H}^*(\bar{x}) > 0$. Hence, $\bar{\bar{x}} \in \bar{R}$. Analogously, we can prove $\bar{x} \in \bar{R}$.

Because $\underset{\sim}{B}^*_i(y)$ are strictly monotone increasing functions on $[m_i, M_i]$, we can obtain by (2)

$$\underset{\sim}{A}_i^*(\bar{\bar{x}}) = \underset{\sim}{B}_i^*(f_i(\bar{\bar{x}})) > \underset{\sim}{B}_i^*(f_i(\bar{x})) = \underset{\sim}{A}_i^*(\bar{x}), \quad i = 1,2,\ldots,p,$$

Thus

$$\underset{\sim}{A}^*(\bar{\bar{x}}) = \bigwedge_{1\leq i\leq p} \underset{\sim}{A}_i^*(\bar{\bar{x}}) > \bigwedge_{1\leq i\leq p} \underset{\sim}{A}_i^*(x) = \underset{\sim}{A}^*(x)$$

By (1) and the assumption, we have

$$\underset{\sim}{H}^*(\bar{\bar{x}}) = \underset{\sim}{A}^*(\bar{\bar{x}}) > \underset{\sim}{A}^*(\bar{x}) = \underset{\sim}{H}^*(\bar{x})$$

which contradicts that $\bar{x}$ is an optimal solution of $(\underset{\sim}{P})$.

Theorem 3.2. Assume that $F(x)$ is a strictly concave vector function and $g(x)$ - a concave vector function. If $\bar{x}$ is an optimal solution of fuzzy programming $(\underset{\sim}{P})$ and the fuzzy feasible set $\underset{\sim}{R}$ is not active, then $\bar{x}$ is an efficient solution on $F = \{x : \underset{\sim}{A}^*(x) \leq \underset{\sim}{R}(x)\}$.

The relation between the efficient and weak-efficient solutions immediately leads to the above conclusion.

We shall now show that under an appropriate assumption a converse statement is also satisfied. That is, if $x$ is an efficient solution or a weak-efficient solution, then we can construct a fuzzy feasible set and a fuzzy optimum point set $\underset{\sim}{A}^*$ of $F(x)$, and the solution $\bar{x}$ is an optimal solution of fuzzy programming $(\underset{\sim}{P})$.

Theorem 3.3. If $\bar{x}$ is a weak-efficient solution of $F(x)$ on $R = \{x : g(x) \geq 0\}$, then there is a fuzzy feasible set $\underset{\sim}{R}$ and a fuzzy optimum point set $\underset{\sim}{A}^*$, such that $\bar{x}$ is an optimal solution of fuzzy programming $(\underset{\sim}{P})$.

Proof. First, we construct the fuzzy constraint by defining the membership function

$$\underset{\sim}{A}_i(x) = \underset{\sim}{B}_i(g_i(x)) = \begin{cases} 0 & \text{for} \quad g_i(x) < 0 \\ 1 & \text{for} \quad g_i(x) \geq 0, \ i=1,2,\ldots,m \end{cases}$$

We define the membership function of a fuzzy feasible set as

$$\underset{\sim}{R}(x) = \bigwedge_{1\leq i\leq p} \underset{\sim}{A}_i(x)$$

Secondly, we construct a fuzzy optimum point set $\underset{\sim}{A}^*$ as follows. Let $0<c<1$. If $f_i(\bar{x}) \neq M_i$ and $f_i(\bar{x}) \neq m_i$, $i \in \{1,2,\ldots,p\}$, then we define the membership function of a fuzzy value set as

$$\underset{\sim}{B}_i^*(y) = \begin{cases} \dfrac{y-m_i}{f_i(\bar{x})-m_i}\,c & \text{for} \quad m_i \leq y < f_i(\bar{x}) \\ \dfrac{y-f_i(\bar{x})}{M_i-f_i(\bar{x})}(1-c)+c & \text{for} \quad f_i(\bar{x}) \leq y \leq M_i \end{cases}$$

If there is some $i_o$, $i_o \in \{1,2,\ldots,p\}$, such that $f_{i_o}(\bar{x}) = M_{i_o}$ or $f_{i_o}(\bar{x}) = m_{i_o}$, then we define the membership function of the fuzzy value set as

$$\underset{\sim}{B}^*_{i_o}(y) = \frac{y-m_{i_o}}{M_{i_o}-m_{i_o}}\, c \qquad \text{for} \qquad m_{i_o} \leqslant y \leqslant f_i(\bar{x})$$

and

$$\underset{\sim}{B}^*_{i_o}(y) = \frac{y-m_{i_o}}{M_{i_o}-m_{i_o}}(1-c)+c \qquad \text{for} \qquad f_i(\bar{x}) \leqslant y \leqslant M_{i_o}$$

respectively, where $m_{i_o}$, $M_{i_o}$ are the infimum and supremum of $f_{i_o}(x)$, respectively. Clearly, $\underset{\sim}{B}^*_i(y)$'s are strictly monotone increasing.

Let $\underset{\sim}{A}^*_i(x) = \underset{\sim}{B}^*_i(f_i(x))$, $i=1,2,\ldots,p$, and

$$\underset{\sim}{A}^*(x) = \bigwedge_{1 \leqslant i \leqslant p} \underset{\sim}{A}^*_i(x)$$

then $\underset{\sim}{A}^*$ is a fuzzy optimum point set and

$$\underset{\sim}{A}^*_i(\bar{x}) = \underset{\sim}{B}^*_i(f_i(\bar{x})) = c, \quad i = 1,2,\ldots,p.$$

Now, we prove that a weak-efficient solution $\bar{x}$ is an optimal solution of fuzzy programming $(\underset{\sim}{P})$.

Obviously, $\underset{\sim}{H}^*(\bar{x}) > 0$. Suppose that $\bar{x}$ is not an optimal solution of $(\underset{\sim}{P})$. Then, there exists an $\bar{\bar{x}} \in R = \{x : g(x) \geqslant 0\}$, such that

$$\underset{\sim}{H}^*(\bar{\bar{x}}) > \underset{\sim}{H}^*(\bar{x})$$

and since

$$\underset{\sim}{H}^*(\bar{\bar{x}}) = \underset{\sim}{A}^*(\bar{\bar{x}}), \qquad \underset{\sim}{H}^*(\bar{x}) = \underset{\sim}{A}^*(\bar{x})$$

then

$$\min_{1 \leqslant i \leqslant p} \underset{\sim}{A}^*_i(\bar{\bar{x}}) > \min_{1 \leqslant i \leqslant p} \underset{\sim}{A}^*_i(\bar{x})$$

Therefore

$$\underset{\sim}{A}^*_i(\bar{\bar{x}}) \geqslant \min_{1 \leqslant i \leqslant p} \underset{\sim}{A}^*_i(\bar{\bar{x}}) > \min_{1 \leqslant i \leqslant p} \underset{\sim}{A}^*_i(\bar{x}) = \underset{\sim}{A}^*_i(\bar{x}) = c, \quad i=1,2,\ldots,p.$$

It follows that

$$\underset{\sim}{B}^*_i(f_i(\bar{\bar{x}})) > \underset{\sim}{B}^*_i(f_i(\bar{x})), \quad i = 1,2,\ldots,p.$$

Because $\underset{\sim}{B}^*_i(y)$ are strictly monotone increasing functions for each i, we can obtain

$$f_i(\bar{\bar{x}}) > f_i(\bar{x}), \qquad i=1,2,\ldots,p.$$

which contradicts that $\bar{x}$ is a weak-efficient solution of F(x) on R.

Corollary 3.1. Assume that $\bar{x}$ is an efficient solution of F(x) on $R=\{x:g(x) \geq 0\}$. Then, there is a fuzzy feasible set $\underset{\sim}{R}$ and a fuzzy optimum point set $\underset{\sim}{A}^*$, such that $\bar{x}$ is an optimal solution of fuzzy programming problem ($\underset{\sim}{P}$).

This completes our analysis of relationships between optimal solutions of fuzzy programming and multiobjective programming.

REFERENCES

Feng Y. (1981). Fuzzy solution of multiple objective problem (in Chinese). Kexue Tongbao 17, 1028-1030.

Feng Y. (1983). A method using fuzzy mathematics to solve the vector-maximum problem. Fuzzy Sets and Syst. 9, 129-136.

Feng Y. and Wei Q. (1982). General form of fuzzy solution in multiobjective programming (in Chinese). J. of Fuzzy Mathematics 2, 29-35.

Kiyotaka, S. (1976). Systems Optimization Theory. Korona, Tokyo.

Wierzbicki, A.P. (1979). A methodological guide to multiobjective optimization. Proc. 9th IFIP Conf. on Optimization Techniques (Warsaw), Springer-Verlag, Berlin.

Zimmermann, H.-J, (1978). Fuzzy programming and linear programming with several objective functions. Fuzzy Sets and Syst. 1, 45-55.

# FUZZY PROGRAMMING AND THE MULTICRITERIA DECISION PROBLEM

J. J. Buckley

Mathematics Department
University of Alabama in Birmingham
Birmingham, AL 35294, USA

Abstract. This paper studies the use of fuzzy programming in determining undominated, and only undominated, solutions to multicriteria decision problems. The multicriteria problem is not fuzzy, and fuzzy programming is employed to generate the set of undominated solutions. Membership functions are defined in the usual way when the objective is to maximize all the objective functions in the multicriteria decision problem. We first consider the product operator as a method of combining the membership functions. We show that the set of solutions to the fuzzy program is the Pareto optimal set for all multicriteria decision problems. We also discuss an interactive application and a solution algorithm for solving the fuzzy program. We next discuss the minimum operator as a procedure for combining the membership functions. We show that the set of solutions to the fuzzy program always contains the set of undominated solutions, but some solutions to the fuzzy program may be dominated. We then study arbitrary methods G of combining the membership functions. We show that the set of solutions to the fuzzy program is the Pareto optimal set for all multicriteria decision problems if and only if G has the dominance and the zero properties. We then apply these results to some new methods of combining membership functions that have recently appeared.

Keywords: fuzzy programming, Pareto set, efficient points, undominated solutions.

## 1. INTRODUCTION

The objective of this paper is to investigate how fuzzy programming may be used to determine the Pareto optimal set (the set of efficient points, the set of undominated solutions) for any multicriteria programming problem.

The multicriteria decision problem is

$$\max_{x} \left(f_1(x), \ldots, f_n(x)\right) \quad \text{subject to: } x \in X \tag{MC}$$

where $n \geq 2$ and the $f_i$ are real-valued functions defined on X. In general, no assumptions will be made about the feasible set X and the objective functions $f_i$ except in Section 2. There we will assume that $\max(f_i(x))$, for $x \in X$, exists and is known for $i = 1,2,\ldots,n$ when we discuss a solution algorithm. In fact, one might consider generalizing to the case where the objective functions take their values in some linearly ordered space, but in this paper we will assume the $f_i$ are real-valued.

Let $F(x) = (f_1(x),\ldots,f_n(x))$, a mapping from X to $R^n$. Define $O = F(X)$, the image of the feasible set under the mapping F. If $x, y \in X$, then x dominates y if and only if $f_i(x) \geq f_i(y)$ for all i and $f_k(x) > f_k(y)$ for some k. The Pareto optimal set P is all the undominated x in X. If $u, v \in R^n$, then $u \leq v$ if and only if $u_i \leq v_i$ for all i. Vector v dominates vector u if and only if $v \geq u$ and $v_k > u_k$ for some k. If A is any subset of $R^n$, then P(A) denotes all the undominated vectors in A. Define $\bar{P} = F(P)$.

Since the feasible set and the objective functions are arbitrary, it may turn out that the Pareto optimal set is empty. If P is empty, then $\bar{P}$ and P(O) are both empty and hence equal. If P is not empty, then it was shown in Buckley (1983) that $\bar{P} = P(O)$ and they are not empty. See Buckley (1983) for sufficient conditions guaranteeing that P is not empty.

Problem (MC) is not fuzzy and we will now employ fuzzy programming as a tool to generate P. We first need to define the membership functions $\mu_i$ for each objective function. If $x \in X$, let $c = F(x)$ where $c = (c_1,\ldots,c_n)$. Also, let $v_i = f_i(x), 1 \leq i \leq n$. For each objective function $f_i$, $1 \leq i \leq n$, define

$$\mu_i(v_i) = \begin{cases} 0, & \text{if} \quad v_i < c_i, \\ h_i(v_i), & \text{if} \quad v_i \geq c_i, \end{cases}$$

where $h_i(c_i) = \delta_i$, $0 < \delta_i < 1$, and the $h_i$ are monotonically increasing on $[c_i,+\infty)$ with $h_i(v_i) < 1$ for all $v_i > c_i$. We will discuss in Section 4 why we have required $h_i(c_i)$ to be strictly positive. We do not need to assume that the functions $h_i$ are continuous or differentiable on $(c_i,+\infty)$. In Section 4 we will add one further condition on the $h_i$ where we will assume that

$$\lim_{v_i \to +\infty} h_i(v_i) = 1.$$

If the maximum value of a $f_i(x)$ on X exists - as assumed earlier - and is known to equal $b_i$, the corresponding membership function $\mu_i$ could be changed to equal one for $v_i \geq d_i$, where $b_i \leq d_i$. The definition of $\mu_i$ would then be

$$\mu_i(v_i) = \begin{cases} 0, & \text{if } v_i < c_i, \\ h_i(v_i), & \text{if } c_i \leq v_i < d_i, \\ 1, & \text{if } v_i \geq d_i, \end{cases}$$

where $c_i < b_i \leq d_i$, $h_i(c_i) = \delta_i$, $0 < \delta_i < 1$, and the $h_i$ are monotonically increasing on $[c_i, d_i)$ with $h_i(v_i) < 1$. If, in addition, $c_i = b_i = d_i$, then we would need to define $\mu_i$ as follows

$$\mu_i(v_i) = \begin{cases} 0, & \text{if } v_i < c_i = b_i = d_i, \\ 1, & \text{if } v_i \geq c_i = b_i = d_i. \end{cases}$$

We will employ this second definition of the $\mu_i$ in Section 2 when we discuss a solution algorithm, otherwise we use the first definition of the membership functions.

Other types of membership functions are used in fuzzy programming. If a goal was to make $f_1(x)$ approximately equal to N, then a triangular membership function centered at M might be appropriate. But, we are not assuming vague, or fuzzy, goal statements. The multicriteria decision problem is crisp, not fuzzy. We are employing fuzzy programming only as a technique to obtain undominated solutions. Therefore, our type of membership function is appropriate if the objective is to maximize all objective functions.

We next need to specify a procedure of combining the membership functions into one objective function. Let $P = [0,1]^n$ and $G: P \rightarrow R$. The fuzzy program is

$$\begin{aligned} &\max\ G(\mu_1(f_1(x)), \ldots, \mu_n(f_n(x))) \\ &\text{subject to: } x \in X. \end{aligned} \tag{G1}$$

Problem (G1) is the same as

$$\begin{aligned} &\max\ G(\mu_1(v_1), \ldots, \mu_n(v_n)) \\ &\text{subject to: } v \in O. \end{aligned} \tag{G2}$$

We will investigate arbitrary methods (G) of combining the $\mu_i$ in Section 4. Two important candidates for G are

$$G(p_1,\ldots,p_n) = \min(p_1,\ldots,p_n),$$

and

$$G(p_1,\ldots,p_n) = \prod_{i=1}^{n} p_i,$$

where $p_i = \mu_i(v_i)$, $1 \leq i \leq n$.

The minimum operator has been the most popular method of combining the $\mu_i$ since it was first introduced in Bellman and Zadeh (1970). The minimum operator has been described as a non-interactive technique while the product operator has been called an interactive procedure (Yager, 1978); see also Bellman and Zadeh, 1970, p. 280; Dyson, 1980). The minimum operator, and the product operator, have not gone uncriticized (Carlsson, 1982; Dyson, 1980; Luhandjula, 1982; Thole, Zimmermann and Zysno, 1979; Zimmermann, 1983; Zimmermann and Zysno, 1980, 1983). We first study solutions to problems (G1) and (G2) in Section 2 when the product operator is used to combine the membership functions. In Section 3 we look at solutions to (G1) and (G2) using the minimum operator. Some authors (Luhandjula, 1982; Yager, 1978; Zimmermann, 1983; Zimmermann and Zysno, 1980, 1983) have suggested other ways to combine the $\mu_i$. We will discuss their methods in Section 4 together with a general result on the structure of G so that solutions to problem (G1) equal P.

Several authors (Buckley, 1983, 1984; Feng, 1983; Hannan, 1979; Leberling, 1981; Zimmermann, 1978) have studied fuzzy programming as a tool to determine P. The initial papers were on linear multicriteria programming using the minimum or product operator. The most general results to date are in Buckley (1983), Ester and Schwartz (1983) and Feng (1983). Both Buckley (1983) and Feng (1983) do assume concave objective functions and place restrictive assumptions on the feasible set in order to obtain some of their major results. They also consider only the minimum or product operator for combining the membership functions.

Finding the Pareto optimal set, without fuzzy programming techniques, has also been an active area of research. In Geoffrion (1968) it is shown that all solutions to

$$\max\left(\sum_{i=1}^{n} \lambda_i f_i(x)\right)$$

$$\text{subject to: } x \in X,$$

for $\lambda_i > 0$, all i, and $\lambda_1 + \ldots + \lambda_n = 1$, are the properly efficient solutions to (MC) when the $f_i$ are concave and X is convex.

See Gal (1983) for a survey of this research area. Recently, some authors (Choo and Atkins, 1983; Ester and Schwartz, 1983) have been able to extend these results to non-convex X, or to other settings.

Before proceeding to the main results of this paper we need to define some notation that will be used in the following sections. Let $\mu = (\mu_1,\ldots,\mu_n)$ and $\delta = (\delta_1,\ldots,\delta_n)$. We usually assume that $\delta$ is fixed and we vary the $\mu_i$ through changing the $c_i$. One changes c by choosing different x in X and computing $c = F(x)$. If v, $\bar{v}$, and $v^*$ are vectors in 0, then $p = \mu(v)$, $\bar{p} = \mu(\bar{v})$, and $p^* = \mu(v^*)$ are their corresponding values in P. We are usually interested in solving (G2) and then translating back to X. Suppose $v^*$ solves (G2). We then need to solve $F(x) = v^*$, for $x^*$, producing a solution to (G1). The solution for $x^*$ involves solving a system of non-linear equations simultaneously. If F is defined on a larger space containing X, then some solutions may not belong to X and must be discarded. In fact, we sometimes wish to solve $F(x) = v$ for x when v is any vector in $R^n$. If F is defined only on X and v is not in O, then $F(x) = v$ has no solution. If F is defined on a space containing X and v is not in 0, then solutions to $F(x) = v$ will not belong to X. Throughout this paper, if v is in 0, then only those x in X solving $F(x) = v$ will be considered solutions to the system of non-linear equations. If v is an arbitrary vector in $R^n$, then we have a test to see if v is not in 0: either $F(x) = v$ has no solution or its solutions are not in X.

## 2. THE PRODUCT OPERATOR

We are interested in solutions to

$$\max\left(\prod_{i=1}^{n} \mu_i(f_i(x))\right) \quad \text{subject to: } x \in X. \tag{P1}$$

Problem (P1) is equivalent to

$$\max\left(\prod_{i=1}^{n} \mu_i(v_i)\right) \quad \text{subject to: } v \in 0. \tag{P2}$$

We need to specify what is meant by $v^*$ or $x^*$ being a solution to (P2) or (P1), respectively. If there is a $v^* \in 0$ so that

$$\prod_{i=1}^{n} \mu_i(v_i) \leq \prod_{i=1}^{n} \mu_i(v_i^*) < +\infty$$

for all v in O, then $v^*$ solves (P2). If $x^*$ is any solution to $F(x) = v^*$, then $x^*$ solves (P1). Conversely, if there is an $x^* \in X$ so that

$$\prod_{i=1}^{n} \mu_i(f_i(x)) \leq \prod_{i=1}^{n} \mu_i(f_i(x^*)) < +\infty$$

for all x in X, then $x^*$ solves (P1) and $v^* = F(x^*)$ solves (P2). Since X and the objective functions are arbitrary, problems (P1) and (P2) may have no solution either because the product is unbounded or the maximum is not attained in X, or O. Naturally, we are always assuming X is not empty or else there is no problem to solve.

Notice also that if $v^* \in \bar{P}$, then any $x^*$ solving $F(x) = v^*$ belongs to P. Conversely, if $x^* \in P$, then $v^* = F(x)$ belongs to $\bar{P}$.

Theorem 1. The set of solutions to (P2) for c in O is $\bar{P}$.

Proof. If problem (P2) has no solution for any c in O, then the set of solutions is empty and a subset of $\bar{P}$. Therefore, let $v^*$ solve (P2) for some c in O. Suppose $v^*$ is not in $\bar{P}$ which implies that there is a $\bar{v}$ in O that dominates $v^*$ with $\bar{v}_k > v_k^*$ for some index k. Domination implies $\mu_i(\bar{v}_i) \geq \mu_i(v_i^*)$ for $i = 1,2,\ldots,n$. From the way the $\mu_i$ were defined we know that no $\mu_i(v_i^*)$ can equal zero because $v^* \geq c$. Therefore, $\mu_k(\bar{v}_k) > \mu_k(v_k^*)$ and it follows that

$$\prod_{i=1}^{n} \mu_i(\bar{v}_i) > \prod_{i=1}^{n} \mu_i(v_i^*).$$

This contradicts $v^*$ solving (P2). Hence $v^* \in \bar{P}$.

If $\bar{P}$ is empty, it is a subset of the set of solutions to (P2), so suppose $v^*$ belongs to P. If $c = v^*$, we claim $v^*$ solves (P2). Choose any $\bar{v}$ in O not equal to $v^*$. Since $v^*$ is in $\bar{P}$, there must be some index k so that $\bar{v}_k < v_k^*$. Then $\mu_k(v_k)$ is zero and

$$\prod_{i=1}^{n} \mu_i(\bar{v}_i) = 0.$$

The product is zero for every $\bar{v}$ in O not equal to $v^*$ and the product evaluated at $v^*$ is positive. Therefore, $v^*$ solves (P2) for $c = v^*$.

Notice $\delta_i > 0$, all i, was used in Theorem 1 especially to show $\bar{P}$ is a subset of the set of solutions to (P2). We will show in Section IV that the fuzzy program (G1) can have dominated solutions when all the $\delta_i$ equal zero.

Corollary 1. The set of solutions to (P1) for x in X, where $c = F(x)$, is P.

The results of Theorem 1 and Corollary 1 are the best possible because they are true for all multicriteria decision problems. They are true for any feasible set and for all real-valued functions defined on X.

An interactive method for finding a "best" compromise solution to (MC) may be constructed for the product operator. An outline of the procedure is as follows:

1. Choose $x_a, x_b, x_c, \ldots$ (randomly) in X and solve (P2) for $c_a, c_b, \ldots$ where $c_a = F(x_a)$, $c_b = F(x_b), \ldots$ . Let the solutions to (P2) be $v_a^*, v_b^*, \ldots$ .

2. Show $v_a^*, v_b^*, \ldots$ to the decision maker. The decision maker rejects some $v_b^*$ and does not reject some other $v_a^*$.

3. Suppose $v_a^*$ is not rejected by the decision maker and let $v^* = (v_{a1}^*, \ldots, v_{an}^*)$ and $c_a = (c_{a1}, \ldots, c_{an})$. The decision maker expresses a desire to see the values of $v_{ai}^*$ increase for $i \in I$, I is some proper subset of $\{1, 2, \ldots, n\}$. Choose $\bar{c}_{ai} > v_{ai}^*$ for i not in I. Solve (P2) for $\bar{c}_a = (\bar{c}_{a1}, \ldots, \bar{c}_{an})$ giving $\bar{v}_a^*$ with $\bar{v}_{ai}^* > v_{ai}^*$ for each i in I. If (P2) has no solution for $\bar{c}_a$, choose smaller increases for $\bar{c}_{ai}$ when i is in I but keep the values at least equal to $v_{ai}^*$.

You never have to translate solutions to (P2) back to X until the end of the process when the decision maker accepts some $v^*$ as the "best" compromise solution. You only need to show the values of the objective functions $v^*$ to the decision maker at each stage. When the decision maker decides on $v^*$ as the solution, then you solve $F(x) = v^*$ for x in P as the "best" compromise solution.

If $\max(f_i(x))$, for $x \in X$, exists and is known to be $b_i$, for some i in I at step (3), then the choice of the $\bar{c}_{ai}$ is much easier. One could use

$$\bar{c}_{ai} = \lambda v_{ai}^* + (1 - \lambda) b_i$$

for $\lambda = 1/4, 1/2, 3/4, 1$. If $v_{ai}^*$ is equal to $b_i$, then we would delete this i from I.

What is needed to implement the interactive procedure is a solution algorithm for problem (P2). For the rest of this section we assume max $f_i(x)$, $x \in X$, exists and is known to equal $b_i$ for $i = 1, 2, \ldots, n$. The results of Theorem 1, and Corollary 1, are independent of the type of function $h_i$ used in the definition of $\mu_i$ and for simplicity we will now make the $h_i$ linear. Also, we will set $\delta_i = \theta$, $1 \leq i \leq n$, where $\theta < \theta < 1$. If $c_i < b_i$, the definition of $\mu_i$ is

$$\mu_i(v_i) = \begin{cases} 0, & \text{if } v_i < c_i \\ m_i(v_i - c_i) + \theta, & \text{if } c_i \leq v_i \leq b_i. \\ 1, & \text{if } v_i \geq b_i, \end{cases}$$

where $m_i = (1 - \theta)/(b_i - a_i)$. If $c_i = b_i$, then $\mu_i(v_i)$ is zero for $v_i < c_i$ and equals one when $v_i \geq c_i$. We will first assume that $c_i < b_i$ for all i. If $b = (b_1, \ldots, b_n)$ belongs to 0, then the objective functions are not conflicting and the solution to (MC) is any x solving $F(x) = b$. We therefore assume b is not in 0. Problem (P2) becomes

$$\max\left(\prod_{i=1}^{n} [m_i(v_i - c_i) + \theta]\right) \qquad \text{subject to: } v \in 0 \text{ and } v \geq c. \tag{P2*}$$

If we set $w_i = m_i(v_i - c_i) + \theta$, $1 \leq i \leq n$ and $w = (w_1, \ldots, w_n)$, then problem (P2*) is equivalent to

$$\max \prod_{i=1}^{n} w_i \qquad \text{subject to: } w \in 0 \text{ and } w \geq \delta, \tag{P3}$$

where $\delta = (\theta, \theta, \ldots, \theta)$ and

$$0 = \{ w \mid w_i = m_i(v_i - c_i) + \theta,\ v \in 0 \}.$$

If $w^*$ solves (P3), then $v^*$ solves ($P2^*$) where $v_i^* = (w_i^*-\Theta)/m_i+c_i$, $1 \leq i \leq n$.

We will outline an algorithm for solving (P3), and hence ($P2^*$) and (P1), when there are two objective functions. The procedure is easily extended to $n \geq 3$. The algorithm depends on an efficient method of testing if v is in 0 given any v in $R^n$. Equivalently, it requires testing if x is in X given any solution x to $F(x) = v$, where $v \in R^n$. Of course, if $F(x) = v$ has no solution for x then v is not in 0. One must solve a system of non-linear equations simultaneously to obtain x given v. There are numerous algorithms available to solve this problem.

Let $\varepsilon$ be a suitably small positive number, set $r = 1 - \varepsilon$, and let K be a suitably large positive integer. The algorithm is:

1. For $j = 0,1,\ldots,K$ let $w_1 = r + j([1-r]/K)$, $w_2 = r/w_1$, $v_i = (w_i - \Theta)/m_i+c_i$ for $i = 1,2$, and $v = (v_1,v_2)$. Test to see if v belongs to 0. If v is in 0, then go to 2. Otherwise, increase j by one if j is less than K. If j equals K, then set $r = r-\varepsilon$ and go back to 1.

2. Refine the estimate of r so that we obtain the desired accuracy on the value of $r^*$ where

$$\left\{w \mid w_1w_2 = r^*\right\} \cap 0 \neq \emptyset$$

but

$$\left\{w \mid w_1w_2 = r\right\} \cap 0 = \emptyset,$$

for any $r > r^*$.

3. Choose $w^* = (w_1^*,w_2^*)$ to be any element in the intersection of $w_1w_2 = r^*$ and 0. Then $v^* = (v_1^*,v_2^*)$ solves ($P2^*$) where $v_i^* = (w_i^* - \emptyset)/m_i + c_i$, $i = 1,2$.

The geometry of problem (P3) is shown in Fig. 1. The first part of the algorithm finds values of w along the curve $w_1w_2 = r$. The idea of the algorithm is to find the first intersection, as r decreases, of the curve $w_1w_2 = r$ and 0. Notice that we need $r < 1$ because (1,1) does not belong to 0 and the curve $w_1w_2 = 1$ will not intersect 0.

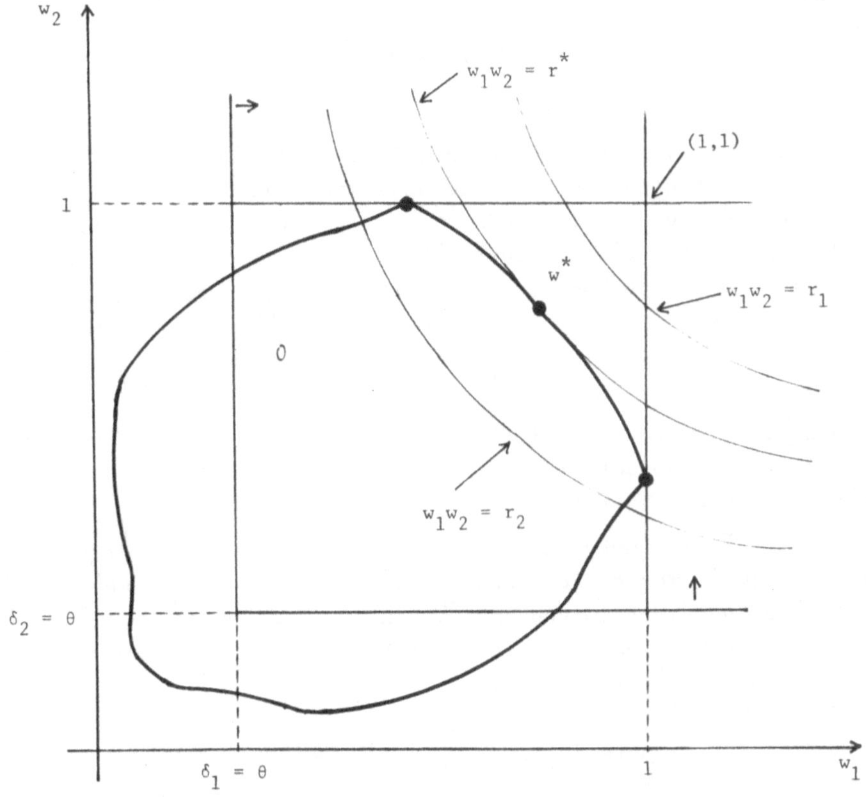


Fig. 1. The geometry of the solution algorithm for two objective functions ($r_2 < r^* < r_1 < 1$)

A $c_i$ value could equal $b_i$ because we have assumed that the maximum value of $f_i$ is known. When this happens we set, and keep, the corresponding $w_i$ to be equal to one, and the procedure obtains the solution where $v_i^*$ equals $b_i$.

## 3. THE MINIMUM OPERATOR

We now study solutions to

$$\begin{aligned}&\max(\min[\mu_1(f_1(x)),\ldots,\mu_n(f_n(x))])\\&\text{subject to: } x \in X.\end{aligned} \quad \text{(M1)}$$

Problem (M1) is equivalent to

$$\max(\min[\mu_1(v_1),\ldots,\mu_n(v_n)])$$

$$\text{subject to: } v \in 0. \qquad \text{(M2)}$$

A vector $v^*$ in 0 solves (M2) if and only if its maximin value is finite and is at least as large as any other maximin value for $v$ in 0. Then $x^*$ solves (M1) if $x^*$ is a solution to $F(x) = v^*$. Conversely, $x^*$ solves (M1) if its maximin value is finite and is greater than or equal to all other maximin values for $x$ in $X$. Then $v^* = F(x^*)$ solves (M2). Of course, both problems may have no solution for a given $c$ in 0.

<u>Theorem 2</u>. The set of solutions to (M2) for $c$ in 0 contains $\bar{P}$. If $S$ is the set of solutions to (M2) for a fixed $c$ in 0, then $P(S)$ is a subset of $\bar{P}$.

<u>Proof</u>. If $\bar{P}$ is empty, then it is a subset of the set of solutions to (M2). Therefore, assume $v^* \in \bar{P}$. If $c = v^*$, then we show that $v^*$ solves (M2). Given any $\bar{v}$ in 0, not equal to $v^*$, then we must have $v_k < v_k^*$, for some k, because $v^*$ belongs to $\bar{P}$. But then $\mu_k(v_k) = 0$ and

$$\min[\mu_1(\bar{v}_1),\ldots,\mu_n(\bar{v}_n)] = 0.$$

The minimum is zero for any $\bar{v}$ in 0 not equal to $v^*$. Since

$$\min(\mu_1(v_1^*),\ldots,\mu_n(v_n^*)) = \min[\delta_i] > 0$$

we see $v^*$ solves (M2) and $\bar{P}$ is a subset of the set of all solutions to (M2).

Fix $c$ in 0 and let $S$ be all solutions to (M2). If $S$ is empty, then its set of undominated vectors is also empty and a subset of $\bar{P}$. Also, if $P(S)$ is empty, it is a subset of $\bar{P}$. Hence, we assume $v^*$ belongs to $P(S)$. If $v^*$ does not belong to $\bar{P}$, then there is a $\bar{v}$ in 0 that dominates $v^*$. Domination implies $\mu_i(\bar{v}_i) \geq \mu_i(v_i^*)$ all i. This implies that $\bar{v}$ also solves (M2) and belongs to $S$. This contradiction says that $v^*$ belongs to $\bar{P}$.

Corollary 2. The set of solutions to (M1) for $x$ in $X$ and $c = F(x)$ contains $P$. If $S$ is all solutions to (M1) for some fixed $c = F(x)$, with $x$ in $X$, then $P(S)$ is contained in $P$.

The following example shows that $S$ may not be contained in $P$. Therefore, the above results are the best possible for the minimum operator. They are true for any feasible set and for all real-valued objective functions.

This example, adopted from Chanas (1985), shows that solutions to (M1) may be dominated. It is a linear multicriteria decision problem which implies that even with very strong assumptions about $X$ and the $f_i$, solutions to (M1) may not belong to $P$.

Example. Let (M1) be:

$$\max(x_1 + x_2 + 1,\ 2x_1 - x_2 + 2,\ x_3 + 1)$$

$$\text{subject to:} \quad \begin{aligned} x_1 &\le 1, \\ x_2 &\le 1, \\ x_3 &\le 1, \\ x_1 + x_2 + x_3 &\le 1.5, \\ x_i &\ge 0. \end{aligned}$$

The variables $x_i$ are real variables so $X$ is a subset of $R^3$. Let $c = (19/10,\ 11/10,\ 16/10)$ which is in $0$. We can easily see that $\max f_i(x)$ exists for $x$ in $X$ and the $b_i$ values are $b_1 = 5/2$, $b_2 = 4$ and $b_3 = 2$. The membership functions $\mu_i$ will all be linear on $[c_i, b_i]$ and their definitions are

$$\mu_1(v_1) = (2/3)v_1 - 2/3, \quad \text{if} \quad 19/10 \le v_1 \le 5/2$$

$$\mu_2(v_2) = (2/15)v_2 + 7/15, \quad \text{if} \quad 11/10 \le v_2 \le 4,$$

$$\mu_3(v_3) = v_3 - 1, \quad \text{if} \quad 16/10 \le v_3 \le 2.$$

Each $\mu_i$ is zero if $v_i < c_i$ and equals one when $v_i \ge b_i$. All the $\delta_i$ values are positive.

In $0$ the solution set to (M2) is

$$S = \left\{ (19/19, v_2, 16/10) \quad 11/10 \le v_2 \le 38/10 \right\},$$

with only $v^* = (19/19, 38/10, 16/10)$ in $\bar{P}$. All other $v$ in $S$

are dominated by this $v^*$. One may check that each $v$ in $S$ produces a maximin value of 3/5 and this is the optimal maximin value for $v$ in $0$. Translating back to $X$, the set of solutions to (M1) is

$$S = \left\{ \left(\frac{1}{3}\lambda - \frac{11}{30}, \frac{38}{30} - \frac{1}{3}\lambda, \frac{6}{10}\right) \mid \frac{11}{10} \leq \lambda \leq \frac{38}{10} \right\}$$

with only $x^* = (9/10, 0, 6/10)$ in $P$.

Therefore, in comparing the product operator and the minimum operator, the product operator is preferable if one wishes to employ fuzzy programming to generate the set of undominated, and only undominated, solutions to multicriteria decision problems.

## 4. THE G OPERATOR

In this section we are interested in finding conditions on $G$ so that the set of solutions to (G2) and (G1) is $\overline{P}$ and $P$, respectively. We will keep the $\delta_i$, $0 < \delta_i < 1$, values fixed and let $\delta = (\delta_1, \ldots, \delta_n)$. We also assume that

$$\lim_{v_i \to +\infty} h_i(v_i) = 1,$$

for $i = 1, 2, \ldots, n$. This will be used in the "only if" part of Theorem 3 below. With $\delta$ fixed, the domain of $G$ is not all of $P = [0,1]^n$. Let $\bar{o} = (0, \ldots, 0)$, $\Gamma = \{ p \mid p \geq \delta,\ p_i < 1 \}$ and $\Lambda = \{ p \mid \text{some } p_i = 0, \text{ the rest } \delta_i \leq p_i < 1 \}$. If $D$ is the union of $\bar{o}$ and $\Gamma$ and $\Lambda$, then $\mu(v)$ is in $D$ for every $v$ in $0$. Therefore, $G$ is only evaluated at those $p$ values in $D$. In this section it is assumed that all the $p$ values always belong to $D$.

<u>Definition 1</u>. $G$ satisfies the dominance (D) condition if and only if $\overline{p}$ dominates $p$ and $G(p) \geq G(\delta)$, then $G(\overline{p}) > G(p)$.

<u>Definition 2</u>. $G$ satisfies the zero (Z) condition if and only if $G(\delta) > G(p)$ whenever $p \in \Lambda$ or $p = \bar{o}$.

The dominance condition is a basic property of any ordinal utility function. Suppose $\overline{v}$ dominates $v$, both in $0$ with $v \geq c$. Any decision maker would prefer $\overline{v}$ to $v$. Then the "utility" of $\overline{v}$ should be greater than the "utility" of $v$. If $\overline{p} = \mu(\overline{v})$ and $p = \mu(v)$, then $\overline{p}$ dominates $p$ and $G(p) \geq G(\delta)$. Hence $G(\overline{p}) > G(p)$ if $G$ has the dominance property. Vectors $v$ not greater than or equal to $c$ are excluded from consideration and cannot solve (G2).

Theorem 3. $\{\forall X, \forall F[v^*$ solves (G2) for some $\mu$ if and only if $v^* \in \bar{P}]\}$ if and only if $\{$ G possesses properties (D) and (Z)$\}$.

Proof. First suppose G satisfies both conditions (D) and (Z).

1. Let $v^*$ belong to $\bar{P}$. We will use the $\mu$ with $c$ equal to $v^*$. If $\bar{v}$ is in 0 and $\bar{v}$ does not equal $v^*$, then $\bar{v}_k < v^*_k$ for some $k$ because $v^*$ is undominated. Hence $p_k = \mu_k(\bar{v}_k) = 0$. If $\bar{p} = \mu(\bar{v})$ and $p^* = \mu(v^*)$, then $\bar{p}$ belongs to $\Lambda$, or $p = \bar{o}$, while $G(p^*) = G(\delta)$. Property (Z) implies $G(p^*) > G(\bar{p})$ for any $\bar{v}$, not equal to $v^*$, in 0. Hence $v^*$ solves (G2). Of course, if $\bar{P}$ is empty, it is a subset of the set of solutions to (G2).

2. Let $v^*$ solve (G2) for some $\mu$. We know $v^* \geq c$ so $G(p^*) \geq G(\delta)$ if $p^* = (v^*)$. If $v^*$ is not in $\bar{P}$, then some $\bar{v}$ in 0 dominates $v^*$. If $\bar{p} = \mu(\bar{v})$, then $\bar{p}$ dominates $p^*$. Condition (D) implies that $G(\bar{p}) > G(p^*)$ contradicting $v^*$ solves (G2). Hence $v^*$ belongs to $\bar{P}$. Again, the case where (G2) has no solutions for all $\mu$ is trivial.

To show the "only if" part of the theorem we will prove the contrapositive. We show if G does not satisfy condition (D) or (Z), then there exists an X and an F so that

$$\{[v^* \text{ does not solve (G2) for any } \mu \text{ and } v^* \text{ belongs to } \bar{P}] \text{ or } [v^* \text{ solves (G2) for some } \mu \text{ and } v^* \text{ is not in } \bar{P}]\}.$$

Our examples will be for two objective functions which may be extended to $n \geq 3$. Also, we choose 0 to be any subset of $R^2$ because X and F are arbitrary. If 0 is a subset of $R^2$, then we can pick $f_i(x) = x_i$ for $i = 1,2,.$ and $X = 0$.

1. First assume G does not possess property (D). The function G may, or may not, have property (Z). Then there exists a $\bar{p}$ and a $p$, $\bar{p}$ dominates $p$ and $G(p) \geq G(\delta)$, but $G(\bar{p}) \leq G(p)$. We will define $\bar{v}$ and $v$, and a third vector $c$, in $R^2$ so that if $0 = \{c, v, \bar{v}\}$ then $\bar{p} = \mu(\bar{v})$, $p = u(v)$, v dominates $v$, and $v$ solves (G2). The vector $v$ must solve (G2) because $G(p) \geq G(p)$ and $G(p) \geq G(\delta)$ where $\delta = \mu(c)$. The vector $v$ will not belong to $\bar{P}$ because it is dominated by $\bar{v}$.

There are a number of cases to consider: $\bar{p}$ and $p$ in $\Gamma$; $\bar{p}$ in $\Gamma$ and $p$ in $\Lambda$; $\bar{p}$ and $p$ in $\Lambda$; $\bar{p}$ in $\Gamma$ and $p=\bar{o}$; and $\bar{p}$ in $\Lambda$ and $p = \bar{o}$. In all cases we can choose $c$ in $R^2$ and the functions $h_i$, $i = 1,2$, so that $\bar{v}$ dominates $v$ and the values of $\mu(\bar{v})$ and $\mu(v)$ are the given $\bar{p}$ and $p$, respectively. For brevity we shall present only two cases in detail.

a. Suppose $\bar{p}$ and $p$ are in $\Gamma$ and assume $\bar{p}_1 > p_1$ and $\bar{p}_2 \geq p_2$. If $p_1 = \delta_1$, then set $c_1 = v_1 = 0$, $\bar{v}_1 = 1$ and choose $h_1$ so that $h_1(1) = \bar{p}_1$. If $p_1 > \delta_1$, then let $c_1 = 0$, $v_1 = 1$, $\bar{v}_1 = 2$ and pick $h_1$ so that $h_1(1) = p_1$ and $h_1(2) = \bar{p}_1$. Now consider the values of $\bar{p}_2$ and $p_2$. If $\bar{p}_2 = p_2 = \delta_2$, then let $c_2 = v_2 = \bar{v}_2 = 0$. If $\bar{p}_2 > p_2 = \delta_2$, then set $c_2 = v_2 = 0$, $\bar{v}_2 = 1$ and have $h_2(1) = \bar{p}_2$. If $\bar{p}_2 > p_2 > \delta_2$, then let $c_1=0$, $v_2 = 1$, $\bar{v}_2 = 2$ and select $h_2$ so that $h_2(1) = p_2$, $h_2(2) = \bar{p}_2$. Finally, if $\bar{p}_2 = p_2 > \delta_2$, then choose $c_2 = 0$, $v_2 = \bar{v}_2 = 1$ and make $h_2(1) = \bar{p}_2 = p_2$. The results are: $c$ is always equal to $(0,0)$ and $\bar{v}$ dominates $v$.

b. Suppose $\bar{p}$ and $p$ are in $\Lambda$ and assume $\bar{p}_1 > p_1$, $\bar{p}_2 = p_2 = 0$. If $p_1 = \delta_1$, then let $c_1 = v_1 = 0$, $\bar{v}_1 = 1$ and $h_1(1) = \bar{p}_1$. If $p_1 > \delta_1$, then set $c_1 = 0$, $v_1 = 1$, $\bar{v}_1 = 2$ and $h_1(1) = p_1$, $h_1(2) = \bar{p}_1$. Also, let $v_2 = \bar{v}_2 = 0$ and $c_2 = 1$. Then $c = (0,1)$ and $v = (0,0)$, $\bar{v} = (1,0)$ or $v = (1,0)$, $\bar{v} = (2,0)$. Therefore, $\bar{v}$ dominates $v$.

We conclude that there is a multicriteria decision problem so that (G2) has dominated solutions if $G$ does not satisfy the dominance condition.

2. Next assume $G$ does not possess the zero property. We may at this point assume that $G$ does have the dominance property. Then there is a $\bar{p}$ in $\Lambda$, or $\bar{p} = \bar{o}$, so that $G(\bar{p}) \geq G(\delta)$. Assume $\bar{p} = (0,\bar{p}_2)$, with $\bar{p}_2 \geq \delta_2$ if $\bar{p} \in \Lambda$ or $\bar{p}_2 = 0$ when $\bar{p} = \bar{o}$. We show there is an $0$ in $R^2$ with $v^*$ in $P$ but $v^*$ does not solve (G2) for every $\mu$.

Let $V = \{(0,v_2) \mid v_2>1\}$, $H = \{(v_1,0) \mid v_1>1\}$ and $0$ is the union of $(1,1)$ and $H$ and $V$. The set $0$ is shown in Fig. 2. Clearly, $v^* = (1,1)$ and $\bar{P}$ consists of only $v^*$. There are three cases.

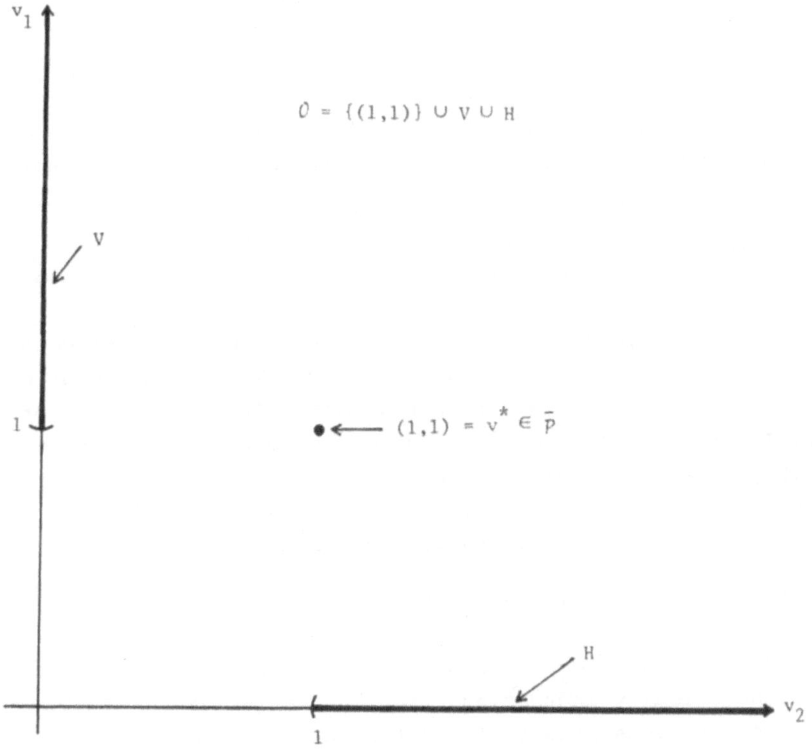


Fig. 2. The second example of a set 0 in Theorem 3 when there are two objective functions

1. Suppose $c = v^*$. There is an $M > 0$ so that $\mu_2(M) > \bar{p}_2$ since we have assumed the $h_2$ values may be made arbitrarily close to one. Let $v = \{(0,\mu_2(M))\}$. Then $\mu(v)$ dominates $\bar{p}$ with $G(\bar{p}) \geq G(\delta)$. The dominance property implies $G(0,\mu_2(M)) > > G(\bar{p})$. Therefore $v^*$ does not solve (G2) because $G(\delta) = = G(\mu(v^*))$.

2. Assume $c$ is in $V$. Then $\mu(v^*) = (\mu_1(1),0)$. If $v^*$ is to solve (G2), then we must have $G(\mu_1(1),0) \geq G(\delta)$. There is an $M > 0$ so that $\mu_1(M) > \mu_1(1)$ because the $h_1$ values may be made arbitrarily close to one. Let $v = (M,0)$. Then $\mu(v)$ do-

minates $(\mu_1(1),0)$ with $G(\mu_1(1),0) \geq G(\delta)$. The dominance property implies that $G(\mu_1(M),0) > G(\mu_1(1),0)$ and $v^*$ cannot solve (G2).

3. Assume $c$ is in $H$. The proof is similar to case 2 above.

We conclude that there is a multicriteria decision problem with an undominated solution not solving (G2) if $G$ does not have the zero property but does satisfy the dominance condition.

Corollary 3. $\{\forall X, \forall F[x^*$ solves (G1) for some $\mu$ if and only if $x^* \in P]\}$ if and only if $\{G$ possesses properties (D) and (Z)$\}$.

The dominance and zero properties are the strongest conditions to be placed on $G$ because we require that the solution set to (G1) is $P$ for all possible multicriteria decision problems. If the $f_i$ are concave and $X$ is convex, then one could possibly obtain a result like Corollary 3 with weaker conditions on $G$.

The zero condition is satisfied for

$$G(p) = \prod_{i=1}^{n} g_i(p_i)$$

if $g_i(0) = 0$ and $g_i(p_i) > 0$ for $p_i > 0$. If each $g_i$ is also monotonically increasing, then $G$ satisfies the dominance condition. Various possibilities for the $g_i$ are

$$g_i(p_i) = p_i^n, \quad n > 0,$$

$$g_i(p_i) = 1 - (1-p_i)^n, \quad n > 0,$$

$$g_i(p_i) = \log(p_i+1),$$

$$g_i(p_i) = \exp(p_i) - 1.$$

The $g_i$ functions in the product do not all have to be the same.

Notice that the minimum operator does not have the dominance property. There have been several articles recently (Luhandjula, 1982; Thole, Zimmermann and Zysno, 1979; Yager, 1978; Zimmermann, 1983; Zimmermann and Zysno, 1983) proposing new methods of combining the membership functions. One such operator is

$$G(p) = [\prod_{i=1}^{n} p_i]^{1-\gamma} \cdot [1 - \prod_{i=1}^{n} (1 - p_i)]^{\gamma},$$

for $0 < \gamma < 1$. This operator does satisfy the dominance and zero conditions and hence may be used to generate undominated, and only undominated, solutions to any multicriteria problem. Another procedure is

$$G(p) = \gamma \min_i p_i + (1 - \gamma)\min(1, \prod_{i=1}^{n} p_i),$$

for $0 < \gamma < 1$. This method satisfies neither the dominance nor the zero property and would be a poor choice to determine the Pareto optimal set for an arbitrary multicriteria problem.

Finally, let us consider what happens if $\delta_i = 0$ for all i. We show by example that (G2), and hence (G1), may have dominated solutions. Let 0, a subset of $R^2$, be $\{(v_1,v_2) | 0 \leq v_1 \leq 1\}$. Then $\overline{P}$ consists of only $v^* = (1,1)$. If $c = (1,1)$, then $G(\mu(v)) = G(0,0)$ for all v in 0 implying that 0 is the solution set. All vectors in 0, not equal to $v^*$, are dominated by $v^*$.

5. SUMMARY

This paper investigated employing fuzzy programming as a tool to obtain undominated, and only undominated, solutions to an arbitrary multicriteria decision problem. We showed that if the product operator is used to combine the membership functions, then the set of solutions to the fuzzy program is the Pareto optimal set for any multicriteria problem. If the minimum operator is employed to combine the membership functions, then solutions to the fuzzy program may be dominated.

The major results pertain to arbitrary methods G of combining membership functions in a fuzzy program. We showed that the set of solutions to the fuzzy program is always the Pareto optimal set if and only if G has a dominance and zero property.

REFERENCES

Bellman, R.E., and L.A. Zadeh (1970). Decision-making in a fuzzy environment. Mang. Sci. 17, 141-164.

Buckley, J.J. (1983). Fuzzy programming and the Pareto optimal set. Fuzzy Sets and Syst. 10, 57-63.

Buckley, J.J. (1986). A reply to 'Note on fuzzy programming and the Pareto optimal set'. Fuzzy Sets and Syst. To appear.

Carlsson, C. (1982). Tackling an MCDM-problem with the help of some results from fuzzy set theory. Eur. J. Op. Res. 10, 270-281.
Chanas, S. (1986). Note on fuzzy programming and the Pareto optimal set. Fuzzy Sets and Syst. To appear.
Choo, E.U., and D.R. Atkins (1983). Proper efficiency on non-convex multicriteria programming. Math. Op. Res. 8, 467-470.
Dubois D., and H. Prade (1980). Fuzzy Sets and Systems: Theory and Applications. Academic Press, New York.
Dyson, R.G. (1980). Maximin programming, fuzzy linear programming and multicriteria decision making. J. Op. Res. Soc. 31, 263-267.
Ester, J., and B. Schwartz (1983). An extended efficiency theorem. Math. Operationsforsch. Statist. Ser. Optim. 14, 331-342.
Feng, Y. (1983). A method using fuzzy mathematics to solve the vectormaximum problem. Fuzzy Sets and Syst. 9, 129-136.
Gal, T. (1983). On efficient sets in vector maximum problems - a brief survey. In P. Hansen (ed.), Essays and Surveys on Multiple Criteria Decision Making. Springer-Verlag, Berlin, 94-114.
Geoffrion, A.M. (1968). Proper efficiency and the theory of vector maximization. J. Math. Anal. Appl. 22, 618-630.
Hannan, E.L. (1979). On the efficiency of the product operator in fuzzy programming with multiple objectives. Fuzzy Sets and Syst. 2, 259-262.
Leberling, H. (1981). On finding compromise solutions in multicriteria problems using the fuzzy min-operator. Fuzzy Sets and Syst. 6, 105-118.
Luhandjula, M.K. (1982). Compensatory operators in fuzzy linear programming with multiple objectives. Fuzzy Sets and Syst. 8, 245-252.
Thole, U., H.-J. Zimmermann, and P. Zysno (1979). On the suitability of minimum and product operators for the intersection of fuzzy sets. Fuzzy Sets and Syst. 2, 167-180.
Zimmermann, H.-J. (1978). Fuzzy programming and linear programming with several objective functions. Fuzzy Sets and Syst. 1, 45-55.
Zimmermann, H.-J. (1983). Using fuzzy sets in operational research. Eur. J. Op. Res. 13, 201-216.
Zimmermann, H.-J., P. Zysno (1980). Latent connectives in human decision making. Fuzzy Sets and Syst. 4, 37-51.
Zimmermann, H.-J., and P. Zysno (1983). Decisions and evaluations by hierarchical aggregation of information. Fuzzy Sets and Syst. 10, 243-260.
Yager, R.R. (1978). Fuzzy decision making including unequal objectives. Fuzzy Sets and Syst. 1, 87-95.

# HIERARCHICAL PROGRAMMING WITH FUZZY OBJECTIVES AND CONSTRAINTS

Yee Leung

Department of Geography
The Chinese University of Hong Kong
Shatin, Hong Kong

Abstract. Procedures of hierarchical programming with fuzzy objectives and constraints are proposed in this paper. Compromise solutions are determined through a sequential optimization procedure by which the objectives are optimized according to their descending order of priorities. In each step, a fuzzy objective function is optimized subject to the fuzzy constraints and the trade-off functions constructed from the permissible trade-offs for the current objective with respect to the higher order objectives. The more stringent the decision makers are on the trade-offs, the more favorable the compromise solution is to the higher order objectives.

Keywords: fuzzy optimization, fuzzy hierarchical programming, compromise solution.

## 1. INTRODUCTION

Decision making within a complex system ordinarily involves a set of conflicting objectives. Resolution of conflicts among objectives is generally accomplished through the search for a compromise solution. System complexity and imprecise cognition often exist in our decision making processes. Programming with fuzzy objectives and constraints is thus pertinent to the modeling of human decision making problems.

Based on a general fuzzy mathematical programming framework proposed by Bellman and Zadeh (1970), optimization with a fuzzy objective and fuzzy constraints has been developed in the last decade or so (see, for example, Tanaka, Okuda, and Asai, 1974; Negoita and Sularia, 1976; Zimmermann, 1976). Several attempts have specifically been made on solving multiobjective linear optimization problems embedded with fuzziness (see, for example, Zimmermann, 1978; Hannan, 1981a, b; Narasimhan, 1980; Rubin and Narasimhan, 1984; and Leung, 1982, 1983, 1984).

One of the basic issues in multiobjective programming is the treatment of priorities. In general, objectives are of varying degrees of importance. Often, a compromise solution is reached by sacrificing, to a certain extent, lower order objectives for the betterment of higher order objectives. To guarantee that objectives are optimized according to their priorities, methods of implicit priorities (Zimmermann, 1978), explicit fuzzy weights (Zeleny, 1973; Hannan, 1981a, b), and composite objective with weighted contributions of individual goals (Narasimhan, 1980; Rubin and Narasimhan, 1984) have been employed.

A common characteristic of these methods is that weighted objectives are simultaneously considered in the optimization process. However, when objectives can be ranked and form a hierarchy, decision makers often consider the objectives one at a time by the descending order of their priorities. The stepwise optimization procedure is executed in such a way that the most important objective is optimized first. The next most important objective is then optimized within a tolerable trade-off specified for the optimal value of the preceding objective. Similarly, the third most important objective is optimized under the restraints imposed by the tolerable trade-offs of the previous optimal solutions. The final solution is then the most appropriate compromise solution favoring the higher order objectives (Waltz, 1967; Nijkamp, 1977). Such a sequential procedure appears to be common in many real-life decision making problems.

The purpose of this paper is to propose some procedures for hierarchical optimization with fuzzy linear objectives and constraints. Two major methods are first discussed. Their variants are then examined.

## 2. PROCEDURES OF HIERARCHICAL OPTIMIZATION

Let the following be a multiobjective fuzzy linear optimization problem:

$$\begin{cases} f_i(x) \gtrsim \bar{z}_i; \underline{z}_i, & i = 1,\ldots,m \\ g_j(x) \geq \bar{b}_j; \underline{b}_j, & j = m+1,\ldots,n \\ x \geq 0 \end{cases} \tag{1}$$

where $f_i$ and $g_j$ represent, respectively, fuzzy objective functions and constraints of the fuzzy "greater than or equal to" type with tolerance intervals $[\bar{z}_i - \underline{z}_i]$ and $[\bar{b}_i - \underline{b}_j]$, $x \in R^n$.

Let the satisfaction functions of $f_i$ and $g_j$ be defined, respectively, by the following membership functions

$$\mu_i(f_i(x)) = \begin{cases} 1 & \text{if } f_i(x) \geq \bar{z}_i \\ 1 - \dfrac{\bar{z}_i - f_i(x)}{z_i - \underline{z}_i} & \text{if } \underline{z}_i < f_i(x) < \bar{z}_i, i=1,\ldots,m \\ 0 & \text{if } f_i(x) \leq \underline{z}_i \end{cases} \tag{2}$$

$$\mu_j(g_j(x)) = \begin{cases} 1 & \text{if } g_j(x) \geq \bar{b}_j \\ 1 - \dfrac{\bar{b}_j - g_j(x)}{\bar{b}_j - \underline{b}_j} & \text{if } \underline{b}_j < g_j(x) < \bar{b}_j, \quad j=m+1,\ldots,n \\ 0 & \text{if } g_j(x) \leq \underline{b}_j \end{cases} \tag{3}$$

Assume that the fuzzy objectives can be ranked in a descending order of priorities as

$$f_1 \succ f_2 \succ \cdots \succ f_m, \tag{4}$$

with $f_1$ being the most important objective and $f_m$ the least important objective.

To obtain a compromise solution which would be in favor of more important objectives, sequential optimization of individual objectives by their descending order of priorities should be carried out. The satisfaction function $\mu_1$ is first optimized subject to the fuzzy constraints $\mu_j$'s. A tolerable level of deviation from the optimal value $f_1^*$ is then employed as a restraint on the optimization of $\mu_2$ in the next step. By the same token, $\mu_3$ is optimized subject to $\mu_j$'s and the tolerable levels of deviation from the optimal values $f_1^*$ and $f_2^*$. The process continues until $\mu_m$ is optimized.

Since trade-offs can be specified in different ways, the sequential procedures can be of varying formats. Two methods and their variants are discussed in the remaining part of this section.

<u>Method 1</u>. Given is the multiobjective fuzzy linear optimization problem defined by (1) - (4).

Step 1. Obtain the optimal solution $(\lambda_1^*, x^*, f_1^*)$ by solving the following single objective optimization problem:

$$\begin{cases} f_1(x) \lesssim \bar{z}_1; \underline{z}_1 \\ g_j(x) \gtrsim \bar{b}_j; \underline{b}_j, \quad j = m+1,\ldots,n \\ x \geq 0 \end{cases} \tag{5}$$

Equivalently, we solve

$$\begin{cases} \max & \lambda \\ \text{s.t.} & \lambda \leq \mu_1(f_1(x)) \\ & \lambda \leq \mu_j(g_j(x)), \quad j = m+1,\ldots,n \\ & \lambda \geq 0, \; x \geq 0 \end{cases} \tag{6}$$

by solving

$$
\begin{cases}
\max \lambda_1 \\
\text{s.t.} \quad \lambda_1 \leqslant \dfrac{f_1(x) - \underline{z}_1}{\bar{z}_1 - \underline{z}_1} \\
\qquad \lambda_1 \leqslant \dfrac{g_i(x) - \underline{b}_j}{\bar{b}_j - \underline{b}_j}, \quad j = 1,\ldots,n \\
\qquad \lambda_1 \geqslant 0, \; x \geqslant 0
\end{cases}
\tag{7}
$$

Step 2. Based on the optimal value $\lambda_1^*$ obtained in Step 1, determine a trade-off coefficient $\beta_1^2$ which indicates the extent to which $\lambda_1^*$ can be compromised in order to best achieve $f_2$ by maximizing $\mu_2$. That is, $\beta_1^2$ can be treated as the tolerance level for the permissible deviation from $\lambda_1^*$ in the process of optimizing $f_2$.

Determine the compromise solution by solving

$$
\begin{cases}
\max \lambda_2 \\
\text{s.t.} \quad \lambda_2 \leqslant \dfrac{f_2(x) - \underline{z}_2}{\bar{z}_2 - \underline{z}_2} \\
\qquad \lambda_2 \leqslant \dfrac{g_j(x) - \underline{b}_j}{\bar{b}_j - \underline{b}_j}, \quad j = m+1,\ldots,n \\
\qquad \beta_1^2 \lambda_1^* \leqslant \dfrac{f_1(x) - \underline{z}_1}{\bar{z}_1 - \underline{z}_1} \\
\qquad \lambda_2 \geqslant 0, \; x \geqslant 0
\end{cases}
\tag{8}
$$

where $0 \leqslant \beta_1 \leqslant \beta_1^2 \leqslant 1$, with $\beta_1$ being the maximal tolerable trade-off of $f_1^*$ for $f_2^*$. For example, $\beta_1$ can be set in a way that the degree of satisfaction of $f_1$ would always be greater than or equal to that of $f_2$.

Step 3. Based on $\lambda_1^*$ and $\lambda_2^*$ obtained in the previous steps, select trade-off coefficients $\beta_1^3$ and $\beta_2^3$ to serve as restraints on the optimization of $f_3$. Then, the next compromise solution is obtained by solving

$$
\left\{
\begin{array}{l}
\max \lambda_3 \\
\text{s.t. } \lambda_3 \leqslant \dfrac{f_3(x) - \underline{z}_3}{\bar{z}_3 - \underline{z}_3} \\
\lambda_3 \leqslant \dfrac{g_j(x) - \underline{b}_j}{\bar{b}_j - \underline{b}_j}, \quad j = m+1,\ldots,n \\
\beta_1^3 \lambda_1^* \leqslant \dfrac{f_1(x) - \underline{z}_1}{\bar{z}_1 - \underline{z}_1} \\
\beta_2^3 \lambda_2^* \leqslant \dfrac{f_2(x) - \underline{z}_2}{\bar{z}_2 - \underline{z}_2} \\
\lambda_3 \geqslant 0, \; x \geqslant 0
\end{array}
\right.
\tag{9}
$$

where $0 \leqslant \beta_1 \leqslant \beta_1^3 \leqslant \beta_1^2 \leqslant 1$ and $0 \leqslant \beta_2 \leqslant \beta_2^3 \leqslant 1$, with $\beta_1$ and $\beta_2$ being respectively the maximal permissible trade-offs of $f_1^*$ and $f_2^*$ for $f_3^*$.

...

Step m. Applying the same method throughout the sequential optimization procedure, the compromise solution in the mth step is obtained by solving

$$
\left\{
\begin{array}{l}
\max \lambda_m \\
\text{s.t. } \lambda_m \leqslant \dfrac{f_m(x) - \underline{z}_m}{\bar{z}_m - \underline{z}_m} \\
\lambda_m \leqslant \dfrac{g_j(x) - \underline{b}_j}{\bar{b}_j - \underline{b}_j}, \quad j = m+1,\ldots,n \\
\beta_i^m \lambda_i^* \leqslant \dfrac{f_i(x) - \underline{z}_i}{\bar{z}_i - \underline{z}_i}, \quad i = 1,\ldots,m-1 \\
\lambda_m \geqslant 0, \; x \geqslant 0
\end{array}
\right.
\tag{10}
$$

where $0 \leqslant \beta_i \leqslant \beta_i^m \leqslant \beta_i^{m-1} \leqslant \ldots \leqslant \beta_i^{i+1} \leqslant 1$, $i = 1,\ldots,m-1$, with

$\beta_i$ being the maximal permissible trade-off of $f_i^*$ for $f_m^*$.

Remarks:

(a) Except for the initial step, in each step of the sequential optimization process a fuzzy objective function is optimized over a decision space delimited by the fuzzy constraints and the trade-off functions (constraints constructed from the permissible trade-offs for a lower order objective with respect to the optimal satisfaction of all higher order objectives). The compromise solution thus derived depends on the extent to which decision makers are willing to trade-off the higher order objectives. The more stringent we are on the trade-offs, the more favorable the compromise solution is to the higher order objectives.

The relation $0 \leqslant \beta_i \leqslant \beta_i^m \leqslant \beta_i^{m-1} \leqslant \ldots \leqslant \beta_i^{i+1} \leqslant 1$ indicates that the trade-offs for more important lower order objectives are more lenient than those of the less important ones. That is, the trade-off extent decreases with decreasing order of priorities of the lower order objectives. Such a specification makes the search for a compromise solution consistent throughout the sequential procedure. The solution obtained from later steps would not reverse or contradict that obtained from the preceding steps.

In (10), if $\beta_i^m = 1$ for all i, the optimization process takes on a non-compromising position and infeasibility may most likely occur. However, if $\beta_i^m = \dfrac{\lambda_m}{\lambda_i^*}$ for all i, the hierarchical optimization process becomes a one-stage optimization process and all objectives are treated equally.

(b) The trade-off functions in (10) are formulated with respect to how much satisfaction of the higher order objective i, $\lambda_i^*$, a decision maker is willing to give up in return for better achievement of a lower order objective m.

Sometimes, decision makers may find it easier to determine the trade-off on the basis of how much of $f_i^*$, instead of how much of $\lambda_i^*$, they are willing to trade for $f_m^*$. Under this situation, we can replace

$$\beta_i^m \lambda_i^m \leqslant \frac{f_i(x) - \underline{z}_i}{\bar{z}_i - \underline{z}_i}, \quad i = 1,\ldots,m-1 \tag{11}$$

in (10) by

$$\alpha_i^m f_i^* \leqslant f_i(x), \quad i = 1,\ldots,m-1 \tag{12}$$

where

$0 \leqslant \alpha_i \leqslant \alpha_i^m \leqslant \alpha_i^{m-1} \leqslant \ldots \leqslant \alpha_i^{i+1} \leqslant 1$, $i = 1,\ldots,m-1$ with $\alpha_i$ being the maximal permissible trade-off of $f_i^*$ for $f_m^*$.

Actually, (11) and (12) are identical if

$$\alpha_i^m f_i^* = \mu_i^{-1}(\beta_i^m \lambda_i^m), \quad i = 1,\ldots,m-1. \tag{13}$$

Specifically,

$$\alpha_i^m f_i^* = \underline{z}_i + (\beta_i^m \lambda_i^m)(\bar{z}_i - \underline{z}_i). \tag{14}$$

(c) Although our formulation is for fuzzy "greater than or equal to" type of objectives, it can easily be extended to hierarchical optimization problems involving fuzzy "less than or equal to" and/or fuzzy "equal to" type of objectives.

(d) To make the hierarchical optimization process more flexible, suitable adjustments of $\beta_i^m$'s should be allowed if the compromise solution obtained in each step is not satisfactory. That is, decision makers could be involved throughout the procedure and the process could be iterative.

Example. Let the following be a simple two-objective fuzzy linear programming problem

$$\begin{cases} f_1 : x \gtrsim 60;20 \\ f_2 : x \lesssim 10;90 \\ g : x \geqslant 50;70 \\ x \geqslant 0 \end{cases} \tag{15}$$

with the priority ordering being $f_1 \succ f_2$ (see Fig. 1).

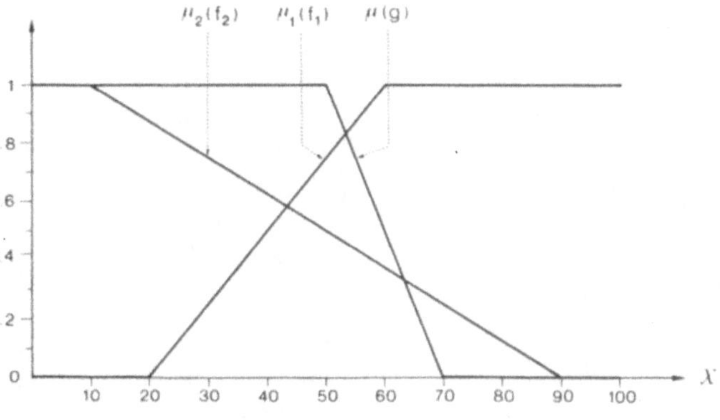


Fig. 1. Membership functions of two fuzzy objectives and one fuzzy constraint

$$\left\{\begin{array}{l} \max \lambda_1 \\ \text{s.t.} \quad \lambda_1 \leqslant \dfrac{x-20}{40} \\ \qquad \lambda_1 \leqslant \dfrac{70-x}{20} \\ \qquad \lambda_1 \geqslant 0, \quad x \geqslant 0 \end{array}\right. \tag{16}$$

Step 2. Obtain $\lambda_2^*$ by solving

$$\left\{\begin{array}{l} \max \lambda_2 \\ \text{s.t.} \quad \lambda_2 \leqslant \dfrac{90-x}{80} \\ \qquad \lambda_2 \leqslant \dfrac{70-x}{20} \\ \qquad (.833)\beta_1^2 \leqslant \dfrac{x-20}{40} \\ \qquad \lambda_2 \geqslant 0, \quad x \geqslant 0 \end{array}\right. \tag{17}$$

Table 1 and Fig. 2 summarize the results of the hierarchical programming procedure for selected values of $\beta_1^2$.

Table 1. Compromise solutions with varying trade-off specifications

| Trade-off coefficients ($\beta_1^2$) | Compromise solutions | | |
|---|---|---|---|
| | $x^*$ | $\lambda_1^*$ | $\lambda_2^*$ |
| 0 | 20 | 0 | 0.875 |
| 0.1 | 23.332 | 0.083 | 0.833 |
| 0.2 | 26.664 | 0.166 | 0.792 |
| 0.3 | 29.996 | 0.250 | 0.750 |
| 0.4 | 33.328 | 0.333 | 0.708 |
| 0.5 | 36.660 | 0.417 | 0.667 |
| 0.6 | 39.992 | 0.500 | 0.625 |
| 0.7 | 43.324 | 0.583 | 0.583 |
| 0.8 | 46.656 | 0.666 | 0.542 |
| 0.9 | 49.988 | 0.750 | 0.500 |
| 1 | 53.333 | 0.833 | 0.458 |

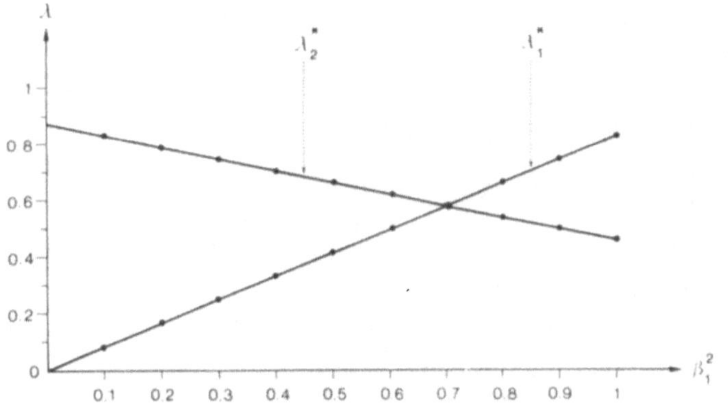


Fig. 2. Trade-offs of objective-achievements

Apparently, not all trade-off specifications favor the optimal achievement of the more important fuzzy objective $f_1$. The critical trade-off coefficient is 0.7. At this level of trade-off, the degrees of satisfaction of both objectives are identical, 0.583. Actually, it is the result of simultaneously optimizing both $f_1$ and $f_2$ by treating them as if they were of equal importance. For $\beta_1^2 < 0.7$, the degree of satisfaction of $f_1$ is lower than that of $f_2$. For $\beta_1^2 > 0.7$, it is the opposite. Therefore, to guarantee that the degree of satisfaction of $f_1$ would always be higher than that of $f_2$, in reaching a compromise, the maximal permissible trade-off of $f_1$ for $f_2$ should be set at a value greater than 0.7. That is, the condition, $0.7 = \beta_1 < \beta_1^2 \leqslant 1$, should be enforced.

Method 2

In method 1, decision makers need to specify trade-offs for all individual lower order objectives with respect to the optimal value of each higher order objective. Often, it might prove to be too taxing an information drill, albeit useful. A less onerous task is to require the determination of an overall permissible trade-off for any lower order objectives.

Should that be the case, the first step of the hierarchical optimization procedure would remain the same, (7), and the second step becomes

$$\begin{cases} \max \lambda_2 \\ \text{s.t. } \lambda_2 \leqslant \dfrac{f_2(x)-\underline{z}_2}{\bar{z}_2-\underline{z}_2} \\ \qquad \lambda_2 \leqslant \dfrac{g_j(x)-\underline{b}_j}{\bar{b}_j-\underline{b}_j}, \quad j = m+1,\ldots,n \\ \qquad \beta_1\lambda_1^* \leqslant \dfrac{f_1(x)-\underline{z}_1}{\bar{z}_1-\underline{z}_1} \\ \qquad \lambda_2 \geqslant 0,\ x \geqslant 0 \end{cases} \tag{18}$$

where $0 \leqslant \beta_1 \leqslant 1$ is the overall trade-off coefficient based on $\lambda_1^*$.

Likewise, the mth step becomes

$$\begin{cases} \max \lambda_m \\ \text{s.t. } \lambda_m \leqslant \dfrac{f_m(x)-\underline{z}_m}{\bar{z}_m-\underline{z}_m} \\ \qquad \lambda_m \leqslant \dfrac{g_j(x)-\underline{b}_j}{\bar{b}_j-\underline{b}_j}, \quad j = m+1,\ldots,n \\ \qquad \beta_i\lambda_i^* \leqslant \dfrac{f_i(x)-\underline{z}_i}{\bar{z}_i-\underline{z}_i}. \quad i = 1,\ldots,m-1 \\ \qquad \lambda_m \geqslant 0,\ x \geqslant 0 \end{cases} \tag{19}$$

where $0 \leqslant \beta_1 \leqslant 1$ is the coefficient of overall trade-off of objective i for any lower order objectives. It remains the same in every step of the sequential optimization procedure.

In both methods 1 and 2, it is assumed that the unconstrained optimal values of the individual objective functions (values obtained when individual objective functions are optimized subject to the fuzzy constraints without the imposition of the trade-off functions of the higher order objectives) are not known. If such information is available, trade-off coefficients can be directly specified on these unconstrained optimal values. The variants of methods 1 and 2 are summarized in method 1′ and 2′, respectively. Without loss of generality, only the mth step is discussed.

Method 1′

Let $\bar{\lambda}_1, \bar{\lambda}_2, \ldots,$ and $\bar{\lambda}_m$ be the unconstrained optimal values of the objective functions $\mu_1, \mu_2, \ldots,$ and $\mu_m$. Then, the mth step, (10), of method 1 becomes

$$\left\{\begin{array}{ll} \max \lambda_m & \\ \text{s.t. } \lambda_m \leq \dfrac{f_m(x)-\underline{z}_m}{\bar{z}_m-\underline{z}_m} & \\ \lambda_m \leq \dfrac{g_j(x)-\underline{b}_j}{\bar{b}_j-\underline{b}_j}, & j = m+1,\ldots,n \\ \beta_i^m \bar{\lambda}_i \leq \dfrac{f_i(x)-\underline{z}_i}{\bar{z}_i-\underline{z}_i}, & i = 1,\ldots,m-1 \\ \lambda_m \geq 0,\ x \geq 0 & \end{array}\right. \tag{20}$$

The difference between the optimization problems in (10) and (20) is that trade-off coefficients in (10) are specified for the suboptimal values $\lambda_i^*$ while those in (20) are specified for the unconstrained optimal values $\bar{\lambda}_i$'s

Method 2'

Similarly, the m-th step in method 2 becomes

$$\left\{\begin{array}{ll} \max \lambda_m & \\ \text{s.t. } \lambda_m \leq \dfrac{f_m(x)-\underline{z}_m}{\bar{z}_m-\underline{z}_m} & \\ \lambda_m \leq \dfrac{g_j(x)-\underline{b}_j}{\bar{b}_j-\underline{b}_j}, & j = m+1,\ldots,n \\ \beta_i \bar{\lambda}_i \leq \dfrac{f_i(x)-\underline{z}_i}{\bar{z}_i-\underline{z}_i}, & i = 1,\ldots,m-1 \\ \lambda_m \geq 0,\ x \geq 0 & \end{array}\right. \tag{21}$$

Remark. In all of the above methods, trade-offs are only specified for the hierarchical fuzzy objective functions. The fuzzy constraints are strictly enforced throughout and are of equal importance with the current objectives. Sometimes, a more satisfactory or pragmatic compromise solution may require the relaxation of constraints also. Under these circumstances, different procedures need to be formulated to keep track of the trade-off and relaxation processes.

## 3. CONCLUSION

Hierarchical optimization with multiple fuzzy objectives and fuzzy constraints has been examined in this paper. Several methods, by no means to be exhaustive, have been formulated. Differing from other single-step methods, the proposed optimization procedures are sequential in structure. Fuzzy objectives are optimized one at a time in the descending order of priorities. The final compromise solution is determined over the decision space delimited by the respective trade-off functions and the fuzzy constraints. Instead of assigning weights, preferential treatment of objectives is provided through the control of the trade-off coefficients.

Although trade-off coefficients can sometimes be difficult to determine, they are at least no more difficult than the determination of explicit or implicit weights required by other methods. In addition, compromise is ordinarily reached by giving up part of the achievement of one objective for the return of another. The hierarchical procedure ties in quite satisfactorily with such a decision making process and should be applicable to many real-life decision making problems with conflicting objectives of varying importances and conflicting interest groups. International disarmament, national economic policies, interregional development, environmental management, growth of a firm and personal investment are just a few of the applicable areas.

Since the hierarchical and the nonhierarchical procedures are constructed on the basis of different rationales, it is then difficult if not impossible to compare their superiorities. Depending on the structures of the decision making problems, one method may be more applicable than the other. However, it may prove to be beneficial to extend the hierarchical procedure to solve large-scale multilevel, multistage problems in a fuzzy environment.

## REFERENCES

Bellman, R.E., and L.A. Zadeh (1970). Decision-making in a fuzzy environment. Mang. Sci. 17, 151-169.

Hannan, E.L. (1981a). On fuzzy goal programming. Dec. Sci. 12, 522-531.

Hannan, E.L. (1981b). Linear programming with multiple fuzzy goals. Fuzzy Sets and Syst. 6, 235-248.

Leung, Y. (1982). Multicriteria conflict resolution through a theory of a displaced fuzzy ideal. In M.M. Gupta and E. Sanchez (eds.), Approximate Reasoning in Decision Analysis. North-Holland, Amsterdam, 381-390.

Leung, Y. (1983). A concept of a fuzzy ideal for multicriteria conflict resolution. In P.P. Wang (ed.), Advances in Fuzzy Sets, Possibility Theory, and Application. Plenum, New York, 387-403.

Leung, Y. (1984). Compromise programming under fuzziness. Control and Cyb. 13 ,203-215.

Narasimhan, R. (1980). Goal programming in a fuzzy environment. Dec. Sci. 11, 325-336.

Negoita, C.V. and M. Sularia (1976). On fuzzy mathematical programming and tolerance in planning. Econ. Comp. and Econ. Cyb. Stud. and Res. 1, 3-14.

Nijkamp, P. (1977). The use of hierarchical optimization criteria in regional planning. J. Reg. Sci. 17, 195-205.

Rubin, P.A. and R. Narasimhan (1984). Fuzzy goal programming with nested priorities. Fuzzy Sets and Syst. 14, 115-129.

Tanaka, H., T. Okuda, and K. Asai (1974). On fuzzy mathematical programming. J. Cyb. 3, 37-46.

Waltz, F.M. (1967). An engineering approach to hierarchical optimization criteria. IEEE Trans. on Aut. Cont. AC-12, 179-180.

Zeleny, M. (1973). Compromise programming. In J.L. Cochrane, and M. Zeleny (eds.), Multiple Criteria Decision Making. Columbia, University of South Carolina, 262-301.

Zimmermann, H-J. (1976). Description and optimization of fuzzy systems. Int. J. Gen. Syst. 2, 209-215.

Zimmermann, H-J. (1978). Fuzzy programming and linear programming with several objective functions. Fuzzy Sets and Syst. 1, 45-55.

AN INTERACTIVE SATISFICING METHOD FOR MULTIOBJECTIVE NONLINEAR PROGRAMMING PROBLEMS WITH FUZZY PARAMETERS

Masatoshi Sakawa* and Hitoshi Yano**

* Department of Systems Engineering, Faculty of Engineering, Kobe University, Kobe 657, Japan

** Department of Information Science, College of Economics, Kagawa University, Kagawa 760, Japan

Abstract. This paper presents a new interactive satisficing method for multiobjective nonlinear programming problems with fuzzy parameters. The fuzzy parameters in the objective functions and the constraints are characterized by fuzzy numbers. The concept of $\alpha$-Pareto optimality is introduced in which the ordinary Pareto optimality is extended based on the $\alpha$-level sets of the fuzzy numbers. In our interactive satisficing method if the decision maker (DM) specifies the degree $\alpha$ of the $\alpha$-level sets and the reference objective values, the augmented minimax problem is solved and the DM is supplied with the corresponding $\alpha$-Pareto optimal solution together with the trade-off rates among the values of the objective functions and the degree $\alpha$. Then by considering the current values of the objective functions and $\alpha$ as well as the trade-off rates, the DM responds by updating his reference objective values and/or the degree $\alpha$. In this way the satisficing solution for the DM can be derived efficiently from among an $\alpha$-Pareto optimal solution set. On the basis of the proposed method, a time-sharing computer program is written and an illustrative numerical example is demonstrated along with the computer outputs.

Keywords: multiobjective nonlinear programming, fuzzy number, interactive optimization, satisficing solution.

## 1. INTRODUCTION

When formulating a multiobjective nonlinear programming problem which closely describes and represents the real decision situation, various factors of the real system should be reflected in the description of the objective functions and the constraints. Naturally, these objective functions and the constraints involve many parameters whose possible values may be assigned by the experts. In the conventional approach, such parameters are fixed at some values in an experimental and/or subjective way through the experts' understanding of the nature of the parameters.

In most practical situations, however, it is natural to consider that the possible values of these parameters are often only ambiguously known to the experts. In this case, it may be more appropriate to interpret the experts' understanding of the parameters as fuzzy numerical data which can be represented by means of fuzzy subsets of the real line known as fuzzy numbers (Dubois and Prade, 1978, 1980). The resulting multiobjective nonlinear programming problem involving fuzzy parameters would be viewed as the more realistic version of the conventional one.

Recently, Tanaka and Asai (1981, 1984) formulated multiobjective linear programming problems with fuzzy parameters. Following the maximizing decision proposed by Bellman and Zadeh (1970), together with traingular membership functions for fuzzy parameters, they considered two types of fuzzy multiobjective linear programming problems; one is to derive a nonfuzzy solution and the other is to derive a fuzzy one.

More recently, Orlovski (1985a,b) formulated general multiobjective nonlinear programming problems with fuzzy parameters. He presented two approaches to the formulated problems by making systematic use of the extension principle of Zadeh (1975) and demonstrated that there exist in some sense equivalent nonfuzzy formulations.

In this paper, in order to deal with the multiobjective nonlinear programming problems with fuzzy parameters characterized by fuzzy numbers, the concept of $\alpha$-Pareto optimality is introduced by extending the usual Pareto optimality on the basis of the $\alpha$-level sets of the fuzzy numbers. Then an interactive satisficing method to derive a satisficing solution of the decision maker (DM) efficiently from among an $\alpha$-Pareto optimal solution set is presented as a generalization of the results obtained in Sakawa (1983), Sakawa and Yano (1984), Sakawa and Yumine (1983), and Sakawa, Yumine and Yano (1984).

## 2. $\alpha$-PARETO OPTIMALITY

In general, the multiobjective nonlinear programming (MONLP) problem is represented as the following vector-minimization problem:

$$\left.\begin{aligned} &\min\ f(x) \triangleq (f_1(x), f_2(x), \ldots, f_k(x))^T \\ &\text{subject to}\quad x \in X = \left\{ x \in E^n \mid g_j(x) \leqq 0,\ j=1,\ldots,m \right\} \end{aligned}\right\} \quad (1)$$

where $x$ is an n-dimensional vector of decision variables, $f_1(x),\ldots,f_k(x)$ are k distinct objective functions of the decision vector x, $g_1(x),\ldots,g_m(x)$ are inequality constraints, and $X$ is the feasible set of constrained decisions.

Fundamental to the MONLP is the Pareto optimality concept, also known as a noninferiority of solution. Qualitatively, a Pareto optimal solution of the MONLP is one where any improvement of one objective function can be achieved only at the expense of another. Mathematically, a formal definition of a

Pareto optimal solution to the MONLP is given below:

Definition 1. (Pareto optimal solution) $x^* \in X$ is said to be a Pareto optimal solution to the MONLP, if and only if there does not exist another $x \in X$ such that $f_i(x) \leqq f_i(x^*)$, $i=1,\ldots,k$, with strict inequality holding for at least one i.

In practice, however, it would certainly be appropriate to consider that the possible values of the parameters in the description of the objective functions and the constraints usually involve the ambiguity of the experts' understanding of the real system. For this reason, in this paper, we consider the following fuzzy multiobjective nonlinear programming (FMONLP) problem involving fuzzy parameters:

FMONLP

$$\left.\begin{aligned} &\min f(x,\tilde{a}) \triangleq (f_1(x,\tilde{a}_1), f_2(x,\tilde{a}_2), \ldots, f_k(x,\tilde{a}_k))^T \\ &\text{subject to } x \in X(\tilde{b}) \triangleq \{x \in E^n \mid g_j(x,\tilde{b}_j) \leqq 0,\ j=1,\ldots,m\} \end{aligned}\right\} \quad (2)$$

where $\tilde{a}_i$ and $\tilde{b}_j$ represent, respectively, a vector of fuzzy parameters involved in the objective function $f_i(x,\tilde{a}_i)$ and the constraint function $g_j(x,\tilde{b}_j)$.

These fuzzy parameters are assumed to be characterized as the fuzzy numbers introduced by Dubois and Prade (1978, 1980). It is appropriate to review here that a real fuzzy number $\tilde{p}$ is a convex continuous fuzzy subset of the real line whose membership function $\mu_{\tilde{p}}(p)$ is defined as:

(1) A continuous mapping from $E^1$ to the closed interval [0,1],

(2) $\mu_{\tilde{p}}(p) = 0$ for all $p \in (-\infty, p_1]$,

(3) Strictly increasing on $[p_1, p_2]$,

(4) $\mu_{\tilde{p}}(p) = 1$ for all $p \in [p_2, p_3]$,

(5) Strictly decreasing on $[p_3, p_4]$,

(6) $\mu_{\tilde{p}}(p) = 0$ for all $p \in [p_4, +\infty)$.

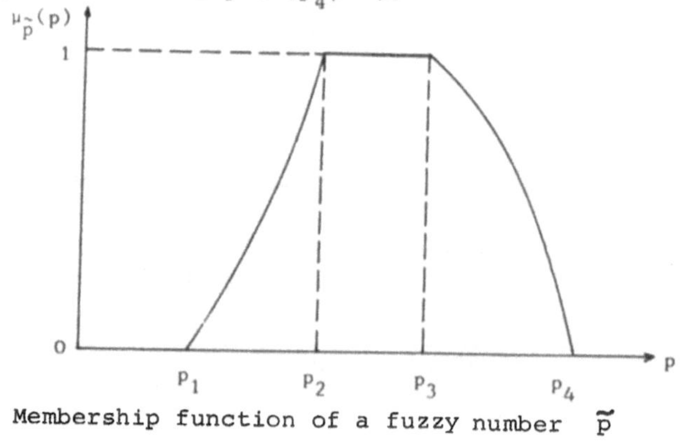


Fig. 1. Membership function of a fuzzy number $\tilde{p}$

Figure 1 illustratesthe graph of a possible shape of a fuzzy number $\tilde{p}$.

We now assume that $\tilde{a}_i$ and $\tilde{b}_j$ in the FMONLP are fuzzy numbers whose membership functions are $\mu_{\tilde{a}_i}(a_i)$ and $\mu_{n_j}(b_j)$, respectively. For simplicity in the notation, define the following vectors:

$$\mu_{\tilde{a}}(a) = (\mu_{\tilde{a}_1}(a_1),\ldots,\mu_{\tilde{a}_k}(a_k)),$$

$$\mu_{\tilde{b}}(b) = (\mu_{\tilde{b}_1}(b_1),\ldots,\mu_{\tilde{b}_m}(b_m)),$$

$$a = (a_1,\ldots,a_k),\ \tilde{a} = (\tilde{a}_1,\ldots,\tilde{a}_k),$$

$$b = (b_1,\ldots,b_m),\ \tilde{b} = (\tilde{b}_1,\ldots,\tilde{b}_m).$$

Then, we can introduce the following $\alpha$-level set or $\alpha$-cut, Dubois and Prade (1980), of the fuzzy numbers $\tilde{a}$ and $\tilde{b}$.

Definition 2. ($\alpha$-level set). An $\alpha$-level set of fuzzy numbers $\tilde{a}$ and $\tilde{b}$ is defined as an ordinary set $L_\alpha(\tilde{a},\tilde{b})$ for which the degree of their membership functions exceeds a level $\alpha$, i.e.

$$L_\alpha(\tilde{a},\tilde{b}) = \left\{(a,b) \mid \mu_{\tilde{a}}(a) \geqq \alpha,\quad \mu_{\tilde{b}}(b) \geqq \alpha\right\} \tag{3}$$

It is clear that the level sets have the following property:

$$\alpha_1 \leqq \alpha_2 \text{ if and only if } L_{\alpha_1}(\tilde{a},\tilde{b}) \supset L_{\alpha_2}(\tilde{a},\tilde{b})\}$$

For a certain degree $\alpha$, the FMONLP (2) can be understood as the following nonfuzzy $\alpha$-multiobjective nonlinear programming ($\alpha$-MONLP) problem:

$\alpha$-MONLP

$$\left.\begin{aligned} &\min\left\{f(x,a) \triangleq (f_1(x,a_1),f_2(x,a_2),\ldots,f_k(x,a_k))^T\right\} \\ &\text{subject to } x\in X(b) \triangleq \left\{x\in E^n \mid g_j(x,b_j) \leqq 0,\ j=1,\ldots,m\right\} \\ &\qquad (a,b) \in L_\alpha(\tilde{a},\tilde{b}) \end{aligned}\right\} \tag{4}$$

It should be emphasized here that in the $\alpha$-MONLP the parameters (a,b) are treated as decision variables rather than constants.

On the basis of the $\alpha$-level sets of the fuzzy numbers, we introduce the concept of $\alpha$-Pareto optimal solutions to the $\alpha$-MONLP.

Definition 3. ($\alpha$-Pareto optimal solution). $x^*\in X(b)$ is said to be an $\alpha$-Pareto optimal solution to the $\alpha$-MONLP(4) if and only if there does not exist another $x\in X(b)$, $(a,b)\in L_\alpha(\tilde{a},\tilde{b})$ such that $f_i(x,a) \leqq f_i(x^*,a^*)$, $i=1,\ldots,k$, with strict inequality holding for at least one i, where the corresponding values of parameters $(a^*,b^*)$ are called $\alpha$-level optimal parameters.

In order to generate a candidate for a satisficing solution which is also α-Pareto optimal, the DM is asked to specify the degree α of the α-level set and reference levels of achievement of the objective functions, called reference levels. Observe that the idea of the reference levels or the reference point was first proposed in Wierzbicki (1979). For the DM's degree α reference levels $\bar{f}_i$, i=1,...,k, the corresponding α-Pareto optimal solution, which is in a sense close to his requirement (or better, if the reference levels are attainable), is obtained by solving the following augmented minimax problem:

$$\min_{\substack{x\in X(b)\\(a,b)\in L(\tilde{a},\tilde{b})}} \left\{ \max_{1\leqq i\leqq k} (f_i(x,a_i) - \bar{f}_i) + \rho \sum_{i=1}^{k} (f_i(x,a_i) - \bar{f}_i) \right\} \tag{5}$$

or, equivalently:

$$\min_{x,v,a,b} \left\{ v + \rho \sum_{i=1}^{k} (f_i(x,a_i) - \bar{f}_i) \right\} \tag{6}$$

subject to:

$$f_i(x,a_i) - \bar{f}_i \leqq v, \; i=1,\ldots,k \tag{7}$$

$$\mu_{\tilde{a}}(a) \geqq \alpha \tag{8}$$

$$\mu_{\tilde{b}}(b) \geqq \alpha \tag{9}$$

$$x \in X(b) \tag{10}$$

The term augmented is adopted because the term $\rho \sum_{i=1}^{k} (f_i(x,a_i) - \bar{f}_i)$ is added to the usual minimax problems, where $\rho$ is a sufficiently small positive scalar. Such an augmented minimax problem can be viewed as a modified version of the augmented weighted Tchebycheff norm problem of Steuer and Choo (1983) or Choo and Atkins (1983).

The relationships between the optimal solutions of the augmented minimax problem and the α-Pareto optimal concept of the α-MONLP can be characterized by the following theorems.

<u>Theorem 1</u>. If $(x^*,a^*,b^*)$ is an optimal solution to the augmented minimax problem for some $\bar{f} = (\bar{f}_1,\ldots,\bar{f}_k)$, then $x^*$ is an α-Pareto optimal solution to the α-MONLP.

<u>Proof</u>. Assume that $x^*$ is not an α-Pareto optimal solution to the α-MONLP, then there exists $x\in X(b)$, $(a,b)\in L_\alpha(\tilde{a},\tilde{b})$, such that $f(x,a) \lneq f(x^*,a^*)$. Then, it holds that

$$\max_{1\leqq i\leqq k} (f_i(x,a_i) - \bar{f}_i) \leqq \max_{1\leqq i\leqq k} (f_i(x^*,a_i^*) - \bar{f}_i)$$

$$\rho \sum_{i=1}^{k} (f_i(x,a_i) - \bar{f}_i) < \rho \sum_{i=1}^{k} (f_i(x^*,a_i^*) - \bar{f}_i)$$

This means that

$$\max_{1 \leq i \leq k} (f_i(x,a_i) - \bar{f}_i) + \rho \sum_{i=1}^{k} (f_i(x,a_i) - \bar{f}_i) <$$

$$< \max_{1 \leq i \leq k} (f_i(x^*,a_i^*) - \bar{f}_i) + \rho \sum_{i=1}^{k} (f_i(x^*,a_i^*) - \bar{f}_i),$$

which contradicts the fact that $(x^*,a^*,b^*)$ is an optimal solution to the augmented minimax problem. Hence $x^*$ is an $\alpha$-Pareto optimal solution to the $\alpha$-MONLP.

Theorem 2. If $x^*$ is an $\alpha$-Pareto optimal solution and $(a^*,b^*)$ is an $\alpha$-level optimal parameter to the $\alpha$-MONLP, then there exists $f = (\bar{f}_1,\ldots,\bar{f}_k)$ such that $(x^*,a^*,b^*)$ is an optimal solution to the augmented minimax problem.

Proof. Assume that $(x^*,a^*,b^*)$ is not an optimal solution to the augmented minimax problem for any $\bar{f}$ satisfying

$$f_1(x^*,a_1^*) - \bar{f}_1 = \ldots = f_k(x^*,a_k^*) - \bar{f}_k.$$

Then there exists $x \in X(b)$, $(a,b) \in L_\alpha(\tilde{a},\tilde{b})$ such that

$$\max_{1 \leq i \leq k} (f_i(x^*,a_i^*) - \bar{f}_i) + \rho \sum_{i=1}^{k} (f_i(x^*,a_i^*) - \bar{f}_i) >$$

$$> \max_{1 \leq i \leq k} (f_i(x,a_i) - \bar{f}_i) + \rho \sum_{i=1}^{k} (f_i(x,a_i) - \bar{f}_i).$$

This implies that

$$\max_{1 \leq i \leq k} (f_i(x,a_i) - f_i(x^*,a_i^*)) + \rho \sum_{i=1}^{k} (f_i(x,a_i) - f_i(x^*,a_i^*)) < 0.$$

Now, if either any $f_i(x,a_i) - f_i(x^*,a_i^*)$ is positive or all $f_i(x,a_i) - f_i(x^*,a_i^*)$, $i=1,\ldots,k$, are zero, this inequality will be violated for sufficiently small positive $\rho$. Hence

$$f(x,a) - f(x^*,a^*) \leq 0$$

must hold which contradicts the fact that $x^*$ is an $\alpha$-Pareto optimal solution and $(a^*,b^*)$ is an $\alpha$-level optimal parameter to the $\alpha$-MONLP, and the theorem is proved.

It is significant to note here that from the property of the $\alpha$-level set, the following relation holds for any two optimal solutions $(x^1,v^1,a^1,b^1)$ and $(x^2,v^2,a^2,b^2)$ to the augmented minimax problems with the same reference levels corresponding to $\alpha^1$ and $\alpha^2$:

$$\alpha^1 \leq \alpha^2 \text{ if and only if } f_i(x^1,a_i^1) \leq f_i(x^2,a_i^2) \quad i=1,2,\ldots,k$$

## 3. TRADE-OFF RATES

Now given the $\alpha$-Pareto optimal solution for the degree $\alpha$ and the reference levels specified by the DM by solving the corresponding augmented minimax problem, the DM must either be satisfied with the current $\alpha$-Pareto optimal solution, or update the reference levels and/or the degree $\alpha$. In order to help the DM express his degree of preference, trade-off information between a standing objective function and each of the other objective functions as well as between the degree $\alpha$ and the objective functions is very useful. Fortunately, such trade-off information is easily obtainable since it is closely related to the strict positive Lagrange multipliers of the augmented minimax problem.

To derive the trade-off information, we first define the Lagrangian function L for the augmented minimax problem (6)-(10) as follows:

$$L(x,v,a,b,\lambda^f,\lambda^a,\lambda^b,\lambda^g,f,\alpha)$$

$$= v + \rho \sum_{i=1}^{k} (f_i(x,a_i) - \bar{f}_i) + \sum_{i=1}^{k} \lambda_i^f (f_i(x,a_i) - \bar{f}_i - v) +$$

$$\sum_{j=1}^{m} \lambda_j^g g_j(x) + \sum_{i=1}^{k} \lambda_i^a (\alpha - \mu_{\tilde{a}_i}(a_i)) + \sum_{j=1}^{m} \lambda_j^b (\alpha - \mu_{\tilde{b}_j}(b_j)) \quad (11)$$

In the following, for notational convenience, we denote the decision variable in the augmented minimax problem (6)-(10) by y = (x,v,a,b) and let us assume that the augmented minimax problem has a unique local optimal solution y* satisfying the following three assumptions.

If all the constraints (7) of the augmented minimax problem are active, namely if $v(\alpha^*) = f_i(x(\alpha^*), a_i(\alpha^*)) - \bar{f}_i$, then the following theorem holds.

<u>Theorem 5</u>.

Let all the assumptions in Theorem 3 be satisfied. Also assume that all the constraints (7) of the augmented minimax problem are active. Then it holds that

$$\left.\frac{\partial f_i(x,a_i)}{\partial \alpha}\right|_{\alpha=\alpha^*} = \frac{1}{1+\rho k} \left( \sum_{i=1}^{k} \lambda_i^{a*} + \sum_{j=1}^{m} \lambda_j^{b*} \right) \quad (13)$$

Regarding a trade-off rate between $f_1(x)$ and $f_i(x)$ for each i=2,...,k, by extending the results in Haimes and Chankong (1979), we can prove that the following theorem holds, Yano and Sakawa (1985).

<u>Theorem 6</u>. Let all the assumptions in Theorem 3 be satisfied. Also assume that the constraints (7) are active. Then it holds that

$$\left.\frac{\partial f_i(x,a_i)}{\partial f_1(x,a_1)}\right|_{\alpha=\alpha^*} = -\frac{\lambda_1^{f*}}{\lambda_i^{f*}}, \quad i=2,\ldots,k \tag{14}$$

It should be noted here that in order to obtain the trade-off rate information from (12) and (13), all the constraints (7) of the augmented minimax problem must be active. Therefore, if there are inactive constraints, it is necessary to replace $\bar{f}_i$ for inactive constraints by $f_i(x^*,a_i^*)$ and solve the correspond-ind augmented minimax problem for obtaining the Lagrange multipliers.

Assumption 1. $y^*$ is a regular point of the constraint of the augmented minimax problem.

Assumption 2. The second-order sufficiency conditions are satisfied at $y^*$.

Assumption 3. There are no degenerate constraints at $y^*$.

Then the following existence theorem, which is based on the implicit function theorem Fiacco (1983), holds.

Theorem 3. Let $y^* = (x^*,v^*,a^*,b^*)$ be a unique local solution of the augmented minimax problem (6)-(10) satisfying assumptions 1,2 and 3. Let $\lambda^* = (\lambda^{f*},\lambda^{a*},\lambda^{b*},\lambda^{g*})$ denote the Lagrange multipliers corresponding to the constraints (7)-(10). Then there exist a continuously differentiable vector-valued function $y(.)$ and $\lambda(.)$ defined on some neighborhood $N(\alpha^*)$, $y(\alpha^*) = y^*$, $\lambda(\alpha^*) = \lambda^*$, where $y(\alpha)$ is a unique local solution of the augmented minimax problem (6)-(10) for any $\alpha \in N(\alpha^*)$ satisfying assumptions 1,2 and 3, and $\lambda(\alpha)$ is the Lagrange multiplier corresponding to the constraints (7)-(10).

In Theorem 3, $\inf_{x,v}\left\{ v + \rho\sum_{i=1}^{k}(f_i(x,a_i)-\bar{f}_i) \mid f_i(x,a_i)-\bar{f}_i \leqq v,\right.$ $\left.(i=1,\ldots,k),\ \mu_{\tilde{a}}(a) \geqq \alpha,\ \mu_{\tilde{b}}(b) \geqq \alpha,\ x\in X(\tilde{b})\right\}$ can be viewed as the optimal value function of the augmented minimax problem (6)-(10) for any $\alpha\in N(\alpha^*)$. Therefore, the following theorem holds under the same assumptions as in Theorem 3.

Theorem 4. If all the assumptions in Theorem 3 are satisfied, then the following relations hold in some neighborhood $N(\alpha^*)$ of $\alpha^*$

$$\frac{\partial\left\{v + \sum_{i=1}^{k}(f_i(x,a_i) - \bar{f}_i)\right\}}{\partial\alpha} = \frac{\partial L}{\partial \alpha} = \sum_{i=1}^{k}\lambda_i^a + \sum_{j=1}^{m}\lambda_j^b \tag{12}$$

## 4. INTERACTIVE ALGORITHM

Following the above discussions, we can now construct an interactive algorithm in order to derive a satisficing solution for the DM from among the $\alpha$-Pareto optimal solutions. The steps marked with an asterisk involve interaction with the DM.

Step 1. Calculate the individual minimum and maximum of each objective function under given constraints for $\alpha=1$.

Step 2*. Ask the DM to select the initial value of $\alpha (0 < \alpha < 1)$ and the initial reference levels $\bar{f}_i$, $i=1,\ldots,k$.

Step 3. For the degree $\alpha$ and the reference levels specified by the DM, solve the augmented minimax problem.

Step 4. The DM is supplied with the corresponding $\alpha$-Pareto optimal solution and the trade-off rates between the objective functions and the degree $\alpha$. If the DM is satisfied with the current objective function values of the $\alpha$-Pareto optimal solution, stop. Otherwise, the DM must update the reference levels and/or the degree $\alpha$ by considering the current values of the objective functions and $\alpha$ together with the trade-off rates between the objective functions and the degree $\alpha$ and return to step 4. Here it should be stressed for the DM that: (1) any improvement of one objective function can be achieved only at the expense of at least one of the other objective functions, and (2) the greater value of the degree $\alpha$ gives worse values of the objective functions for some fixed reference levels.

## 5. NUMERICAL EXAMPLE

Based on the proposed method, we have developed a new interactive computer program. Here we demonstrate the interactive processes using our computer program by means of an illustrative example which is designed to test the program. Consider the following three-objective nonlinear programming problem with fuzzy parameters:

$$\left.\begin{array}{l} \min\left\{f_1(x,\tilde{a}_1) = \tilde{a}_{11}x_1^2 + (x_2+5)^2 + 2(x_3-\tilde{a}_{12})^2\right\} \\ \min\left\{f_2(x,\tilde{a}_2) = (x_1+\tilde{a}_{21})^2 + \tilde{a}_{22}(x_2-55)^2 + 3(x_3+20)^2\right\} \\ \min\left\{f_3(x,\tilde{a}_3) = \tilde{a}_{31}(x_1-55)^2 + \tilde{a}_{32}(x_2+\tilde{a}_{33})^2 + (x_3+20)^2\right\} \\ \text{subject to: } g_1(x,\tilde{b}_1) = \tilde{b}_{11}x_1^2 + \tilde{b}_{12}x_2^2 + \tilde{b}_{13}x_3^2 \leqq 100 \end{array}\right\} \quad (15)$$

$$0 \leqq x_i \leqq 10, \quad i = 1,2,3.$$

The membership functions for the fuzzy numbers $\tilde{a}_1$, $\tilde{a}_2$, $\tilde{a}_3$ and $\tilde{b}_1$ in this example are explained in Table 1 where L and E represent respectively linear and exponential membership functions.

```
COMMAND:
=GO

INPUT SUFFICIENTLY SMALL POSITIVE SCALAR FOR AUGEMENTED TERM:
=0.00001

---------------------< ITERATION 1 >---------------------

INPUT YOUR REFERENCE VALUES F(I) (I=1,3):
=4925 5640 6042

INPUT THE DEGREE ALFA OF THE ALFA LEVEL SETS
FOR THE FUZZY PARAMETERS:
=0.5

( KUHN-TUCKER CONDITIONS SATISFIED )

ALFA-PARETO OPTIMAL SOLUTION
TO THE AUGEMENTED MINIMAX PROBLEM
FOR INITIAL REFERENCE VALUES
------------------------------
     OBJECTIVE FUNCTION
------------------------------
F(1) =         6030.7663
F(2) =         6745.7663
F(3) =         7147.7663
------------------------------------------------------------
   X( 1) =            6.4142    X( 2) =              6.8306
   X( 3) =            4.8060
------------------------------------------------------------
TRADE-OFFS AMONG FUNCTIONS
 -DF(2)/DF(1) =             1.2226
 -DF(3)/DF(1) =             1.4022
------------------------------------------------------------
TRADE-OFFS BETWEEN ALFA AND FUNCTIONS
 DF/DALFA =            549.4475
------------------------------------------------------------

ARE YOU SATISFIED WITH THE CURRENT OBJECTIVE VALUES OF
THE ALFA-PARETO OPTIMAL SOLUTION ?
=NO

---------------------< ITERATION 2 >---------------------

INPUT YOUR REFERENCE VALUES F(I) (I=1,3):
=5500 7500 6500

INPUT THE DEGREE ALFA OF THE ALFA LEVEL SETS
FOR THE FUZZY PARAMETERS:
=0.6

        ........................................
```

Fig. 2 Interactive Satisficing Processes

Table 1. Fuzzy numbers

| $\tilde{p}$ | $(p_1,$ | $p_2,$ | $p_3,$ | $p_4)$ | TYPE left | TYPE right |
|---|---|---|---|---|---|---|
| $\tilde{a}_{11}$ | (3.8, | 4.0, | 4.0, | 4.3) | L | E |
| $\tilde{a}_{12}$ | (57.0, | 59.0, | 60.0, | 63.5) | E | L |
| $\tilde{a}_{21}$ | (18.0, | 19.5, | 20.0, | 22.5) | E | E |
| $\tilde{a}_{22}$ | (1.75, | 2.0, | 2.0, | 2.25) | E | L |
| $\tilde{a}_{31}$ | (2.3, | 2.5, | 2.5, | 2.75) | L | E |
| $\tilde{a}_{32}$ | (1.25, | 1.4, | 1.5, | 1.7) | L | L |
| $\tilde{a}_{33}$ | (17.5, | 20.0, | 20.0, | 22.0) | L | E |
| $\tilde{b}_{11}$ | (0.9, | 1.0, | 1.0, | 1.1) | E | E |
| $\tilde{b}_{12}$ | (0.8, | 0.95, | 1.0, | 1.2) | E | E |
| $\tilde{b}_{13}$ | (0.85, | 1.0, | 1.0, | 1.15) | E | L |

In Fig. 2, the interaction processes using the time-sharing computer program under TSS of the ACOS-1000 digital computer in the computer center of Kobe University in Japan are explained expecially for the first iteration through the aid of some of the computer outputs. α-Pareto optimal solutions are obtained by solving the augmented minimax problem using the revised version of the generalized reduced gradient (GRG) (Lasdon, Fox and Ratner, 1974) program called GRG2 (Lasdon, Warren and Ratner, 1980).

In this example, in the 4th iteration, the DM's satisficing solution is derived and the values of the objectives and decision variables are shown in Fig. 3. The whole interactive pro-

```
THE FOLLOWING VALUES ARE YOUR SATISFICING SOLUTION :

-----------------------------
     OBJECTIVE FUNCTION
-----------------------------

F(1) =          5624.2467
F(2) =          8324.2467
F(3) =          6724.2467
-----------------------------------------------------------
   X( 1) =               7.5887    X( 2) =          0.9739
   X( 3) =               6.7895
-----------------------------------------------------------

COMMAND:
=STOP
```

Fig. 3. Satisficing Solution for the DM

cesses are summarized in Table 2. CPU time required in this interaction process was 5.046 seconds and the example session takes about 10 minutes.

Table 2. Interactive processes

| Iteration | 1 | 2 | 3 | 4 |
|---|---|---|---|---|
| $\bar{f}_1$ | 5025 | 5500 | 5500 | 5300 |
| $\bar{f}_2$ | 5650 | 7500 | 7900 | 8000 |
| $\bar{f}_3$ | 6062.5 | 6500 | 6300 | 6400 |
| $\alpha$ | 0.5 | 0.6 | 0.6 | 0.65 |
| $f_1$ | 6030.77 | 5769.10 | 5763.16 | 5624.25 |
| $f_2$ | 6745.77 | 7769.10 | 8163.16 | 8324.25 |
| $f_3$ | 7147.77 | 6769.10 | 6563.16 | 6724.25 |
| $x_1$ | 6.41 | 7.64 | 8.08 | 7.59 |
| $x_2$ | 6.83 | 3.11 | 1.25 | 0.97 |
| $x_3$ | 4.81 | 6.18 | 6.21 | 6.79 |
| $\partial f_2/\partial f_1$ | -1.22 | -1.88 | -2.49 | -2.84 |
| $\partial f_3/\partial f_1$ | -1.40 | -1.22 | -1.08 | -1.19 |
| $\partial f/\partial\alpha$ | 549.45 | 561.49 | 560.61 | 547.38 |

## 6. CONCLUSIONS

In this paper, we have proposed an interactive satisficing method using the augmented minimax problems in order to deal with the multiobjective nonlinear programming problem with fuzzy parameters characterized by fuzzy numbers. Through the use of the concept of the α-level sets of the fuzzy numbers, a new solution concept called the α-Pareto optimality has been introduced. In our interactive scheme, the DM's satisficing solution can be derived by updating the reference levels and/or the degree α based on the current values of the membership functions and α together with the trade-off rates between the objective functions and the degree α. Furthermore, α-Pareto optimality of the generated solution in each iteration is guaranteed. Based on the proposed method, the time-sharing computer program has been written to facilitate the interaction processes. An illustrative numerical example demonstrated the feasibility and efficiency of both the proposed method and its interactive computer program by simulating the responses of the hypothetical DM. However, further applications must be carried out in cooperation with a person actually involved in decision making. From such experiences the proposed method and its computer program must be revised.

## REFERENCES


Bellman, R.E., and L.A. Zadeh (1970). Decision making in a fuzzy environment. Mang. Sci. 17, 141-164.

Choo, E.U., and D.R. Atkins (1983). Proper efficiency in nonconvex multicriteria programming. Math. Oper. Res. 8, 467-470.

Dubois, D., and H. Prade (1978). Operations on fuzzy numbers. Int. J. Systems Sci. 9, 613-626.

Dubois, D., and H. Prade (1980). Fuzzy Sets and Systems: Theory and Applications. Academic Press, New York.

Fiacco, A.V. (1983). Introduction to Sensitivity and Stability Analysis in Nonlinear Programming. Academic Press, New York.

Haimes, Y.Y., and V. Chankong (1979). Kuhn Tucker multipliers as trade-offs in multiobjective decision-making analysis. Automatica 15, 59-72.

Lasdon, L.S., R.L. Fox, and M.W. Ratner (1974). Nonlinear optimization using the generalized reduced gradient method. Revue Francaise d'Automatique, Informatique et Research Operationnelle 3, 73-103.

Lasdon, L.S., A.D. Waren and M.W. Ratner (1980). GRG2 User's Guide. Technical Memorandum. University of Texas.

Orlovski, S.A. (1983a). Problems of Decision-Making with Fuzzy Information. IIASA Working Paper WP-83-28. Laxenburg, Austria.

Orlovski, S.A. (1983b). Multiobjective Programming Problems with Fuzzy Parameters. IIASA Working Paper. Laxenburg, Austria.

Sakawa, M. (1983). Interactive computer programs for fuzzy linear programming with multiple objective. Int. J. Man-Machine Stud. 18, 489-503.

Sakawa, M., and T. Yumine (1983). Interactive fuzzy decision-making for multiobjective linear fractional programming problems. Large Scale Syst. 5, 105-114.

Sakawa, M., T. Yumine, and H. Yano (1984). An Interactive Fuzzy Satisficing Method for Multiobjective Nonlinear Programming Problems. IIASA Collaborative Paper CP-84-18. Laxenburg, Austria.

Sakawa, M., and H. Yano (1984). An interactive fuzzy satisficing method using penalty scalarizing problems, Proc. Int. Computer Symposium. Tamkang, Univ. Taiwan, 1122-1129.

Steuer, R.E., and E.U. Choo (1983). An interactive weighted Tchebycheff procedure for multiple objective programming. Math. Prog. 26, 326-344.

Tanaka, H., and K. Asai (1981). A formulation of linear programming problems by fuzzy function. (in Japanese) Syst. and Cont. 25, 351-357.

Tanaka, H., and K. Asai (1984). Fuzzy linear programming problems with fuzzy numbers. Fuzzy Sets and Syst. 13, 1-10.

Wierzbicki, A.P. (1979). The Use of Reference Objectives in Multiobjective Optimization - Theoretical Implications and Practical Experiences. IIASA Working Paper WP-79-66. Laxenburg, Austria.

Yano, H., and M. Sakawa (1985). Trade-off rates in the weighted Tchebycheff norm method. (in Japanese) Trans. S.I.C.E. 21, 248-255.

Zadeh, L.A. (1975). The concept of a linguistic variables and its application to approximate reasoning-1. Inf. Sci. 8, 199-249.

# INTERACTIVE POLYOPTIMIZATION FOR FUZZY MATHEMATICAL PROGRAMMING

Cs. Fabian, Gh. Ciobanu and M. Stoica

Academy of Economic Studies
Bucharest, Romania

Abstract. An optimization model with non-linear (deterministic and/or fuzzy) constraints and several objective functions is formulated and transformed into a problem with an objective synthesis function. The paper refers to the situation when there is no feasible solution, this case being frequently met in practice. A resolution method is indicated further on. The algorithm is illustrated by an example. A concrete model of loading up the capacities with fuzzy constraints and with several objective functions is also given.

Keywords: fuzzy constraints, multicriteria decision making, nonlinear programming, global optimization, cluster analysis, interactive algorithms, capacity loading.

## 1. INTRODUCTION

The study of modelling and solving mathematical programming problems with several objective functions and fuzzy constraints is a necessity resulting from economical, technical and social practice. To this effect, we set forth a mathematical optimization model and a relevant solution method.

This paper is concerned with problems which may be said to be at the confluence of several fields, with each of them being dealt with separately, or sometimes together. These fields are multicriteria optimization, fuzzy mathematical programming, global optimization, interactive solving of decision making problems and application of mathematical programming in production scheduling.

As far as multicriteria optimization is concerned, we refer to the papers of Roy (1970), Boldur-Stancu Minasian (1971), Marusciac (1970) and to the excellent works of Zeleny (1974), Zionts (1978) and Stadler (1979). The interactive approach to multicriterial optimization problems appears in Zionts and Wallenius (1976), Leclercq (1982), Sakawa (1984) and Warfield (1984).

The analysis of optimization problems by means of fuzzy sets appears in the works of, e.g., Dubois and Prade (1984), Negoita and Sularia (1976), and Zimmermann (1978).

Global optimization is also a field very much debated. We mention among the respective works one by Boender (1982) and an excellent synthesis by Dixon and Szegö (1978). We will refer in this paper to the Törn (1977) clustering method and to the directed simulation method of Hartman (1973).

The interactive approach to decision making problems is a mode of solving them which is now widespread thanks to computing methods and microcomputers.

In this paper we present a fuzzy optimization model with non-linear constraints and several objective functions.

The optimization problem with several objective functions and several (deterministic and/or fuzzy) constraints is formulated and transformed into a problem with a synthesis objective function; the concept of an M-efficient solution is introduced; the mode of associating the membership degrees to constraints and objective functions is described, laying stress on determining the parameters of the membership function and on the method of composing several optimum functions into a synthesis function.

A resolution method is indicated further on. When there is a feasible solution, the Sakawa (1984) method is recommended.

The main section of this paper refers to the situation when there is an infeasible solution, this case frequently being met in practice. For this case, a global optimization algorithm is suggested which is founded on the combination and modification of the Törn (1977), Hartman (1973) and Box (1965) methods. The operation of the algorithm is illustrated by means of an example.

We conclude the paper with a concrete model of loading up capacities with fuzzy constraints and with several objective functions.

## 2. MULTIOBJECTIVE OPTIMIZATION PROBLEM

### 2.1. Statement of the problem

The optimization problem with several objective functions is represented as follows:

$$\min_x \left\{ f(x) := (f_1(x), f_2(x), \ldots, f_p(x)) \right\} \quad (1)$$

subject to

$$x \in D = \left\{ x \mid x \in R^n, 0 \leqslant x \leqslant d, g_j(x) \leqslant 0,\ j=1,2,\ldots,m \right\}$$

where: $x$ is an n-dimensional vector of the decision making variables with nonnegative components $x_1, x_2, \ldots, x_n$ ($x_i \geq 0$, $i=1,2,\ldots,n$); $d$ is an n-dimensional vector with the components $d_1, d_2, \ldots, d_n$ upper bounding the decision variables ($x_i \leqslant d_i$, $i=1,2,\ldots,n$); $f_1, f_2, \ldots, f_p$ are p distinct objective functions of the decision vector $x$,

$$g_j(x) \leq 0,\ j = 1,2,\ldots,m \quad (2)$$

are constraints on the decision variables and $D$ is the feasible set of constraints (2) in the hyperparallelepiped $0<y<d$

One assumes that in the practical problems the decision maker knows at least approximatively the range $d=(d_1,d_2,\ldots,d_n)$ of decision variable $x = (x_1,x_2,\ldots,x_n)$.

The concept of the Pareto optimum, also known as a non-inferior solution, is fundamental to problem (1). Qualitatively, a Pareto optimal solution to problem (1) is one where any improvement of one objective function can be achieved only at the expense of another one. Usually, Pareto optimal solutions form an infinite set of points from D. The choice of a preferred solution among the Pareto optimal solutions means the solving of the following problem:

$$\max_x \eta(f_1(x),f_2(x),\ldots,f_p(x)) \tag{3}$$

subject to $x \in D^P$

where: $D^P$ is the set of Pareto optimal solutions of problem (1) and $\eta(.)$ is an aggregating function of the functions $f_1(x),f_2(x),\ldots,f_p(x)$, defined as a utility function on $D^P$. As a rule, these utility functions $\eta(.)$ are not explicitly known by the decision makers.

It is difficult to obtain such Pareto optimal solutions of problem (3), as the set $D^P$ is not "visible". The practical requirements of the problem compel the decision makers to present the desired (aspiration) values of the objective functions

$$f_1(x),f_2(x),\ldots,f_p(x)$$

as bounding values $M_1,M_2,\ldots,M_p$. We may thus consider p constraints of the form

$$f_i(x) \le M_i, \qquad i = 1,2,\ldots,p \tag{4}$$

Let $D^M = \{x|x\in R^n,\ f_i(x) \le M_i,\ i = 1,2,\ldots,p\}$. If $D^M \neq \emptyset$, let $D^{PM} = D^P \cap D^M$. One observes that if $D^{PM} \neq \emptyset$, then $D^P \neq \emptyset$ and $D \neq \emptyset$, respectively. As a rule, the converse is not true, since the set $D^{PM}$ may be empty due to $M_1,M_2,\ldots,M_p$ selected by the decision maker as his or her aspiration levels. But this is not known a priori at the moment when the decision maker selects the values $M_1,M_2,\ldots,M_p$.

It may be noticed that when $D^{PM} \neq \emptyset$, this is the set of Pareto solutions sought by the decision maker.

We mention that Sakawa (1984) deals with the case when $D^M \neq \emptyset$ and gives a method which, in certain conditions, leads to the selection of a solution from $D^{PM}$.

Sometimes, the sets of solutions $D^M$ and $D^{PM}$, respectively, are empty, due to the set of values $M_1, M_2, \ldots, M_p$ supplied by the decision maker and to their inconsistence with the data of the problem. Thus, we consider the values of the objective functions $f_i(x)$, $i=1,2,\ldots,p$, to be around $M_i$, $i=1,2,\ldots,p$, hence constraints (4) are fuzzy constraints

$$f_i(x) \lesssim M_i, \; i = 1,2,\ldots,p \tag{5}$$

In many practical cases (for instance, in production scheduling and programming), constraints (2) yield an empty set ($D = \emptyset$), the decision maker asks for the solution of problem (1), disregarding the fact whether D is an empty set or not. In order to solve problem (1) in this case also, we allow the violation of some constraints (2), considering them as fuzzy constraints

$$g_j(x) \lesssim 0, \quad j = 1,2,\ldots,m \tag{6}$$

2.2. Fuzzy multiobjective model

We associate with the fuzzy constraints (5) and (6) the membership functions $\mu_{f_i}(x)$, $i=1,2,\ldots,p$, concerning the degree of feasibility of the levels $M_i$, $i=1,2,\ldots,p$, desired by the decision maker, and $\mu_{g_j}(x)$, $j=1,2,\ldots,m$, concerning the degree of fulfillment of the fuzzy constraints $g_j(x) \lesssim 0$, $j=1,2,\ldots,m$. Thus, we may consider problem (1) as a multicriteria maximization problem, the objective functions being membership functions, and the constraints remaining those of variable boundings, i.e.

$$\max_x \Big\{ \mu(x) = (\mu_{f_1}(x), \mu_{f_2}(x), \ldots, \mu_{f_p}(x), \mu_{g_1}(x), \mu_{g_2}(x), \ldots, \mu_{g_m}(x)) \Big\} \tag{7}$$

subject to

$$x \in X = \Big\{ x \in R^n \,|\, 0 \leqslant x \leqslant d \Big\}$$

Let us denote by $X^{PM}$ the set of Pareto optimal solutions for problem (7); a solution $x \in X^{PM}$ will be referred to as an M-efficient solution.

Having in view the analogy between problems (1) and (3), for determining an optimal Pareto solution of problem (7), we solve the following problem

$$\max_{x} \chi(x) \tag{8}$$

subject to

$$x \in X^{PM}$$

where: $\chi(x)$ is a synthesis membership function obtained by applying some algebraic composition operators on the membership functions $\mu_{f_i}(x)$, $i=1,2,\ldots,p$, and $\mu_{g_j}(x)$, $j=1,2,\ldots,m$,

$$\chi(x) = \chi(\mu_{f_1}(x),\ldots,\mu_{f_p}(x),\ \mu_{g_1}(x),\ldots,\mu_{g_m}(x))$$

Problem (3) makes sense if and only if problem (1) has a non-empty solution set ($D \neq \emptyset$). This also implies the fact that the set of Pareto optimal solutions is non-empty ($D^P \neq \emptyset$). In many concrete cases $D = \emptyset$, hence $D^P = \emptyset$ also, and this is why problems (7) and (8) have been considered.

Let us further consider the problem

$$\max_{x \in X} \chi(x) \tag{9}$$

and let $x^*$ be an optimal solution.

Under some resonable assumptions (Dubois and Prade, 1984) for $\mu_{f_i}(x)$, $i=1,2,\ldots,p$, and $\mu_{g_j}(x)$, $j=1,2,\ldots,m$ (x is preferred to x′ if and only if $(\forall) i \leq p$, $\mu_{f_i}(x) \geq \mu_{f_i}(x')$ and $(\forall) j \leq m$, $\mu_{g_j}(x) \geq \mu_{g_j}(x')$) and for $\chi(x)$ ($(\forall)(x_1,\ldots,x_n)$, $(y_1,\ldots,y_n)$ if $(\forall) k \leq n$, $x_k \geq y_k$, then $\chi(x_1,\ldots,x_n) \geq \chi(y_1,\ldots,y_m)$); it is obvious that $x^* \in X^{PM}$ and $x^* \in D^P$. In other words, an M-efficient solution is also a Pareto optimal solution for problem (1).

## 3. ASSOCIATED MEMBERSHIP FUNCTIONS

### 3.1. On objective functions and constraints

We define the membership functions associated with constraints (5) and (6). The membership function for a constraint $h(x) \lessapprox \alpha$ ($\alpha > 0$ given), may be defined, e.g., as follows:

$$\mu_h(x) = \begin{cases} 1 & \text{if } h(x) \leq \alpha \\ \exp(-qh(x)), & \text{if } h(x) > \alpha \end{cases}$$

where: q is a positive scalar factor expressing the relative importance of the constraint $h(x) \leq \alpha$ ; $\mu_h(x)$ depends on the order of magnitude of h(x) which imposes a normalization. To this effect, the decision maker should specify an additional value β, with $\beta > \alpha$, by which a maximum violation of the constraint $h(x) \leq \alpha$ is allowed. The membership function $\mu_h(x)$ may be defined as follows

$$\mu_h(x) = \begin{cases} 1 & \text{if } h(x) \leq \alpha \\ \exp\left[- q \dfrac{h(x) - \alpha}{\beta - \alpha}\right] & \text{if } \alpha < h(x) \leq \beta \\ 0 & \text{if } h(x) > \beta \end{cases} \tag{10}$$

Likewise, for the case of our problem, we may also define other membership functions which have to satisfy the following conditions:

$\mu_h(x)$ - monotonously decreasing for $\alpha<h(x)<\beta$, and
$\mu_h(x) = 1$ if $h(x) \leq \alpha$;
$0 < \mu_h(x) < 1$ if $\alpha < h(x) \leq \beta$ , and
$\mu_h(x) = 0$ if $h(x) > \beta$.

The introduction of the membership functions $\mu_h(x)$ complicates the solving of the problems stated in the preceding section, however, in the cases $D = \emptyset$ and/or $D^M = \emptyset$ we are able to supply a solution, associated with a degree of membership regarding the satisfaction of the constraints or of the bounds $M_1, M_2, \ldots, M_p$ set by the decision maker.

As a rule, it is difficult for the decision maker to supply (decide) the values of the parameters q, α and β . Meeting the decision maker half way, we suggest the following procedure:

- the points $x^1, x^2, x^3, \ldots, x^c$ from $X = \{ x \in R^n \mid 0 \leq x_k \leq d_k, k=1,2,\ldots,n \}$ are uniform randomly generated;

- the $h(x^1), h(x^2), h(x^3), \ldots, h(x^c)$ values are evaluated and put in an increasing order, i.e.

$h(x^1) \leq h(x^2) \leq h(x^3) \leq \ldots \leq h(x^c)$

- knowing this variation range of the function h(x), the decision maker can decide on the selection of the values α and β;

- for determining the parameter, q, the decision maker should evaluate the values of the membership function relevant to $h(x^1), \ldots, h(x^c)$, completing, for instance, the following table for the points $x^k$ with $\alpha < h(x^k) < \beta$

| h(x) | α | $h(x^u)$ | ... | $h(x^v)$ | β |
|---|---|---|---|---|---|
| $\mu_h(x)$ | 1 | . | ... | . | 0 |

thus, the value of the parameter q from the membership function (10) can be statistically determined.

With our problem, according to constraints (4) [$f_i(x) \leq M_i$, $i=1,2,\ldots,p$], the membership functions (10) are

$$\mu_{f_i}(x) = \begin{cases} 1 & \text{if } f_i(x) \leq M_i \\ \exp[-r_i \dfrac{f_i(x)-M_i}{M_i' - M_i}], & \text{if } M_i<f_i(x)\leq M_i', I=1,2,\ldots,p \\ 0 & \text{if } f_i(x) > M_i \end{cases} \quad (11)$$

where: $r_i$, $i=1,2,\ldots,p$, are positive scalar factors, corresponding to the factor q, expressing the relative importance of the objective functions $f_1(x),\ldots,f_p(x)$; $M_i$ and $M_i'$, $i=1,2,\ldots,p$, are the aspiration levels $\alpha$ and $\beta$ related to each objective function $f_i(x)$, $i=1,2,\ldots,p$.

Corresponding to constraints (2) [$g_j(x) \leq 0$, $j=1,2,\ldots,m$], the associated membership functions (10) are

$$\mu_{g_j}(x) = \begin{cases} 1 & \text{if } g_j(x) \leq 0, \\ \exp(-s_j g_j(x)/b_j), & \text{if } 0<g_j(x)\leq b_j, j=1,2,\ldots,m \\ 0 & \text{if } g_j(x) > b_j \end{cases} \quad (12)$$

where $s_j$, $j=1,2,\ldots,m$, are positive scalar factors expressing the relative importance of the constraints $g_j(x)\leq 0$, $j=1,2,\ldots,m$, In this case, $\alpha = 0$ and $\beta = b_j$ are considered for each constraint.

The parameters $r_i$, $s_j$, $M_i$, $M_i'$, $b_j$, $i=1,2,\ldots,p$, $j=1,2,\ldots,m$, entering the expression of the membership functions are determined in the same manner as the parameters q, $\alpha$ and $\beta$ from the general membership function (10).

In order to determine the $r_i$ and $s_j$ one may also resort to a group of decision makers. Let us assume that, when the importance of the constraint $g_j(x) \leq 0$ is greater than that of $g_k(x) \leq 0$, we have $s_j \geq s_k$, and, likewise, when the objective function $f_u(x)$ is more important than the objective function $f_v(x)$, we have $r_u \geq r_v$.

The parameters $r_i$ and $s_j$ may be determined in case of a group of decision makers, as follows:

- concrete values are given to the parameters $r_i$ and $s_j$ by each decision maker of the group;

- when no qualitative inconsistencies exist (the importance order is identical and only the values are different) the arithmetic mean is computed on each parameter;

- when there are qualitative inconsistencies, these should be analysed and argued, and the parameter values should be brought into accord.

### 3.2. Synthesis membership functions

Similarly to the existing methods for solving multicriteria problems or decision problems, we define a series of $\chi(x)$ operators for problem (8) by composing the membership functions associated to the objective functions $\mu_{f_i}(x)$, i=1,2,...,p,and to the constraints $\mu_{g_j}(x)$, j=1,2,...,m, respectively.

These composition operators $\chi(x) = \chi(\mu_{f_i}(x),\dots,\mu_{f_p}(x), \mu_{g_1}(x),\dots,\mu_{g_m}(x))$ are synthesis membership functions if they satisfy the following two axioms, according to Dubois and Prade (1984):

A1. $\chi(1,1,\dots,1) = 1$

This condition requires the membership function $\chi$ to take the value 1 if $\mu_{f_i}(x) = 1$, i=1,2,...,p, and $\mu_{g_j}(x) = 1$, j=1,2,...,m. Obviously, the dual requirement is also very natural, i.e.

$\chi(0,0,\dots,0) = 0$

A2. The second axiom will be derived from a compatibility condition with respect to the Pareto approach to multigoal decision problem: $x \in X$ is preferred to $y \in X$ if and only if

$$\mu_{f_i}(x) \geq \mu_{f_i}(y), \quad i=1,2,\dots,p$$

$$\mu_{g_j}(x) \geq \mu_{g_j}(y), \quad j=1,2,\dots,m$$

imply

$$\begin{aligned}&\chi(\mu_{f_1}(x),\dots,\mu_{f_p}(x),\mu_{g_1}(x),\dots,\mu_{g_m}(x)) \geq \\ &\geq \chi(\mu_{f_1}(y),\dots,\mu_{f_p}(y),\mu_{g_1}(y),\dots,\mu_{g_m}(y))\end{aligned} \qquad (13)$$

Such composition operators satisfying axioms $A_1$ and $A_2$ are, for example, the following:

$$\lambda_1(x) = \min(\mu_{f_1}(x),\dots,\mu_{f_p}(x),\mu_{g_1}(x),\dots,\mu_{g_m}(x))$$

$$\lambda_2(x) = \max(\mu_{f_1}(x),\dots,\mu_{f_p}(x),\mu_{g_1}(x),\dots,\mu_{g_m}(x))$$

$$\chi_3(x) = (\sum_i \mu_{f_i}(x) + \sum_j \mu_{g_j}(x))/(p+m)$$

$$\chi_4(x) = \min[a \min_i \mu_{f_i}(x) + (1-a) \max_i \mu_{f_i}(x),$$

$$b \min_j \mu_{g_j}(x) + (1-b) \max_j \mu_{g_j}(x)]$$

where $a\in[0,1]$ and $b\in[0,1]$.

While solving problem (8), the decision maker may choose a certain composition function $\chi(x)$. Thus, in case of a pessimistic attitude (Wald's principle), the decision maker can select the function $\chi_1(x)$. When the decision maker has in view a maximum caution, problem (8) may be modified to the effect that $\chi_1(x)$ is minimized: $[\min \chi(x), x\in X]$. This overcarefulness may also be accepted and used when the obtained optimal version meets the minimal requirements of the decision maker who may select the function $\chi_2(x)$.

If constraints (5) and (6) are considered equally important (the Bayes-Laplace principle), the decision maker may select the function $\chi_3(x)$. The idea of maximum uncertainty and indeterminism regarding the degree of fulfillment of constraints (5) and (6) is thus suggested.

By adopting the idea of a compromise between the maximum and the minimum values of the membership functions corresponding to (5) and (6) (Hurwicz principle), the decision maker may select the function $\chi_4(x)$ and the constants a and b being previously established by him on the basis of his intuition, experience and art. From this criterion, we can derive the pessimistic caution criterion for a = b = 1, and the superoptimism criterion for a = b = 0, respectively.

## 4. RESOLUTION METHOD

Thanks to the mode of building the synthesis membership functions, by solving problem (9) instead of problem (8) a Pareto optimal solution of problem (7) will be obtained.

Therefore, let us consider the following problem:

$$\begin{cases} \max_x \chi(x) \\ \text{subject to} \\ x\in X = \{x \mid x\in R^n,\ 0 \leqslant x \leqslant d\} \end{cases} \tag{14}$$

where $\chi(x)$ is a synthesis membership function, and d the upper bound vector.

The problem (14) is a global optimization problem for whose resolution Törn (1977) suggests a cluster analysis method. The method suggested by Törn consists in generating N uniformly distributed points $x^i$ $(i=1,2,\ldots,N)$. By using a local optimizer, some iterations are carried out with the view of forming some clusters, reaching the points $y^i$ $(i=1,2,\ldots,N)$ from the points $x^i$. When only a cluster can be formed, a local optimizer is further applied until the desired accuracy is obtained. When several clusters are formed, a representative is selected in every cluster and a local optimizer applied. In the end, the global optimum is selected.

Another method for global optimization is indicated by Hartman (1973) and based on a directed simulation according to the marginal distribution of some random variables built up by means of the function to be optimized and of a generator of uniformly distributed random numbers.

The Hartman algorithm consists in the generation of N points by means of a uniform marginal distribution $\pi_j$. The function to be optimized is evaluated in every point generated $x^k$ $(k=1,2,\ldots,N)$, and the obtained sequence is decreasingly arranged for maximum criteria, and increasingly arranged for minimum criteria. The first $[\alpha N]$ values from the sequence $(\alpha = 0.20 - 0.25)$, considered to be "success" points, are preserved. The intervals $[0,d_j]$, $j = 1,2,\ldots,n$, are divided into $m_j$ subintervals $(m_j \geq 2)$ and the absolute frequencies $M_{jt}$, $t=1,2,\ldots,m_j$, of occurrence of the "success" points in the t-th subinterval are computed. The new density $\pi_j'$ is defined by means of them.

The following steps result in a resolution method for problem (14) founded on the modified Törn and Hartman methods.

Let us consider the marginal density $\pi_j$ for the component $x_j$ $(j=1,2,\ldots,n)$ of the initially uniform variable x.

Step 0. The number of points to be generated is determined by means of the relation

$$N = \text{entier}\left(\prod_{j=1}^{n}(d_j + 1)/v\right)$$

where: v is a value specified by the decision maker and by means of which he assumes the risk of having $D \neq \emptyset$ non-identified, or to find an $\bar{x}\in X^{PM}$ which is not in $D^P$, respectively. In practice, the decision maker can supply the acceptable tolerances on the variables, $\Delta x_i$. In this case, we have

$$v = \prod_{j=1}^{n} \Delta x_j$$

Step 1. N points $x^k = (x_1^k,\ldots,x_n^k)$, $k=1,\ldots,N$, distributed with densities $\pi_j$, $j=1,\ldots,n$, in the hyperparallelepiped $0 \leqslant x \leqslant d$ are generated.

If there exists an $x^k$ for which $\mu_{g_i}(x^k) = 1$, $i=1,2,\ldots,m$, then, e.g., the method given by Sakawa (1984) may be applied; otherwise go to step 2.

The Sakawa (1984) method consists of:

- Selecting the levels $M_i$ for p-1 constraints so that the system $f_i(x) \leq M_i$, $i=1,\ldots,p$, be consistent.

- Optimizing the p-th function, a Pareto optimal point is determined as are the corresponding Lagrange multipliers. If all the M-constraints are active, we pass to the following step; otherwise, the inactive constraint values $M_i$ are replaced by $f_i(x^*(M_i))$ and the new Pareto optimal solution and the corresponding multipliers are determined.

- Appraising the marginal rate of substitution of the decision makers in a fuzzy form.

- When improvements are no longer possible, the Pareto optimal solution of the decision maker has been obtained so that one goes to the next step. The direction of improvement is determined by means of the substitution rate and of the Lagrange multipliers.

- The preferred form of the function for local optimization is selected and its parameters are determined with the help of the decision maker.

- By modifying the step size, a local optimum is determined in the neighbourhood of the starting point and the method is resumed with the determination of another Pareto optimal point.

Step 2. We keep $[\alpha N]$ $(0 < \alpha < 1)$ points $x^k$ and the values $\chi(x^k)$ which have led to the greatest values $\chi(x^k)$ (Hartman, 1973). We suggest for $\alpha$ the values 0.5; 0.035 or 0.25.

Step 3. We apply a local optimizer for the function $\chi(x)$ with the initial points $x^k$ for which $\chi(x^k) \neq 1$, $k=1,2,\ldots,s$, $s \leq [\alpha N]$.

Due to the difficulties of establishing the convexity and differentiability of the function $\chi(x)$, we recommend the application of a method that does not use the derivative, for example that of Box (1965).

The Box method consists in determining n+1 feasible points where the function to be optimized is evaluated. The lowest value is determined and eliminated and the centroid of the remaining n points is determined. The symmetric of the point relevant to the lowest value relative to the centroid is built up and the process goes on.

With the previously determined number N, we divide the hyperparallelepiped $0 \leqslant x \leqslant d$ in N spheres, whose volume is equal to $\prod_{j=1}^{n} d_j / N$.

The radius of a sphere is

$$r = \sqrt[n-1]{\frac{\Gamma(n/2) \prod_{j=1}^{n} d_j}{2\pi^{n/2} N}}$$

where $\Gamma(x)$ is the well-known gamma function.

We generate n+1 points, uniformly distributed on a sphere with radius r (Rubinstein, 1982). Let these generated points be $x_i^k$, $i=1,\dots,n+1$, $0 \leq x_i^k \leq d_i$, and a new $x^k$ ( $y^k$) is calculated according to

$$y^k = \arg\max \left\{ \chi(x_i),\ i=1,\dots,n+1 \right\}$$

or by applying the Box method until $\chi(x_{j+1}^k) < \min\left\{\chi(x_{j-2}^k), \chi(x_{j-1}^k), \chi(x_j^k), \dots\right\}$.

The points $y^k$ are given by

$\chi(y^k) = \max\left\{\chi(x_{j-2}^k), \chi(x_{j-1}^k), \chi(x_j^k), \dots\right\}$ and we have

$\chi(x^k) \leq \chi(y^k)$, $k=1,\dots,s$

The generation of the n+1 initial feasible points in an arbitrary domain is difficult for the original Box method. In our case, as we have the simple constraints $0 \leq x \leq d$, the generation of these initial feasible points meets no difficulties.

With the Box method, only the most unfavourable point is eliminated with each step and thus we obtain a weak convergence. We can speed up the convergence by eliminating, for instance, $[\beta(n+1)]$ from the most disadvantageous points, where $0 < \beta < 1$, and by generating $[\beta(n+2)]$ new points with each step.

When there exists a $y^k$ for which $\chi(y^k) = 1$, the process may be stopped or continued for the more complete description of the set of optimal solutions.

<u>Step 4</u>. We divide the intervals $0 \leq x_j \leq d_j$, $j=1,\dots,n$ into $m_j$ subintervals (see Hartman, 1973) and calculate the values $M_{jt}$ by counting how many times the component $y_j^k$ of the $[\alpha N]$ points has entered the t-th interval.

We calculate

$$p_{jt} = M_{jt} / \sum_{l=1}^{m_j} M_{jl}$$

and the new marginal density $\pi_j'$ is characterized by $p_{jt}$, $j=\hat{1},\dots,m_j$.

We consider $\pi_j = \pi_j'$. When $p_{jt}=0$, the t-th interval can be eliminated from the domain of component j.

As far as the parameters $m_j$ are concerned, we may proceed at first with lower values $m_j$ and go on with the subdividing process in the fields of interest. We also may consider $m_j = [d_j]$ when $d_i$ are not great values or we may resort to a dichotomization process with $m_j = 2$.

Step 5. We count the groups of remaining elementary hyperparallelepipeds. These will give the number of clusters. When we have a single cluster or a small number of clusters, in order to increase the exactness, the algorithm may be successively applied on the new hyperparallelepiped $d' \leq x \leq d''$ from the obtained clusters.

Step 6. The representatives of the clusters are selected or calculated (for instance by the arithmetic mean) and the point $x^k$ which realized the greatest $\chi(x^k)$ is considered to be the solution of the problem. It is obvious that, from this point, the algorithm may be resumed from step 1, when a greater accuracy is desired in determining the global optimum, or one may stop if $|\chi(x^{k+1}) - \chi(x^k)| \leq \varepsilon$ and $|x^{k+1} - x^k| \leq \delta$.

We conclude these considerations by mentioning that the suggested method should be included into an interactive system wherein the decision maker intervenes during the solving of real problems.

The decision maker supplies some information, for instance for $M_i$, N, $\alpha$, $m_j$ and, analyzing $\mu_{f_i}(x^k)$, and $\mu_{g_j}(x^k)$, and $\chi(x^k)$, respectively, intervenes on the previously given parameters until a solution is obtained, which, from the practical point of view, satisfies to a greater degree than those previously found, or he stops if the new attempts to obtain an improvement of the objectives considered do not end with a significant favourable by-result.

Example. Let us consider the following problem (Fig. 1):

$$\min\{f_1(x) = x_1 - x_2\}$$

$$\max\{f_2(x) = x_1\}$$

$$g_1(x) = (x_1 + 2x_2)^2 - 4(1+x_1x_2) \leq 0$$

$$g_2(x) = (x_1-x_2)^2 - 6(x_1+x_2) + 15 \leq 0$$

$$x_1 \geq 0, \; x_2 \geq 0$$

By considering

$M_1=0$, $M_1'=2$, $M_2=0$, $M_2'=0.5$, $b_1=16$, $b_2=15$, $r_1=2$, $r_2=0.5$, $s_1=16$, $s_2=15$, the membership functions (10) and (11) are defined as follows:

$$\mu_1(x) = \begin{cases} 1, & x_1-x_2 \leq 0 \\ \exp(-(x_1-x_2)), & 0 < x_1-x_2 \leq 2 \\ 0 & x_1-x_2 > 2 \end{cases}$$

$$\mu_2(x) = \begin{cases} 1 & x_1 \geq 0.5 \\ \exp(-(-x_1+0.5)), & 0 < x_1 < 0.5 \\ 0 & x_1 \leq 0 \end{cases}$$

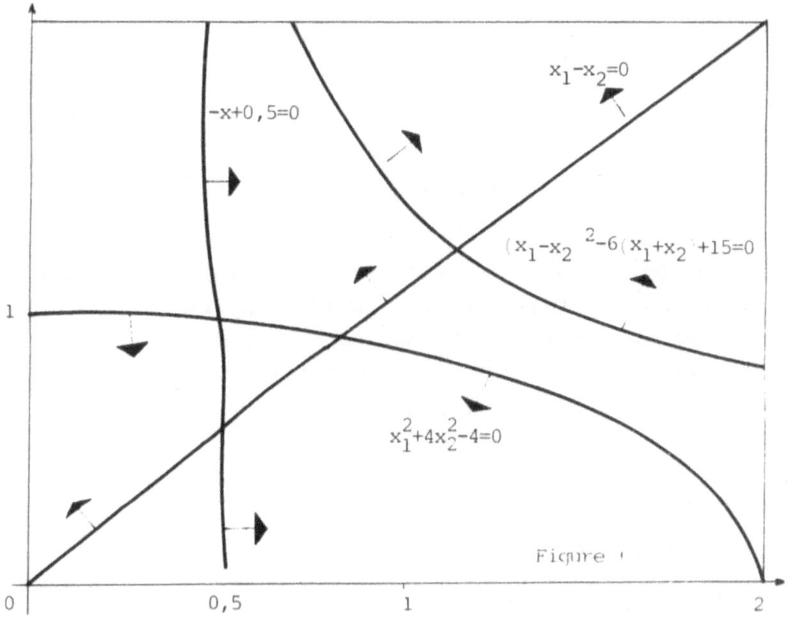


Figure 1

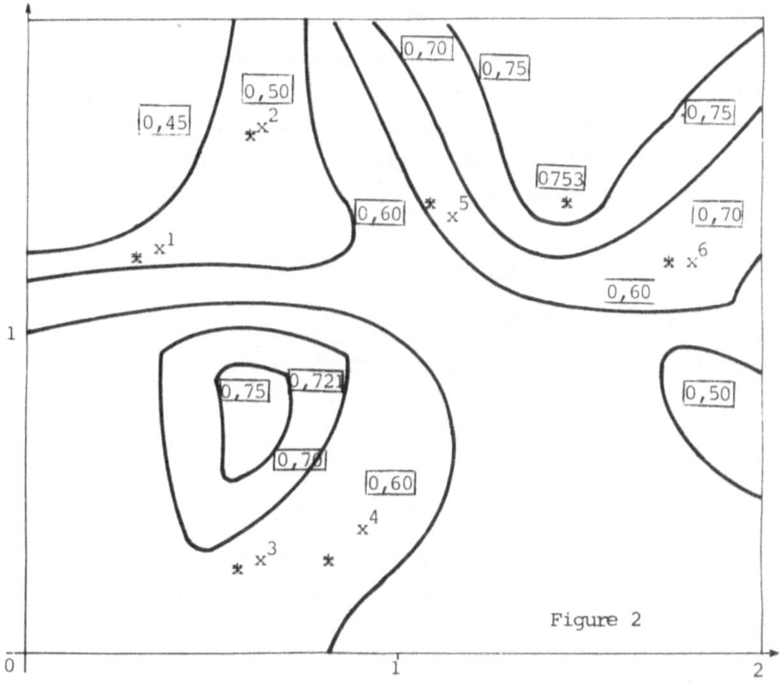


Figure 2

$$\mu_3(x) = \begin{cases} 1 & g_1(x) \le 0 \\ \exp(-g_1(x)), & 0 < g_1(x) \le 16 \\ 0 & g_1(x) > 16 \end{cases}$$

$$\mu_4(x) = \begin{cases} 1 & g_2(x) \le 0 \\ \exp(-g_2(x)), & 0 < g_2(x) \le 15 \\ 0 & g_2(x) > 15 \end{cases}$$

and the global membership function, according to the operator $\chi_3(x)$, is $\chi(x) = (\mu_1(x)+\mu_2(x) + \mu_3(x) + \mu_4(x))/4$.

Further on we also consider: $d_1=d_2=2$, $N=6$, and $\alpha=0.5$.

When step 1 is carried out, we find:

$x^1 = (0.4;\ 1.2)$, $\chi(x^1) = 0.513$;

$x^2 = (0.6;\ 1.6)$, $\chi(x^2) = 0.516$;

$x^3 = (0.6;\ 0.2)$; $\chi(x^3) = 0.668$;

$x^4 = (1.2;\ 0.4)$, $\chi(x^4) = 0.613$;

$x^5 = (1.0;\ 1.4)$, $\chi(x^5) = 0.619$;

$x^6 = (1.6;\ 1.2)$, $\chi(x^6) = 0.671$;

These points and the level curves of the function $\chi(x)$ are illustrated in Fig. 2.

The best $[\alpha N] = 3$ values for $\chi(x)$ are attained in $x^6$, $x^3$ and $x^5$. Starting from these points, by using a local optimizer, and with the radius $r = 0.46$, we proceed from $x_0^6 = x^6$ through:

$x_1^6 = (1.72;\ 1.58)$, $\chi(x_1^6) = 0.72$,

$x_2^6 = (1.21;\ 0.91)$, $\chi(x_2^6) = 0.57$,

$x_3^6 = (1.41;\ 1.63)$, $\chi(x_3^6) = 0.75$.

to

$y^1 = (1.41;\ 1.5)$, $\chi(y^1) = 0.75$;

from $x_0^3 = x^3$, through $x_1^3 = (0.22;\ 0.08)$, out,

$x_2^3 = (0.74;\ 0.63)$, $\chi(x_2^3) = 0.72$,

$x_3^3 = (0.12;\ 0.20)$, $\chi(x_3^3) = 0.67$,

to $y^2 = (0.74;\ 0.63)$, $\chi(y^2) = 0.72$;

from $x_0^5 = x^5$ through $x_1^5 = (0.82;\ 0.97)$, $\chi(x_1^5) = 0.66$,

$x_2^5 = (1.37;\ 1.63)$, $\chi(x_2^5) = 0.75$,

$x_3^5 = (0.85;\ 1.85)$, $\chi(x_3^5) = 0.75$,

to $y^3 = (1,37;\ 1,63)$, $\chi(y^3) = 0.753$.

The local optimizer stops at the next iteration since the stopping condition given in step 3 of the algorithm is satisfied.

By using $m_1=m_2=2$, we obtain $p_{12}=p_{22}=0$ and two elementary clusters: $C_1 = [0;1]\times[0;1]$ and $C_2 = [1;2]\times[1;2]$.

We may notice that the best value (0.753) is attained in the cluster $C_2$ at the point $y^3 = (1.37;\ 1.63)$.

## 6. A MODEL OF LOADING CAPACITIES

The loading of production capacities consists in the distribution of the manufacturing tasks and operations to be carried out over the groups of equipment of a section or enterprise.

Here is a mathematical model for loading production capacities by considering several objective functions (e.g., cost, price and profit).

Let P be the set of products to be obtained in a certain time period:

- $O_i$ - the set of all operations required for realizing product $i\in P$;
- $G_{ij}$ - the set of groups of equipment on which operation $j\in O_i$ regarding product $i\in P$ may be carried out;
- $G$ - the set of all groups of equipment;
- $x_{ijk}$ - the amount of product $i\in P$ to be processed through operation $j\in O_i$ with the group of equipment $k\in G_{ij}$;
- $t_{ijk}$ - a time norm relevant to the group of equipment k for carrying out operations j for product i;
- $T_k$ - the available time of the group of equipment k in a given time period;
- $Q_i$ - the required amount from the product $i\in P$;
- $e^s_{ijk}$ - the coefficients of the objective functions: cost, price (s=1), profit (s=2) with the decision makers' aspiration levels $N_1$ and $N_2$, respectively;

A model of the loading problem of the production capacities is as follows:

$$\text{optimize}\left\{\sum_{i\in P}\ \sum_{j\in Q_i}\ \sum_{k\in G_{ij}} e^s_{ijk}\, x_{ijk}\right\}$$

subject to:

$$\sum_{i\in P}\sum_{j\in Q_i} t_{ijk}\, x_{ijk} \leq T_k,\ k \in G; \tag{15}$$

$$\sum_{k\in G_{ij}} x_{ijk} \geq Q_i,\ j \in Q_i,\ i \in P$$

$$x_{ijk} \geq 0,\ i \in P,\ j \in Q_i,\ k \in G_{ij}.$$

Since, frequently, the set of solutions of the model is empty, it is necessary to transform the system into a permanently consistent system, obtained by making the capacity constraints and those for the demand of products fuzzy.

Formally, this transformation consists in the substitution of the "$\leq$" signs with "$\lesssim$" and "$\geq$" with "$\gtrsim$".

A method for resolving thus obtained model, suggested by Ciobanu and Stoica (1981), consists in supplementing the model by some variables $\lambda_k$ - the capacity surplus, $\mu_i$ - the product deficit and $\eta_s$ - deviations of the values of the objective functions from the aspiration levels.

The model (15) is transformed into the following linear programming model:

$$\sum_{i\in P}\sum_{j\in Q_i}\sum_{k\in G_{ij}} e^1_{ijk}\, x_{ijk} \leq N_1 + \eta_1$$

$$\sum_{i\in P}\sum_{j\in Q_i}\sum_{k\in G_{ij}} e^2_{ijk}\, x_{ijk} \leq N_2 - \eta_2$$

$$\sum_{i\in P}\sum_{j\in Q_i} t_{ijk}\, x_{ijk} \leq T_k + \lambda_k,\ k \in G$$

$$\sum_{k\in G_{ij}} x_{ijk} \geq Q_i - \mu_i,\ j \in Q_i,\ i \in P$$

$$x_{ijk} \geq 0,\ i \in P,\ j \in Q_i,\ k \in G_{ij}$$

$$\eta_s \geq 0,\ s=1,2,\ \lambda_k \geq 0,\ k \in G,\ \mu_i \geq 0,\ i \in P$$

$$\min\left\{\sum_{s=1}^{2} a_s\,\eta_s + \sum_{k\in G} b_k\lambda_k + \sum_{i\in P} c_i\,\mu_i\right\}$$

where $a_s$, $b_k$ and $c_i$ are the corresponding importance coefficients.

The variables $\eta_s$, $\lambda_k$ and $\mu_i$ may possibly also have upper bounds to be interactively determined by the decision maker.

Let $X^* = (x^*_{ijk}, \eta^*_s, \lambda^*_k, \mu^*_i)$ $s=1,2$, $i\in P$, $j\in Q_i k\in G_{ij}$, be the solution of the transformed model.

If $\eta_s^* = 0$, $s = 1,2$, $\lambda_k^* = 0$, $k \in G$, $\mu_i^* = 0$, $i \in P$, the constraints system is consistent and the solution $X^*$ is accepted from the point of view of the objective functions also.

If some components $\eta_s^*$, $\lambda_k^*$, $\mu_i^*$ are non-zero, a membership degree of the solution $X^*$ may be calculated for the corresponding constraints of the system by means of membership functions (Dumitru and Luban, 1982) of the following form

$$f(x) = \frac{x}{x+\varepsilon}, \quad x > 0, \quad \varepsilon \geq 0$$

From these individual membership functions, a synthesis membership function $F(\eta,\lambda,\mu)$ is built up.

For the capacity constraints we define

$$\varphi_k(\lambda_k) = \frac{T_k}{T_k - \lambda_k} = \frac{1}{1 + \frac{\lambda_k}{T_k}}, \quad k \in G$$

where $\lambda_k << T_k$.

Also, for the product demand constraints

$$\psi_i(\mu_i) = \frac{Q_i - \mu_i}{Q_i - \mu_i + \mu_i} = \frac{1}{1 + \frac{\mu_i}{Q_i - \mu_i}} = 1 - \frac{\mu_i}{Q_i}, \quad i \in P$$

where $\mu_i << Q_i - \mu_i$.

Analogously, we define the membership functions for the constraints derived from the objective function.

If we consider a composition operator of the multiplication type, we obtain a synthesis membership function of the form

$$F(\eta, \lambda, \mu) = [1/(1+\eta_1/N_1)]\cdot[1/(1-\eta_2/N_2)]\cdot \prod_{k \in G} \varphi_k(\lambda_k)\cdot \prod_{i \in P} \psi_i(\mu_i)$$

Taking into consideration the form of the functions $\varphi_k(\lambda_k)$ and $\psi_i(\mu_i)$, respectively, because of the fact that $\lambda_k/T_k<<1$ and $\mu_i/(Q_i-\mu_i) << 1$ (the sign << should be read "much smaller than") we have (Ciobanu and Stoica, 1981)

$$F(\eta, \lambda, \mu) = \frac{1 - \frac{\eta_2}{N_2} - \sum_{i \in P} \frac{\mu_i}{Q_i}}{1 + \frac{\eta_1}{N_1} + \sum_{k \in G} \frac{\lambda_k}{T_k}}$$

Aiming at obtaining a high membership degree F, we consider

$$\min\left(\frac{\eta_1}{N_1} + \sum_{k \in G} \frac{\lambda_k}{T_k} + \frac{\eta_2}{N_2} + \sum_{i \in P} \frac{\mu_i}{Q_i}\right)$$

In conjunction with model (16) it follows that, for the beginning, we may select the coefficients

$a_s = 1/N_s$, $s=1,2$, $b_k=1/T_k$, $k \in G$, and $c_i = 1/Q_i$, $i \in P$.

It is of importance that the model remains linear with this approach to a linear problem with several objective functions and fuzzy conditions by means of the membership functions, and that this is a facility taken into consideration in regards to resolving the model.

## 7. FINAL CONCLUSIONS

The method suggested in this paper is different from other methods in the clustering side, as it is founded on the hyper-parallelepiped type clusters, not on spheres (Törn, 1977) or ellipsoids (Boender et al., 1982).

This method also differs regarding the local optimizer, in instead of using a gradient or a quasi-Newton method, both supposing differentiability or the existence of the Hessian, it is based on the Box method, modified in the selection of the initial points and converegence acceleration.

## REFERENCES

Boender, C.G.E., A.H.G. Kan Rinnooy, G.T. Timmer, and L.A. Stougis (1982). Stochastic method for global optimization. Math. Progr. 22, 125-140.

Boldur, G., I. Stancu-Minasian, (1971). Methods of solution of some multicriteria linear programming problem (in French). Rev. Roumaine de Math. Pures et Appl. 16, 313.327.

Box, N.I. (1965). A new method of constrained optimization and a comparison with other methods. Comp. J. 8,

Ciobanu, G., and M. Stoica (1981). Production scheduling in fuzzy conditions. Ec. Comp. and Ec. Stud. and Res. 3, 67-79.

Dixon, L.C.W., and G.P. Szegö (eds.) (1978). Towards Global Optimization Part 2, North-Holland, Amsterdam.

Dumitru, V., and F. Luban (1982). Membership functions, some mathematical programming models and production scheduling. Fuzzy Sets and Syst. 9, 19-33.

Dubois, D., and H. Prade (1984). Criteria aggregation and valueing of alternatives in the framework of fuzzy set theory. In H.J. Zimmermann, L.A. Zadeh and B.R. Gaines (eds.), Fuzzy Sets and Decision Analysis. North-Holland, Amsterdam.

Fabian, C., G. Ciobanu, and M. Stoica (1983). A general method for resolving multicriterial and fuzzy problems (in Rumanian). Ec. Comp. and Ec. Cyb. and Res. 1, 47-54.

Fabian, C., and M. Stoica (1984). Fuzzy integer programming. In H.J. Zimmermann, L.A. Zadeh, and B.R. Gaines (eds.), Fuzzy Sets and Decision Analysis. North-Holland, Amsterdam.

Hartman, I.K. (1973). Some experiments in global optimization. Naval Res. Log. Quart. 20.

Leclercq, I.P. (1982). Stochastic Programming: a interactive multicriteria approach. Eur. J. Oper. Res. 10, 33-41.

Marusciac, I., and M. Radulescu (1970). General linear programming problem with multiple economic functions (in French). Studia Universitatis Babes-Bolyai 2 ,55-65.

Negoita, C.V., and D.A. Ralescu (1975). Application of Fuzzy Sets to Systems Analysis. Birkhauser-Verlag, Basel.

Negoita, C.V., and M. Sularia (1976). Fuzzy linear programming and tolerances in planning. Econ. Comp. Econ. Cybernet. Stud. Res. 1, 3-15.

Rubinstein, R.I. (1982). Generating random vectors uniformly inside and on the surface of different regions. Eur. J. Op. Res. 10, 205-209.

Roy, B. (1970). Problems and Methods with Multiple Objective Functions. SEMA, Note de travail 60.

Sakawa, M. (1984). Interactive multiobjective decision-making by the fuzzy sequential proxy optimization technique-FSPOT. In H.J. Zimmermann, L.A. Zadeh and B.R. Gaines (eds.), Fuzzy Sets and Decision Analysis. North-Holland, Amsterdam.

Stadler, W. (1979). A survey of multicriteria optimization on the vector maximum problem. Part I: 1776-1960. J. Opt. Theory Appl. 29, 1-52.

Stancu-Minasian, I.M. (1980). Stochastic Programming with Several Objective Functions (in Rumanian). Editura Academiei, Bucharest.

Törn, A. (1977). Cluster analysis as a tool in a global optimization model. In Modern Trends in Cybernetics and Systems. Springer-Verlag, Berlin.

Warfield, I.N. (1984). Progress in interactive management. Proc. the 6th Int. Congress of Cybernetics and Systems, 10-14 Sept. 1984, AFCET, Paris, 147-152.

Zeleny, M. (1974). Linear Multiobjective Programming. Springer-Verlag, Berlin.

Zimmermann, H.J. (1978). Fuzzy programming and linear programming with several objective functions. Int. J. Gen. Syst. 1, 45-55.

Zionts, S. (1978). Multiple Criteria Problem Solving. Springer-Verlag, Berlin.

Zionts, S., and J. Wallenius (1976). An interactive linear programming method for solving the multiple criteria problem. Mang. Sci. 22, 8652-8663.

# A CONCEPT OF RULE-BASED DECISION SUPPORT SYSTEMS

V. Rajkovic[*,**], J. Efstathiou[***] and M. Bohanec[*]

* School of Organisational Science, Kranj, Yugoslavia

** J. Stefan Institut, Ljubljana, Yugoslavia

*** Queen Mary College, London, England

Abstract. We present an outline of rule-based decision theory where decision knowledge is represented and handled as logical rules, with probability and/or fuzziness. The theory is based on the fundamental belief that people are able to express their opinion on preferences using rules. The paper presents some practical results on obtaining, representing, aggregating and verifying rule-based decision knowledge for a decision support system. We emphasise the differences between traditional decision theory and the rule-based theory.

Keywords: multi-attribute decision-making, decision support systems, rule-based knowledge representation, consistency checking, knowledge elicitation.

## 1. INTRODUCTION

Traditional approaches to decision analysis, using weights, scores and utilities, have been useful in decision support. However, we feel that it is not the mathematical model of the process which is useful in itself, but the structure which the formalism imposes upon the decision-making process that is useful. Unfortunately, this approach means that some of the processes which are important for effective decision-making are made difficult, such as knowledge elicitation, learning, group communication and explanation. It is our belief that many of these problems can be solved by better knowledge representation methods.

Our practical experiences have shown that rule-based knowledge representation can lead towards more effective techniques for decision analysis and support. Therefore, we propose a rule-based methodology which has been implemented on a computer as a mixed initiative decision support system and has been used in practical trials (Bohanec, Bratko and Rajkovic, 1983; Efstathiou and Rajkovic, 1979).

Recent interest in Artificial Intelligence and Expert Systems have drawn attention to rule-based knowledge representation (Fox, 1983). We also believe that more emphasis must be placed on man-machine interaction, enabling the user to acquire the knowledge for a decision.

Rules are a natural method for communication. Accepting rules as the method of knowledge representation means that man-man and man-machine interaction is easier and more effective in terms of:

- knowledge elicitation,
- learning,
- knowledge verification,
- explanation of decision process,
- handling'soft' knowledge (Michie, 1979),
- improving effectiveness of computer usage.

## 2. RULE-BASED MULTI-ATTRIBUTE DECISION-MAKING

Multi-attribute decision-making proceeds by selecting a set of attributes which can describe the options. Each option may be scored against the attributes. The attributes may be weighted to reflect their relative importance. An aggregation formula combines the weights and scores to provide an overall figure of merit for each option, subject to independence criteria. The best option may be selected on the basis of these figures.

Rule-based multi-attribute decision-making also uses attributes to describe the options. Each attribute is described by a vocabulary, e.g., security: (unsafe, tolerable, secure, very secure). The scores are now presented as a semantically meaningful description of the option (Zadeh, 1975). For complex decisions, attributes are usually arranged hierarchically, as a tree. This structures the user's decision space (Bohanec, Bratko and Rajkovic, 1983).

Weighting of the attributes and aggregation of the scores are replaced by rules. A rule is a combination of values of the attributes for which the decision-maker states a utility value, for example:

IF the security is unsafe and the price is low
THEN the utility of the system is unacceptable

The syntax of the rules accepted by the existing system (Bohanec, Efstathiou and Rajkovic, 1983) is presented in Fig.1.

The simplicity of the rules is due to the effectiveness of the knowledge elicitation, using an interactive computer dialogue. Rules were accepted by decision makers from different levels and backgrounds as a natural way to express opinions and understand the explanation of decisions. Following our work with rule-based expert systems for decision making (Bohanec, Bratko and Rajkovic, 1983), we formulated the postulate that people are able to express their opinion on preferences using rules in a way that is realistic and satisfactory for decision support systems.

In practice, a decision may require many rules, but a decision maker is usually capable of providing them, with the assistance of a computer.

Let us take a simplified example with two attributes: SECURITY and PRICE, and illustrate it by Fig. 2. Every rule

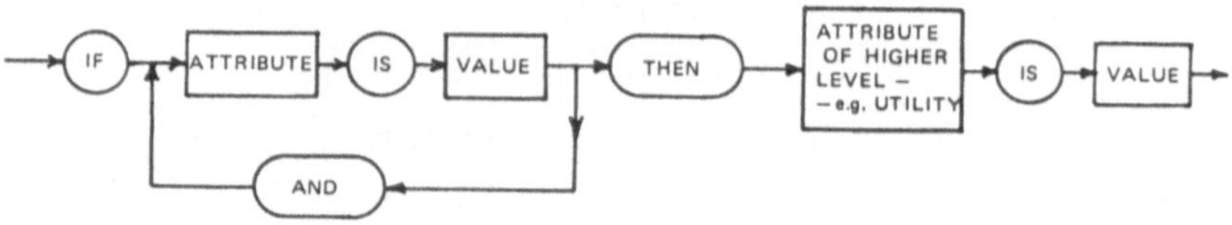


Fig. 1. Syntax of rules for expressing an aggregated opinion on preferences

defines a point on the UTILITY knowledge surface. The rule mentioned above is illustrated by dotted lines in Fig. 2. In this case, 12 rules are needed for the complete definition of the decision knowledge. It is not necessary for the decision-maker to supply all the rules because they can also be generated by computer. A new rule, i.e. point on the "utility" surface, can be interpolated from the points in the neighbourhood. The computer may suggest rules for modification by the user or may ask the user in an easy, structured way to provide the rules. The construction of the knowledge base (utility surface in Fig. 2) of rules is, therefore, by a mixed initiative between man and machine. But it has to be emphasized that every rule is completely under the control of the decision-maker.

The decision-maker can check the knowledge-base for consistency. Once the rule-base has been constructed, it may be used in several ways:

- evaluation of options
- explanation of evaluations
- option generation

By using rules, we avoid some of the difficulties with the traditional approach.

Our emphasis on the decision-maker and his own learning process together with a better interface between man and computer means that the decision analyst's burden of acting as an interface between man and technique is alleviated. Therefore, he can focus on decision-aming as a creative, learning process, leading to better decisions.

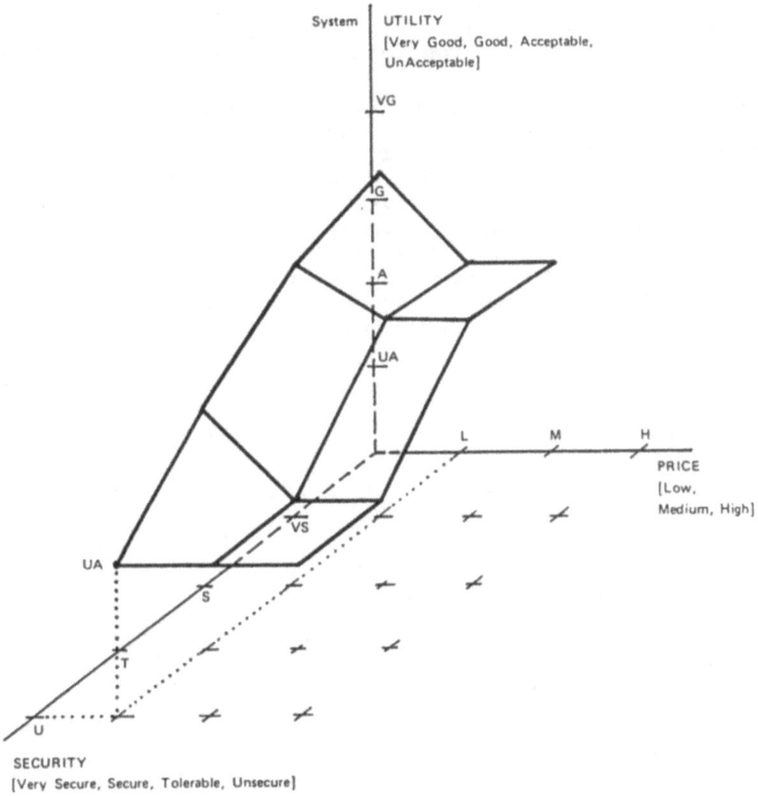


Fig. 2. Graphical illustration of aggregated UTILITY function; rule-based knowledge representation surface

## 3. CONSISTENCY CHECKING OF KNOWLEDGE-BASE AND EXPLANATION OF EVALUATION

The rule-base will have been constructed piecewise, by examining a small area of the decision space at once. Some rules are interpolated by the computer and they need to be checked. The global consistency of the knowledge-base must be verified. In most cases, inconsistency means that the overall utility decreases towards the more preferred end of the attribute scale.

The consistency can be checked by holding the values of all the attributes except one constant and varying its values to see how the overall utility is affected. If the terms on the scale of the attribute have been ordered preference-wise, then the graph should show, in general, a non-decreasing trend.

The graphs represent cross-sections through the multi-dimension decision space (see Fig. 3). They are displayed rapidly on the screen and one key only is pressed to bring up the next. In this way, the consistency of the knowledge-base can be checked visually and very quickly. Deviations can be marked for modification by the user.

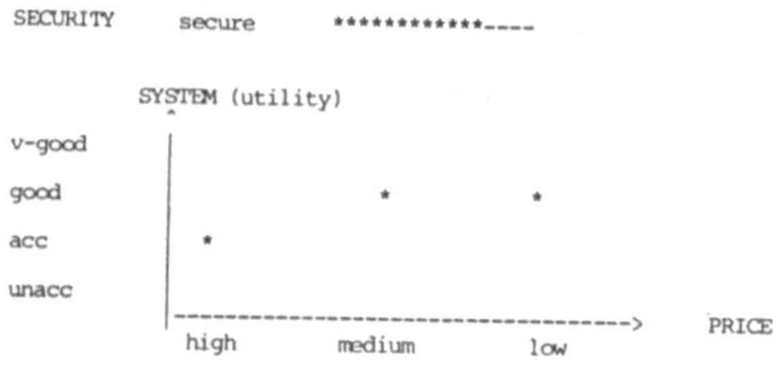


Fig. 3. Example of a consistency checking graph. Note the display of the constant attributes at the top of the picture, with attribute name, value and bar graph representation of the value

Once the user is satisfied with the consistency of the rule-base, then it may be used for evaluation. If an option is described according to the attributes, its utility is derived from the rules. An explanation of the evaluation is useful for the decision-maker to compare options more deeply and to check for mistakes in the description of the options and the rules. The explanation is particularly useful when several interest groups are involved in the decision and a consensus must be reached.

There are two kinds of explanation, a full explanation and a summary. The full explanation displays all the rules that were invoked in the evaluation of the option. The summary explanation uses only the rules that had an important effect in obtaining the evaluation. It lists the good points of the option, its bad points and the compensations (Bohanec, Efstathiou and Rajkovic, 1983).

## 4. EXPERT SYSTEMS AND DECISION SUPPORT SYSTEMS

Many current expert systems use similar rules but are applied to quite different problems, such as fault diagnosis. These expert systems are different from multi-attribute decision support systems, such as that described herein.

Expert systems are used to record the existing, well-established knowledge of an expert in a particular domain. This knowledge may be transmitted to novices through their use of the expert system. A decision support system (Alter, 1980) should enable a decision-maker to obtain insight into his own decision-making process, using the knowledge elicitation process to learn. The construction of the knowledge base is a dynamic, interactive, learning process for the decision-maker. This process of decision support is the main part of any decision analysis.

The rules which are obtained from the user to describe his knowledge will be quite simple, because of the structure of the knowledge base and because it has been created by the user. This is in contrast with other expert systems, which are not useful unless they are complex. A simple expert system would not be needed because it would be easily learnt. However, the complexity means that explanations of the expert system's decisions are unwieldy and difficult to understand. Since the emphasis of decision support systems is on analysing and learning, then simple rules and easily understood explanations are essential.

Rule-based expert systems have the problem of incompleteness. The absence of rules can only be detected in use. With the multi-attribute rule-based approach, the boundaries of the domain are defined by the attributes and their values. Let us look at Fig. 2 again. The definition domain (PRICE - SECURITY) of the UTILITY function is known. Therefore, we know that 12 rules are needed and what those rules are. The missing rules may be detected and can be added by the decision-maker or computer.

## 5. PRACTICAL EXPERIENCE

Software to support rule-based decision making has been written and used in several different problems (Bohanec, Bratko and Rajkovic, 1983; Bohanec, Efstathiou and Rajkovic, 1983). It has been written in Pascal and runs on DEC-10 under TOPS-10 and VAX under VMS and on PDP under RSX and RT-11. A PC version is in preparation. The practical applications may be divided into three main groups : consultancy support, multi-option ranking and personal decision making.

Under consultancy support, the approach has been used mainly for hardware and software evaluation and selection. The decision analysis phase was quick and the development of the knowledge base was easy. The transparency to the user was used as more than an explanation in another software engineering project. The status of the project was analysed and it was evaluated as no longer feasible. The software manager, using the knowledge base, varied some values of attributes until a feasible solution was achieved. He discovered which aspects of the project needed to be changed to make the project successful in future. In this case, the knowledge representation method was significant.

In a multi-option ranking problem, about two thousand application forms had to be rank ordered according to a stated policy. The policy was built into the knowledge base and applied consistently to every applicant. This task could have been done by a traditional method, but the advantage was that an explanation of the decision for each case could be supplied to the appointment committee and to each applicant. The time needed to carry out the ranking process was reduced to about 20% of the previous effort. Furthermore, the project led to a new, better application form.

This software was tested by a group of students on individual multi-attribute decision making problems. They were not assisted and were able easily to define a decision tree and rule base. The consistency of their knowledge bases varied and reflected their understanding of the problem. This suggests that the software could be used as a check on students' comprehension of a subject.

## 6. CONCLUSION

We conclude that rule-based decision making can be useful and effective. Software to support this has been written and tested extensively in real applications. The technique relies upon mixed initiative dialogue between decision maker and computer. The methodology has several new features some of which have been described above, but in particular we point out that the computer itself plays a fundamental role. Just as Monte Carlo techniques were known, they could not be widely used until computers were available. This method also relies upon the power of the computer to enhance the human's decision-making and learning processes.

There are similarities between this approach and existing expert systems. However, we point out the differences which derive from the specific features of the decision making problem and argue that these differences make this effective as a decision support system. Further research and development is required, but we believe that this approach already provides a toolbag of techniques that could provide the foundation for a fifth generation decision support environment.

ACKNOWLEDGEMENTS

We acknowledge the financial support of the British Council and ZAMTES, Ljubljana to enable visits between QMC and JSI. We also want to thank Prof. E.H. Mamdani at QMC for his support.

REFERENCES

Alter, S. (1980). Decision Support Systems: Current Practice and Continuing Challenges. Addison Wesley, Reading, Mass.

Bohanec, M., I. Bratko, and V. Rajkovic (1983). An expert system for decision making. In H.G. Sol (ed.), Processes and Tools for Decision Support. North-Holland, Amsterdam.

Bohanec, M., J.Efstathiou, and V. Rajkovic (1983). Rule-based Decision Support. Software: User's Manual V 1.0. Josef Stefan Institut, Ljubljana, Internal Report DP-3192.

Efstathiou, J., and V. Rajkovic (1979). Multiattribute decision making using a fuzzy heuristic approach. IEEE Trans. Syst., Man Cyber. SMC-9, 326-333.

Fox, J. (1983). Formal and knowledge-based methods in decision technology. Proc. of the 9th Conf. on Subjective Probability, Utility and Decision Making. Groningen.

Michie, D. (1979). Expert Systems in the Microelectronic Age. Edinburgh University Press, Edinburgh.

Zadeh, L.A. (1975). The concept of a linguistic variable and its application to approximate reasoning - Parts I,II,III. Inf. Sci. 8 and 9, 199-251, 301-357, 43-80.

## II.4. Fuzzy Network Optimization, Location, Transportation and Resource Allocation Models

# FUZZY OPTIMIZATION IN NETWORKS

Stefan Chanas

Institute of Production Engineering and Management
Technical University of Wrocław
50-370 Wrocław, Poland

Abstract. Some selected problems of fuzzy optimization in networks are analysed. The maximum-flow problem in a network with fuzzy arc capacities is considered in the first part of the paper. The fuzzy arc capacities are described by possibility distributions for feasible flows in the arcs. The criterion of maximization of the flow value is replaced in the problem with a fuzzy goal (a fuzzy set in the real line). The next problem considered concerns network analysis of a project with fuzzy activity times (fuzzy variables with given possibility distributions). Determination of the project's completion time is the main topic. The third part of the paper is devoted to critical discussion on possible approaches to the shortest path problem in a network with fuzzy arc lengths.

Keywords: network optimization, fuzzy optimization, maximum flow, network planning, shortest path.

## 1. INTRODUCTION

From among classic mathematical programming problems, a class of network optimization problems may be singled out. A network as the main element of a problem description is a common feature of all problems in this class. This fact is utilized to a considerable degree in construction of special solution algorithms which are more efficient than the general mathematical programming methods that could also be used to solve the problems. For example, the following problems: the max-flow problem, the min-cost flow problem, the shorstest route problem and others (see, e.g., Boffey, 1982), may also be stated as linear programming problems. But, applying the simplex method to solve them would be very inefficient. That is why some special "network" algorithms have been developed separately for those problems.

One may expect a similar situation among fuzzy mathematical programming problems. Also here a class of fuzzy network optimization problems can be distinguished, and, as in the classic case, the network description of problems may be utilized to construct solution algorithms.

In this paper we review some problems of this type and present solution algorithms based on a network representation of these problems. We lay special stress on the formulation of these problems, explanation of a solution concept, interpretation of membership functions and the like, as they are not uniquely understood and clear. This results from the existing fuzzy parameters in the models of problems.

In section 2 some problems of choosing a proper flow in a network with fuzzy capacity constraints in arcs are presented. In section 3 a problem of network analysis of a project with fuzzy activity times is discussed. Section 4 is devoted to the shortest route problem in a network with fuzzy arc lengths. Some results presented there are also pertiment in such problems as the traveling salesman problem, the minimal spanning tree problem, etc.

## 2. FLOW PROBLEMS IN A NETWORK WITH FUZZY ARC CAPACITIES

In this section we present and analyse some results selected from Chanas (1982), Chanas and Kołodziejczyk (1982, 1984, 1985).

### 2.1. Max-flow problem - a fuzzy formulation

Let $S = \langle N,A \rangle$ be a directed network, where N denotes the set of vertices and $A \subset C \times N$ is the set of arcs. Two vertices, a source $s \in N$ and a sink $t \in N$ are specified in S. A fuzzy interval $C_{ij}$ is associated with each arc $(i,j) \in A$. The membership function, $\mu_{ij} : R^+ \to [0,1]$ ($R^+$ is the nonnegative part of the real line) of $C_{ij}$ is of the form presented in Fig. 1. In a particular case, the $\mu_{ij}$ function may be linear on $[c_{ij}, \bar{c}_{ij}]$. It is also admissible that $c_{ij} = \bar{c}_{ij}$ and in this case $C_{ij} = [0, c_{ij}]$ is a crisp interval.

The $C_{ij}$ is a non-sharply defined capacity of arc (i,j) or, more precisely, a fuzzy interval of feasible flow in arc (i,j).

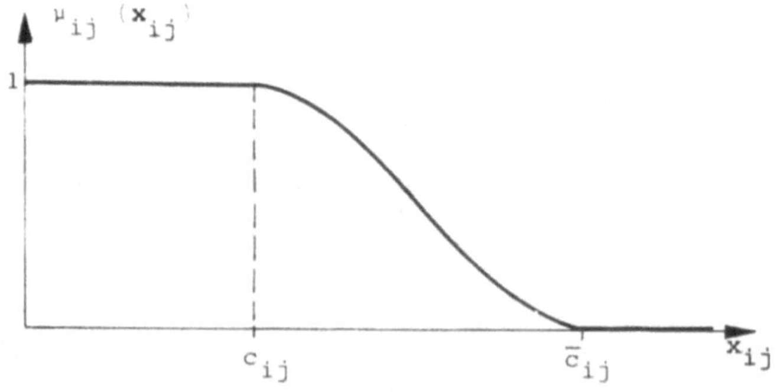


Fig. 1. The general form of the membership function $\mu_{ij}$.

The considered problem consists of finding a flow $x_v = \{x_{ij} \in R^+ \mid (i,j) \in A\}$ (a set of arc flows $x_{ij}$) from the source s to the sink t such that

$$"v \in G" \rightarrow \max \tag{1}$$

$$\sum_i x_{ij} - \sum_k x_{jk} = \begin{cases} -v & \text{for } j = s \\ 0 & \text{for } j \neq s,t \\ v & \text{for } j = t \end{cases} \tag{2}$$

$$"x_{ij} \in C_{ij}" \rightarrow \max, \ (i,j) \in A, \tag{3}$$

where $v$ is a value of the flow $x_v$ and $G$ is a fuzzy goal being a fuzzy interval of flow values accepted by the decision maker. The admissible shape of the membership function $\mu_G$ is presented in Fig. 2. It is possible that $v_0 = v_1$ which would mean that a decision maker requires a flow $x_v$ of value $v \geqslant v_0$.

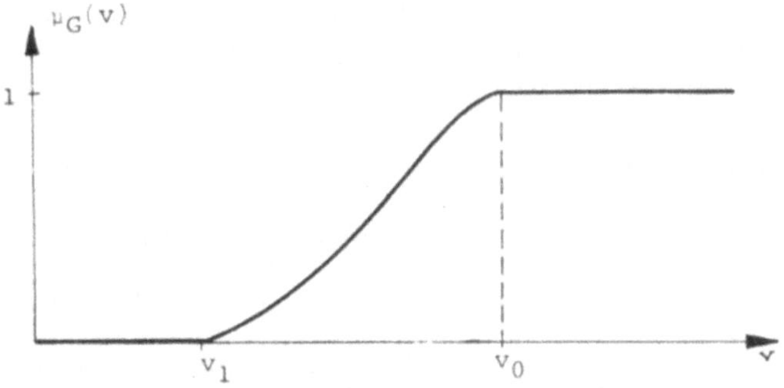


Fig. 2. The admissible form of the membership function $\mu_G$.

Verbally, problem (1) - (3) may be stated as follows: find a flow $x_v$ fulfiling the conservation constraints (2), and satisfying the fuzzy goal (1) and the fuzzy capacity constraints (3) to the maximum degree. It is natural to assume the values $\mu_G(v)$ and $\mu_{ij}(x_{ij})$ as degrees to which $x_v$ satisfies fuzzy conditions (1) and (3), respectively. One can easily notice that problem (1) - (3) in case of linear functions $\mu_G$ and $\mu_{ij}$ is an example of a partially fuzzy linear programming problem as formulated in Zimmermann (1978). But, using the method proposed there would not be efficient here because of the special form of the problem.

Using Bellman-Zadeh's (1970) approach we define a fuzzy set, D, in the set X of flows $x_v$ satisfying the conservation condition (2), called a fuzzy decision. The membership function of D is

$$\mu_D(x_v) = \mu_G(v) \wedge \mu_C(x_v) \tag{4}$$

where

$$\mu_C(x_v) = \bigwedge_{(i,j)\in A} \mu_{ij}(x_{ij}) \tag{5}$$

and "$\wedge$" stands for the minimum operation.

It is natural to consider a flow $x_v$ which belongs to D to the maximum degree, as a proper final choice. Thus, problem (1) - (3) may be reduced now, according to Bellman and Zadeh's methodology, to the following mathematical programming problem: find $x_v$, such that

$$\mu_D(x_v) \to \max, \qquad \text{subject to (2)} \tag{6}$$

Before we present solution algorithms for this problem we will present some remarks and comments concerning a possible interpretation of the membership functions $\mu_{ij}$ and $\mu_G$. It is closely related to the problem of selection of a proper operation in the definition of the fuzzy decision D in (4) and (5).

### 2.2. Discussion on interpretation of membership functions $\mu_{ij}$ and $\mu_G$

The first possible interpretation is closely connected with the notion of tolerance. The arc capacities $c_{ij}$ and the flow value $v_o$ are considered to be standards established beforehand which may be violated in given ranges of tolerance - maximally to $\bar{c}_{ij}$ and $v_1$, respectively. However, a decision-maker is interested in minimization of these deviations. In such a case the values $1 - \mu_{ij}(x_{ij})$ and $1 - \mu_G(v)$ may be regarded as degrees of deviations from the respective standards and problem (6) is reduced to the choice of a flow for which the largest of the deviations is minimal. With such an interpretation the linear form of the membership function $\mu_{ij}$ and $\mu_G$ seems to be sensible and also the min operation used in (4) and (5) is interpretable. Another approach, closer to the fuzzy set theory, consists in regarding the membership functions $\mu_{ij}$ and $\mu_G$ as possibility distributions of fuzzy variables in the sense of Zadeh (1978). The $\mu_{ij}(x_{ij})$ means then "the possibility" of the flow $x_{ij}$ performance through the arc (i,j) or, in other words, the possibility that the (i,j) arc capacity is at least equal to or greater than $x_{ij}$. Similarly, the $\mu_G(v)$ means "the possibility" of the objective accomplishment with a v-value flow, the possibility that a demand for the flow value in the sink will be satisfied.

There are two different interpretations of possibility. In the first, a "physical" interpretation, possibility is treated as a degree of easiness of a system performance when a concrete value of a parameter, described by a fuzzy variable, is assumed. The second interpretation, an epistemic one, is similar to the interpretation of probability. Possibility is a measure of uncertainty connected with an event. Usually, the possibility distribution is a subjective characterization resulting from the meaning of fuzzy propositions as, e.g., expert opinions expressed in a natural language (Zadeh, 1978).

In the case of the physical interpretation, functions $\mu_{ij}$ and $\mu_G$ may be defined subjectively but, in some situations, they can be determined as strict dependences on measurable physical parameters. The min operation used in (4) and (5) is acceptable in this case and can be interpreted. In the case of the epistemic interpretation of the possibility distribution, the problem is more complicated and the form of membership functions $\mu_{ij}$ and $\mu_G$, as well as the operation in (4) and (5), should be selected separately in every case according to the nature and structure of information used to define the membership functions. Formally, any operation from the class of triangular norms (see, e.g., (Alsina and Trillas, 1980; Klement, 1980) could be used in (4) and (5).

A triangular norm, briefly a t-norm, is a binary operation on $[0,1]$ that satisfies:

(i) $T(a,b) \leqslant T(c,d)$, for all $a,b,c,d \in [0,1]$ such that $a \leqslant c$, $b \leqslant d$,
(ii) $T(a,b) = T(b,a)$ for all $a,b \in [0,1]$,
(iii) $T(a,1) = a$ for all $a \in [0,1]$,
(iv) $T(T(a,b),c) = T(a,T(b,c))$ for all $a,b,c \in [0,1]$.

The min operation is just a particular case of t-norm. Also the product operation belongs to this class.

Presented in Chanas (1984) is the reliable flow problem which may be treated formally as a special case of problem (1)-(3) and in which the use of the product operation is more justified.

Let us assume that a flow value in problem (1) - (3) is fixed and therefore condition (1) may be omitted. Thus, problem (2) - (3) consists of finding a v-value flow $x_v$ that satisfies the capacity constraints to the maximum degree, i.e. such that

$$\mu_C(x_v) = \bigwedge_{(i,j)\in A} \mu_{ij}(x_{ij}) \rightarrow \max \tag{7}$$

Let the arc capacities (upper bounds on the arc flows) be independent random variables $B_{ij}$, $(i,j)\in A$, with distribution functions $F_{ij}(x_{ij})$. With each arc $(i,j)\in A$ we can now associate a random interval $[0,B_{ij}]$ of "possible" flows in the arc. Next, the random interval $[0,B_{ij}]$ can be used to define a fuzzy interval $C_{ij}$ characterized by its membership function $\mu_{ij}$ defined as follows:

$$\mu_{ij}(x_{ij}) = \text{Prob}(x_{ij} \in [0, B_{ij}]) = 1 - F_{ij}(x_{ij}) \tag{8}$$

The membership function now has a concrete interpretation - it is the reliability function (see, Jiang, 1983; Wang and Sanchez, 1982). The $\mu_{ij}(x_{ij})$ means the reliability of the flow $x_v$ with respect to the capacity constraint on the arc $(i,j) \in A$. Naturally, the use of the min operation in (7) is still sensible in the case considered but the product operation seems to be more justified. Problem (7) with the product operation resolves itself into maximization of the reliability of $x_v$ with respect to all the arc capacity constraints, i.e.

$$\mu_C(x_v) = \prod_{(i,j) \in A} \mu_{ij}(x_{ij}) = \prod_{(i,j) \in A} (1 - F_{ij}(x_{ij})) \to \max \tag{9}$$

It is shown in Chanas (1984) that problem (9) is equivalent to the convex min-cost flow problem for a wide class of distribution functions $F_{ij}$. Thus it can be solved with the aid of one of the methods adapted for this problem (see, e.g., Weintraub, 1974).

At the end of this subsection we want the reader to notice that problem (1) - (3) when (1) is replaced with "$\mu_G(v) \to \max$" and (3) with "$\mu_{ij}(x_{ij}) \to \max$, $(i,j) \in A$" may be treated as a multi-objective programming problem. Thus, the methods used in the multi-objective programming (see, e.g., Hwang and Masud, 1979) could be adapted to analyse problem (1) - (3). But then the problem arises of the proper choice of a compromise solution concept. It is as difficult as the problem of choosing a proper operation in (4) and (5).

2.3. Solution algorithms for real-valued flows

In Chanas and Kołodziejczyk (1984) the fuzzy max-flow problem for real flows in a network with two-sided fuzzy capacity constraints is exhaustively analysed. Here, for simplicity, we present only some selected results concerning the problem (1) - (3) which is a particular case of the problem considered in Chanas and Kołodziejczyk (1984).

First, we formulate a theorem which is a direct generalization of the Ford-Fulkerson max-flow min-cut theorem (Ford and Fulkerson, 1962). Next, we present the solution algorithm and an example of calculations.

At the end of this subsection we will show, quoting some results from Chanas (1984), that a parametric approach may be used to analyse problem (1) - (3) when the $\mu_{ij}$'s are linear.

Definition 1. A fuzzy set $V$ in $R^+$ with the membership function $\mu_V(v) = \max_{x_v} \mu_C(x_v)$ is called a fuzzy capacity of the network S.

The value $\mu_V(v)$ may be interpreted, according to its definition, as the possibility of reaching the v-value flow in the network S in the face of the existence of the fuzzy capacity constraints (3).

Definition 2. A cut in the network S is a partition of the set of vertices N into two sets X, $\bar{X}$ such that the source belongs to X and the sink to $\bar{X}$.

Definition 3. The fuzzy capacity of the cut $(X,\bar{X})$ is a fuzzy set in $R^+$ defined in the following way:

$$C(X,\bar{X}) = \bigoplus_{\substack{i\in X\\ j\in\bar{X}}} C_{ij} \qquad (10)$$

Obviously, the sign $\oplus$ appearing in (10) means the extended addition operation on fuzzy numbers, i.e. given by

$$\mu_{A\oplus B}(x) = \sup_{\substack{x_1,x_2\in R\\ x_1\ x_2=x}} \min\left\{\mu_A(x_1),\ \mu_B(x_2)\right\}.$$

Let $\mu_{ij}$, $(i,j)\in A$, be functions of the same type (see Dubois and Prade (1980)) on the intervals $[c_{ij},\infty)$, i.e.

$$\mu_{ij}(x_{ij}) = R\left(\frac{x_{ih} - c_{ij}}{\alpha_{ij}}\right) \quad \text{for} \quad x_{ij} \geqslant c_{ij},\ \alpha_{ij} > 0, \qquad (11)$$

where R(x) is any function continuous and decreasing on $[0,\infty)$ and taking on 1 at 0. For example, it may be $R(x) = \max\{0,1-|x|\}$, $\alpha_{ij} = \bar{c}_{ij} - c_{ij}$ and then $\mu_{ij}(x_{ij})$ is linear on $[c_{ij},\bar{c}_{ij}]$.

If R is fixed for all $C_{ij}$, $(i,j)\in A$, then $C_{ij}$ can be uniquely identified with the parameters $c_{ij}$ and $\alpha_{ij}$ and the addition in (10) can be performed very easily. Let us denote $C_{ij} = (c_{ij},\ \alpha_{ij})_R$. It follows from Dubois and Prade (1980) that $C(X,\bar{X})$ is a fuzzy interval of the same type as $C_{ij}$'s:

$$C(X,\bar{X}) = \bigoplus_{\substack{(i,j)\in A\\ i\in X,\ j\in\bar{X}}} (c_{ij},\alpha_{ij})_R = (c,\alpha)_R, \qquad (12)$$

where

$$c = \sum_{\substack{(i,j)\in A\\ i\in X,\ j\in\bar{X}}} c_{ij} \quad \text{and} \quad \alpha = \sum_{\substack{(i,j)\in A\\ i\in X,\ j\in\bar{X}}} \alpha_{ij}. \qquad (13)$$

Theorem 1. Let W denote the set of all cuts in the network S. Then, the following relation is valid

$$V = \bigotimes_{(X,\bar{X})\in W} C(X,\bar{X}). \qquad (14)$$

Under the accepted assumption on the form of $C_{ij}$, the extended minimum operation, $\textcircled{\wedge}$, in (14) is equivalent to the intersection operation of fuzzy sets and therefore:

$$V = \bigcap_{(X,\bar{X})\in W} C(X,\bar{X}), \tag{15}$$

$$\mu_V(v) = \bigwedge_{(X,\bar{X})\in W} \mu_{C(X,\bar{X})}(v).$$

Theorem 1 is an analogue to the Ford-Fulkerson theorem and it states a relation between the possible flow value in the network (that is, the network capacity) and the cut capacities.

Now, we are prepared to present the solution algorithm for problem (1) - (3).

Algorithm

Step 1. Determine using any known algorithm the maximum real flow $x_w$ and the respective minimal cut $(X,\bar{X})$ in the network with arc capacities equal to $c_{ij}$. If $\mu_G(w) = 1$, then $x_w$ is optimal. Otherwise go to Step 2.

Step 2. Calculate the fuzzy interval $C(X,\bar{X})$ and coordinates $(v,r)$ of the intersection point of the functions $\mu_{C(X,\bar{X})}$ and $\mu_G$. If $r = 0$, then STOP; the problem is infeasible. Otherwise, go to Step 3.

Step 3. Determine the maximum flow $x_w$ and the respective minimal cut $(X,\bar{X})$ in S with arc capacities equal to $c^r_{ij}$, where $c^r_{ij} = \sup\{x_{ij}\ \mu_{ij}(x_{ij}) \geq r\}$, $(i,j)\in A$. If $\mu_D(x_w) = r$ and $w = v$ then STOP - the flow $x_w$ is optimal. Otherwise go to Step 2.

Numerical example

Let us consider a sample network as presented in Fig. 3. Assume that $C_{ij} = (c_{ij}, \bar{c}_{ij} - c_{ij})_R$, where $R(x) = \max\{0, 1 - |x|\}$, i.e. $\mu_{ij}$ is linear on $[c_{ij}, \bar{c}_{ij}]$, $(i,j)\in A$. Let it be: $C_{s1} = (4,16)_R$, $C_{s2} = (5,40)_R$, $C_{12} = (4,4.5)_R$, $C_{1t} = (7.5, 1.5)_R$, $C_{2t} = (8.5, 4)_R$. Let $\mu_G$ be linear on $[v_1,v_o]$ (see Fig. 2) and $v_1 = 18$, $v_o = 20$.

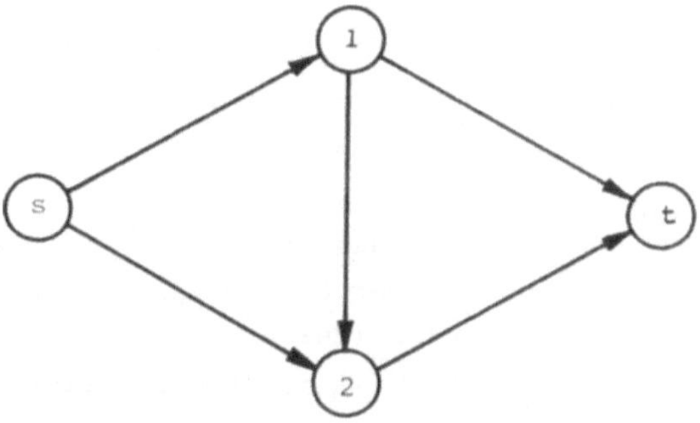


Fig. 3. A sample network

The algorithm runs as follows:

Step 1. $x_w = x_9$ and $x_{s1} = 4$, $x_{s2} = 5$, $x_{12} = 0$, $x_{1t} = 4$, $x_{2t}=5$, $(X,\bar{X}) = (\{s\}, \{1,2,t\})$.
$\mu_G(9) = 0 \neq 1$. Go to Step 2.

Step 2. $C(\{s\}, \{1,2,t\}) = (9,56)_{R'}$
$(v,r) = (19.62, 0.81)$. $r = 0.81 \neq 0$. Go to Step 3.

Step 3. $c^r_{ij}$ s: $c^r_{s1} = 7.04$, $c^r_{s2} = 12,6$, $c^r_{12} = 4.86$, $c^r_{1t} = 7.78$, $c^r_{2t} = 9.26$. $w = 16.3$ and $x_w$ : $x_{s1} = 7.04$, $x_{s2} = 9.26$, $x_{12} = 0$, $x_{1t} = 7.04$, $x_{2t} = 9.26$. $(X,\bar{X}) = (\ s,2\ ,\ 1,t\ )$.
1,t ).

$\mu_D(x_w) = 0 \neq r = 0.81$ and $w = 16.3 \neq v = 19.62$.

Step 2. $C(\{s,2\},\{1,t\}) = (12.5, 20)_{R'}$
$(v,r) = (19.32, 0.66)$, $r = 0.66 \neq 0$. Go to Step 3.

Step 3. $c^r_{ij}$ 's: $c^r_{s1} = 9,44$, $c^r_{s2} = 13.6$, $c^r_{12} = 5.53$, $c^r_{1t} = 8.01$, $c^r_{2t} = 9.86$. $w = 17.87$ and $x_w$:
$x_{s1} = 8.01$, $x_{s2} = 9.86$, $x_{12} = 0$, $x_{1t} = 8.01$, $x_{2t} = 9.86$
$(X,\bar{X}) = (\{s,1,2\},\{t\})$. $\mu_D(x_w) = 0 \neq r = 0.66$, and $w = 17.87 \neq$
$\neq v = 19.32$. Go to Step 2.

Step 2. $C(\{s,1,2\}, \{t\}) = (16, 5.5)_{R'}$
$(v,r) = (18.93, 0.467)$, $r = 0.467 \neq 0$. Go to Step 3.

<u>Step 3.</u> $c^r_{ij}$'s: $c^r_{s1} = 12.54$, $c^r_{s2} = 26.32$, $c^r_{12} = 6.4$, $c^r_{1t} = 8.3$, $c^r_{2t} = 10.63$. $w = 18.93$ and $x_w$ : $x_{s1} = 8.3$, $x_{s2} = 10.63$, $x_{12} = 0$, $x_{1t} = 8.3$, $x_{2t} = 10.83$. $(X,\bar{X}) = (\{s,1,2\}, \{t\})$. $\mu_D(x_w) = 0.467 = r$ and $w = 18.93 = v$.
STOP. The current $x_w$ is optimal.

Now we will show that in the case of linear $\mu_{ij}$ functions ($\mu_G$ does not need to be linear) a parametric max-flow procedure (see, e.g., Ruhe, 1985) may be used (Chanas, 1984).

Let us associate with problem (1) - (3) the following parametric max-flow problem:

$$v \to \max$$
$$\sum_i x_{ij} - \sum_k x_{jk} = \begin{cases} -v & j = s, \\ 0 & j \neq s,t, \\ v & j = t, \end{cases} \tag{16}$$
$$0 \leqslant x_{ij} \leqslant c_{ij} + t(\bar{c}_{ij} - c_{ij}),\quad (i,j)\in A,\ t\in[0,1],$$

where $t\in[0,1]$ is a parameter of variation.
By solving problem (16), for instance using the method presented in Ruhe (1985), we obtain an analytically expressed set of solutions $\{x_{v(t)} \mid t\in[0,1]\}$ explicity depending on t.
The $x_{v(t)} = \{x_{ij}(t) \mid (i,j)\in A\}$ for any $t\in[0,1]$ fulfils the capacity constraints at least to the degree of 1-t (i.e. $\mu_{ij}(x_{ij}(t)) \geqslant 1-t$, $(i,j)\in A$) and simultaneously maximizes the flow value $v(t)$ (thereby $\mu_G(v(t))$) by this condition. For any $t\in[0,1]$ the value $\mu_C(x_{v(t)})$ (see (5)) is equal to 1-t and it can not be improved provided $v = v(t)$. This results from the fact that $x_{v(t)}$ is a maximum flow, $\mu_{ij}(x_{ij}(t)) \geqslant 1-t$, $(i,j)\in A$, and there are arcs in the network (arcs of the minimal cut) for which $x_{ij}(t) = c_{ij} + t(\bar{c}_{ij} - c_{ij})$ (so $\mu_{ij}(x_{ij}(t)) = 1-t$). By finding the value of parameter t for which

$$\mu_D(x_{v(t)}) = \mu_G(v(t)) \wedge (1-t) \to \max \tag{17}$$

we identify a maximizing solution of (6). The $\mu_D(x_{v(t)})$, $t\in[0,1]$ also provides us with information about solutions "close" to the maximizing alternative and it may be regarded as a fuzzy solution of the initial problem (1) - (3).

<u>Example</u>. Let us return to the example already considered (see Fig. 3) in this paragraph. As the assumed $\mu_{ij}$ and $\mu_G$ are linear functions, we can also use the parametric technique to solve the

problem. Reducing the problem to the parametric max-flow problem (16) and applying the method from Ruhe (1985) we obtain the following results:

for $t \in [0, 0.0972222]$:

$x_{s1}(t) = x_{1t}(t) = 4 + 16t$, $x_{s2}(t) = x_{2t}(t) = 5 + 40t$,

$x_{12}(t) = 0$, $v(t) = 9 + 56t$;

for $t \in [0.0972222, 0.2413793]$:

$x_{s1}(t) = x_{1t}(t) = 4 + 16t$, $x_{s2}(t) = x_{2t}(t) = 8.5 + 4t$,

$x_{12}(t) = 0$, $v(t) = 12.5 + 20t$;

for $t \in [0.2413793, 1]$:

$x_{s1}(t) = x_{1t}(t) = 7.5 + 1.5t$, $x_{s2}(t) = x_{2t}(t) = 8.5 + 4t$,

$x_{12}(t) = 0$, $v(t) = 16 + 5.5t$.

In Fig. 4 the functions $\mu_C(x_v(t)) = 1-t$, $\mu_G(v(t))$ and $\mu_D(x_{v(t)})$ (the membership function of the fuzzy solution are presented. Obviously, the maximizing flow is the same as before and is obtained for $t = 0.5333333$.

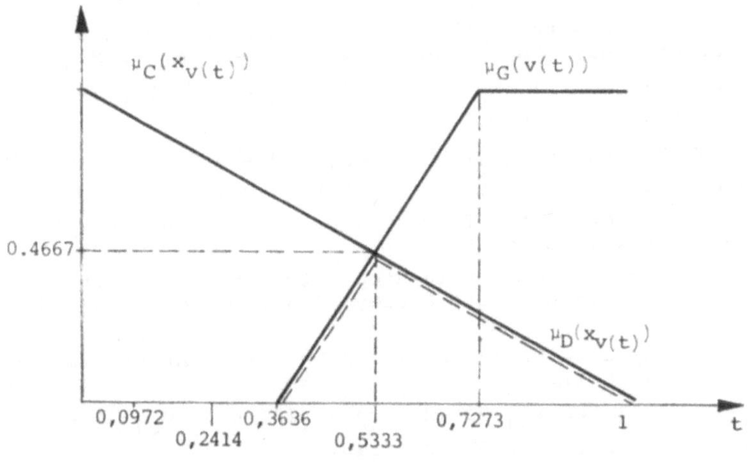


Fig. 4. Fuzzy solution for the example

2.4. <u>Solution algorithm for integer flows</u>

Let us extend problem (1) - (3) (and problem (6) associated with it) with the additional condition: $x_{ij}$ has to be an integer for any $(i,j) \in A$. Naturally, $C_{ij}$ and G are fuzzy sets in $I^+$ now

($I^+$ - the set of nonnegative integers). We assume that membership functions $\mu_{ij}$ and $\mu_G$ have the similar form as before, i.e. $\mu_{ij}(x_{ij}) = 1$ for $x_{ij} = 0,1,\dots,c_{ij}$ and $\mu_{ij}$ is decreasing on $[c_{ij},\infty)$. Similarly, $\mu_G(v) = 1$ for $v \geqslant v_0$ and $\mu_G$ is increasing on $[0, v_0]$.

Further on we present the solution algorithm which is a slightly improved version of the method proposed in Chanas and Kołodziejczyk (1982).

Algorithm.

Step 1. Find a maximum flow $x_v$ in the classic sense in the network S assuming arc capacities equal to $c_{ij}$, $(i,j)\in A$. Evidently, $\mu_C(x_v) = 1$. If $\mu_G(v) = 1$, then STOP, $x_v$ is optimal. Otherwise, go to Step 2.

Step 2. Let $x_v$ be a current flow. Determine a path r leading from s to t and maximizing the value of the expression

$$\mu^r(x_v) = \bigwedge_{(i,j)\in\vec{r}} \mu_{ij}(x_{ij} + 1) \wedge \bigwedge_{(i,j)\in\overleftarrow{r}} \mu_{ij}(x_{ij}-1) \to \max \qquad (18)$$

where $\vec{r}$ and $\overleftarrow{r}$ are sets of the forward and backward arcs in the path r, respectively. Additionally, we assume that if $x_{ij} = 0$, then $\mu_{ij}(x_{ij}-1) = -1$. If $\mu_D(x_v) > \mu^r(x_v)$, then STOP; $x_v$ is optimal. Otherwise, go to Step 3.

Step 3. Set $x_v := x_{v+1}$, where $x_{v+1}$ is the v+1 - value flow obtained by increase of the flow $x_v$ with a unit on the path r (adding a unit flow to the forward arcs and subtracting it from the backward arcs). Go to Step 2.

To determine the path r in Step 2 the shortest path algorithm of Dijkstra (see, e.g., Boffey, 1982) may be easily adapted.

Example. Let the network as in Fig. 3 be given. Assume values of $\mu_{ij}$ and $\mu_G$ as follows:

$\mu_{s1}(x_{s1}) = 1$ for $x_{s1}=0,1,2$, and $\mu_{s1}(3)=0.7$, $\mu_{s1}(4)=0.4$, $\mu_{s1}(5)=0.2$, $\mu_{s1}(x_{s1})=0$ for $x_{s1}\geqslant 6$;

$\mu_{s2}(x_{s2})=1$ for $x_{s2}=0,\dots,5$, and $\mu_{s2}(6)=0.8$, $\mu_{s2}(7)=0.5$ $\mu_{s2}(8)=0.1$, $\mu_{s2}(x_{s2})=0$ for $x_{s2}\geqslant 9$;

$\mu_{1t}(x_{1t})=1$ for $x_{1t}=0,1$, and $\mu_{1t}(2)=0.9$, $\mu_{1t}(3)=0.4$, $\mu_{1t}(x_{1t})=0$ for $x_{1t}\geqslant 4$;

$\mu_{2t}(x_{2t})=1$ for $x_{2t}=0,\dots,6$, and $\mu_{2t}(t)=0.7$, $\mu_{2t}(8)=0.4$, $\mu_{2t}(9)=0.1$, $\mu_{2t}(x_{2t})=0$ for $x_{2t}\geqslant 10$;

$\mu_{12}(x_{12})=1$ for $x_{12}=0,\ldots,2$, and $\mu_{12}(3)=0.7$, $\mu_{12}(4)=0.3$,
$\mu_{12}(x_{12})=0$ for $x_{12}\geqslant 5$;
$\mu_G(v)=1$ for $v\geqslant 10$, $\mu_G(9)=0.8$, $\mu_G(8)=0.6$,
$\mu_G(7)=0.2$, $\mu_G(v)=0$ for $v\leqslant 6$.

The algorithm runs as follows:

Step 1. $x_v : x_{s1} = 2$, $x_{s2} = 5$, $x_{12} = x_{1t} = 1$, $x_{2t} = 6$. $v = 7$, $\mu_G(7) = 0.2 \neq 1$. Go to Step 2.

Step 2. $r = (s,2,1,t)$, $\mu^r(x_v) = 0.8>\mu_D(x_v) = 0.2$. Go to Step 3.

Step 3. $x_v : x_{s1} = x_{1t} = 2$, $x_{s2} = x_{2t} = 6$, $x_{12} = 0$. Go to Step 2.

Step 2. $r = (s,1,2,t)$, $\mu^r(x_v) = 0.7>\mu_D(x_v) = 0.6$. Go to Step 3.

Step 3. $x_v : x_{s1} = 3$, $x_{s2} = 6$, $x_{12} = 1$, $x_{1t} = 2$, $x_{2t} = 7$. Go to Step 2.

Step 2. $r = (s,2,1,t)$, $\mu^r(x_v) = 0.4<\mu_D(x_v) = 0.7$. STOP. The current flow $x_v$ is optimal.

## 3. NETWORK ANALYSIS OF A PROJECT WITH FUZZY ACTIVITY TIMES

We shortly review some main results from (Chanas and Kamburowski, 1981; Dubois and Prade, 1978; Kamburowski, 1983) (see also Chanas, 1982). Next we propose one more approach to the fuzzy network analysis of a project. A fuzzy number is used in this approach to induce a proper probability distribution for duration of an activity. If the fuzzy activity times are replaced in the network with random variables, then a probabilistic method of network analysis may be used.

### 3.1. Fuzzy network planning

In the sequel we assume that a project can be represented as a directed, compact and acyclic network $\langle X,A\rangle$. Assume that $X = \{1,2,\ldots,n\}$ is the set of nodes (events), 1 - the single start node, n - the single terminal node, and $A\subset X\times X$ is the set of arcs (activities). The events of the project are of "the conjunction type", i.e. all the activities "starting" from a fixed node i are to be executed and their execution can be started at the time when all the activities "entering" node i are completed.
For convenience we assume that the events are labelled from 1 to n in such a way that $i<j$ for each activity $(i,j)\in A$. Associated with each activity $(i,j)$ is a fuzzy number $T_{ij}$ which is a normal, bounded and convex fuzzy set in $R^+$ (or $I^+$) with a membership function denoted by $\mu_{ij}$. The $T_{ij}$ is regarded as a fuzzy duration time of activity $(i,j)$.

Determination of the earliest time at which a project can

be completed is one of the main problems of network analysis. A natural approach to this problem, in the case considered here, consists in a direct extension of Ford's algorithm used in CPM (Critical Path Method) by replacing the addition and maximum operations in the algorithm with proper extended operations, i.e.

$$T_j = \bigotimes_{i \in \Gamma_j^-} (T_i \oplus T_{ij}), \quad j = 2,\ldots,n, \tag{18}$$

where $\Gamma_j^- = \{ i \in X \mid (i,j) \in A \}$ and $T_1$ (fuzzy or crisp) is given a priori. $T_j$ is the earliest (fuzzy) time at which event i may occur.

Naturally, the realization of formula (18) in a general case by direct utilization of the definition of $\circledvee$ and $\oplus$ is very troublesome. Therefore Dubois and Prade (1978, 1980) propose to use for the representation of $T_{ij}$'s fuzzy numbers of a special form - the L-R type fuzzy numbers. Assuming the L-R type, $T_{ij}$'s can be represented by parameters only, i.e. $T_{ij} = (m_{ij}, \alpha_{ij}, \beta_{ij})_{LR}$. The extended operations reduce themselves to the usual operations on the parameters. However, it should be stressed that from the two operations used on (18) only the addition $\oplus$ may be executed precisely in this way. The maximum operation $\circledvee$ frequently provides us with a result out of the class of the L-R type numbers and the outcome has to be approximated. This may considerably distort the final result, i.e. the project completion time $T_n$. The necessity of representing all the activity times with the fuzzy numbers of the same type is also a disadvantage of this approach.

In Chanas and Kamburowski (1981) an other way of performing formula (18) is presented. In the method given there, the properties of the r-cuts of fuzzy numbers are utilized. The r-cut of a fuzzy number $T_{ij}$, $r \in [0,1]$, is an interval defined as

$$T_{ij}^r = \{ t \mid \mu_{ij}(t) \geq r \} = [\underline{t}_{ij}^r, \bar{t}_{ij}^r] \tag{19}$$

Since the extended operations preserve the interval operations on the r-cuts of the arguments, the following formula gives the precise result for any $r \in [0,1]$

$$T_j^r = [\underline{t}_j^r, \bar{t}_j^r] = [\max_{i \in \Gamma_j^-} (\underline{t}_i^r + \underline{t}_{ij}^r), \max_{i \in \Gamma_j^-} (\bar{t}_i^r + \bar{t}_{ij}^r)], \tag{20}$$

$$j = 2,\ldots,n,$$

where $T_1^r = [\underline{t}_1^r, \bar{t}_1^r]$ is given beforehand.

For being able, in a general case, to precisely identify the membership function, $\mu_i$, $i = 1,2,\ldots,n$, one has to perform formula (20) for many values of r. It seems that knowledge of the r-cuts of $T_i$; $i = 1,\ldots,n$, for a chosen few values or r (e.g., $r = 0.1, 0.2,\ldots,1$) will be sufficient for the decision maker. Of course, the formula (20) is also correct for integer activity times $T_{ij}$.

Example. Let a network as presented in Fig. 5 be given. Assume that all the fuzzy activity times are of the triangular form (see Fig. 6): $T_{11} = (5,2,4)$, $T_{13} = (6,5,2)$, $T_{24} = T_{34} = (3,2,1)$. The activity (2,3) is dummy and $T_{23} = (0,0,0)$. It is assumed that $T_1 = (0,0,0)$.

Table 1 contains $T_i^r$, $i = 1,\dots,4$ for $r = 0.1, 0.2,\dots,1$.

In Fig. 7 the membership function $\mu_4$ of the project completion time $T_4$ is presented. It is not a triangular fuzzy number although all the $T_{ij}$'s are triangular.

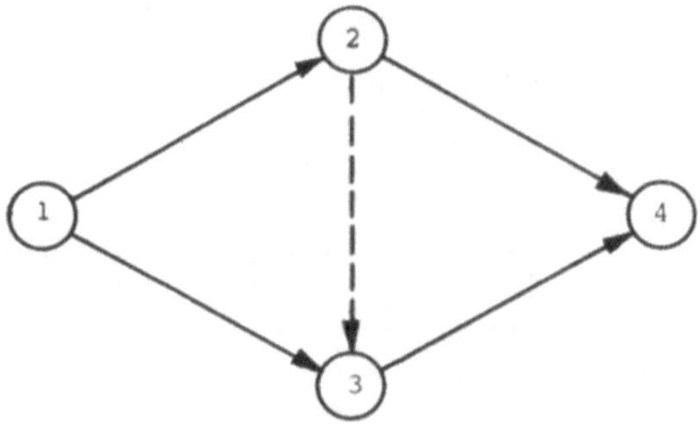


Fig. 5. A sample of project network

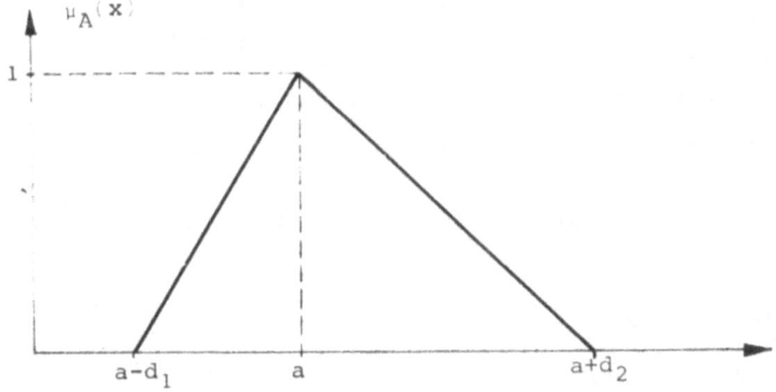


Fig. 6. A membership function for a triangular fuzzy number $A = (a, d_1, d_2)$.

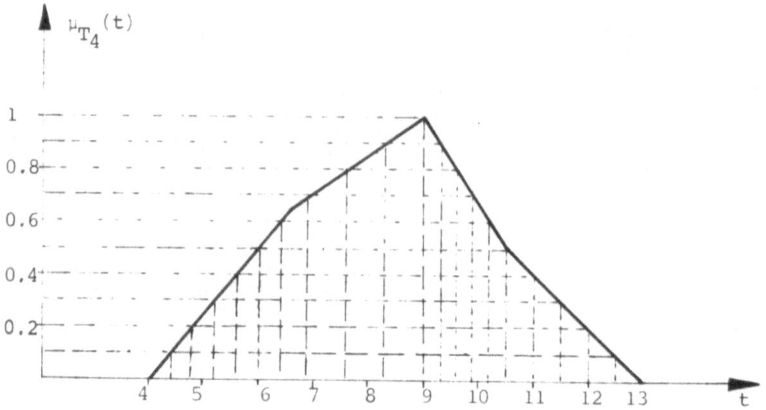


Fig. 7. The membership function $\mu_4$ of the project completion time $T_4$.

Table 1. The r-cuts of $T_i$

| r | $T_1$ | $T_2$ | $T_3$ | $T_4$ |
|---|---|---|---|---|
| 0.1 | [0,0] | [3.2, 8.6] | [3.2, 8.6] | [4.4, 12.5] |
| 0.2 | [0,0] | [3.4, 8.2] | [3.4, 8.2] | [4.8, 12 ] |
| 0.3 | [0,0] | [3.6, 7.8] | [3.6, 7.8] | [5.2, 11.5] |
| 0.4 | [0.0] | [3.8, 7.4] | [3.8, 7.4] | [5.6, 11 ] |
| 0.5 | [0,0] | [4, 7 ] | [4, 7 ] | [6, 10.5] |
| 0.6 | [0,0] | [4.2, 6.6] | [4.2, 6.8] | [6.4, 10.2] |
| 0,7 | [0,0] | [4.4, 6.2] | [4.5, 6.6] | [6.9, 9.9 ] |
| 0.8 | [0,0] | [4.6, 5.8] | [5, 6.4] | [7.6, 9.6 ] |
| 0.9 | [0,0] | [4.8, 5.4] | [5.5, 6.2] | [8.3, 9.3 ] |
| 1 | [0,0] | [5, 5 ] | [6, 6 ] | [9, 9 ] |

Both in Dubois and Prade (1983) and in Kamburowski (1983), a similar definition of the critical path is proposed. Namely, a criticality value of any path p from 1 to n is determined as

$$\mu(p) = \text{Poss}(T_1^n(p) \text{ is } T_n) = \text{hgt}(T_1^n(p) \cap T_n) \tag{21}$$

where

$$T_1^n(p) = \bigoplus_{(i,j)\in p} T_{ij} \tag{22}$$

is a fuzzy length of path p.

Of course, one may imagine other measures of path criticality than (21). Anyway, the problem of proper choice of a criticality measure depends on the accepted interpretation of the fuzzy activity times $T_{ij}$'s (more precisely, of the membership function $\mu_{ij}$'s).

In the following paragraph we propose a methodologically quite different utilization of fuzzy numbers in the network model of a project.

### 3.2. A new approach to fuzzy network planning

The membership function $\mu_{ij}$ of a fuzzy activity time $T_{ij}$ may be treated as a possibility distribution of a fuzzy variable. As we said, the notion of possibility may be interpreted in two ways (see subsection 2.2).

In the case of the physical interpretation of possibility, procedure (18) seems to be sensible. But, $T_n$ should be treated as a fuzzy restriction on the possible termination time of the project (not as an evaluation of the unknown time) - being the result of fuzzy restrictions $T_{ij}$ on the activity times. Naturally, if there are no additional criteria or restrictions, then the decision maker should choose a time $t_o$, such that $\mu_n(t_o)=1$, as a proper termination time.

The problem becomes more complicated when $\mu_{ij}$ is treated as a possibility distribution in the epistemic sense. In this case $T_{ij}$ is a fuzzy variable which is realized similarly as a random variable. In such a situation, the idea of replacing the fuzzy variable with a random variable, whose probability distribution is "consistent" in a certain sense with the possibility distribution, seems be natural.

The problem of approximation of a fuzzy variable by a random variable or, in other words, the problem of random generation of single values of a fuzzy variable is considered thoroughly in Chanas and Nowakowski (1985). We will use some of their results concerning the continuous possibility distribution.

Let us assume that a fuzzy number, bounded in R, induces by an evidence "X is A" a fuzzy variable X with the possibility distribution (see Zadeh, 1978).

$$\text{Poss}(X = x) = \pi_X(x) = \mu_A(x), \qquad x \in R \tag{23}$$

The following random variable, $X_A$, may be associated with X (or A)

$$X_A = \underline{a}(T) + S(\bar{a}(T) - \underline{a}(T)), \tag{24}$$

where T and S are independent random variables uniformly distributed over (0,1] and $\underline{a}(t)$, $\bar{a}(t)$ are the ends of an interval being the t-cut of A, i.e. $A^t = [\underline{a}(t), \bar{a}(t)]$.

One may make an attempt at determining the probability distribution of $X_A$ (in Chanas and Nowakowski, 1985 the distribution function of $X_A$ is found for a triangular fuzzy number A). It is however not necessary if we are interested only in the distribution parameters of $X_A$. For example, it may be easily shown that the expected value of $X_A$ can be obtained by the following formula

$$E(X_A) = \int_0^1 \frac{\underline{a}(t) + \bar{a}(t)}{2} dt \tag{25}$$

If $A = (a, d_1, d_2)$ is a triangular fuzzy number (Fig. 6), then

$$E(X_A) = a + \frac{d_2 - d_1}{4} \tag{26}$$

If $X_A$ and $X_B$ are random variables induced by fuzzy numbers A and B, respectively, then

$$E(X_{A \oplus A}) = E(X_A) + E(X_B). \tag{27}$$

Let us return to the main topic of our consideration. If we replace fuzzy activity times $T_{ij}$ in the network with random variables $X_{T_{ij}}$, we will obtain a problem for which a probabilistic method should be used. Such methods have been developed intensively during the last years. For example, estimation of distribution parameters of a project completion time is a separate problem (see Kamburowski, 1985, and its references).

Obviously, the operations "v" and "+" on random variables $X_{T_{ij}}$ are not equivalent to the proper operations $\textcircled{v}$ and $\oplus$ on $T_{ij}$ s. It means that the random variable $X_{T_n}$ generated by the fuzzy time $T_n$ need not be the same as the random completion time obtained from the probabilistic analysis of the network. Only in one case we get such an equivalence - namely, when we assume a strong (functional) dependency between the random variables $X_{T_{ij}}$, $(i,j) \in A$, i.e. when T and S are the same random variables for all $X_{T_{ij}}$´s, i.e.

$$X_{T_{ij}} = \underline{t}_{ij}^{T} + S(\bar{t}_{ij}^{T} - \underline{t}_{ij}^{T}), \quad (i,j) \in A. \tag{28}$$

Such a way of simultaneous generating values of several fuzzy variables was called in Chanas and Nowakowski (1985) the generation according to the extension principle.

Analysis of the network model by the assumption of same other relations among $X_{T_{ij}}$ is also an interesting problem. For example, one may assume that $X_{T_{ij}}$, $(i,j) \in A$, are independent variables. It is the case when variables T i S used in (28) are different and independent for each $(i,j) \in A$. Such a way of simultaneous generating random values of several fuzzy variables was called independent generating in Chanas and Nowakowski (1985). However, we can not elaborate on this subject here because of space limitations.

## 4. THE SHORTEST ROUTE PROBLEM IN A NETWORK WITH FUZZY LENGTHS OF ARCS

In this section we make a critical appraisal of various approaches to the fuzzy shortest path problem and of the resulting consequences for solution algorithms. Most of the remarks put forward here also remain valid for other similar problems such as: the shortest spanning tree problem, traveling salesman problem and the like. All these problems have one common feature: the estimate of the alternative (of the path, spanning tree, route for the salesman, etc.) is a fuzzy number being the sum of several fuzzy numbers (lengths of arcs).

### 4.1. On the Dubois and Prade approach

Dubois and Prade (1978, 1980) and Prade (1980) touch several times upon the problem of choosing a shortest path. However, they do it superficially, without solving conclusively all the questions connected with the problem. What we mean by this is an unequivocal formulation of a solution concept as well as of a solution algorithm for the problem. They propose to substitute in the Ford algorithm (for acyclic networks) and in that of Floyd (in a more general case) for the operations "+" and "min" the respective extended operations $\oplus$ and $\textcircled{\wedge}$. The realization of the algorithms with the operations changed in this manner yields for a fixed couple of vertices i and j a fuzzy number $\tilde{l}_{ij}$, being a fuzzy equivalent of the length of the shortest path between i and j. Because the $\textcircled{\wedge}$ operation of several fuzzy numbers does not necessarily yield one of those numbers, it is possible that no path has fuzzy length $\tilde{l}_{ij}$. For this reason the authors do not make use of these parts in Ford's and Floyd's algorithms which serve to identify merely an optimal path (the labeling procedure in Ford's algorithm and the procedure of calculating the table of predecessors in Floyd's algorithm). Having the fuzzy quantity $\tilde{l}_{ij}$ defined, Dubois and Prade introduce the notion of criticality of a path, defining for each path k between i and j the value of criticality as equal to $\mathrm{hgt}(\tilde{l}_k \cap \tilde{l}_{ij})$, where $\tilde{l}_k$ is the fuzzy length of the path k (i.e. the extended sum of the fuzzy lengths of the arcs of k). The value of $\mathrm{hgt}(\tilde{l}_k \cap \tilde{l}_{ij})$ can be treated (as stated in Prade, 1980) as a membership function of the "fuzzy set of the shortest paths between i and j". Let us add from ourselves that, analogously as in Bellman and Zadeh (1970), one could now treat in turn this fuzzy set as a fuzzy decision (fuzzy solution) for the problem in question.

But the problem of the definitive choice of a proper single route still remain. It seems natural, again by analogy to Bellman and Zadeh (1970), to take the route of the maximal criticality for the final solution (maximizing alternative). Dubois and Prade would interpret the final solution in this way although they do not state it clearly anywhere.

However, it is easy to notice that it follows from the properties of fuzzy numbers and from those of the operations

$\wedge$ and $\oplus$ that there always exists a path between i and j of the criticality degree equal to 1 (hence of the maximal one) and in order to identify it, it is not necessary to calculate the quantities $\tilde{l}_{ij}$ - it is enough to realize one of the classical algorithms, assuming determined lengths of arcs, equal to the modal values $m_{ij}$ of the fuzzy lengths of arcs (such that $\mu_{ij}(m_{ij}) = 1$. Thus, we could say, the problem ceases to be fuzzy.

## 4.2. On the Chanas and Kamburowski approach

In Chanas and Kamburowski (1983) an attempt has been made to differently formulate the concept of the solution of the problem and to find a means of extending the classical algorithms to the case of fuzzy data. We are going to shortly state this concept.

Suppose that a fuzzy nonstrict preference relation $\mu$ on the set of the nonnegative fuzzy numbers $F(R^+)$ is given. $\mu(A,B), A,B \in F(R^+)$, is interpreted as the degree to which the preference $A \leqslant B$ is true. Having defined the relation $\mu$, we can construct the fuzzy strict preference relation as follows (Orlovski, 1978):

$$\mu^s(A,B) = \max\left\{\mu(A,B) - \mu(B,A),\ 0\right\}. \tag{29}$$

$\mu^s(A,B)$ may be treated as the degree to which $A < B$ (A is strictly less than B).

Let $X = \left\{A_1, A_2, \ldots, A_n\right\}$ be a finite set of fuzzy numbers. The fuzzy subset of the nondominated elements of X, according to $\mu$, may be defined by the following membership function:

$$\mu^{ND}(A_j) = 1 - \max_{A_i \in X} \mu^s(A_i, A_j),\ A_j \in X, \tag{30}$$

i.e. $\mu^{ND}(A_j)$ describes the degree to which fuzzy number $A_j$ is strictly dominated by none of the numbers from X. It is natural to regard as "the least" fuzzy number in X the one for which the value of the $\mu^{ND}$ function is maximal, i.e. the maximal nondominated element in X.

Let us return to our main problem. Denote by $P(i,j)$ the set of all paths between vertices i and j. The maximal nondominated path in $P(i,j)$, i.e. the path for which

$$\mu^{ND}(\tilde{l}_k) = 1 - \max_{d \in P(i,j)} \mu^s(\tilde{l}_d, \tilde{l}_k) \rightarrow \max \tag{31}$$

is regarded as the shortest path between i and j.

Now arises the problem of constructing a suitable algorithm for (31). In Chanas and Kamburowski (1985) algorithms for a certain class of the relation $\mu$ are proposed. Unfortunately, the algorithms suggested there, which are a direct adaptation of Ford's and Dijkstra's algorithms, are not fully correct - they may lead in some cases to nonoptimal solutions, though close to the optimal ones. Consequently, they can be treated only as approximate algorithms. In order to eliminate that defect in one of the theorems, one would have to complicate those algorithms, as in Kołodziejczyk (1984), where he applied an identical approach to the shortest spanning tree problem.

However, it appears that problem (31) for some definite relations (satisfying the conditions required in Chanas and Kamburowski, 1985 and Kołodziejczyk, 1984) becomes decidedly simpler and does not necessitate constructing separate algorithms. For example, it is easy to prove that if $\mu$ is a relation of Baas and Kwakernaak (1977), then the solution of problem (31) will be arrived at by solving the classical shortest route problem with arcs lengths (like in the Dubois and Prade approach treated in the preceding paragraph) to the modal values $m_{ij}$ of the fuzzy arc lengths.

Now we are going to show that also for the relation introduced by Kołodziejczyk (1984) problem (31) (as well as the shortest spanning tree problem considered there) resolves itself into the classical problem and does not require a new algorithm.

Kołodziejczyk (1984) defines the preference relation as

$$\mu(A_i,A_j) = \frac{d(A_i^l \circledvee A_j^l, A_i^l) + d(A_i^p \circledvee A_j^p, A_i^p) + d(A_i \cap \overline{A_j}, 0)}{d(A_i^l, A_j^l) + d(A_i^p, A_j^p) + 2d(A_i \cap A_j, 0)}, \tag{32}$$

where

$$d(A,B) \overset{df}{=} \int_R |\mu_A^{(x)} - \mu_B^{(x)}| \, dx, \tag{33}$$

is the measure of the Hamming distance, and

$$\mu_A^l(x) \overset{df}{=} \begin{cases} \mu_A(x) & \text{for } x \leqslant z, \\ 1 & \text{for } x \geqslant z, \end{cases}$$
$$\mu_A^p(x) \overset{df}{=} \begin{cases} \mu_A(x) & \text{for } x \geqslant z, \\ 1 & \text{for } x \leqslant z, \end{cases} \tag{34}$$

where $z$ is such that $\mu_A(z) = 1$.

The following theorem is suprising and, at the same time, interesting, inasmuch as it relates the relation (32) with the generative value of the fuzzy number (see (25)) introduced in Chanas and Kamburowski (1985). (A ranking function which is equivalent to the generative mean value, $E : F([0,1]) \to R$, mapping each fuzzy set in the unit interval into the real line was earlier introduced by Yager, 1981).

<u>Theorem 2</u>. The following relation holds

$$\mu(A_i,A_j) \geqslant \tfrac{1}{2} \Leftrightarrow E(A_i) \leqslant E(A_j),$$

where $\mu$ is defined by (32) and $E(.)$ by (25).

<u>Proof</u>. We are going to bring forward only the main idea of the proof: It is easy to notice that

$$\mu(A_i, A_j) = \frac{S_L^+ + S_P^- + S}{S_L^+ + S_l^- + S_P^+ + S_P^- + 2S} \tag{35}$$

where $S_L^+$, $S_L^-$, $S_P^+$, $S_P^-$ and S are areas consisting of subareas as marked in Fig. 8.

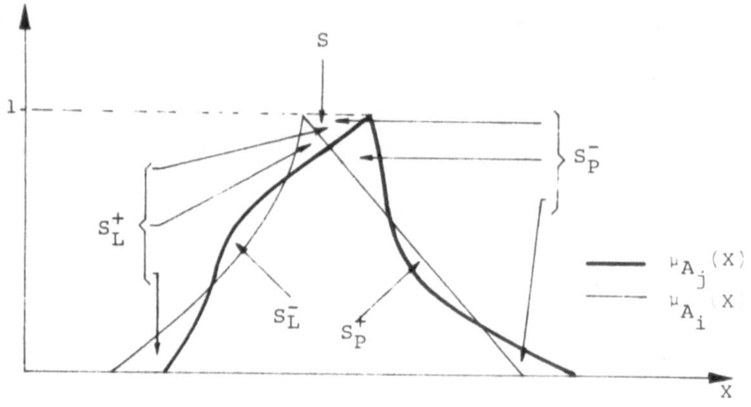


Fig. 8. The areas used in formulae (35) and (36)

Further, using (25) one may show that

$$E(A_i) = E(A_j) - \frac{1}{2}(A_L^+ - S_L^- + S_P^- - S_P^+) \tag{36}$$

And the theorem follows immediately from (35) and (36).

It is from theorem 2 and the additivity property of the generative mean value (see (27)), as well as the property of the relation (32), $\mu(A,B) + \mu(B,A) = 1$, that in order to solve problem (31) with the preference relation (32), it suffices to solve the classical problem with the lengths of arcs equal to $E(T_{ij})$. Needless to say that this remark remains valid also for the spanning tree problem considered in Kołodziejczyk (1984) and with such a preference relation as (32) the algorithms worked out there are redundant.

## 5. FINAL REMARKS

It was not our objective to review all the problems connected with network optimization which may be met in the literature. For example, we have not considered here the min-cost flow problem, Chanas and Machaj (1985), and related problems (e.g., Chanas, Kołodziejczyk and Machaj (1984)). We rather wanted to present and analyse more deeply some chosen typical fuzzy network problems - such as the maximum flow problem, the network analysis of a project and the shortest path problem. At the same time while presenting each of these problems, we tried to lay stress on two questions: the mere formulation of the problem in its fuzzy version as well as solution algorithms in which specific features of the network representation of the problem are used extensively. We tried not to hide the fact that there may be different approaches to all the problems considered depending on how fuzzy input data are interpreted in the model of the problem. Therefore, for example, we presented two completely different ways of the use of fuzzy sets in the network analysis of a project. Different interpretation and meaning of the membership function of a fuzzy number cause, for example, that the problem of ordering fuzzy numbers cannot be uniquely solved (see, e.g., Bortolan and Degani, 1985). That is also why a solution concept for the discrete optimization problems as, for example, the shortest path problem cannot be uniquely stated. We also wanted to stress the parts where we were making a critical review of different approaches to the fuzzy shortest path problem in Section 4.

REFERENCES

Alsina, C., E. Trillas, and L. Valverde (1980). On non-distributive logical connectives for fuzzy set theory. Busefal 2, 18-29.

Bass, S.M., and H. Kwakernaak (1977). Rating and ranking of multi-aspect alternatives using fuzzy sets. Automatica 13, 47-58.

Barlow, R.E., and F. Proschan (1975). Statistical Theory of Reliability and Life Testing. Holt, New York (1970).

Bellman, R.E., and L. A. Zadeh (1970). Decision-making in a fuzzy environment. Mang. Sci. 17, B141-B164.

Boffey, T.B., (1982). Graph Theory in Operations Research. Macmillan, London.

Bortolan, G., and R. Degani (1985). A review of some methods for ranking fuzzy subsets, Fuzzy Sets and Syst. 16, 1-19.

Chanas, S., (1984). Reliable Flows in Networks. Institute of Production Engineering and Management, Technical University of Wrocław, Report 36.

Chanas, S. (1982). Fuzzy sets in few classical operational research problems. In M.M. Gupta and E. Sanchez (eds.), Approximate Reasoning in Decision Analysis. North-Holland, 351-363.

Chanas, S., and J. Kamburowski (1981). The use of fuzzy variables in PERT. Fuzzy Sets and Syst. 1, 11-19.

Chanas, S., and J. Kamburowski (1985). The fuzzy shortest route problem. Proc. Polish Symp. Interval and Fuzzy Math. Poznań, 35-41.

Chanas, S., and W. Kołodziejczyk (1982). Maximum flow in a network with fuzzy arc capacities, Fuzzy Sets and Syst. 8, 165-173.

Chanas, S., and W. Kołodziejczyk (1984). Real-valued flows in a network with fuzzy arc capacities, Fuzzy Sets and Syst.

13, 139-151.
Chanas, S., and W. Kołodziejczyk (1986). Integer flows in network with fuzzy capacity constraints. Networks. To appear.
Chanas, S., W. Kołodziejczyk, and A. Machaj (1984). A fuzzy approach to the transportation problem. Fuzzy Sets and Syst., 13, 211-221.
Chanas, S., and A. Machaj (1985). A Parametric Version of the Out-of-kilter Method and its Application to the Fuzzy Network Flow Problems. Institute of Production Engineering and Management, Technical University of Wrocław, Report 59.
Chanas, S., and M. Nowakowski (1985). Single Value Simulation of Fuzzy Variable. Institute of Production Engineering and Management, Technical University of Wrocław, Report 25.
Dubois, D., and H. Prade (1978). Shortest path algorithms with fuzzy data (in French), R.A.I.R.O. Op. Res. 12, 214-227.
Dubois, D., and H. Prade (1979). Operations on fuzzy numbers. Int. J. Syst. Sci. 9, 613-626.
Dubois, D., and H. Prade (1980). Fuzzy Sets and Systems: Theory and Applications. Academic Press, New York.
Ford, L.R., and D.R. Fulkerson (1962), Flows in Networks. A RAND Corporation Research Study. Princeton Univ. Press.
Hwang, C.L., and A.S.M. Masud, in collaboration with S.R. Paidy and K. Yoon (1979). Multiple Objective Decision Making. Methods and Application: A State - of the - Art Survey. Springer Verlag, Berlin.
Jiang, Pei-rong (1983). Distribution of fuzzy subset projected from random intervals. Busefal 15, 64-71.
Kamburowski, J., (1983). Fuzzy activity duration times in critical path analysis. Int. Symp. on Project Mang. New Delhi, 194-199.
Kamburowski, J. (1985). An upper bound on the expected project completion time in PERT networks. Eur. J. Op. Res. 21, 206-212.
Klement, E.P. (1980). Fuzzy sigma algebras and fuzzy measures with respect to t-norms, Round Table on Fuzzy Sets. Lyon.
Kołodziejczyk, W. (1984). The Shortest Spanning Tree Problem in a Network with Fuzzy Parameters. Institute of Production Engineering and Management, Technical University of Wrocław, Report 44.
Kołodziejczyk, W. (1986). Orlovsky s concept of decision-making with fuzzy preference relation - further results. Fuzzy Sets and Syst. To appear.
Nakamura, K. (1979). Preference Relations Between Fuzzy Outcomes. The Working Group on Fuzzy Systems. Report 5. Tokyo, Japan.
Orlovsky, S.A. (1978). Decision-making with a fuzzy preference relation. Fuzzy Sets and Syst. 1, 155-167.
Prade, H. (1980). Operations research with fuzzy data. In P.P. Wang, and S.K. Chang (eds.), Fuzzy Sets. Theory and Application to Policy Analysis and Information Systems. Plenum Press, New York, 155-170.
Ruhe, G.,(1985). Characterization of ali optimal solutions and parametric maximal flows in networks, Optim. 16, 51-61.
Wang Pei-zhuang, and E. Sanchez (1982). Treating a fuzzy subset as a projectable random subset. In M.M. Gupta and E. Sanchez (eds.), Fuzzy Information and Decision Processes.

North-Holland, Amsterdam, 213-219.
Weintraub, A. (1974). A primal algorithm to solve network flow problems with convex costs. Mang. Sci. 21, 87-97.
Yager, R.R. (1981). A procedure for ordering fuzzy subsets of the unit interval. Inf. Sci. 24, 143-161.
Zadeh, L.A. (1978). Fuzzy sets as a basis for theory of possibility. Fuzzy Sets and Syst. 1, 3-28.
Zimmermann, H.J. (1978). Fuzzy programming and linear programming with several objective functions. Fuzzy Sets and Syst. 1, 45-55.

# ON FUZZY LOCATION MODELS

John Darzentas

Department of Mathematical Sciences and Computing
Polytechnic of the South Bank
London SE1 OAA, UK

Abstract. Typically, in location problems "optimisation" is judged according to criteria which are mainly deterministic and crisply defined. In a great many location problems, however, the policy and decision makers are not always in a position to determine crisply either the criteria on the basis of which optimisation will be assumed or the aims and objectives to be achieved or even the restrictions imposed on the overall project. This paper presents some fuzzy location models and solution methodologies which can be applied to such situations. The emphasis here is on the application of fuzzy discrete location models on i) The general mixed integer programming formulation of the discrete model, commonly known as "the simple plant location model", and ii) the set covering and set partitioning 0-1 pure integer programming formulations.

Keywords: location problem, fuzzy location problem, fuzzy set covering, fuzzy set partitioning.

## 1. INTRODUCTION

The great volume of papers which has appeared in the last twenty years on location analysis reflects the tremendous interest shown by the researchers. This interest is due to the theoretical rewards and also the realisation that locational analysis is an area of optimization with direct impact to real life problems such as those faced by operational researchers, management scientists, planners, etc. These problems range from minimization of transport costs to maximization of social welfare.

The locational models which have appeared in the literature can be broadly classified into three main categories: continuous (planar), network, and discrete.

Typically, a location problem would be to optimally locate a number of new facilities within an area. These are usually assumed to be dimension-less points in space. The new facilities could be expected to be located either anywhere in the plane (continuous problem), or anywhere on a transport network (network problem), or at specific points within the area (discrete problem).

Planar models require more assumptions and approximations of the distance involved, hence they are sometimes less realistic, while the discrete problems need less assumptions and they are the most realistic of the three, but at the cost of tractability since they are NP-hard. Network models are more easily analysed and solved when they contain no cycles, i.e. they are "trees". However, network problems can usually be transformed into equivalent discrete models.

The optimization is judged according to criteria which are mainly deterministic and crisply defined such as distances, fixed costs, travelling time and cost of travelling. Some more broadly defined criteria which take into consideration characteristics of the population involved (Koenig, 1981) could be used in socially orientated studies, nevertheless these are also crisply quantified.

In a great many locational problems, however, the policy makers and planners are not always in a position to determine crisply either the criteria on the basis of which optimisation will be assumed or the aims and objectives to be achieved, or even the restrictions imposed on the overall project. Forcing the policy/decision makers to obtain, say, specific numerical values for the constraints or to come up with crisp criteria may result in masking feasible alternative solutions which may be much more within the policy lines than those an inflexible model would provide. In these cases, the fuzzy sets approach can be successfully used.

This paper presents some fuzzy location models which can be applied to situations where the constraints and objectives may be expressed as, for instance: "the optimum must be within the policy lines" or "it must satisfy that sector of the population and be acceptable to the rest", or "the maximum number of people served by each facility must not exceed a certain number by very much", etc. The emphasis here is on the application of fuzzy discrete location models, in particular on a) the general mixed integer programming formulation of the discrete model, commonly known as "the simple plant location model" (SPLP), and b) the set covering and set partitioning 0-1 pure integer programming formulations. Both a) and b) are introduced in the next section. Fabian and Stoica's (1984) approach to the fuzzy integer programming problem, modified to describe the mixed integer formulation of the SPLP model, is presented in Section 3. The same section discusses the possible use of Zimmermann and Pollatschek's (1984) approach to formulating and solving the 0-1 fuzzy integer programming problem to describe the discrete location problem. Next, Section 4 presents an approach based on the set covering/set partitioning 0-1 formulation with fuzzy criteria, and Section 5 a solution method where a simple implicit enumeration algorithm is used to obtain the set of covers and their membership values according to a membership function. Finally, Section 6 discusses the consequences of the assumptions about the membership function introduced in these models and their applicability in general. The level of fuzziness of the problems and its impact on the choice of model used to solve them are discussed during the examination of the above models.

## 2. NETWORK AND DISCRETE LOCATION MODELS

Because of the widespread interest in location models, there are a number of excellent reviews (Francis and McGinnis, 1983; Krarup and Pruzan, 1983), and for one of the most popular models, the simple plant location problem (SPLP), a comprehensive review of papers on the subject up to 1982 is presented in Krarup and Pruzan's (1983) paper.

As far as the network location problems are concerned, the main interest focuses on the p-centre/p-median type of problem. In the p-centre problem new facilities are to be located with respect to m locations in order to minimise a maximum of weighted distances between each existing location and its closest centre. In the p-median problem the difference is in the objective function where the p new facilities are located so as to minimise a sum of weighted distances between each existing facility and its closest median. The existing locations are situated on the vertices of the network, while the new facilities can be anywhere on the network. An array of algorithms, backed by theoretical development exists especially for the cases where the network is a tree since it usually implies convexity. However, Krarup and Pruzan (1983) have shown that the p-centre p-median problems are transformable to SPLP. They also established relationships between SPLP and the set covering problems in integer programming, emphasizing that way the importance of SPLP.

The general SPLP problem which is fully treated in Krarup and Pruzan (1983) can be described by the following mixed integer programming formulation:

$$\text{Minimise } Z = \sum_{i \in I} \sum_{j \in J} (c_i + t_{ij}) d_{ij} + \sum_{i \in I} f_i y_i \qquad (1)$$

Subject to:

$$\sum_{i \in I} d_{ij} \geq l_j, \quad j \in J \qquad (2)$$

$$k_i y_i - \sum_{j \in J} d_{ij} \geq 0, \quad i \in I \qquad (3)$$

$$d_{ij} \geq 0, \ i \ I, \ j \in J \qquad (4)$$

$$y_i \geq 0, \ y_i \geq 1 \quad \text{and integer} \qquad (5)$$

where:

- $i \in I$ represents the potential locations for new facilities; $I = 1(1)m$,
- $j \in J$ represents the locations to be served by the new facilities; $J = 1(1)n$,
- $f_i$ is the fixed cost of opening the facility i,
- $c_i$ is the operating cost per unit for facility i,
- $l_j$ is the demand by j,
- $t_{ij}$ is the cost of transporting a unit from i to j,

- $k_i$ is a constant greater than the maximum production potential of i.

And the variables:

$d_{ij}$ are the units to be transported from i to j,

$y_i$ = 1, if a facility is located at i, 0 otherwise.

The objective function (1) is the standard fixed charge problem function where the opening of a facility i ($y_i$=1) implies the fixed cost $f_i$, and constraints (3) guarantee that if $d_{ij} > 0$ for facility i, then $y_i$=1, since $k_i$ can be chosen as a constant greater than the maximum production of i, usually each $k_i$ is replaced by the sum of $l_j$'s (i.e. the total demand). Constraints (2) guarantee that the demand at j is satisfied.

Discrete location models have also been successfully formulated as zero-one pure integer programming problems in the form of the set covering, set partitioning problem. Typically, the general problem in this case can be stated as the optimal location of a number of facilities at some points out of a set of candidate points S so as to serve the whole membership of S. In other words, let $S = \{1,2,\ldots,i,\ldots,n\}$ be the set of n potential locations and $K = \{\{l_{ij}, j\in J\}, i\in I\}$ be a class of subsets of S where $l_{ij} \in K$ locates a facility at i to serve the rest of the points in the subset (points in $l_{ij}$).

Let:

$$y_{ij} = \begin{cases} 1 & \text{if } l_{ij} \text{ is in the cover} \\ 0 & \text{otherwise} \end{cases}$$

$$\text{Minimise} \quad Z = \sum_{i\in I} \sum_{j\in J} a_{ij} y_{ij} \tag{6}$$

Subject to:

$$\sum_{\text{for}\{l_{ij}, j\in J\}} y_{ij} \geq 1 \qquad i\in I \tag{7}$$

$$y_{ij} \geq 0, \; y_{ij} \leq 1 \quad \text{and integer} \tag{8}$$

The objective function (6) will minimise the "cost" of subsets of S (members of K) necessary to cover the members of S, and thus obtain the minimum cost facilities needed, according to certain criteria quantified by $a_{ij}$. The coefficients $a_{ij}$ represent the "cost" of serving the group of points of the subset

$l_{ij}$ by a facility located at i. The constraints (7) guarantee that every $i \in S$ will be covered by at least one $l_{ij} \in K$, since for every i there is a constraint forcing at least one $l_{ij}$ to be in the cover ($y_{ij} = 1$).

In the case where (7) are equalities, the above formulation is known as the set partitioning type of formulation.

Several specialised algorithms, as well as the standard integer programming ones, can be used to solve these types of problems. However, their combinatorial nature (NP-hard) imposes limitations on the size of the problem which can be solved.

## 3. THE FUZZY MODELS

A number of papers appeared recently dealing with the case of mathematical programming problems in fuzzy environments (Fabian and Stoica, 1984; Zimmermann and Pollatschek, 1984). Although the fuzzification may occur at many stages in the formulation and through a number of elements of the problem, such as the constants or the variables, it is mainly considered that the objective function(s) and some or all the constraints are fuzzy sets and in most cases both are treated as restrictions to be satisfied simultaneously according to the principles discussed in Bellman and Zadeh's (1970) paper. The suggested solution methods are based on transforming the fuzzy models into deterministic non-fuzzy equivalents which can be solved using the standard algorithms.

The SPLP could be fuzzified to represent a real life problem in a more pragmatic way than its non-fuzzy counterpart. For instance

$$Z_1 \underset{\sim}{\simeq} \min Z \tag{9}$$

subject to:

$$\sum_{i \in I} d_{ij} \underset{\sim}{\geq} l_j, \quad j \in J \tag{10}$$

$$k_i y_i - \sum_{j \in J} d_{ij} \underset{\sim}{\geq} 0, \quad i \in I \tag{11}$$

where "$\underset{\sim}{\simeq}$" may mean almost optimal, and "$\underset{\sim}{\geq}$" almost satisfied. The actual constraints can be labelled as, for example, "the demand is almost satisfied", in the case of (10), and the fuzzification of (10) and (11) may imply that if the total amount produced at i and delivered to j is very small, then the $y_i$ may be allowed to be zero. In other words, the fixed cost may not occur because it is more beneficial to use some of the stock, or even buy it from elsewhere just to maintain the customers.

Alternatively, if $k_i$ is defined as being a specific maximum outflow from location i, then for $y_i = 1$, $d_{ij}$ may be allowed to go a little over that maximum, for policy reasons.

Several more logical constraints can be incorporated in general into the SPLP, and these can be fuzzified if more appropriate. The fuzzy objectives and constraints, in most cases considered interchangeable, are the course characterized by their membership functions. The fuzzy mixed integer programming formulations suggested so far by various researchers (Fabian and Stoica, 1984; Zimmermann and Pollatschek, 1984) use specific types of membership functions which lead to a non-fuzzy equivalent problem.

In Fabian and Stoica's (1984) approach, for example, the membership function of the objective function would be defined as follows:

let $X_o$ be the non-integer solution to the non-fuzzy problem, then

$$\mu_Z(X) = \begin{cases} Z(X)/Z(X_o) & \text{if} \quad Z(X) < a \\ 1 & \text{if} \quad Z(X_o) \geq Z(X) \geq a \\ \exp(-p_o(Z(X)-Z(X_o))) & \text{if} \quad Z(X) > Z(X_o) \end{cases} \tag{12}$$

where a is a minimum threshold defined by the decision maker(s) and $p_o$ is a membership parameter.

A membership function for the constraints (10) may be

$$\mu(d_{ij}, y_j) = \begin{cases} 1 & \text{if} \quad \Sigma d_{ij} \geq l_i \\ \exp(-p_i \; \Sigma d_{ij}) & \text{if} \quad \Sigma d_{ij} < l_i \end{cases} \tag{13}$$

and similarily for (11)

$$\mu(d_{ij}, y_i) = \begin{cases} 1 & \text{if } k_i y_i - \Sigma d_{ij} \geq 0 \\ \exp(-p_i(k_i y_i - \Sigma d_{ij})) & \text{if } k_i y_i - \Sigma d_{ij} < 0 \end{cases} \tag{14}$$

In the same paper, Fabian and Stoica (1984) also show that assuming the membership functions (12) to (14) the problem (9) to (11) is equivalent to a non-fuzzy one with non-linear objective function and linear constraints which they solve using simulation methods.

Furthermore, Zimmermann and Pollatschek (1984) suggested non-fuzzy equivalent formulations for the 0-1 integer problem with fuzzy right hand side. Their approach could be applied to discrete location problems where the set S, class K and consequently the set of covers C are all crisp, but the restrictions and goals are fuzzy, for example, consider (6) and (7), in their fuzzy form they would be:

$$\Sigma\Sigma a_{ij} \, y_{ij} \lesssim Z_o \tag{15}$$

$$- \Sigma y_{ij} \lesssim -1 \tag{16}$$

respectively, where the "$\lesssim$" means that the constraints can be violated up to a tolerance level $d_{ij}$, and $Z_o$ may be a level defined by the decision maker(s). One further assumption is that the membership functions of the objective and constraints are linear functions of $d_{ij}$, in this case:

$$\mu_i(y_{ij}) = \begin{cases} 1 & \text{if } -\Sigma a_{ij}y_{ij} \leq -1 \\ 1-(-\Sigma a_{ij}y_{ij}+1)/d_{ij} & \text{if } -1 \leq \Sigma a_{ij}y_{ij} \leq d_{ij}-1 \\ 0 & \text{if } -\Sigma a_{ij}y_{ij} \geq -1+d_{ij} \end{cases} \tag{17}$$

The membership function of the decision is the intersection of the membership functions of the constraints and the objective(s) using an intersection operator such as the minimum. To obtain the $y_{ij}$ which will yield that minimum, the authors solve a crisp 0-1 problem.

The use of the above approaches to solve the discrete location problem applied in this paper would depend on:

a) the importance the decision maker attaches to the restriction of the specific (decreasing) membership functions for the constraints and the objectives,

b) his ability to define working tolerance levels, and

c) the assumption that the class K of the subsets of the facility points in the case of the set covering approach can be crisply defined.

## 4. AN ALTERNATIVE FUZZY APPROACH

The approach presented in this section is based on the 0-1 set covering, set partitioning type of formulations discussed in the previous sections.

In a deterministic non-fuzzy problem the subsets (members of K) can be either homogenous, which means that all $a_{ij}=1$ in (6), or they can have different values attached to them, for example the location of fire stations or other emergency services where the objective is to reach their destination within a maximum time period, the location of warehouses, etc. These problems can be tackled by the formulation (6) to (8) presented in Section 2 subject to the problem size.

On the other hand, there are many location problems which are associated with social policies, such as decentralisation policies, where major funding is made available for locating public services in certain areas in order to improve the standards of living and to boost local economies, etc.

The decision makers' problem here is the identification and evaluation of criteria on the basis of which an optimum will be obtained. In these cases, one of the major tasks, assuming that a set S of potential facility points can be crisply defined, is the construction of the class K of the subsets.

of S. The choice of specific location points and of the rest of the points these should serve can only be based on questions like:

How far should people travel to reach a service point?
How important are bad and good roads and public transport?
Is homogeneity of class and income within a subset important?
Is it very unfair to locate two major facilities in one point?

It becomes apparent that in the way the above problem is expressed, the labels which may be used to describe the aims of a policy cannot always (if ever) be crisply defined.

The fuzzy nature of the problem can be accepted and introduced at various stages in the analysis. To begin with, the members of K can be regarded as fuzzy sets, with the construction of their membership functions being the major difficulty, in other words, the evaluation of "nearness" or "accessibility" of a service point i from the other points.

Let S′ be the non-fuzzy set of $x_{im}$(i.e. linked pairs members of S such that i serving m) generated by S.

Consider a point i as a potential facility location, then a fuzzy subset may be expressed as follows:

$$l_{im} = \left\{ x_{im},\ \mu_1(x_{im}) \mid x_{im} \in S', \text{ where m is served by centre i, } i \in S \right\} \quad (18)$$

$\mu_1(x_{im})$ measures the "serviceability" of i to m, and may depend on how far m is from i, how many people m has, what is the maximum number of people, approximately, that $l_{im}$ should have (i.e. maximum number of people i should serve), whether i is in competition with m in any way, etc.

Alternatively, the class K of subsets of S can be regarded as a fuzzy set

$$K = \left\{ l_{im}\ \mu_k(l_{im}) \mid l_{im} \text{ is one of p subsets which consist of points which are served by i, } i \in S \right\} \quad (19)$$

where $l_{im}$'s can be non-fuzzy subsets generated according to a procedure, e.g., containing up to a maximum total population or consisting of points m which are up to a maximum distance away from i, etc. Also $l_{im}$'s can be the fuzzy sets as defined in (18) with $\mu_k(l_{im})$ being the intersection of the $\mu_1(x_{im})$ for every m in $l_{im}$.

The $\mu_k(l_{im})$ may measure the acceptableness of $l_{im}$ as a member of K, i.e., how acceptable and useful $l_{im}$ would be as a member of a cover, and can also be evaluated on the basis of the same criteria used for the membership of (18).

Nevertheless, the individual membership functions of $l_{im}$'s do not necessarily represent the membership value of a cover. It is very likely that a cover will "look" and be valued differently than an aggregate value which is based on the membership functions of the subsets (members of k) it consists of.

Thus one may prefer to accept the fuzziness of the problem at the stage where the covers are examined

$$A = \left\{ c_n, \ \mu_A(c_n)/c_n \right\} \in C \tag{20}$$

where C is the finite set of covers and $\mu_A(c_n)$ the membership function of $c_n$. Possible labels for the evaluation of $\mu_A(c_n)$ are: "fair cover", "a cover within the policy lines", etc.

The next section discusses a solution method for this case.

## 5. A METHOD FOR THE EVALUATION OF FUZZY COVERS

The following example is used to illustrate the approach:

Consider the road network shown in Fig. 1 which is part of a real road network.

The points 1-4 represent villages whose populations are given in Table 1a. The distances in kilometres between the villages are given in Table 1b.

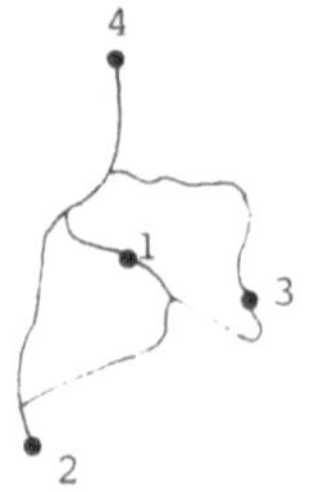


Fig. 1

Table 1a

| | |
|---|---|
| 1 | 1.100 |
| 2 | 650 |
| 3 | 1.350 |
| 4 | 730 |

Table 1b

| | 1 | 2 | 3 | 4 |
|---|---|---|---|---|
| 1 | - | 11 | 7 | 9 |
| 2 | 11 | - | - | 14 |
| 3 | 7 | - | - | - |
| 4 | 9 | 14 | - | - |

A possible problem is to optimally locate three facilities in order to serve (cover) each village by only one facility. Thus this problem in its non-fuzzy form can be formulated as a set partitioning problem.

Nevertheless, whatever the approach, the first major task is the construction of the class K of subsets of the set $S = \{1,2,3,4\}$. The hypotheses and assumptions set out in Section 4 could be considered when making up K. However, for simplicity, and without loss of generality, the assumption here is that each member of K should not have more than 2.000 total population, and the distances between villages should not be more than 15 km. This, of course, is a non-fuzzy restriction an alternative fuzzy label could be: Each member of K should not have much more than 2.000 total population and the distances between villages should not be much more than 15 km. In this case, the support S(K) of K could be obtained according to a membership function or empirically calculated membership values of the members of K.

Table 2 gives the members of K for the example and, in addition, all the information needed for the non-fuzzy formulation (6) - (8) presented in Section 2.

Table 2

| members of K | $l_j$ | total population $p_j$ | total distance $\Sigma d_{ij}$ | $y_{ij}$ | $a_{ij}=\Sigma d_{ij}/p_j$ | $a_{ij} \times 1000$ |
|---|---|---|---|---|---|---|
| 1 | 1 | 1,100 | 0 | $<_{11}$ | 0 | 0 |
| 2 | 1,2 | 1,750 | 11 | $y_{12}$ | 0.006 | 6 |
| 3 | 1,4 | 1,830 | 9 | $y_{13}$ | 0.005 | 5 |
| 4 | 2 | 650 | 0 | $y_{24}$ | 0 | 0 |
| 5 | 2,1 | 1,750 | 11 | $y_{25}$ | 0.006 | 6 |
| 6 | 2,4 | 1,380 | 14 | $y_{26}$ | 0.010 | 10 |
| 7 | 3 | 1,350 | 0 | $y_{37}$ | 0 | 0 |
| 8 | 4 | 730 | 0 | $y_{48}$ | 0 | 0 |
| 9 | 4,1 | 1,830 | 9 | $y_{49}$ | 0.005 | 5 |
| 10 | 4,2 | 1,380 | 14 | $y_{4\ 10}$ | 0.010 | 10 |

The crisp formulation is:

$$\min Z = 6y_{12} + 5y_{13} + 6y_{25} + 10y_{26} + 5y_{49} + 10y_{4\ 10} \tag{21}$$

Subject to:

$$y_{11}+y_{12}+y_{13}+y_{25}+y_{49} = 1 \quad (22)$$

$$y_{12}+y_{24}+y_{25}+y_{26}+y_{4\,10} = 1 \quad (23)$$

$$y_{13}+y_{26}+y_{48}+y_{49}+y_{4\,10} = 1 \quad (24)$$

$$y_{37} = 1 \quad (25)$$

$$y_{11}+y_{12}+y_{13}+y_{24}+y_{25}+y_{26}+y_{37}+y_{48}+y_{49}+y_{4\,10} = 3 \quad (26)$$

To provide a solution to a fuzzy version of the example according to the treatment presented in the previous section, the set of covers (partitions) is needed assuming that either a membership function or calibrated membership values of the members of the set of covers (partitions) can be obtained.

A simple branch-and-bound based enumeration algorithm can be used to obtain the covers. The particular nature of the set partitioning problem provides a number of working bounds, hence there are a number of special algorithims. In the case of the example, the bounds are provided by the specific number of new facilities on one hand and, of course, the equality constraints on the other, which impose the restriction that each village should only be covered once. Hence two bounds can be constructed: Let V = 3 be an upper bound for the number of villages, and M = 0 if no village is covered more than once, and 1 otherwise.

As an example, see Fig. 2 which shows a part of the search tree.

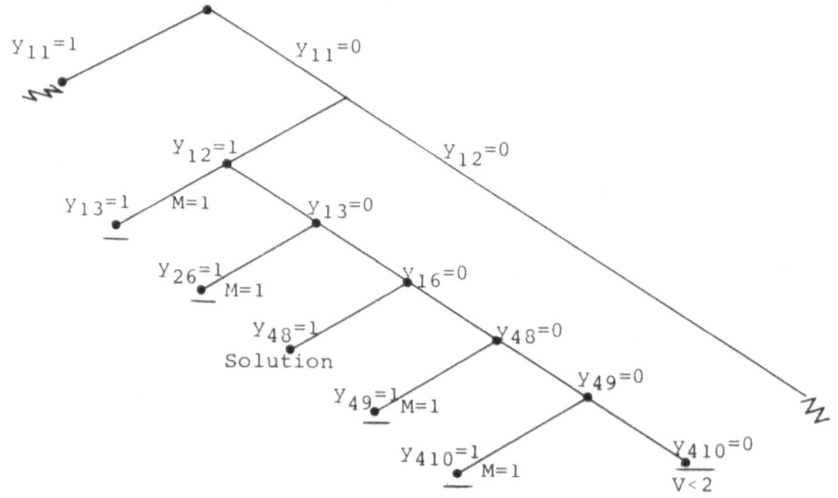


Fig. 2

The search for the example is from left to right in the table of constraints (22) to (26). When $y_{ij} = 1$, then row i and column j are removed, the number of $y_{ij}$'s $\neq 0$ in the j column are however recorded for the benefit of M. Also for the example the matrix can be reduced since $y_{37}$ should be 1, that is because 3 can only be served by itself (the only non-zero element in its column and row).

The enumeration gives the following feasible covers:

1) $y_{12}, y_{48}$ 2) $y_{13}, y_{24}$ and 3) $y_{25}, y_{48}$

Fig. 3 gives the locations of the facilities and the villages they serve (cover).

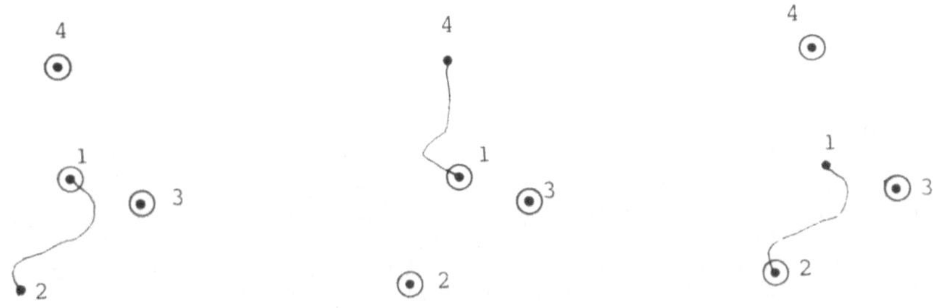


Fig. 3

According to the crisp formulation the optimum (minimizing Z) is 5.

However, the fuzzy version of the problem can be tackled on the basis of the membership values these covers take in relation to the fuzzy constraints and objectives treated here symmetrically according to Bellman and Zadeh (1970).

For example, the problem may be structured as in Table 3 where $c_1, c_2, c_3$ are the covers (feasible solutions). The numerical values in the table are the membership values of each cover in relation to each label (constraint). The membership values of the decision are obtained as the intersection of the corresponding values using the minimum as the operator.

This maximum approach, if the minimum operator is adopted, is very sensitive to the extreme membership values. However, other intersection operators which may take into consideration "compensation" could be used (Thole, Zimmermann and Zysno, 1979).

For our example, the membership values for the decision are (0.9, 0.6, 0.5) and if a straight optimum is sought, then the maximum of 0.9 is taken.

Table 3

| | $c_1$ | $c_2$ | $c_3$ |
|---|---|---|---|
| If possible, separate 1 and 4: | 0.9 | 0.6 | 0.9 |
| It is preferable not to locate at 2: | 0.9 | 0.9 | 0.5 |
| It is better policy to locate this type of facility in villages with high population: | 0.9 | 0.8 | 0.7 |
| Membership values of the decision: | 0.9 | 0.6 | 0.5 |

## 6. DISCUSSION

The justification for suggesting, let alone adopting, the fuzzy approaches to the discrete location problem should be allocated the major effort in this section. Indeed, in the previous sections, the fuzzy location problems were described in such a way that it is difficult to deny that they are real, and in most cases the decisions are part of policies where there are elements which go beyond the stochastic level, i.e. the policy/decision makers cannot identify and define their problems crisply. However, the argument may be that sensitivity analysis applied to an array of crisp formulations will provide equally good solutions if not better looking ones. The counter argument is: why should one produce a good-looking solution to a problem which is not recognizable? Nevertheless, it is beyond the scope of this paper to justify the existence of fuzzy sets theory. Experience has shown, though, that the "client" gives more information about his problem and even understands it better if he or she is not forced to quantify everything.

Looking back through the stages of the suggested approach, possible drawbacks may be identified:

- The assumption about the membership functions for the objective(s) and constraints in formulations (9) to (11) and (15) to (17);

- The examination of possible covers in the approach presented in Sections 4 and 5;

- The adoption of the appropriate intersection operator.

As for the specific membership functions, it is true they may be restrictive but one could expect that in real life empirical membership functions can be described by decreasing mathematical functions. If, however, one feels that the real fuzzy nature of the problem could be overlooked by such assumptions, the approach presented in Sections 4 and 5 may be more appropriate.

In the context of a comparison between the non-fuzzy integer programming formulation (6) to (8), and the fuzzy approach suggested here, the restrictions imposed upon the size of the problem by the algorithms used to solve the non-fuzzy one, is at least as great as the restriction imposed by a total implicit enumeration which is necessary in order to construct the fuzzy set A and the membership of the constraints.

As to the intersection operator, several researchers have suggested various operators (see, Thole, Zimmermann and Zysno 1979) and it should be acknowledged that each study should be treated individually with an open mind towards the various operators. Data availability on, say, preferences expressed by users, would obviously help a great deal in testing the value of the operator, but in addition to that a calibration of the personal perception of the concept of "and" by the decision and policy makers would certainly be a worthwhile exercise.

## REFERENCES

Bellman, R., and L.A. Zadeh (1970). Decision making in a fuzzy environment. Mang. Sci. 17B, 141-164.

Fabian, C., and M. Stoica (1984). Fuzzy integer programming. In H.J. Zimmermann, L.A. Zadeh, and B.R. Gaines (eds.), Fuzzy Sets and Decision Analysis. North Holland, Amsterdam, 123-132.

Francis, R.L., L.F. McGinnis, and J.A. White (1983). Locational analysis. Eur. J. Op. Res. 12, 220-252.

Koenig, J.G. (1981). Indicators of urban accessibility: theory and application. Transportation 9, 145-172.

Krarup, J., and P.M. Pruzan (1983). The simple plant location problem: survey and synthesis. Eur. J. Op. Res. 12, 36-81.

Thole, V., H.J. Zimmermann, and P. Zysno (1979). On the suitability of minimum and product operators for the intersection of fuzzy sets. Fuzzy Sets and Syst. 2, 167-180.

Zimmermann, H.J., and M.A. Pollatschek (1984). Fuzzy 0-1 linear programs. In H.J. Zimmermann, L.A. Zadeh, and B.R. Gaines (eds.), Fuzzy Sets and Decision Analysis, North-Holland, Amsterdam, 133-145.

FUZZY TRANSPORTATION PROBLEMS: A GENERAL ANALYSIS

M. Delgado, J.L. Verdegay and M.A. Vila
Departamento de Estadistica Matematica
Facultad de Ciencias
Universidad de Granada
18071 Granada, Spain

Abstract. The transportation problems have a recognized importance. Their range of applications can be enlarged when some fuzziness in its formulation is accepted. This paper is devoted to the study of a resolution method for fuzzy transportation problems. In order that this may be done, in accordance with the decomposition theorem for fuzzy sets, a formulation as the transshipment problems cut by cut is done. Feasibility or unfeasibility of the former fuzzy problem is analyzed on these cuts by means of four functions. These are straightforwardly defined from the membership functions of the fuzzy parameters involved in the starting formulation. In order to find a fuzzy solution using an auxiliary problem, a parametric type algorithm is proposed. This one is shown to be more efficient than others existing in the current literature because of the lower dimensionality of the mentioned auxiliary problem.

Keywords: fuzzy transportation problem, transportation problem, fuzzy mathematical programming.

## 1. INTRODUCTION

A great part of practical applications of linear programming falls into the field of network flow problems. The transportation problem (TP) is of special interest. As is well known, TP is formulated as follows: A product is to be transported from each of m sources to any of n destinations. The sources are production facilities (supply points) characterized by available capacities $a_1,\ldots,a_m$. The destinations are consumption facilities characterized by required levels of demand $b_1,\ldots,b_n$. There is a cost $c_{ij}$ associated with transporting a unit of the given product from the i-th source to the j-th destination. One must determine the amounts $x_{ij}$ of the product to be transported from all sources i to all destinations j so that the total operation cost will be minimized.

It is usual to impose $\sum_{i=1}^{m} a_i = \sum_{j=1}^{n} b_j$ and with this assumption TP can be formulated as the following linear programming problem:

$$\sum_{i=1}^{m} \sum_{j=1}^{n} c_{ij} x_{ij} \to \min_{x_{ij}}$$

s.t:

$$\sum_j x_{ij} = a_i, \; i \in I = \{1, \ldots, m\}$$

$$\sum_i x_{ij} = b_j, \; j \in J = \{1, \ldots, n\}$$

$$x_{ij} \geq 0$$

This formulation has a clear economic interpretation from which the great interest in this type of problems arises. On the other hand, methods exist (the stepping stone or the Hungarian one) which permit the problem to be solved more easily than by the simplex algorithm (in fact the stepping stone method is a version of the simplex method adjusted to the special features of TP). These reasons justify, in part, extraordinary importance TP has for operations research. Moreover, there are many problems, not being exactly TP, which can be solved in the same way because of their mathematical formulation. On the other hand, general network flow problems contain TP as subproblems.

All these areas of interest in TP can be enlarged when it is assumed that some parameters taking part in the formulation of the problem are fuzzy. Thus, the attempt is to solve problems, such as those mentioned above, assuming that some fuzziness is present in them. In this case we deal with the so-called fuzzy transportation problem (FTP).

FTP was first studied by Prade (1980); Oheigeartaigh (1982), Verdegay (1983) and Delgado and Verdegay (1984) should also be mentioned. Recently, a parametric approach to solve FTP has been proposed by Chanas, Kołodziejczyk and Machaj (1984).

## 2. PROBLEM FORMULATION

Suppose we have an FTP with a fuzzy supply and demand, and nonfuzzy objective function given as usually (see, e.g., Chanas, Kołodziejczyk and Machaj, 1984; Oheigeartaigh, 1982; Delgado and Verdegay, 1984), i.e.

$$\sum_{i=1}^{m} \sum_{j=1}^{n} c_{ij} x_{ij} \to \min_{x_{ij}}$$

s.t:

$$\sum_{j=1}^{n} x_{ij} = \underset{\sim}{a}_i \;, \; i \in I$$

$$\sum_{i=1}^{m} x_{ij} = \underset{\sim}{b}_j, \; j \in J \qquad (1)$$

$$x_{ij} \geq 0, \; (i,j) \in I \times J$$

where $\underset{\sim}{a}_i$ and $\underset{\sim}{b}_j$ denote nonnegative fuzzy numbers, $i \in I$ and $j \in J$. Some of them can be crisp numbers. However, we suppose there exists at least one noncrisp number.

For any $\{x_{ij}, (i,j) \in I \times J\}$ the value of $\mu_{\underset{\sim}{a}_i}(\sum_{j=1}^{n} x_{ij})$, $(\mu_{\underset{\sim}{b}_j}(\sum_{i=1}^{m} x_{ij}))$ is interpreted as a feasibility (compatibility) degree of the solution $\{x_{ij}, (i,j) \in I \times J\}$ with respect to the ith (jth) constraint.

Thus, this model can also be used to describe the situation in which the right-hand-sides of the constraints are crisp, with the equality relation being fuzzy.

## 3. A PARAMETRIC APPROACH

According to the decomposition theorem (Negoita and Ralescu, 1975) and using the parametric approach to solve fuzzy mathematical programming problems (Verdegay, 1982) (particularly its application to FTP in Delgado and Verdegay, 1984; Verdegay, 1983), problem (1) can be changed into

$$\sum_{i=1}^{m} \sum_{j=1}^{n} c_{ij} x_{ij} \rightarrow \min_{x_{ij}}$$

s.t:

$$\mu_{\underset{\sim}{a}_i}(\sum_j x_{ij}) \geq \alpha, \quad i \in I$$

$$\mu_{\underset{\sim}{b}_j}(\sum_i x_{ij}) \geq \alpha, \quad j \in J \qquad (2)$$

$$\alpha \in (0,1], x_{ij} \geq 0, \; (i,j) \in I \times J$$

If $\underset{\sim}{a}_i$, $i \in I$, $\underset{\sim}{b}_j$, $j \in J$, are LR fuzzy numbers (Dubois and Prade, 1980) or, more particularly, trapezoidal fuzzy numbers, as in Chanas, Kołodziejczyk and Machaj (1984), problem (2) becomes:

$$\sum_{i=1}^{m} \sum_{j=1}^{n} c_{ij} x_{ij} \rightarrow \min_{x_{ij}}$$

s.t:

$$\sum_j x_{ij} \in [a_i(\alpha), A_i(\alpha)], \quad i \in I$$

$$\sum_i x_{ij} \in [b_j(\alpha), B_j(\alpha)], \quad j \in J \qquad (3)$$

$$x_{ij} \geq 0, \quad \alpha \in (0,1], \ (i,j) \in I \times J$$

where $[a_i(\alpha), A_i(\alpha)]$ $([b_j(\alpha), B_j(\alpha)])$ $\alpha \in (0,1]$ is the $\alpha$-cut of level $\alpha$ of $\underset{\sim}{a}_i$ $(\underset{\sim}{b}_j)$, $i \in I$ $(j \in J)$.

As each constraint in (3) can be stated by means of two inequalities, it is obvious that (3) is a parametric linear programming problem. Thus, from a theoretical point of view it can be solved by using the simplex method, obtaining the family $\{x_{ij}(\alpha), a \in (0,1], (i,j) \in I \times J\}$ of solutions, and, in accordance with the applied approach (Verdegay, 1982), this family shall be considered the fuzzy solution to the original fuzzy problem.

However, as in the case of classical TP, we ask about a way to solve (3) so that, taking into account the special structure of the problem, it allows us to obtain $\{x_{ij}, \alpha \in (0,1], (i,j) \in I \times J\}$ faster and more easily than the direct application of the simplex method. The following sections are devoted to this subject.

## 4. A FORMULATION AS A TRANSSHIPMENT PROBLEM

In economic terms, for any $\alpha \in (0,1]$, (3) may be interpreted as a specific TP on a bipartite network as follows. There exist n destinations (demand points) $D_1, \ldots, D_n$ each of them demanding an amount of a product. The special feature is that the demand of jth destination must be between $b_j(\alpha)$ and $B_j(\alpha)$ (there is a minimal and a maximal demand), $j \in J$.

To supply that demand, there exist m sources (supply points) each of them having an amount of the said product. As before, the stock of $0_i$ is between $a_i(\alpha)$ and $A_i(\alpha)$ ($0_i$ must supply at least $a_i(\alpha)$ but not more than $A_i(\alpha)$), $i \in I$.

The unit transportation cost from $0_i$ to $D_j$ is $c_{ij}$ ($c_{ij} = \infty$ if such transportation is impossible). As is usual, we will assume $c_{ij} \geq 0$, $\forall (i,j) \in I \times J$.

It is required to find a flow $\{x_{ij}(\alpha), (i,j) \in I \times J\}$ $(\alpha \in (0,1])$, having a minimal total transportation cost.

Following a usual method in network flow theory, this problem can be changed into a new transshipment problem on a network with a single source and a single destination obtained as follows:

1) By adding a new source 0 with infinite supply, joined to each original source $0_i$ by an arc $(0, 0_i)$ with null cost,

lower capacity equal to $a_i(\alpha)$ and upper capacity equal to $A_i(\alpha)$, $i \in I$.

2) By adding a new destination D, joined to each original destination $D_j$ by an arc $(D_j, D)$ with null cost, lower capacity equal to $b_j(\alpha)$ and upper capacity equal to $B_j(\alpha)$, $j \in J$.

3) Original sources and destinations are considered as transit points (null demand or supply points).

Now we look for a flow from 0 to D being compatible with the capacity constraints of the arcs and having minimal total cost of transportation.

This transshipment problem (which will be denoted $T1(\alpha)$) can be solved for every $\alpha \in (0,1]$ by using well known techniques (see, e.g., Simmonard, 1973).

On the other hand, the dimensions of $T1(\alpha)$ can be reduced by taking into account the values $a_i(\alpha)$, $A_i(\alpha)$, $b_j(\alpha)$, $B_j(\alpha)$, $i \in I$, $j \in J$, $\alpha \in (0,1]$, and their relations. Define:

$$a(\alpha) = \sum_i a_i(\alpha) \qquad b(\alpha) = \sum_j b_j(\alpha)$$

$$\alpha \in (0,1]$$

$$A(\alpha) = \sum_i A_i(\alpha) \qquad B(\alpha) = \sum_j B_j(\alpha)$$

Obviously $a(\alpha) < A(\alpha)$ and $b(\alpha) < B(\alpha)$ for any $\alpha$ in $(0,1)$. When there is some nonunimodal fuzzy number in each of the sets $\{\tilde{a}_i;\ i \in I\}$, $\{\tilde{b}_j;\ j \in J\}$, then $a(1) < A(1)$ and $b(1) < B(1)$.

According to Hoffman's theorem, $T1(\alpha)$ has no solution if

$$B(\alpha) < a(\alpha) \quad \text{or} \quad A(\alpha) < b(\alpha)$$

In the original transportation problem (3), these conditions have the following interpretation:

i) When $B(\alpha) < a(\alpha)$, the problem can not have a solution because the constraints of lower supply and upper demand are impossible to verify: the sources must send $a(\alpha)$ at least, but the destinations can receive $B(\alpha)$ at most, with $B(\alpha) < a(\alpha)$.

ii) When $A(\alpha) < b(\alpha)$, the constraints of lower demand and upper supply can not be verified: we must satisfy a demand equal or greater than $b(\alpha)$ but we have a maximal total supply equal to $A(\alpha)$ with $A(\alpha) < b(\alpha)$.

Since the transportation costs are nonnegative when $T1(\alpha)$ is feasible, its optimal solution must be such that the amount sent from 0 to D (from sources to destinations in the original transportation problem) is the minimal one for which the constraints are satisfied. Thus for every $\alpha$ we can distinguish three cases:

A) $a(\alpha) < b(\alpha)$. In such a situation the minimal feasible amount is $b(\alpha)$ and thus the optimal solution must satisfy

$$\sum_i \sum_j x_{ij}(\alpha) = b(\alpha)$$

We can obtain it from the following modification of (3):

$$\sum_{i=1}^{m} \sum_{j=1}^{n} c_{ij} x_{ij} \to \min_{x_{ij}}$$

$$\text{s.t.:} \quad \sum_{j=1}^{n} x_{ij} \in [a_i(\alpha), A_i(\alpha)] \qquad i \in I$$

$$\sum_{i=1}^{m} x_{ij} = b_j(\alpha) \qquad j \in J \qquad (4)$$

$$x_{ij} \geq 0 \qquad (i,j) \in I \times J$$

This problem is equivalent to a transshipment problem obtained from the original TP as follows:

A.1) By adding a new source 0 with supply $b(\alpha)$ and joined to each $0_i$ by means of an arc $(0,0_i)$ with null cost, upper capacity equal to $A_i(\alpha)$ and lower capacity equal to $a_i(\alpha)$, $i \in I$.

A.2) By fixing the demand at destination $D_j$ exactly equal to $b_j(\alpha)$, $j \in J$.

A.3) Original sources are considered as transit points (null supply points).

This problem will be denoted $T2(\alpha)$.

B) $a(\alpha) > b(\alpha)$. In such a case the minimal feasible flow is $a(\alpha)$; thus the optimal solution must satisfy

$$\sum_{i=1}^{m} \sum_{j=1}^{n} x_{ij}(\alpha) = a(\alpha)$$

and we can obtain it from the following modification of problem (3):

$$\sum_{i=1}^{m} \sum_{j=1}^{n} c_{ij} x_{ij} \to \min_{x_{ij}}$$

$$\text{s.t.:} \quad \sum_{j=1}^{n} x_{ij} = a_i(\alpha) \qquad i \in I$$

$$\sum_{i=1}^{m} x_{ij} \in [b_j(\alpha), B_j(\alpha)] \qquad j \in J \qquad (5)$$

$$x_{ij} \geq 0 \qquad (i,j) \in I \times I$$

This version can be formulated as a transshipment problem on the network obtained from the original as follows:

B.1) By adding a new destination D with demand $a(\alpha)$ and joined to each $D_j$ by means of an arc $(D_j D)$ with null cost, upper capacity equal to $B_j(\alpha)$ and lower capacity equal to $b_j(\alpha)$, $j \in J$.

B.2) By setting the supply of $0_i$ exactly equal to $a_i(\alpha)$

B.3) Original destinations are considered transit points (null demand points).

This problem will be denoted $T3(\alpha)$.

C) $a(\alpha) = b(\alpha)$. In this situation the optimal solution must satisfy

$$\sum_{i=1}^{m} \sum_{j=1}^{n} x_{ij}(\alpha) = a(\alpha) = b(\alpha),$$

being possible to obtain it from:

$$\sum_{i=1}^{m} \sum_{j=1}^{n} c_{ij}\, x_{ij} \rightarrow \min_{x_{ij}}$$

$$\text{s.t.:} \quad \sum_{j=1}^{n} x_{ij} = a_i(\alpha) \qquad i \in I$$

$$\sum_{i=1}^{m} x_{ij} = b_j(\alpha) \qquad j \in J \tag{6}$$

$$x_{ij} \geq 0 \qquad (i,j) \in I \times J$$

which is a classical TP on the original bipartite network with supply $a_i(\alpha)$ in $0_i$, $i \in I$, and demand $b_j(\alpha)$ in $D_j$, $j \in J$.

Since the feasibility and the kind of problem associated with each $\alpha$ (each $\alpha$-cut of the fuzzy problem (1)) depend on the relative position of $a(\alpha)$, $b(\alpha)$, $A(\alpha)$, $B(\alpha)$, it is very important to study behaviour of these functions in $(0,1]$.

We shall suppose $a_i$, $b_j$ $i \in I$, $j \in J$, are convex fuzzy numbers of L-R type, i.e. (see Dubois and Prade, 1980):

i) their membership functions are continuous on R (on their support).

ii) L and R are strictly increasing and decreasing, respectively. With these hypotheses we can assure that:

a) $a_i(\alpha)$, $A_i(\alpha)$, $b_j(\alpha)$, $B_j(\alpha)$, $i \in I$, $j \in J$, are continuous on $\alpha \in (0,1]$. So, $a(\alpha)$, $A(\alpha)$, $b(\alpha)$, $B(\alpha)$ are continuous, too.

b) $a_i(\alpha)$, $b_j(\alpha)$, $i \in I$, $j \in J$, are strictly increasing, and so are $a(\alpha)$, $b(\alpha)$.

c) $A_i(\alpha)$, $B_j(\alpha)$, $i \in I$, $j \in J$, are strictly decreasing and so are $A(\alpha)$ and $B(\alpha)$.

For any $\alpha \in (0,1]$ the feasibility of the problem depends upon the relative position of $a(\alpha)$, $B(\alpha)$ and $b(\alpha)$, $A(\alpha)$ as discussed above. However, the relation between $a(\alpha)$ and $B(\alpha)$ is not independent of the relation between $b(\alpha)$, $A(\alpha)$. For instance, when $B(\alpha) < a(\alpha)$ (there is no solution), $b(\alpha) < A(\alpha)$ must be.

On the other hand we can assure:

1) If $B(\hat{\alpha}) \leq a(\hat{\alpha})$ $(A(\hat{\alpha}) \leq b(\hat{\alpha}))$ for some $\hat{\alpha}\varepsilon(0,1]$, then

$B(\alpha) < a(\alpha)$ $(A(\alpha) \leq b(\alpha))$, $\forall\alpha > \hat{\alpha}$

2) If $B(\hat{\alpha}) \geq a(\hat{\alpha})$ $(A(\hat{\alpha}) \geq b(\hat{\alpha}))$ for some $\hat{\alpha}\varepsilon(0,1]$, then

$B(\alpha) \geq a(\alpha)$ $(A(\alpha) \geq b(\alpha))$, $\forall\alpha < \hat{\alpha}$.

Taking into account the above comments, we can analize the feasibility (with $\alpha$) as follows:

I. Consider the equation $B(\alpha) = a(\alpha)$ in $(0,1]$.

I.1. If it has no solution, it may be:

I.1.1. $a(\alpha) > B(\alpha)$, $\alpha\varepsilon(0,1]$. We can conclude the problem is unfeasible for any $\alpha\varepsilon(0,1]$. Hence, according to our approach, we assert that the fuzzy problem has no solution.

I.1.2. $a(\alpha) < B(\alpha)$, $\alpha\varepsilon(0,1]$. In such a case the study of a relation between $A(\alpha)$ and $b(\alpha)$ is needed (see II below).

I.2. If there is a solution $\alpha^*\varepsilon(0,1]$ for $a(\alpha) = B(\alpha)$, then it must be

$$B(\alpha) > a(\alpha),\ \alpha < \alpha^*;\ B(\alpha) < a(\alpha),\ \alpha > \alpha^*$$

Thus (3) is infeasible for $\alpha\varepsilon(\alpha^*,1]$.

When $\alpha\varepsilon(0,\alpha^*]$, we have

$$A(\alpha) \geq A(\alpha^*) \geq a(\alpha^*) = B(\alpha^*) \geq b(\alpha^*) \geq b(\alpha)$$

and thus we can assure (3) is feasible in $(0,\alpha^*]$.

II. Solve the equation $A(\alpha)=b(\alpha)$.

II.1. When it has no solution, then:

II.1.1 $A(\alpha) < b(\alpha)$, $\alpha\varepsilon(0,1]$. Problem (3) is infeasible in $(0.1]$. Thus we assert that problem (1) has no solution.

II.1.2 $A(\alpha) > b(\alpha)$, $\alpha\varepsilon(0,1]$. In this case the problem is feasible for every $\alpha\varepsilon(0,1]$.

II.2 The equation $A(\alpha) = b(\alpha)$ has a solution $\alpha^{**}\varepsilon(0,1]$. We can finally assure that problem (3) is feasible for every $\alpha\varepsilon(0,\alpha^{**}]$.

In the following we shall consider (3) is feasible in $(0,\alpha]$ where $\alpha$ may be $0,1,\alpha^*$ or $\alpha^{**}$.

For any $\alpha\varepsilon(0,\hat{\alpha}]$ the type and structure of problem (3) depend upon the relative position of $a(\alpha)$ and $b(\alpha)$ just as we have described in the previous sections.

Let us consider the sets:

$$A^{>} = \{\alpha\varepsilon(0,\hat{\alpha}]/a(\alpha) > b(\alpha)\}$$
$$A^{=} = \{\alpha\varepsilon(0,\hat{\alpha}]/a(\alpha) = b(\alpha)\}$$
$$A^{<} = \{\alpha\varepsilon(0,\hat{\alpha}]/a(\alpha) < b(\alpha)\}$$

According to our assumptions about $a_i$, $i \in I$, and $b_j$, $j \in J$, we have:

i) $A^<$ and $A^>$ are the union of open or halfopen intervals,

ii) $A^=$ is the union of closed intervals (some of them can be a single point).

From a theoretical point of view, given $\alpha \varepsilon (0,\alpha]$, we have:

- For $\alpha \varepsilon A^<$ we must solve (4) or T2($\alpha$),
- For $\alpha \varepsilon A^>$ we must solve (5) or T3($\alpha$),
- For $\alpha \varepsilon A^=$ we must solve (6).

In practice, we may solve (4) on $A^< \cup A^= (A^<)$ and (5) on $A^> (A^> \cup A^=)$. This simplification is specialy useful when $A^=$ is a finite union of isolated points.

Remark 1. - In the classical TP, it is usually assumed that the total supply ($\Sigma a_i$) is equal to the global demand ($\Sigma b_j$). In FTP one can think about a similar condition, i.e. to impose that the fuzzy addition of $a_i$ must be equal to the fuzzy addition of $b_j$ ($\Sigma \underset{\sim}{a}_i = \Sigma \underset{\sim}{b}_j$). In this case, obviously:

$$a(\alpha) = b(\alpha),\ A(\alpha) = B(\alpha),\ \forall \alpha \varepsilon (0,1]$$

Thus, for any $\alpha \varepsilon (0,1]$ (3) is feasible and it can be solved by means of (6), i.e., a classical TP satisfying the condition of equality between supply and demand.

However, the condition $\Sigma\, a_i = \Sigma\, b_j$ seems to be too strong in practice. In fact (due to the properties of fuzzy addition), statements about $\underset{\sim}{a}_i$ and $\underset{\sim}{b}_j$, $i \in I$ and $j \in J$, satisfying such equality, cannot be established.

A condition weaker than the above one is

$$\sum_i \hat{a}_i = \sum_j \hat{b}_j$$

with $\hat{a}_i$ and $\hat{b}_j$ being modes of $\underset{\sim}{a}_i$ and $\underset{\sim}{b}_j$, $i \in I$ and $j \in J$. In this case it is easy to prove that

$$[a(\alpha),A(\alpha)] \cap [b(\alpha),B(\alpha)] \neq \emptyset \quad \forall \alpha \varepsilon (0,1]$$

and thus it can be neither $a(\alpha) > B(\alpha)$ nor $b(\alpha) > A(\alpha)$, for any $\alpha \varepsilon (0,1]$. Thus (3) is feasible for all $\alpha$ in (0,1]. However, we can say nothing about the relative position of $a(\alpha)$ and $b(\alpha)$.

Remark 2. - When $\underset{\sim}{a}_i$, $i \in I$, and $\underset{\sim}{b}_j$, $j \in J$, are trapezoidal fuzzy numbers, it is obvious that $a(\alpha)$ and $b(\alpha)$ are linear. Thus the equation $a(\alpha) = b(\alpha)$ may have, at most, a single solution, i.e., $A^{=}$ is empty or reduced to a single point.

If $A^{=} = \emptyset$, then $A^{>} = (0,\alpha)$ and $A^{<} = \emptyset$ or $A^{>} = \emptyset$ and $A^{<} = (0,\hat{\alpha})$. If $A^{=} = \{\bar{\alpha}\}$, then $A^{>} = [0,\bar{\alpha}]$ and $A^{>} = (\bar{\alpha},\hat{\alpha})$ or $A^{<} = (\bar{\alpha},\hat{\alpha})$ and $A^{>} = (0,\bar{\alpha}]$

## 5. RESOLUTION METHODS FOR FUZZY TRANSPORTATION PROBLEM

According to our approach, $\{x_{ij}(\alpha),\ \alpha\varepsilon(0,\hat{\alpha}],\ i\in I,\ j\in J\}$ shall be considered the fuzzy (optimal) solution to the FTP (1), with $x_{ij}(\alpha)$ being the optimal solution of a transshipment problem on a particular network such as we have previously discussed.

In general we can solve these problems using a parametric technique. However, transshipment problems have a laborious resolution (of course, much less so than the simplex method) and thus, we ask for the existence of an easier way of resolution.

A first idea may be to transform the transshipment problems into TP though an increase of dimension arises. In Chanas, Kołodziejczyk and Machaj (1984) this method is applied starting from (3) directly. Then, a TP is obtained as follows:

1) If $D_1,\dots,D_n$ are the original destinations, then n new destinations $D_1^-,\dots,D_n^-$ are added. It is assumed $D_i,\dots,D_n$, $D_1^-,\dots,D_n^-$ have demands $b_1(\alpha),\dots,b(\alpha)$, $d_1(\alpha) = B_1(\alpha) - b_1(\alpha),\dots,d_n(\alpha) = B_n(\alpha) - b_n(\alpha)$, respectively. Thus the total demand is equal to $B(\alpha)$.
2) If $0_1,\dots,0_m$ are the original sources, then m new sources $0_1^-,\dots,0_m^-$ are added. The supply of $0_1,\dots,0_m$, $0_1^-,\dots,0_m^-$ is supposed to be equal to $a_1(\alpha),\dots,a_m(\alpha)$, $d_1^-(\alpha) = A_1(\alpha) - a_1(\alpha),\dots,d_m^-(\alpha) = A_m(\alpha) - a_m(\alpha)$, respectively. Thus the total supply is equal to $A(\alpha)$.
3) If $c_{ij}$ is the original unit transportation cost from $0_i$ to $D_j$, then in the new network this value is assigned to the arcs $(0_i,D_j)$, $(0_i^-,D_j)$, $(0_i,D_j^-)$ and $(0_i^-,D_j^-)$, $i\in I$ and $j\in J$.
4) A fictious source OF with supply equal to $B(\alpha)-b(\alpha)$ and a fictious destination DF with demand $A(\alpha)-b(\alpha)$ are introduced. The unit cost of transportation from $0_i$ to DF or from OF to $D_j$ is set to be equal to M, a large value. The unit cost from $0_i^-$ to DF or from OF to $D_j^-$ is made equal to zero, $i\in I$, $j\in J$.

In this way, the transportation table (Table 1) is obtained.

Table 1

| | $D_1 \ldots D_n$ | $D_1^- \ldots D_n^-$ | DF | |
|---|---|---|---|---|
| $0_1$<br>.<br>.<br>.<br>$0_m$ | $c_{ij}$ | $c_{ij}$ | M<br>.<br>.<br>.<br>M | $a_1(\alpha)$<br>.<br>.<br>.<br>$a_m(\alpha)$ |
| $0_1^-$<br>.<br>.<br>.<br>$0_m^-$ | $c_{ij}$ | $c_{ij}$ | 0<br>.<br>.<br>.<br>0 | $d_1^-(\alpha)$<br>.<br>.<br>.<br>$d_m^-(\alpha)$ |
| OF | M ... M | 0 ... 0 | | $B(\ )-b(\alpha)$ |
| | $b_1(\alpha) \ldots b_n(\alpha)$ | $d_1(\alpha) \ldots d_n(\alpha)$ | $A(\alpha)-b(\alpha)$ | |

The net flow from $0_i$ to $D_j$ in the original problem is the sum of the flows in the four entries with index $(i,j)$ of this table, $(i,j) \in I \times J$.

The basic idea of the transformation is to satisfy the lower demands $b_j(\alpha)$ with the lower supplies $a_i(\alpha)$. If this is not possible, the differences $d_i^-(\alpha)$ or $d_j(\alpha)$ are used. As usual, OF and DF are to absorb the disarrangement between supply and demand. The large cost M blocks the possibility of a fictious flow involving lower supplies or demands.

Within the framework of our formulation it is easy to prove that an optimal solution gives non null-flow from $0_i$ to DF (from OF to $D_j$) for some i (for some j) when $a(\alpha) > b(\alpha)$ ($b(\alpha) > A(\alpha)$), i.e., when (3) is infeasible for such $\alpha$. If not, the optimal solution must transport a flow equal to $\mathrm{Max}(a(\alpha), b(\alpha))$.

Let us note that a fuzzy problem with m sources and n destinations needs a parametric transportation table with (m+n+1) rows and (m+n+1) columns. On the basis of our analysis from the last section, we propose a considerable reduction of this table:

A) When $a(\alpha) < b(\alpha)$, the optimal solution of (3) may be obtained from (4) or $T2(\alpha)$, for any $\alpha \in (0,\alpha]$. In their turn these problems may be solved by means of a TP obtained as follows:

A.1) $D_1,\ldots,D_n$ are considered destinations with demands $b_1(\alpha),\ldots,b_n(\alpha)$, respectively.

A.2) If $0_1,\ldots,0_m$ are the original sources, m new sources $0_1^-,\ldots,0_m^-$ are added.

It is assumed $0_i$ has supply equal to $a_i(\alpha)$ and $0_i$ has supply equal to $d_i^-(\alpha) = A_i(\alpha) - a_i(\alpha)$, $i \in I$. The transportation cost from both $0_i$ and $0_i^-$ to $D_j$ is set to be equal to $c_{ij}$, $i \in I$, $j \in J$.

A.3) A fictious destination with demand $A(\alpha)-b(\alpha)$ is added. The transportation cost from $0_i$ to DF is set to be equal to M, a large value, and the cost from $0_i^-$ to DF is set to be equal to zero, $i \in I$.

The justification of this transformation is the same as in the above. The transportation table is now as given in Table 2.

Table 2

| | $D_1$ ... $D_n$ | DF | |
|---|---|---|---|
| $0_1$ | | M | $a_1(\alpha)$ |
| . | | . | . |
| . | $c_{ij}$ | . | . |
| . | | . | . |
| $0_m$ | | M | $a_m(\alpha)$ |
| $0_1^-$ | | 0 | $d_1^-(\alpha)$ |
| . | | . | . |
| . | $c_{ij}$ | . | . |
| . | | . | . |
| $0_m^-$ | | 0 | $d_m^-(\alpha)$ |
| | $b_1(\alpha),\ldots,b_n(\alpha)$ | $A(\alpha)-b(\alpha)$ | |

This table has 2m rows and n+1 columns. Thus, this is, aproximately, one half of the preceding one.

Solving such a TP, the sum of the flow in the two entries of index (i,j) gives the flow from $0_i$ to $D_j$ in (3), $(i,j)\in I\times J$.

Remark 3. In the feasibility of DF lies the feasibility of (3). When $A(\alpha)<b(\alpha)$, (3) has no solution; in its turn, DF is a destination with negative demand.

B) When $a(\alpha)>b(\alpha)$, the optimal solution of (3) may be obtained from (5) or from $T3(\alpha)$ for every $\alpha\in(0,\hat{\alpha}]$. In their turn, these problems may be changed into a TP with:

B.1) The m sources $0_1,\dots,0_m$ with supplies $a_1(\alpha),\dots,a_m(\alpha)$.

B.2) If $D_1,\dots,D_n$ are the original destinations, a new destinations $D_1^-,\dots,D_n^-$ are added. We suppose $D_j$ has demand equal to $b_j(\alpha)$ and $D_j^-$ has demand equal to $d_j(\alpha)=B_j(\alpha)-b_j(\alpha)$, $j\in J$. The transportation cost from $0_i$ to both $D_j$ and $D_j^-$ is set to be equal to $c_{ij}$, $(i,j)\in I\times J$.

B.3) A fictious source OF with supply $B(\alpha)-a(\alpha)$ is added. The transportation cost from OF to $D_j$ is taken to be equal to M, a large value, and the cost from OF to $D_j^-$ is set to be equal to zero, $j\in J$.

The justification of this transformation is the same as in the former ones. The transportation table is as in Table 3.

Table 3

| | $D_1 \dots D_n$ | $D_1^- \dots D_n^-$ | |
|---|---|---|---|
| $0_1$ . . . $0_m$ | $c_{ij}$ | $c_{ij}$ | $a_1(\alpha)$ . . . $a_m(\alpha)$ |
| OF | $b_1$ M ... M | 0 ... 0 | $B(\alpha)-a(\alpha)$ |
| | $b_1(\alpha) \dots b_n(\alpha)$ | $d_1(\alpha) \dots d_n(\alpha)$ | |

Now the table has m+1 rows and 2n columns, i.e. it is again approximately a half of the one proposed in Chanas,

Kołodziejczyk and Machaj (1984).

The flow from $O_i$ to $D_j$ in (3) is the sum of the flow in the two entries of index $(i,j)$ of the transformed problem, $(i,j) \in I \times J$.

Remark 4. - Again, the feasibility of (3) is equivalent to the feasibility of OF. When $B(\alpha) < a(\alpha)$ for some $\alpha$, (3) is infeasible for such $\alpha$ and OF is a source with negative supply.

## 6. A PARAMETRIC ALGORITHM FOR FUZZY TRANSPORTATION PROBLEMS

Summarizing the above analysis, we propose the following algorithm to solve fuzzy transportation problems:

Step 1. By means of $a(\alpha)$, $b(\alpha)$, $A(\alpha)$, $B(\alpha)$, establish the interval $(0, \hat{\alpha}]$ where (3) is feasible.

Step 2. Determine $A^<$, $A^=$, $A^>$ according to their definition.

Step 3. Solve (8) and (9) for any $\alpha \in A^< \cup A^=$ and any $\alpha \in A^>$, respectively.

Alternatively:

Step 3˙. Solve (8) and (9) for any $\alpha \in A^<$ and any $\alpha \in A^> \cup A^=$, respectively.

For step 3 (3˙˙) we may use a parametric technique (see, Gal, 1979). In this way the family $\left\{ x_{ij}(\alpha), i \in I, j \in J, \alpha \in (o, \alpha] \right\}$ is obtained. According to our approach it will be considered the fuzzy (optimal) solution of the FTP (1).

As proved in Verdegay (1982), when there exists a fuzzy goal with membership function $\mu_G( \sum_i \sum_j c_{ij} x_{ij} )$, the optimal solution (with Bellman-Zadeh´s criterion) is $\left\{ x_{ij}(\hat{r}), i \in I, j \in J \right\}$, $\hat{r}$ being a fixed point of $g(\alpha) = \mu_G(\sum_{ij} c_{ij} x_{ij}(\alpha))$, i.e. a solution to the equation $g(\alpha) = \alpha$.

According to remark 2, this algorithm is more efficient than the one proposed in Chanas, Kołodziejczyk and Machaj (1984) when $\underset{\sim}{a}_i$ and $\underset{\sim}{b}_j$ are trapezoidal fuzzy numbers $\forall i, \forall j$, or, in general, when $A^<$, $A^=$ and $A^>$ constitute a "regular" partition of $(0, \hat{\alpha}]$.

## 7. EXAMPLE

Consider the FTP proposed in Chanas, Kołodziejczyk and Machaj (1984). It has two sources and three destinations with the unit transportations costs:

| | $D_1$ | $D_2$ | $D_3$ |
|---|---|---|---|
| $0_1$ | 10 | 20 | 30 |
| $0_2$ | 20 | 50 | 60 |

Supplies and demands are triangular (trapezoidal) fuzzy numbers with membership functions (the value is given on their support only):

$$\mu_{a_1}(x) = \begin{cases} (x-5)/5 & \text{if } x\in[5,10] \Rightarrow a_1(\alpha) = 5+5\alpha \\ (15-x)/5 & \text{if } x\in[10,15] \Rightarrow A_1(\alpha) = 15-5\alpha \end{cases}$$

$$\mu_{a_2}(x) = \begin{cases} (x-11)/5 & \text{if } x\in[11,16] \Rightarrow a_2(\alpha) = 11+5\alpha \\ (21-x)/5 & \text{if } x\in[16,21] \Rightarrow A_2(\alpha) = 21-5\alpha \end{cases}$$

$$\mu_{b_1}(x) = \begin{cases} (x-5)/5 & \text{if } x\in[5,10] \Rightarrow b_1(\alpha) = 5+5\alpha \\ (15-x)/5 & \text{if } x\in[10,15] \Rightarrow B_1(\alpha) = 15-5\alpha \end{cases}$$

$$\mu_{b_2}(x) = \begin{cases} (x-5)/4 & \text{if } x\in[5,9] \Rightarrow b_2(\alpha) = 5+4\alpha \\ (13-x)/4 & \text{if } x\in[9,13] \Rightarrow B_2(\alpha) = 13-4\alpha \end{cases}$$

$$\mu_{b_3}(x) = \begin{cases} x & \text{if } x\in[0,1] \Rightarrow b_3(\alpha) = \alpha \\ 2-x & \text{if } x\in[1,2] \Rightarrow B_3(\alpha) = 2-\alpha \end{cases}$$

Thus:

$$a(\alpha) = (5+5\alpha)+(11+5\alpha) = 16+10\alpha$$
$$A(\alpha) = (15-5\alpha)+(21-5\alpha) = 36-10\alpha$$
$$b(\alpha) = (5+5\alpha)+(5+4\alpha)+\alpha = 10+10\alpha$$
$$B(\alpha) = (15-5\alpha)+(13-4\alpha)+(2-\alpha) = 30-10\alpha.$$

From these values it is easy to prove that

$a(\alpha) \leq B(\alpha)$ if $\alpha\in(0,0.7]$, and $a(\alpha) > B(\alpha)$ otherwise,

$b(\alpha) \leq A(\alpha)$, for any $\alpha$ belonging to $(0,1]$,

hence the parametric version of this FTP is feasible for any $\alpha\in(0,0.7]$.

On the other hand

$$a(\alpha) = 16+10\alpha \geq 10+10\alpha = b(\alpha) \quad \forall\alpha\in(0,0.7]$$

and thus, for all $\alpha$ in $(0,0.7]$, problem (3) may be changed into (5) or T3($\alpha$). These problems, in their turn, can be solved by means of a TP like (8). According to the above values, the transportation table is:

| | $D_1$ | $D_2$ | $D_3$ | $D_1^-$ | $D_2^-$ | $D_3^-$ | |
|---|---|---|---|---|---|---|---|
| $O_1$ | 10 | 20 | 30 | 10 | 20 | 30 | $5+5\alpha$ |
| $O_2$ | 20 | 50 | 60 | 20 | 50 | 60 | $11+5\alpha$ |
| OF | M | M | M | 0 | 0 | 0 | $14-20\alpha$ |
| | $5+5\alpha$ | $5+4\alpha$ | $\alpha$ | $10-10\alpha$ | $8-8\alpha$ | $2-2\alpha$ | |

Applying a parametric resolution method (see Gal, 1979), we obtain the following optimal solution:

| | $D_1$ | $D_2$ | $D_3$ | $D_1^-$ | $D_2^-$ | $D_3^-$ | |
|---|---|---|---|---|---|---|---|
| $O_1$ | | $5+4\alpha$ | $\alpha$ | | | | |
| $O_2$ | $5+5\alpha$ | | | 6 | | | |
| OF | | | | $4-10\alpha$ | $8-8\alpha$ | $2-2\alpha$ | $\alpha\in(0,0.4]$ |

| | $D_1$ | $D_2$ | $D_3$ | $D_1^-$ | $D_2^-$ | $D_3^-$ | |
|---|---|---|---|---|---|---|---|
| $O_1$ | | $5+4\alpha$ | $\alpha$ | | | | |
| $O_2$ | $5+5\alpha$ | | | $10-10\alpha$ | $-4+10\alpha$ | | |
| OF | | | | | $12-18\alpha$ | $2-2\alpha$ | $\alpha\in[0.4,2/3]$ |

| | $D_1$ | $D_2$ | $D_3$ | $D_1^-$ | $D_2^-$ | $D_3^-$ | |
|---|---|---|---|---|---|---|---|
| $O_1$ | | | $\alpha$ | | $17-14\alpha$ | $-12+18\alpha$ | |
| $O_2$ | $5+5\alpha$ | $5+4\alpha$ | | $10-10\alpha$ | $-9+6\alpha$ | | |
| OF | | | | | | $14-20\alpha$ | $\alpha\in[2/3,0.7]$ |

Hence, the optimal (fuzzy) solution for the original FTP, may be represented by the following tables:

| | $D_1$ | $D_2$ | $D_3$ | |
|---|---|---|---|---|
| $O_1$ | | $5+4\alpha$ | $\alpha$ | |
| $O_2$ | $11+5\alpha$ | | | $\alpha\in(0,0.4]$ |

| | $D_1$ | $D_2$ | $D_3$ | |
|---|---|---|---|---|
| $O_1$ | | $5+4\alpha$ | $\alpha$ | |
| $O_2$ | $15-5\alpha$ | $-4+10\alpha$ | | $\alpha\in[0.4,2/3]$ |

| | $D_1$ | $D_2$ | $D_3$ | |
|---|---|---|---|---|
| $O_1$ | | $-17-14\alpha$ | $-12+19\alpha$ | |
| $O_2$ | $15-5\alpha$ | $-4+10\alpha$ | | $\alpha\in[2/3,0.7]$ |

Let us note that as by means of the analysis of the relations between a $(\alpha)$, $A(\alpha)$, $b(\alpha)$ and $B(\alpha)$, we can easily establish the interval of feasibility for the problem and to use a TP with a table being one half of (7).

REFERENCES


Chanas, S., W. Kołodziejczyk, and A. Machaj (1984). A fuzzy approach to the transportation problem. Fuzzy Sets and Syst. 13, 211-221.

Delgado, M., and J.L. Verdegay (1984). Resolution of a fuzzy transportation problem with generalized triangular membership functions (in Spanish). Actas del XIV Congreso Nacional de Estadistica, Investigación Operativa e Informatica, Granada (Spain), 748-758.

Dubois, D., and H. Prade (1980). Fuzzy Sets and Systems. Theory and Applications. Academic Press, New York.

Gal, T. (1979). Postoptimal Analysis, Parametric Programming and Related Topics. Mc Graw Hill, New York.

Hamacher, H., H. Leberling and H.J. Zimmermann (1978). Sensitivity analysis in fuzzy linear programming. Fuzzy Sets and Syst. 1, 269-281.

Negoita, C.V., and D. Ralescu (1975). Applications of Fuzzy Sets to Systems Analysis. Birkhauser Verlag, Basel.

Oheigeartaigh, M. (1982). A fuzzy transportation algorithm. Fuzzy Sets and Syst. 8, 235-243.

Prade, H. (1980). Operations research with fuzzy data. In P.P. Wang and S.K. Chang (eds.), Fuzzy Sets. Theory and Applications to Policy Analysis and Information Systems. Plenum Press, New York, 155-170.

Simonnard, M. (1973). Linear Programming, vol. 2. Extensions. Dunod, Paris.

Verdegay, J.L. (1983). Transportation problem with fuzzy parameters (in Spanish). Revista de la Real Academia de Ciencias Matematicas, Fisico Quimicas y Naturales de Granada, II, 47-56.

Verdegay, J.L. (1982). Fuzzy mathematical programming. In M.M. Gupta and E. Sanchez (eds.), Fuzzy Information and Decision Processes. North-Holland, Amsterdam, 231-237.

Zimmermann, H.-J. (1976). Description and optimization of fuzzy systems, Int. J. Gen. Syst. 2, 209-215.

# FUZZY PARAMETERS IN OPTIMAL ALLOCATION OF RESOURCES

Jaroslav Ramik[*] and Josef Řimánek[**]

*Research Institute for Economic and Social Development
Revolucni 19, 701 65 Ostrava, Czechoslovakia

**Mining and Metallurgical University
Osvoboditelû  33, 701 00 Ostrava, Czechoslovakia

Abstract. Inequality relations between resources, represented by fuzzy numbers, are investigated in the first part of this paper. The concept of an R-relation and L-relation is introduced. An optimal allocation problem with fuzzy parameters is then stated and by the use of the extension principle a fuzzy optimal solution is defined and investigated. For trapezoidal fuzzy numbers the problem can be transformed into a linear programming one as is demonstrated on a simple example.

Keywords: resource allocation, fuzzy optimization, fuzzy number.

## 1. INTRODUCTION

When dealing with problems of resource allocation, two types of resources may be encountered. The first one, the expended resource (sometimes called used resource or consumption source) will be denoted by "a", while the second one, the resource at disposal (capacity resource or supply level), by "b". The following inequality should take place

$$a \leqslant b \tag{1}$$

In many cases a and b are not known precisely. To deal quantitatively with such imprecision due to the observer (decision maker), one can use the concepts and techniques of fuzzy sets theory. The imprecisely known resources a and b can be expressed quantitatively by means of fuzzy numbers. By a <u>fuzzy number</u>, a normalized convex fuzzy subset of the real line $E^1$ is meant. A fuzzy set $\underset{\sim}{a}$ on $E^1$ is said to be <u>normalized</u> if its membership function $\mu_{\underset{\sim}{a}} : E^1 \to [0,1]$ attains its maximal value 1. By the symbol $[\underset{\sim}{a}]_\tau$ $0 \leqslant \tau \leqslant 1$, we denote the <u>$\tau$-level</u> set of a fuzzy number $\underset{\sim}{a}$, namely

$$[\underset{\sim}{a}]_\tau = \left\{ x \in E^1 ; \mu_{\underset{\sim}{a}}(x) \geqslant \tau \right\}. \tag{2}$$

Let us recall that the convexity of a fuzzy number $\underset{\sim}{a}$ means that the set $[\underset{\sim}{a}]_\tau$ is convex for each $\tau \in [0,1]$. If $\underset{\sim}{a}$ is a fuzzy number expressing an imprecisely known resource, then (2) is the set of all possible values of this resource whose possibility of occurrence is not less than $\tau$. The set of all (normal) fuzzy numbers will be denoted here by $N(E^1)$.

When modelling the resources by fuzzy numbers, $\underset{\sim}{a}$, $\underset{\sim}{b} \in N(E^1)$, the question arises immediately how to compare $\underset{\sim}{a}$, $\underset{\sim}{b}$. It seems natural to consider a fuzzy preference relation between fuzzy numbers, i.e. a relation $\rho: N(E^1) \times N(E^1) \rightarrow [0,1]$. Then, $\rho(\underset{\sim}{a},\underset{\sim}{b}) = \sigma$ means that the fuzzy number $\underset{\sim}{a}$ is "less than or equal to" the fuzzy number $\underset{\sim}{b}$ with grade $\sigma$. The question is what type of fuzzy preference relation should be adopted for $\rho$.

The most direct way is the application of the extension principle to the ordinary relation of inequality "$\leqslant$". Denoting for $a,b \in E^1$

$$\rho_{\leqslant}(a,b) = 1 \quad \text{if} \quad a \leqslant b$$
$$= 0 \quad \text{otherwise} \tag{3}$$

we obtain an ordinary relation on $E^1$ which can be extended to a fuzzy relation on $N(E^1)$ according to the extension principle (see, e.g., Zadeh, 1985; or Orlovski, 1981) by

$$\rho_{\underset{\sim}{\leqslant}}(\underset{\sim}{a},\underset{\sim}{b}) = \sup_{x \leqslant y} \left\{ \min(\mu_{\underset{\sim}{a}}(x), \mu_{\underset{\sim}{b}}(y)) \right\} \tag{4}$$

Consider, for instance, the fuzzy numbers $\underset{\sim}{a}$, $\underset{\sim}{b}$, $\underset{\sim}{c}$, $\underset{\sim}{d}$, $\underset{\sim}{e}$, $\underset{\sim}{f}$ with membership functions depicted in Fig. 1.

According to definition (4) we have

$$\rho_{\underset{\sim}{\leqslant}}(\underset{\sim}{a},\underset{\sim}{b}) = \rho_{\underset{\sim}{\leqslant}}(\underset{\sim}{b},\underset{\sim}{a}) = 1$$

$$\rho_{\underset{\sim}{\leqslant}}(\underset{\sim}{c},\underset{\sim}{d}) = \rho_{\underset{\sim}{\leqslant}}(\underset{\sim}{d},\underset{\sim}{c}) = 1$$

$$\rho_{\underset{\sim}{\leqslant}}(\underset{\sim}{e},\underset{\sim}{f}) = 1, \qquad \rho_{\underset{\sim}{\leqslant}}(\underset{\sim}{f},\underset{\sim}{e}) = h < 1$$

It is evident that the partial ordering on the set of fuzzy members introduced by relation (4) does not have the properties one would believe it should. Thus, for instance, the fuzzy number $\underset{\sim}{c}$ should intuitively be less than $\underset{\sim}{d}$. Still, according to relation (4) $\underset{\sim}{d}$ is "less than or equal to" $\underset{\sim}{c}$ with grade 1. Such a definition of an inequality relation does not seem to be convenient when comparing resources characterized by fuzzy numbers. In the following we present a definition of two different types of fuzzy preference relations which order fuzzy numbers more delicately and which have a natural interpretation in the context of resource allocation.

## 2. FUZZY PREFERENCE RELATIONS

We shall start with the definition of a fuzzy preference relation.

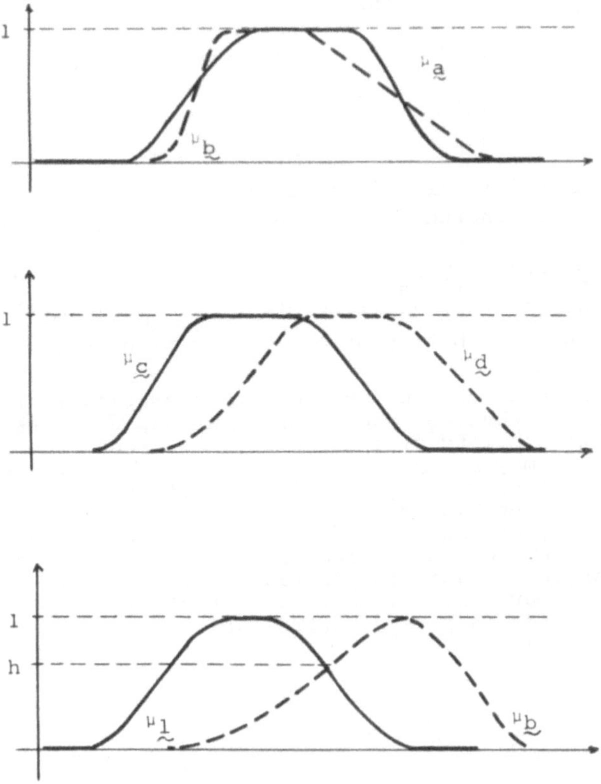


Fig. 1

Definition 1. Let $\underset{\sim}{a}, \underset{\sim}{b} \in N(E^1)$. Then

$$\rho_R(\underset{\sim}{a}, \underset{\sim}{b}) = 0 \quad \text{if} \quad \sup[\underset{\sim}{a}]_1 > \sup[\underset{\sim}{b}]_1$$
$$= 1-\inf\{\tau;\ 0<\tau<1, \forall\lambda \geq \tau :$$
$$\sup[\underset{\sim}{a}]_\lambda < \sup[\underset{\sim}{b}]_\lambda\} \quad \text{otherwise} \qquad (5)$$

The fuzzy preference relation $\rho_R$ is called the right-hand -inequality fuzzy relation (R-relation).

Further

$$\varrho_L(\underset{\sim}{a},\underset{\sim}{b}) = 0 \quad \text{if} \quad \inf[\underset{\sim}{a}]_1 > \inf[\underset{\sim}{b}]_1$$
$$= 1-\inf\left\{\tau \; ; \; 0\leqslant\tau\leqslant 1, \forall\lambda \geqslant \tau \; : \; \inf[\underset{\sim}{a}]_\lambda \leqslant \inf[\underset{\sim}{b}]_\lambda\right\} \quad \text{otherwise.} \tag{6}$$

The fuzzy preference relation $\varrho_L$ is called the <u>left-hand -inequality fuzzy relation (L-relation)</u>.

To put it another way, a fuzzy number $\underset{\sim}{a}$ is "right-hand-less or equal" than a fuzzy number $\underset{\sim}{b}$ with grade $(1-\tau)$ $(0\leqslant\tau\leqslant 1)$ if for any $u\in E^1$ with $\mu_{\underset{\sim}{a}}(u) \geqslant \tau$ there is a $v\in E^1$, $u \leqslant v$, such that the grade of possibility of $v$ in $\underset{\sim}{b}$ is at least as great as the possibility of occurrence of u in $\underset{\sim}{a}$, i.e. $\mu_{\underset{\sim}{a}}(u) \leqslant \mu_{\underset{\sim}{b}}(v)$.

There is a straightforward interpretation in the language of resources. Whatever the value of an expended resource $\underset{\sim}{a}$ with the possibility of occurrence (the grade of possibility) at least $\tau$ may be, there exists a sufficient value of the resource at disposal, $\underset{\sim}{b}$, with at least as great a grade of possibility, see Fig. 2. In such a case the number $\tau$ may serve as an acceptable level of risk or aspiration level on which the decision maker makes his decisions. The above approach can be suitable in practical situations in which the consumed resource is not controllable at all, whereas the capacity resource is partly controllable.

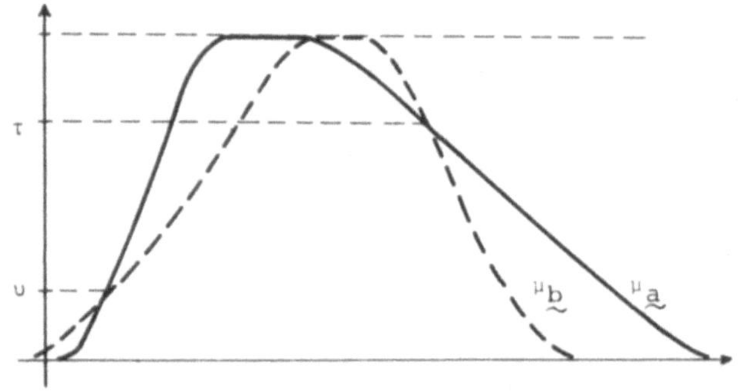


Fig. 2

On the other hand, a fuzzy number $\underset{\sim}{a}$ is "left-hand-side less or equal" than a fuzzy number $\underset{\sim}{b}$ with the grade of possibility $(1-\vartheta)$, if for any $v^-\in E^1$ with the grade of possibility of $v^-$ in $\underset{\sim}{b}$ greater than or equal to $\vartheta$, i.e. $\mu_{\underset{\sim}{b}}(v^-) \geqslant \vartheta$, there exists a $u^-\in E_1$ with $u^- \leqslant v^-$, such that the grade of possibility of $u^-$ in $\underset{\sim}{a}$ is at least as high as that of $v^-$ in $\underset{\sim}{b}$, i.e. $\mu_{\underset{\sim}{a}}(u^-) \geqslant \mu_{\underset{\sim}{b}}(v^-)$ (see Fig. 2).

Whatever the value of a resource at disposal $\underset{\sim}{b}$ on the aspiration level $\nu$ should be, there exists a sufficiently small value of consumption $\underset{\sim}{a}$ of this resource with at least as great a possibility of occurrence. Such a relation of inequality could be adopted in the case of uncontrollable capacity and partly controllable expenditure.

It is possible and it may even be reasonable to consider a combination of the two previous fuzzy preference relations $\rho_R$ and $\rho_L$. Such a combination could take, e.g., one of the following two forms:

$$\rho(\underset{\sim}{a},\underset{\sim}{b}) = \min\left\{ \rho_L(\underset{\sim}{a},\underset{\sim}{b}),\ \rho_R(\underset{\sim}{a},\underset{\sim}{b}) \right\} \tag{8}$$

$$\rho(\underset{\sim}{a},\underset{\sim}{b}) = \max\left\{ \rho_L(\underset{\sim}{a},\underset{\sim}{b}),\ \rho_R(\underset{\sim}{a},\underset{\sim}{b}) \right\} \tag{9}$$

The interpretation of these fuzzy preference relations is evident. For another approach to fuzzy preference relations, see Dubois and Prade (1983).

## 3. NONDOMINATED ELEMENTS

Considering a problem of optimal resource allocation we also come across the necessity of modelling imprecise knowledge about the outcome of a decision, e.g., a profit or a production expense. Expressing, for instance, the profit of two given decisions by fuzzy numbers we encounter once more the problem of comparing two fuzzy quantities.

Let $\rho$ be a fuzzy preference relation on $N(E^1)$, i.e. $\rho : N(E^1) \times N(E^1) \to [0,1]$, and let $\mathcal{X} \subset N(E^1)$ be a set of fuzzy numbers. We shall investigate the problem of finding an element of the set $\mathcal{X}$ (a fuzzy number), which, in a certain sense, presents a rational choice with respect to the fuzzy relation $\rho$.

First, we shall define the strong fuzzy preference relation $\rho^S : N(E^1) \times N(E^1) \to [0,1]$ by the formula

$$\rho^S(\underset{\sim}{a},\underset{\sim}{b}) = \max\left\{ 0,\ \rho(\underset{\sim}{a},\underset{\sim}{b}) - \rho(\underset{\sim}{b},\underset{\sim}{a}) \right\} \tag{10}$$

For any fixed fuzzy number $\underset{\sim}{b}$ the function $\rho^S(\underset{\sim}{a},\underset{\sim}{b})$ given by (10) describes a fuzzy set of fuzzy numbers which are strongly dominated by $\underset{\sim}{b}$ (cf. Orlovski, 1985).

<u>Definition 2</u>. A fuzzy number $\underset{\sim}{z}^* \in \mathcal{X} \subset N(E^1)$ is said to be a <u>$\rho$-nondominated element of</u> $\mathcal{X}$ ($\rho$ - ND element) if there is no $\underset{\sim}{z} \in \mathcal{X}$ with the property

$$\rho^S(\underset{\sim}{z}^*,\ \underset{\sim}{z}) > 0 \tag{11}$$

In the last part of this section we investigate the problem of existence of $\rho$-ND elements, primarily in connection with the fuzzy preference relations $\rho_L$ and $\rho_R$ from Definition 1. The following proposition gives the necessary conditions for the existence of $\varrho_R$ - and $\varrho_L$ - ND elements of $\mathcal{X} \subset N(E^1)$.

Proposition 1. Let $\mathcal{X} \subset N(E^1)$, $\underset{\sim}{z}^* \in \mathcal{X}$. We have:

(a) If $\underset{\sim}{z}^*$ is a $\varrho_R$-ND element of $\mathcal{X}$, then

$$\sup[\underset{\sim}{z}]_1 \leqslant \sup[\underset{\sim}{z}^*]_1 \text{ for each } \underset{\sim}{z} \in \mathcal{X} \tag{12}$$

Conversely, if $\underset{\sim}{z}^*$ is a unique fuzzy number from $\mathcal{X}$ with property (12), then it is a $\varrho_R$-ND element of $\mathcal{X}$.

(b) If $\underset{\sim}{z}^*$ is a $\varrho_L$-ND element of $\mathcal{X}$, then

$$\inf[\underset{\sim}{z}]_1 \leqslant \inf[\underset{\sim}{z}^*]_1 \text{ , for each } \underset{\sim}{z} \in \mathcal{X} \tag{13}$$

Conversely, if $\underset{\sim}{z}^*$ is a unique fuzzy number from $\mathcal{X}$ with the property (13), then $\underset{\sim}{z}^*$ is a $\varrho_L$ - ND element of $\mathcal{X}$.

Proof. Part (a) of the proposition will be proven, the proof of the rest is analogous.
Suppose the existence of a $\bar{\underset{\sim}{z}} \in \mathcal{X}$ with

$$\sup[\underset{\sim}{z}^*]_1 < \sup[\bar{\underset{\sim}{z}}]_1 .$$

According to (5) and the assumption of convexity of $\bar{\underset{\sim}{z}}$ and $\underset{\sim}{z}^*$, we have $\varrho_R^S(\underset{\sim}{z}^*, \bar{\underset{\sim}{z}}) > 0$, a contradiction to (11).
Conversely, let $\underset{\sim}{z}^*$ be a unique element of $\mathcal{X}$ with inequality (12) holding for each $\underset{\sim}{z} \in \mathcal{X}$. Suppose that there is a $\bar{\underset{\sim}{z}} \in \mathcal{X}$ with $\varrho_R^S(\underset{\sim}{z}^*, \bar{\underset{\sim}{z}}) > 0$, i.e.

$$\varrho_R(\underset{\sim}{z}^*, \bar{\underset{\sim}{z}}) > \varrho_R(\bar{\underset{\sim}{z}}, \underset{\sim}{z}^*) \geqslant 0 \tag{14}$$

Due to the uniqueness of $\underset{\sim}{z}^*$, it holds

$$\sup[\underset{\sim}{z}^*]_1 > \sup[\bar{\underset{\sim}{z}}]_1$$

Considering (5), we obtain $\varrho_R(\underset{\sim}{z}^*, \bar{\underset{\sim}{z}}) = 0$, which contradicts (14). Q.E.D.

## 4. FORMULATION OF OPTIMAL ALLOCATION PROBLEM WITH FUZZY PARAMETERS

The problem we shall deal with is the maximization of the payoff (objective ) function

$$\sum_{j=1}^{n} f_j(c_j, x_j) \tag{15}$$

subject to the given constraints

$$\sum_{j=1}^{n} g_{ji}(a_{ij}, x_j) \leqslant b_i \qquad i=1,\ldots,M \tag{16}$$

Here, $f_j(c_j, x_j)$ is the payoff in the j-th cell (independent area in $E^1$) for the employed strategy $x_j$, and $g_{ij}(a_{ij}, x_j)$ is for each $i=1,\ldots,M$ the amount of the i-th cell correspondind to the stragegy $x_j$ employed in this cell, $c_j$, $a_{ij}$ and $b_i$ being vector parameters.

Now, let $\underset{\sim}{c}_j$, $\underset{\sim}{a}_{ij}$, $\underset{\sim}{b}_i$ be fuzzy quantities (fuzzy vectors) describing imprecision of the parameters due to the decision maker, i.e. imperfect knowledge of $c_j$, $a_{ij}$ and $b_i$. For an arbitrary decision vector $x = (x_1,\ldots,x_N) \in E^N$ the extension principle allows the functions $f_j$ abd $g_{ij}$ to be extended for the fuzzy parameters $\underset{\sim}{c}_j$, $\underset{\sim}{a}_{ij}$ and $\underset{\sim}{b}_i$. As a result we obtain fuzzy sets on both sides of the inequality (16) and in (15) denoting the corresponding functions by $\underset{\sim}{f}_j$ and $\underset{\sim}{g}_{ij}$.

Along with problems (15) and (16) with fuzzy parameters we now consider two types of fuzzy preference relations:

$\rho_0 : N(E^1) \times N(E^1) \to [0,1]$ is the payoff fuzzy preference relation,

$\rho_i : N(E^1) \times N(E^1) \to [0,1]$ is the i-th constraint fuzzy preference relation, $i=1,\ldots,M$.

Using the payoff fuzzy preference relation $\rho_0$ we compare fuzzy goals, i.e. fuzzy payoffs or profits, whereas for comparison of fuzzy resources the constraint fuzzy preference relations $\rho_i$ are used. In the part of $\rho_i$, $i=0,1,\ldots,M$, either $\rho_R$ or $\rho_L$ defined in Definition 1 may be used.

Denoting

$$\underset{\sim}{a}_i = \underset{\sim}{g}_{i1}(\underset{\sim}{a}_{i1}, x_1) + \ldots + \underset{\sim}{g}_{iN}(\underset{\sim}{a}_{iN}, x_N) \tag{17}$$

the membership function $\mu_i$ of the fuzzy set of alternatives satisfying the constraint $i$ may be written in the following form

$$\mu_i(x) = \rho_i(\underset{\sim}{a}_i, \underset{\sim}{b}_i) \tag{18}$$

This membership function assigns to each decision $x \in E^N$ the degree to which this decision satisfies constraint $i$.

Now it is natural to accept that $x$ satisfies all the constraints to degree

$$\mu_{\underset{\sim}{C}}(x) = \min_{i=1,\ldots,M} \mu_i(x)$$

or, using (17) and (18),

$$\mu_{\underset{\sim}{C}}(x) = \min_{i=1,\ldots,M}\left\{\boldsymbol{\rho}_i\left(\sum_{j=1}^{n} \underset{\sim}{g}_{ij}(\underset{\sim}{a}_{ij},x_j),\ \underset{\sim}{b}_i\right)\right\}. \tag{19}$$

To put it differently, the fuzzy set $\underset{\sim}{C}$ of all feasible decisions may be defined as the intersection of the fuzzy sets of decisions satisfying the respective constraints.

In what follows, our approach to problems of fuzzy optimization (see Ramik, 1983, 1985; Ramik and Rimanek, 1985) is utilized. For $x = (x_1,\ldots,x_N) \in E^N$ and a crisp set $X \subset E^N$, the following notation is introduced

$$\underset{\sim}{f}(x) = \sum_{j=1}^{N} \underset{\sim}{f}_j(\underset{\sim}{c}_j,x_j) \tag{20}$$

$$\underset{\sim}{f}(X) = \left\{\underset{\sim}{z}\in N(E^1);\ \underset{\sim}{z} = \underset{\sim}{f}(x),\ x\in X\right\} \tag{21}$$

Evidently, $\underset{\sim}{f}(x)\in N(E^1)$, $\underset{\sim}{f}(X)\subset N(E^1)$. Define a multifunction $G : \mathcal{P}(E^N) \to \mathcal{P}(E^N)$, $\mathcal{P}(E^N)$ being a (strict) set of all strict subsets of $E^N$. Let $X \subset E^N$, and

$$G(X) = \left\{ x\in X,\ \underset{\sim}{f}(x) \text{ is a } \boldsymbol{\rho}_0 - \text{ND element of } \underset{\sim}{f}(X)\right\} \tag{22}$$

<u>Definition 3</u>. Let $\boldsymbol{\rho}_0$, respectively, $\boldsymbol{\rho}_i$ $(i=1,\ldots,M)$ be a payoff fuzzy preference relation, respectively, a constraint fuzzy preference relations to problems (15) and (16). Then a <u>fuzzy optimal solution of problems (15) and (16)</u> is a fuzzy set $\underset{\sim}{X}^{OPT}$ defined by the membership function

$$\mu^{OPT}(x) = \max\left\{ 0,\ \sup\left\{\boldsymbol{\gamma}\in [0,1];\ x\in G([\underset{\sim}{C}]_{\boldsymbol{\gamma}})\right\}\right\} \tag{23}$$

where $[\underset{\sim}{C}]_{\boldsymbol{\gamma}}$ is the $\boldsymbol{\gamma}$-level set of fuzzy set $\underset{\sim}{C}$ of all feasible solutions defined by (19).

If $x^* \in E^N$, $\mu^{OPT}(x^*) = \tau$, then according to (20), (21), and (22), the fuzzy number $\sum_{j=1}^{N} \underset{\sim}{f}_j(\underset{\sim}{c}_j,x_j^*)$ is a $\boldsymbol{\rho}_0 - $ND element of the set

$$\left\{ \underset{\sim}{z};u = \sum_j \underset{\sim}{f}_j(\underset{\sim}{c}_j,x_j),\ \mu_{\underset{\sim}{C}}(x) \geqslant \tau,\ x\in E^N\right\} \tag{24}$$

Choosing decision $x^*$, the decision maker knows that his decision is the "best" one on the aspiration level $\tau$ , i.e. it is the most rational decision among all decisions with the grade of feasibility higher than or equal to $\tau$ . When solving the allocation problem, the decision maker chooses a suitable type of fuzzy preference relation $\rho_i$, $i=0,1,\ldots,M$, for instance $\rho_L$ or $\rho_R$, then successively chooses aspiration levels $\tau \in (0,1]$ and solves a sequence of mathematical programming problems the parameters of which are no longer fuzzy. Such a procedure will be demonstrated in the following part of the paper for the case of the so-called trapezoidal fuzzy numbers.

## 5. TRAPEZOIDAL FUZZY NUMBERS

In this section a certain type of fuzzy numbers, called trapezoidal fuzzy numbers, will be dealt with.

<u>Definition 4</u>. A fuzzy number $\tilde{a} \in N(E^1)$ is said to be a <u>trapezoidal fuzzy number</u> if there are four real numbers $m$, $n$, $\alpha$ and $\beta$, $m \leqslant n$, $\alpha \geqslant 0$, $\beta \geqslant 0$, such that

$$\begin{aligned} \mu_{\tilde{a}}(t) &= 1 && \text{for} \quad m \leqslant t \leqslant n, \\ &= 0 && \text{for} \quad t < m-\alpha \quad \text{or} \quad t > n+\beta \\ &= \frac{n+\beta-t}{\beta} && \text{for} \quad n \leqslant t \leqslant n+\beta \ , \quad \beta > 0 \\ &= \frac{t-m+\alpha}{\alpha} && \text{for} \quad m-\alpha \leqslant t \leqslant m, \quad \alpha > 0 \end{aligned} \tag{25}$$

Thus, any trapezoidal fuzzy number $\tilde{a}$ is fully determined by a quadruple of real numbers $m$, $n$, $\alpha$ and $\beta$ . This will be denoted by

$$\tilde{a} = (m,\ n,\ \alpha,\beta).$$

Different types of trapezoidal fuzzy numbers are depicted in Fig. 3. The set of all trapezoidal fuzzy numbers, denoted by symbol $T(E^1)$, is a subset of the set of all (convex and normal) fuzzy numbers $N(E^1)$.

In problems of optimal allocation of resources, parameters and preferences are often evaluated by the help of consulting experts who frequently do not have a fully clear idea in respect to these values. Trapezoidal fuzzy numbers may serve as a flexible and more adequate form of representation of information than traditional crisp numbers. On the other hand, the trapezoidal fuzzy numbers have a comparatively simple structure acceptable to human reasoning. Experience indicates that a more complex structure of fuzzy numbers is usually hardly acceptable by experts. Moreover, the trapezoidal fuzzy numbers include crisp numbers, interval numbers and triangular numbers (see Fig. 3). Due to these facts, the trapezoidal fuzzy numbers are of extreme importance when modelling reality by fuzzy quantities.

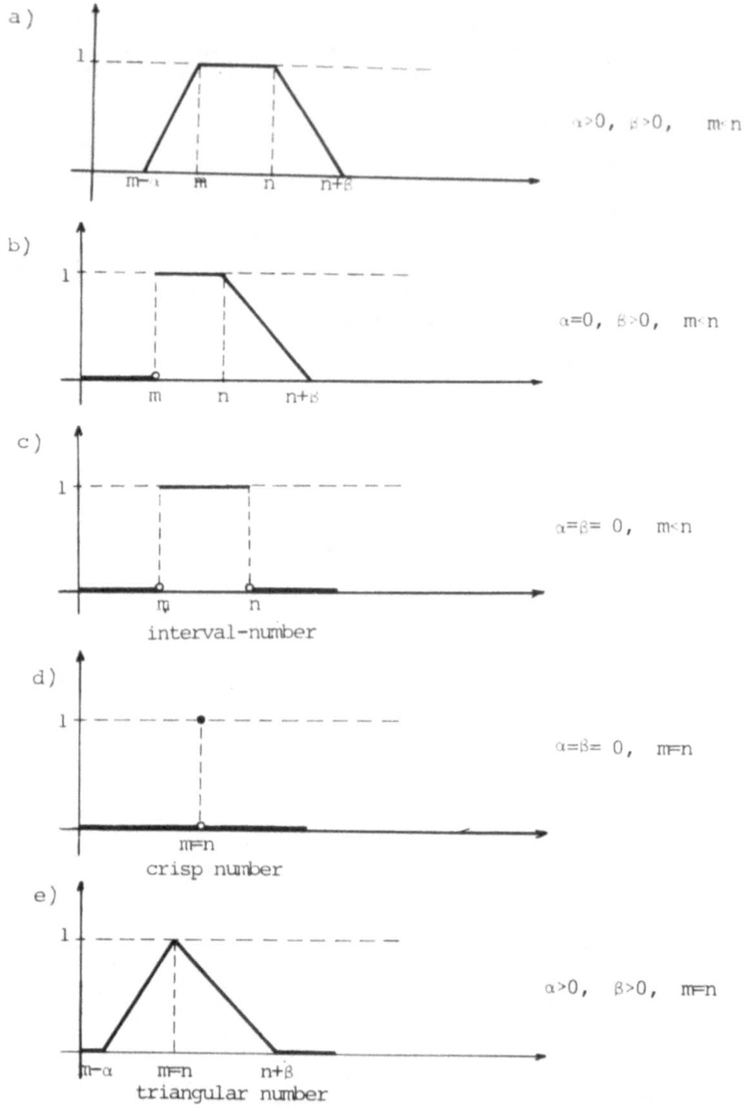
a)
1
m−α
m
n
n+β
α>0, β>0, m<n
b)
1
m
n
n+β
α=0, β>0, m<n
c)
1
m
n
interval-number
α=β= 0, m<n
d)
1
m=n
crisp number
α=β= 0, m=n
e)
1
m−α
m=n
n+β
triangular number
α>0, β>0, m=n

Fig. 3.

In the case of the set of all trapezoidal fuzzy numbers, $T(E^1)$, the fuzzy preference relations $\varrho_R$ and $\varrho_L$ can be expressed in a simple way as the following proposition shows.

Proposition 2. Let $\underset{\sim}{a}$ and $\underset{\sim}{b}$ be trapezoidal fuzzy numbers, $\underset{\sim}{a} = (m_a, n_a, \alpha_a, \beta_a)$, $\underset{\sim}{b} = (m_b, n_b, \alpha_b, \beta_b)$. Then

$$\varrho_R(\underset{\sim}{a},\underset{\sim}{b}) = 0 \quad \text{if} \quad n_a > n_b,$$

$$= \frac{n_b - n_a}{\beta_a - \beta_b} \quad \text{if} \quad n_a \leqslant n_b \quad \text{and} \quad n_a + \beta_a >$$

$$> n_b + \beta_b, \qquad (26)$$

$$= 1 \quad \text{otherwise},$$

$$\varrho_L(\underset{\sim}{a},\underset{\sim}{b}) = 0 \quad \text{if} \quad m_a > m_b,$$

$$= \frac{m_b - m_a}{\alpha_b - \alpha_a} \quad \text{if} \quad m_a \leqslant m_b \quad \text{and} \quad m_a - \alpha_a >$$

$$> m_b - \alpha_b, \qquad (27)$$

$$= 1 \quad \text{otherwise}.$$

The proof follows directly from Definition 1 and is omitted here.

## 6. OPTIMAL ALLOCATION PROBLEM WITH TRAPEZOIDAL FUZZY PARAMETERS

Problems (15) and (16) with trapezoidal fuzzy parameters $\underset{\sim}{c}_j$, $\underset{\sim}{a}_{ij}$ and $\underset{\sim}{b}_i$ will be investigated. To be more specific, we define the concept of a fuzzy vector and introduce some assumptions and notation.

Definition 5. A fuzzy k-vector $\underset{\sim}{a}$ is an ordered k-tuple of fuzzy numbers, i.e. $\underset{\sim}{a} = (\underset{\sim}{a}_1,\ldots,\underset{\sim}{a}_k)$ with $\underset{\sim}{a}_i \in N(E^1)$. The set of all fuzzy k-vectors will be denoted by $N(E^k)$. A fuzzy vector $\underset{\sim}{a} = (\underset{\sim}{a}_1,\ldots,\underset{\sim}{a}_k)$ is said to be trapezoidal if $\underset{\sim}{a}_i \in T(E^1)$, the set of all trapezoidal fuzzy k-vectors denoted by $T(E^k)$.

Evidently, $T(E^k) \subset N(E^k)$. If $\underset{\sim}{a} = (\underset{\sim}{a}_1,\ldots,\underset{\sim}{a}_k)$ and $\underset{\sim}{a}_i = (m_i, n_i, \alpha_i, \beta_i)$, then $\underset{\sim}{a}$ will be denoted by $\underset{\sim}{a} = (m, n, \alpha, \beta)$, $m, n, \alpha, \beta \in E^k$, with $m = (m_1,\ldots,m_k)$, $n = (n_1,\ldots,n_k)$, $\alpha = (\alpha_1,\ldots,\alpha_k)$ and $\beta = (\beta_1,\ldots,\beta_k)$.

From now on the following linear structure of problem (15) and (16) will be considered

$$\underset{\sim}{f}_j(\underset{\sim}{c}_j, x_j) = \underset{\sim}{c}_j f_j(x_j) \qquad (28)$$

$$\underset{\sim}{g}_{ij}(\underset{\sim}{a}_{ij}, x_j) = \underset{\sim}{a}_{ij} \; g_{ij}(x_j) \tag{29}$$

where $\underset{\sim}{c}_j$ and $\underset{\sim}{a}_{ij}$ are trapezoidal fuzzy k-vectors, and for $i = 1, \ldots, M$, $\underset{\sim}{b}_i$ in (16) is a trapezoidal fuzzy number, i.e.

$$\underset{\sim}{c}_j = (m_{cj}, \; n_{cj}, \; \alpha_{cj}, \; \beta_{cj}) \in T(E^k)$$

$$\underset{\sim}{a}_{ij} = (m_{ij}, \; n_{ij}, \; \alpha_{ij}, \; \beta_{ij}) \in T(E^k)$$

$$\underset{\sim}{b}_i = (m_{bi}, \; n_{bi}, \; \alpha_{bi}, \; \beta_{bi}) \in T(E^l)$$

The vector functions

$$f_j : E^1 \to E^k \quad \text{and} \quad g_{ij} : E^1 \to E^k$$

are generally nonlinear.

Such a linear structure of the optimal allocation problem (15) and (16) enables elements of fuzzy optimal solution with the prescribed membership grade to be obtained. To prove our final result the following two lemmas will be needed.

<u>Lemma 1</u>. For $s = 1, \ldots, p$, $p \geqslant 1$, let $\underset{\sim}{a}_s = (m_s, \; n_s, \; \alpha_s, \; \beta_s)$ be trapezoidal fuzzy numbers, $g_s \in E^1$. Then

$$\underset{\sim}{a} = \underset{\sim}{a}_1 g_1 + \underset{\sim}{a}_2 g_2 + \ldots + \underset{\sim}{a}_p g_p \tag{30}$$

is a trapezoidal fuzzy number such that $\underset{\sim}{a} = (m, \; n, \; \alpha, \beta)$ with

$$m = \sum_{s=1}^{p} m_s \; g_s, \qquad n = \sum_{s=1}^{p} n_s \; g_s \tag{31}$$

$$\alpha = \sum_{s=1}^{p} (\alpha_s \; \max\{0, g_s\} - \beta_s \; \min\{0, g_s\}) \tag{32}$$

$$\beta = \sum_{s=1}^{p} (\beta_s \; \max\{0, g_s\} - \alpha_s \; \min\{0, g_s\}) \tag{33}$$

The proof of Lemma 1 can be obtained by a straight forward application of the extension principle. Therefore, it is omitted here.

<u>Lemma 2</u>. Let $0 < \tau < 1$, $\rho_0 = \rho_1 = \ldots = \rho_M = \rho_R$. Then $x \in E^N$ satisfies

$$\sum_{j=1}^{N}\Big\{ n_{ij}q_{ij}(x_j) + \tau\big[\beta_{ij}\max\{0,g_{ij}(x_j)\} - \alpha_{ij}\min\{0,g_{ij}(x_j)\}\big]\Big\} \leqslant n_{bi} + \tau\beta_{bi} \qquad (34)$$

and

$$\sum_{j=1}^{N} n_{ij}\, q_{ij}(x_j) \leqslant n_{bi}, \quad i=1,\ldots,M \qquad (35)$$

if and only if

$$\rho_i\Big(\sum_{j=1}^{N} \tilde{a}_{ij}\, q_{ij}(x_j), \tilde{b}_i\Big) \geqslant \tau\ , \quad i=1,\ldots,M \qquad (36)$$

<u>Proof.</u> Part 1. Let (34) and (35) be valid. Then

$$n_{bi} - \sum_j n_{ij}\, q_{ij}(x_j) \geqslant \tau\Big[\sum_j \beta_{ij}\max\{0,g_{ij}(x_j)\} - \alpha_{ij}\min\{0, g_{ij}(x_j)\}\Big] - \beta_{bi} \qquad (37)$$

for i=1,...,M. Assume that for i=1,...,M

$$\sum_j n_{ij}g_{ij}(x_j) + \sum_j \big[\beta_{ij}\max\{0,g_{ij}(x_j)\} - \alpha_{ij}\min\{0,g_{ij}(x_j)\}\big] > n_{bi} + \beta_{bi} \qquad (38)$$

Applying (35) we obtain

$$\sum_j \big[\beta_{ij}\max\{0,g_{ij}(x_j)\} - \alpha_{ij}\min\{0,g_{ij}(x_j)\}\big] - \beta_{bi} > 0, \quad i=1,\ldots,M \qquad (39)$$

Due to this fact, (37) implies

$$\frac{n_{bi} - \sum_j n_{ij}\cdot g_{ij}(x_j)}{\sum_j \big[\beta_{ij}\max\{0,g_{ij}(x_j)\} - \alpha_{ij}\min\{0,g_{ij}(x_j)\}\big] - \beta_{bi}} > \tau\ , \quad i=1,\ldots,M. \qquad (40)$$

The last inequality along with Proposition 2 and Lemma 1 gives (36).

If (38) does not hold for a certain i, then Proposition 2 indicates that

$$\rho_R\Big(\sum_j \tilde{a}_{ij}\, g_{ij}(x_j), \tilde{b}_j\Big) = 1$$

i.e. inequalities (36) are satisfied too.

Part 2. Let (36) hold and suppose that (39) takes place. By Proposition 2 and Lemma 1, we have (40), and consequently (34) and (35) hold. On the other hand, if the inequality in (39) is not valid for a certain i, then

$$\sum_j \alpha_{ij}\, g_{ij}(x_j) \leqslant \beta_{bi}$$

and

$$\sum_j n_{ij}\, g_{ij}(x_j) + \sum_j [\beta_{ij} \max\left\{0,\, g_{ij}(x_j)\right\} - $$

$$- \alpha_{ij} \min\left\{0,\, g_{ij}(x_j)\right\}] \leqslant n_{bi} + \beta_{bi}$$

Consequently, (34) and (35) are also satisfied. Q.E.D.

Our final result gives the possibility of computing elements of the fuzzy optimal solution set with a prescribed membership grade. Solving a parametric mathematical programming problem, an approximation of the fuzzy optimal solution is obtained which can serve as a basis for a real (nonfuzzy) decision.

Proposition 3. Let $0 < \tau \leqslant 1$, $\varrho_o = \varrho_1 = \ldots = \varrho_M = \varrho_R$. Let $x^* = (x_1^*,\ldots,x_n^*) \in E^N$ be the unique optimal solution of the following mathematical programming problem:

$$\text{maximize} \quad \sum_{j=1}^{N} n_{cj}\, f_j(x_j) \tag{41}$$

subject to (34) and (35). Then

$$\mu^{OPT}(x^*) \geqslant \tau \quad , \tag{42}$$

where $\mu^{OPT}$ is the membership function of the fuzzy optimal solution of the problem (15) and (16).

Proof. Using Lemma 2, the constraints (34) and (35) of the optimization problem (41), can be expressed in the form (36). The rest of the proof follows from Proposition 1, from the observation that

$$\sum_j n_{cj}\, f_j(x_j^*) = \sup\ [\sum_j \underset{\sim}{c}_j\, f_j(x_j^*)]_1$$

and from relations (23), (22), (20) and (21). Q.E.D.

An analogon of Proposition 3 can be formulated for the fuzzy preference relation $\varrho_L$.

## 7. ILLUSTRATIVE EXAMPLE

In the last section we present a simple illustrative example which reflects some characteristic features of the real application of optimal allocation of production of metal in a metallurgical plant on two individual metal producing devices with respect to their individual state. Profit of the overall production is being maximized, parameters of the problem are trapezoidal fuzzy numbers; for more details, see Ramik (1985). The problem is formulated as follows:

$$\text{Maximize}\left\{\underset{\sim}{c}_{11}x_1^2 + \underset{\sim}{c}_{12}x_1 + \underset{\sim}{c}_{21}x_2^2 + \underset{\sim}{c}_{22}x_2\right\} \tag{43}$$

$$\text{subject to: } \underset{\sim}{a}_1x_1 + \underset{\sim}{a}_2x_2 \leqslant \underset{\sim}{b}, \tag{44}$$

$$x_i \geqslant 0, \quad i=1,2$$

where:

$\underset{\sim}{c}_{11} = (-0.3,\ -0.25,\ 0,\ 0)$

$\underset{\sim}{c}_{12} = (\ 0.8,\ 1.0,\ 0,\ 0)$

$\underset{\sim}{c}_{21} = (-0.5,\ -0.44,\ 0,\ 0)$

$\underset{\sim}{c}_{22} = (\ 1.0,\ 1.33,\ 0,\ 0)$

$\underset{\sim}{a}_1 = (2.5,\ 3.0,\ 0.3,\ 0.3)$

$\underset{\sim}{a}_2 = (0.8,\ 1.0,\ 0.3,\ 3.0)$

$\underset{\sim}{b} = (2.0,\ 3.0,\ 1.0,\ 0.3)$

are trapezoidal fuzzy numbers.
Applying Proposition 3, fuzzy optimal solution of problem (43) and (44) is calculated, see Fig. 4, and Table 1 for fuzzy preference relation $\varrho_R = \varrho_o = \varrho_1$. The last column in Tab. 1 contains the right mean values of the respective trapezoidal fuzzy profits.

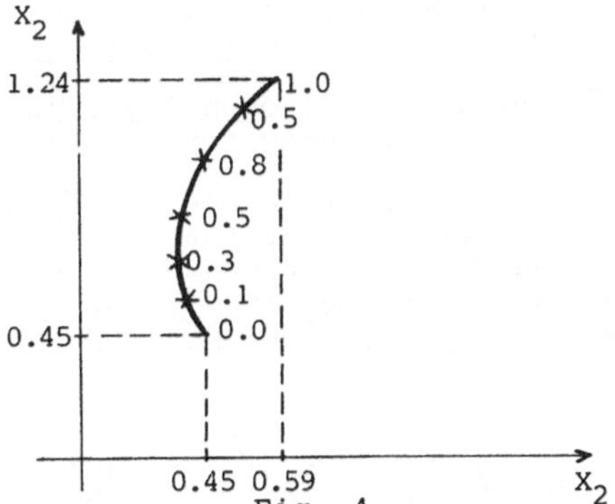


Fig. 4

Table 1

| | $x_1$ | $x_2$ | z |
|---|---|---|---|
| 0.0 | 0.45 | 0.45 | 0.91 |
| 0.1 | 0.43 | 0.50 | 0.94 |
| 0.3 | 0.39 | 0.63 | 1.02 |
| 0.5 | 0.38 | 0.78 | 1.11 |
| 0.8 | 0.45 | 1.05 | 1.31 |
| 1.0 | 0.59 | 1.24 | 1.47 |

REFERENCES


Dubois, D., H. Prade, (1983). Ranking fuzzy numbers in the setting of possibility theory. Inf. Sci. 30, 183-224.

Orlovski, S.A. (1978). Decision-making with a fuzzy preference relation. Fuzzy Sets and Syst. 1, 155-167.

Orlovski, S.A. (1981). Problems of Decision Making with Fuzzy Information (in Russian). Nauka, Moscow.

Orlovski, S.A. (1984). Mathematical Programming Problems with Fuzzy Parameters. Working Paper, IIASA, Laxenburg.

Ramik, J., J. Rimánek (1981). Inequality relation between fuzzy numbers and its use in fuzzy optimization. Fuzzy Sets and Syst. 16, 549-564.

Ramik, J. (1983). Extension principle and fuzzy-mathematical programming. Kybernetika 19, 516-525.

Ramik, J. (1985). An application of fuzzy optimization to optimum allocation of production. Proc. of the International Workshop on Fuzzy Set Applications. Akademie-Verlag, Berlin.

Ramik, J. (1986). Extension principle in fuzzy optimization. Fuzzy Sets and Syst. To appear.

Zadeh, L.A. (1973). The Concept of a Linguistic Variable and its Application to Approximate Reasoning. American Elsevier, New York.

# III. APPLICATIONS

# ANALYSIS OF WATER USE AND NEEDS IN AGRICULTURE THROUGH A FUZZY PROGRAMMING MODEL

Jan W. Owsiński, Sławomir Zadrożny and Janusz Kacprzyk

Systems Research Institute, Polish Academy of Sciences
ul Newelska 6, 01-447 Warsaw, Poland

Abstract. A regional agricultural system is represented and optimized via a two-level linear programming model of significant dimensions. One of its major purposes is to assess the role of water resources in the system with special emphasis on the feasibility of irrigation. Some of the model's data are assumed fuzzy because of the specifics of the problem, i.e. lack of precise knowledge and an appropriate statistical basis, with a simultaneous clear interest in attaining or not exceeding certain predefined levels. The paper presents the model and how fuzziness is represented in it. Results of several runs are shown and commented, related to the use of water, to its significance for the system's overall performance, as well as to the interrelations of water with other crucial resources such as, e.g., capital. The conclusions refer mainly to the features of the system and to the technical and interpretational aspects of the ways fuzziness is represented and manipulated.

Keywords: fuzzy linear programming, fuzzy optimization, agricultural modeling, irrigation.

## 1. INTRODUCTION

This paper is a result of the authors' long involvement in modeling agricultural (mainly regional) systems by using mathematical (mainly linear) programming tools (see e.g., Albegov, Kacprzyk, Orchard-Hays, Owsiński and Straszak, 1982). Part of these efforts (see, e.g., Kacprzyk, and Owsiński, 1984; Kacprzyk, Owsiński and Zadrożny, 1985; Owsiński, Kacprzyk and Zadrożny, 1986) is related to the introduction of fuzziness into the above mentioned linear programming models to better reflect the inherent imprecision of data and relations in agriculture which cannot be adequately dealt with by probabilistic and statistical means due to, e.g., lack of data or subjectivity in the experts' assessments.

In this paper we consider the problem of how to assess the role of water resources in the agricultural system under consideration with particular emphasis on the feasibility and economic justification of irrigation. We assume fuzzy constraint parameters and right hand sides (RHS's) in the capital and water related constraints. We use Wierzchoń, Kindler and Tyszewski's (1986) model of linear programming with fuzzy coefficients and RHS's that make it possible to obtain both a pessimistic and optimistic optimal solution.

In Section 2 we present the model of the considered agricultural system with more rationale for introducing fuzziness. In Section 3 we briefly sketch the approach to fuzzy linear programming to be employed. In Section 4 we present a summary of computational results with the main emphasis on interpretation of the results. In Section 5 some indications are given pertaining to the water use policy in the agricultural system considered.

## 2. A TWO-LEVEL MODEL FOR PLANNING IRRIGATION EXPANSION

The model, whose runs are given as an illustration in this paper, forms a part of a two-level structure of LP models, called SEMORA - see, e.g., Owsiński and Zadrożny (1985), Owsiński and Hołubowicz (1985) - which is in turn a derivation of a one-level LP model for regional agricultural planning, called GRAM (Albegov, Kacprzyk, Orchard-Hays, Owsiński and Straszak, 1982). The two-level structure contains on its lower level a number of LP models which are meant to optimize the production and trade structure in a subregional agriculture. The upper level model coordinates a number of lower-level models, thereby representing a sort of a regional centre, see Fig. 1. In the particular application, whose very partial results are quoted here, the model system was built in order to

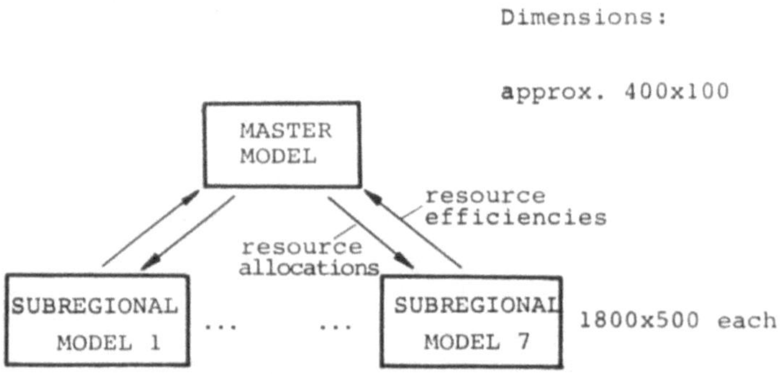


Fig. 1. Structure of the two-level model

assess the economic rationality, dimensions, and ways of utilization of the potential additional water supply obtained via construction of a series of new water system components (reservoirs, pumping stations, irrigation systems, etc.), because the region suffered from inadequate water supply due to a large industrial undertaking within its boundaries.

The large industrial undertaking which is influencing conditions of agricultural development is related to mining and power generation (Owsiński and Hołubowicz, 1985). Its main influences are: diversion of the labour force from agriculture, land use transformation (both of which are not taken up explicitly in this paper), groundwater level decrease, disappearance of certain surface water bodies, and finally, pollution. On the other hand, due to proper regulations, regional agriculture is capable of obtaining large damage compensations which can be used mainly for capital investment purposes. Thus, in view of agriculture under stress and additional financial resources, the following questions arise: How to allocate additional investments? and: Is it rational to replace disappearing water with irrigation?

The present paper highlights a portion of the study meant to answer these and more detailed questions.

The main objective function maximized on both levels is the total net agricultural income. With the help of lower level models, the optimal characteristics with regard to this objective are formed as functions of the allocatable resources. On the basis of these characteristics the upper level model performs a final allocation, thereby defining its rules.

The results given in the paper refer to one of the subregional, i.e. lower level, models. Each such model has approx. 1800 variables and 500 constraints, with an approx. 3% density. These dimensions are related to the necessity of a fairly good description of the system. Hence, a large quantity of items is simultaneously distinguished in each model, e.g., 8 types of crops, 7 types of animals, 5 soil qualities, or 5 farm types. The last example is important for the runs quoted in which a relation between water demand and use and the capital investment availability was analysed. Water used and capital consumed are constrained by a series of constraints, each separately for all farms belonging to a farm type. It is assumed that capital investments shall come from both own resources and from state-given bank credit. The latter position is less certain than the previous one. Besides that there is, albeit of different nature, an uncertainty related to capital intensiveness of activities resulting from moving costs and prices. Similarly, additional water supply is subject to decision-bound uncertainty, while unit water consumptions are climate-dependent.

Farm types (p) distinguished are: p=1; state farms, p=2; cooperative farms, p=3: small private farms, p=4: medium size private farms and p=5: bigger private farms.

The model distinguishes also three types of crop technologies: s=1: traditional technologies with which the present yields are obtained: s=2: intensive technologies: s=3: intensive technologies with irrigation. Livestock breeding can be conducted with two technologies: s'=1: traditional, and s'=2: intensive, i.e. more resource consuming.

The analyses reported here refer primarily to capital for for capital investments and water resource constraints, both seen in their totals and for the 5 farm types distinguished.

In conditions of economic and resource-wise threat to regional agriculture, capital was deemed to be the crucial factor in facilitating changes that could make this agriculture viable. The changes envisaged were of a technological nature, also involving investments directed towards water availability for livestock and for irrigation.

The main reasons behind employment of the fuzzy-set-theoretic approaches were as follows:

a. significant uncertainty as to possibly available outside financing of local capital investments,
b. uncertainty as to possible water resources other than resulting from precipitation,

none of which could be adequately represented in probabilistic or statistical terms, and:

c. cost and price changes which entail uncertainty as to the capital investment requirements for particular activities,
d. in conditions of disappearing rural water resources (in-field ponds, small streams) an uncertainty emerges as to the additional water necessary for particular activities.

Thus, the analysis was performed in two directions:

I. With respect to capital RHS's and capital requirement parameters.
II. With respect to capital and water RHS's and respective requirement parameters.

This division will also be followed in Section 4 which presents the results obtained.

## 2.1. Contents of the lower-level models

The lower-level models are models of subregions defined as contiguous clusters of basic administrative units. The clusters were determined so as to correspond as best as possible to areas in which water resources and more generally - environment - are subject to similar degrees of stress.

The upper-level, i.e. master model, and the coordination procedure shall not be presented here since this paper addresses a different set of issues, methodological and substantial. For more information on the two-level model and its working, see Owsiński and Zadrożny (1986).

All the subregion models have an identical structure. With approx. 3% density and dimensions of 1800 x 500, each such model contains approx. 25000 - 30000 non-zero coefficients, of which some 30% change from submodel to submodel, along with the right hand side values.

The variables distinguished in the subregion models can be classified into the following groups:

* crop raising, i.e. areas under various crops,
* crop product sales,
* crop product consumption,
* crop use for forage,
* crop product purchase,
* livestock breeding, i.e. numbers of particular livestock types,
* livestock product sales,
* livestock product consumption,
* livestock product purchase.

The values that variables contained in the above groups can take are constrained by a number of balances which can be classified into the following groups:

** land availability, total, according to soil qualities, according to crop rotation requirements and second crops,
** crop product balances,
** forage balances,
** consumption balances,
** herd balances,
** livestock product balances,
** resource balances, i.e. availability of:
  - water, annual and for two peak ten-day periods,
  - fertilizers,
  - pulling power, and
  - labour force,
** sales and purchase balances in kind,
** financial balances.

It is out of these groups of constraints that coordination variables are taken to then be included in the coordination procedure, leading to creation and solution of the upper-level model. This aspect, however, shall not be discussed in the present paper.

Thus, it can easily be seen that water availability and consumption questions appear in the model mainly through the water balances and possibility of application of irrigated crop technologies. Since the latter are connected with appropriate parameter values of resource consumption, including current costs and capital investments necessary, the influence and significance of water application spreads over the whole system, thus tainting the solutions obtained.

As mentioned before, a number of fuzziness-generating circumstances influence the water-capital relation. One of the main goals of application of the two-level model is, besides identification of optimal and stable production structures, and place of irrigation therein, also determination of an in-farm capital investment program which would have to parallel the water-system investments into reservoirs, dams, pumps etc. That is why it was deemed important to assess the water-capital relation, with special emphasis on uncertainties which turn out to be of a non-probabilistic and/or of a non-statistical nature, and with distinction of differences among various farm types.

## 3. LINEAR PROGRAMMING WITH FUZZY CONSTRAINT COEFFICIENTS AND RIGHT HAND SIDES

In this section we will briefly present an approach to dealing with fuzzy coefficients and right hand sides in linear programs.

We consider a linear program is the form

$$\{z = f(x)\} \rightarrow \max_{x}$$

subject to:

$$Ax \leqslant b$$

$$x \geqslant 0 \qquad (1)$$

where $x = (x_1,...,x_n)^T \in R^n$, $b = (b_1,...,b_m)^T \in R^m$ and A is an $n \times m$ real matrix.

Since in reality this general formulation may be too rigid, its "softening" has been strongly recommended. As a promising approach, the use of fuzzy sets has been advocated. The first fuzzification, due to Zimmermann, consisted of a fuzzification of strict requirements to "exactly" maximize the objective function and satisfy the constraints. Then some approach appeared (e.g., due to Dubois and Prade) that tried, not always efficiently enough for practical purposes, to account for fuzzy coefficients in the objective function and/or constraints, and in the right hand sides. These latter approaches are more relevant here. For a survey and references, see Kacprzyk and Orlovski's paper earlier in this volume.

In this paper we consider the following LP problem:

$$\begin{cases} cx \longrightarrow \max_{x} \\ \text{subject to:} \\ A_i x \leqslant R_i \qquad i = 1,...,m \\ x \geqslant 0 \end{cases} \qquad (2)$$

where $A_i$ is a row vector of n fuzzy numbers (fuzzy coefficients) corresponding to the particular variables in the i-th constraint, $R_i$ is a fuzzy number representing the fuzzy right hand side of the i-th constraint, and $x = (x_1,...,x_n)^T$. Notice that the objective function is nonfuzzy, and fuzziness is only in the coefficients and right hand sides of the constraints.

We assume the fuzzy numbers to be triangular and use Wierzchoń, Kindler and Tyszewski's (1986) approach to solving problem (2). This approach defines two types of optimal solutions to problem (2), which are feasible to the degree not less than t, namely:

- an optimistic optimal solution to (2) which is obtained by solving

$$\begin{cases} cx \rightarrow \max \\ \text{subject to:} \end{cases}$$

$$\begin{cases} \underline{A}_i x + t(A_i - \underline{A}_i)x \leqslant R_i - t(R_i - \underline{R}_i), \quad i=1,\dots,m \\ x \geqslant 0, \quad t \in [0,1] \end{cases} \tag{3}$$

- a pessimistic optimal solution to (2) which is obtained by solving

$$\begin{cases} cx \to \max \\ \text{subject to:} \\ \overline{A}_i x + t(\overline{A}_i - A_i)x \leqslant \overline{R}_i - t(\overline{R}_i - R_i), \quad i=1,\dots,m \\ x \geqslant 0, \quad t \in [0,1] \end{cases} \tag{4}$$

where the triangular fuzzy numbers are given as $A_i = (\underline{A}_i, A_i, \overline{A}_i)$ and $R_i = (\underline{R}_i, R_i, \overline{R}_i)$. Details on this approach may be found in Wierzchoń, Kindler and Tyszewski (1986), while for some other approaches that are close in spirit, see, e.g., Tanaka, Ichihashi and Asai (1984) or Verdegay (1984).

The two equivalent problems (3) and (4) for determining the solutions sought may be quite efficiently solved by using the parametric programming option in any commercial LP package. This is a serious advantage in view of the large size of our models.

## 4. SUMMARY OF COMPUTATIONAL RESULTS

As indicated before, the results presented here pertain primarily to interrelations between capital investment use and availability on the one hand, and water use and water needs, as, e.g., expressed by dual values, on the other hand. First, only the capital investment constraints are fuzzified, reflecting uncertainty as to both availability of capital (possible credits, own resources) and capital intensity of activities (capital costs of future activity changes). These results are relatively easily interpretable. Besides, capital-related conditions influence not only water use and demand but also a number of other aspects of the local agricultural system.

Thus, having these results facilitates the analysis performed for simultaneous fuzzification of capital and water constraint groups.

In accordance with the methodology adopted, in order to gain a better insight into the field of possible outcomes, runs were carried out for "optimistic" and "pessimistic" formulations of fuzzified constraints.

### 4.1. <u>Fuzzy RHS's and parameters in capital investment constraints</u>

#### 4.1.1. Pessimistic case

Tables 1.A.1 to 1.A.3 illustrate the results of parametrization runs of the model with respect to the constraint satisfaction index α ranging from 1.0, i.e. full satisfaction, through 0.5, i.e. constraint at the limit of satisfaction, down to 0.0, i.e. complete violation, where the constraints analysed were the group of capital investment constraints for various farm types. Tables 1.A.3 and 1.B.3 show how capital availability for and intensity of investments in regional agriculture are related to water use and marginal water value.

Table 1.A.1. Objective function value and total capital investments vs. satisfaction of constraints (see Fig. 2.A.1). Pessimistic case.

| Constraint satisfaction index | Objective function value | Total capital investment disbursements |
|---|---|---|
| 1.Ø* | 3.228.10$^{9}$ zl/year* | 532.49.10$^{6}$ zl/year* |
| Ø.8 | 3.499.10$^{9}$ zl/year | 453.61.10$^{6}$ zl/year |
| Ø.6 | 3.581.10$^{9}$ zl/year | 477.27.10$^{6}$ zl/year |
| Ø.4 | 3.635.10$^{9}$ zl/year | 475.Ø2.10$^{6}$ zl/year |
| Ø.2 | 3.662.10$^{9}$ zl/year | 479.26.10$^{6}$ zl/year |
| Ø.Ø | 3.675.10$^{9}$ zl/year | 481.98.10$^{6}$ zl/year |

* infeasible solution

Table 1.A.2. Capital investments and their duals for particular farm types vs. satisfaction of constraints index (see Fig. 2.A.2). Pessimistic case.

| Constraint satisfaction index | Capital investments/duals for farm types | | | | |
|---|---|---|---|---|---|
| | Farm types | | | | |
| | 1 | 2 | 3 | 4 | 5 |
| 1.0* | 24.12/.19* | 74.95/-5.5* | 129.03/-.6* | 188.98/1.35* | 115.41/.0* |
| 0.8 | 25.57/1.28 | 79.45/.22 | 136.77/2.33 | 200.32/2.33 | 11.50/.0 |
| 0.6 | 27.02/1.25 | 83.95/.12 | 144.51/2.24 | 211.66/2.23 | 10.13/.0 |
| 0.4 | 28.46/.15 | 61.82/.0 | 152.25/1.41 | 223.00/0.94 | 9.94/.0 |
| 0.2 | 17.09/.0 | 59.21/.0 | 159.99/.61 | 234.34/.4 | 8.63/.0 |
| 0.0 | 12.08/.0 | 48.24/.0 | 167.73/.29 | 245.68/.08 | 8.25/.0 |

Capital investment disbursements in $10^{6}$ zl/year

* infeasible solution

Table 1.A.3. Annual water uses and their duals vs. satisfaction of constraints index (see Fig. 2.A.3). Pessimistic case

| onstraint tisfaction index | Water uses in $10^3$ cu. m./duals | | | | | |
|---|---|---|---|---|---|---|
| | Farm types | | | | | |
| | 1 | 2 | 3 | 4 | 5 | Total |
| 1.0* | 33.0/.0* | 184.5/.0* | 479.7/.0* | 326.0/.0* | 427.5/-73.2* | 1450.7* |
| 0.8 | 36.5/.0 | 211.9/77.6 | 211.0/.0 | 519.1/.0 | 62.9/.0 | 1101.4 |
| 0.6 | 60.4/.0 | 211.9/85.0 | 473.2/.0 | 679.0/.0 | 63.2/.0 | 1487.7 |
| 0.4 | 77.2/83.6 | 211.9/78.1 | 504.9/.0 | 712.5/.0 | 63.1/.0 | 1569.6 |
| 0.2 | 77.2/60.0 | 211.9/65.0 | 548.4/.0 | 782.1/21.4 | 63.4/.0 | 1683.0 |
| 0.0 | 77.2/60.0 | 211.9/59.3 | 571.5/26.3 | 782.1/44.7 | 63.4/.0 | 1706.1 |

* infeasible solution

4.1.2. Optimistic case

Tables 1.B.1. to 1.B.3 illustrate the results of parametrization runs of the model for the constraint satisfaction index ranging from 1.0 to 0.0 but assuming the optimistic case

Table 1.B.1. Objective function value and total capital investments vs. constraint satisfaction index (see Fig. 1.B.1). Optimistic case

| Constraint satisfaction index | Objective function value, $10^9$ zl/year | Total capital investment disbursements, $10^6$ zl/year |
|---|---|---|
| 1.0 | 3.674 | 370.75 |
| 0.8 | 3.678 | 355.50 |
| 0.6 | 3.679 | 335.50 |
| 0.4 | 3.679 | 312.63 |
| 0.2 | 3.679 | 289.75 |
| 0.0 | 3.679 | 266.89 |

Table 1.B.2. Capital investments and their duals for particular farm types vs. constraint satisfaction index. optimistic case.

| Constraint satisfaction index | Capital investments in $10^6$ zl/year / duals | | | | |
|---|---|---|---|---|---|
| | Farm types | | | | |
| | 1 | 2 | 3 | 4 | 5 |
| 1.0 | 9.29/.0 | 37.11/.0 | 129.03/.38 | 188.98/.10 | 6.34/.0 |
| 0.8 | 8.37/.0 | 31.46/.0 | 136.77/.15 | 172.94/.0 | 5.96/.0 |
| 0.6 | 7.84/.0 | 29.44/.0 | 130.74/.0 | 161.90/.0 | 5.58/.0 |
| 0.4 | 7.30/.0 | 27.44/.0 | 121.83/.0 | 150.86/.0 | 5.20/.0 |
| 0.2 | 6.77/.0 | 25.43/.0 | 112.91/.0 | 139.82/.0 | 4.82/.0 |
| 0.0 | 6.23/.0 | 23.43/.0 | 104.0/.0 | 128.79/.0 | 4.44/.0 |

Table 1.B.3. Annual water uses and their duals vs. satisfaction of constraints index. Optimistic case

| Constraint satisfaction index | Water uses in $10^3$ cu. m./duals | | | | | |
|---|---|---|---|---|---|---|
| | Farm types | | | | | |
| | 1 | 2 | 3 | 4 | 5 | total |
| 1.0 | 77.2/60.0 | 211.9/59.3 | 571.5/26.3 | 782.1/44.7 | 63.4/0.0 | 1706.1 |
| 0.8 | " / " | " /56.1 | " /36.9 | " /45.6 | " / " | " |
| 0.6 | " / " | " /55.6 | " /46.2 | " /44.5 | " / " | " |
| 0.4 | " / " | " / " | " / " | " / " | " / " | " |
| 0.2 | " / " | " / " | " / " | " / " | " / " | " |
| 0.0 | " / " | " / " | " / " | " / " | " / " | " |

It is important to see the differences of reactions to changes in capital supply and use with respect to water for various farm types. The importance stems from the fact that regional agricultural policies should be shaped differently for those farm types. Data from the tables shown here are plotted in Figs. 2.A.1 (Tables 1.A.1 and 1.B.1), 2.A.2 (Table 1.A.2) and 2.A.3 (Table 1.A.3). Tables 1.B.2 and 1.B.3 were not plotted because of their near-trivial contents. This is due to the fact that the optimistic (B.) formulation yields conditions in which one would have the constraints fully satisfied and simultaneously have enough capital for utilizing efficiently all the water available, in all farm types except for p=5. Note that the passage from optimistic to pessimistic formulation changes this image entirely. Namely in the pessimistic case (A.) only for $\alpha$ nearing 0.0 does there occur full utilization of available water resources in farm types p=1 to 4. It is interesting to see that for $\alpha \geqslant 0.5$ (constraint satisfied) no farm type except p=2 can secure full utilization of available water resources. Thus, some farms in the area may simply not encounter and/or have economic conditions for using additional water.

Note that observation of Fig. 2.A.3 togegher with 2.A.1 conveys quite a significant indication of which way to take in terms of water economy in agriculture in view of the uncertain situation as to the availability and intensity of use of capital. This information can be used in a sort of multiobjective (3 objectives) micro-exercise, which, however, does not have to be necessarily formalized. Indeed, the choice of α can be made on an intuitive basis, with the help of some pre-given precepts. When a certain α, and therefore water use, is chosen, then an optimal production and exchange structure corresponding to it must be checked for conditions of increasing α in order to see its margin of feasibility.

Besides the information as to α, information arises on the marginal values of capital and water and their relations for various farm types. This helps in establishing the water investment programmes and in indicating the possible existence of other factors limiting utilization of water. In the case considered, the foremost example of that is given by farm type p=5 where, evidently, some other type of constraint severely limits the use of water and the propensity to use water. From some other considerations it can be concluded that the constraint in question is labour force availability, as witnessed further on by Figs. 4.1 and 4.2.

## 4.2. Fuzzy RHS's and parameters in capital investment and water availability constraints

Having analysed the intuitively simpler case of fuzzy capital investment constraints let us turn to the case of simultaneous fuzzification of those and the water availability constraints. The optimistic formulation was not looked at because of the already indicated near-triviality of the results then obtained.

Table 2.1. Objective function, total capital investments and total water use vs. constraint satisfaction index (see Fig. 3.1)

| Constraint satisfaction index | Objective function value $10^9$ zl/year | Total capital investment disbursements $10^6$ zl/year | Total water use, annual $10^6$ cu.m. | Total water use, peak 10 days $10^3$ cu.m. |
|---|---|---|---|---|
| 0.5 | 3.595 | 466.86 | 1.979 | 146.49 |
| 0.6 | 3.566 | 471.31 | 1.932 | 142.17 |
| 0.7 | 3.531 | 465.89 | 1.825 | 132.38 |
| 0.8 | 3.489 | 453.61 | 1.539 | 105.73 |
| 0.9 | 3.448 | 443.51 | 1.359 | 88.74 |
| 1.0 | 3.403 | 445.33 | 1.264 | 81.26 |

Table 2.2 Capital investments and their duals for particular farm types vs. constraint satisfaction idex (see Fig. 3.2)

| Constraint satisfaction index | Capital investment disbursements*/their duals for farm types: 1 | 2 | 3 | 4 | 5 |
|---|---|---|---|---|---|
| 0.5 | 27.74/0.20 | 63.72/0.0 | 148.38/1.39 | 217.33/0.98 | 9.69/0.0 |
| 0.6 | 27.02/0.58 | 78.23/0.0 | 144.51/1.98 | 211.66/1.19 | 9.89/0.0 |
| 0.7 | 26.29/1.30 | 81.70/0.19 | 140.64/2.38 | 205.99/2.38 | 11.27/0.0 |
| 0.8 | 25.57/1.31 | 79.45/0.22 | 136.77/2.36 | 200.32/2.36 | 11.50/0.0 |
| 0.9 | 24.85/1.32 | 77.20/0.24 | 132.90/2.35 | 194.65/2.35 | 13.91/0.0 |
| 1.0 | 24.12/2.23 | 74.95/1.12 | 129.03/2.76 | 188.98/2.76 | 28.25/0.0 |

* in $10^6$ zl/year

Table 2.3 Annual water uses, and their duals for particular farm types vs. constraint satisfaction index (see Fig. 3.3)

| Constraint satisfaction index | Annual water uses, in $10^3$ cu. m./their duals for farm types: 1 | 2 | 3 | 4 | 5 |
|---|---|---|---|---|---|
| 0.5 | 88.0/58.3 | 243.6/59.8 | 657.2/54.9 | 899.4/85.6 | 91.5/0.0 |
| 0.6 | 86.5/44.0 | 237.3/43.9 | 640.1/22.7 | 876.0/80.6 | 93.4/0.0 |
| 0.7 | 71.4/0.0 | 230.9/112.9 | 576.2/0.0 | 852.5/0.0 | 95.0/0.0 |
| 0.8 | 56.2/0.0 | 224.6/127.7 | 419.7/0.0 | 741.2/0.0 | 96.9/0.0 |
| 0.9 | 67.6/0.0 | 218.2/126.5 | 339.4/0.0 | 635.6/0.0 | 98.0/0.0 |
| 1.0 | 59.8/0.0 | 211.8/90.2 | 258.7/0.0 | 518.7/0.0 | 213.7/0.0 |

In spite of all the differences between this and the previous case, certain similarities of results can be observed caused by the features of the system analysed. This concerns primarily the behaviour of farm type 5.

Let us first note that the fuzziness of water availability constraints is of a somewhat different nature than that of the capital investment ones. In fact, water availability and use is of a more probabilistic character than capital which is subject to decisions being more often than not made outside of the reach of agricultural producers. The water availability/use relations have their probabilistic (precipitation, temperature etc.) and decision-bound (system construction and current control of within this system) components. Since the latter is to a large degree dependent on investments, it may seem that the first case analysed could be sufficient for making appropriate decisions. The second case, however, where both relations are accounted for via fuzzy representations, is needed primarily for corroboration of results of the first one. When $\alpha$ tends from 1 to the limit of acceptability, i.e. $\alpha=0.5$, not only the RHS's of both groups of constraints become more relaxed but

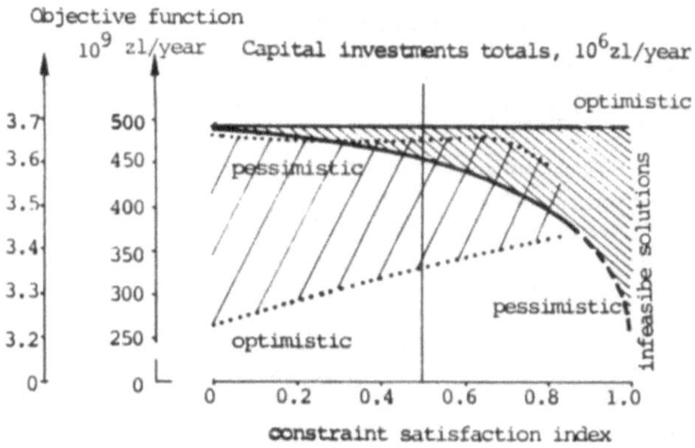


Fig. 2.A.1.

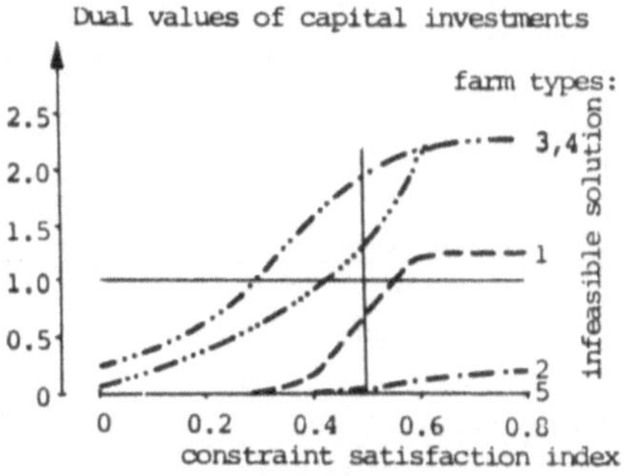


Fig. 2.A.2.

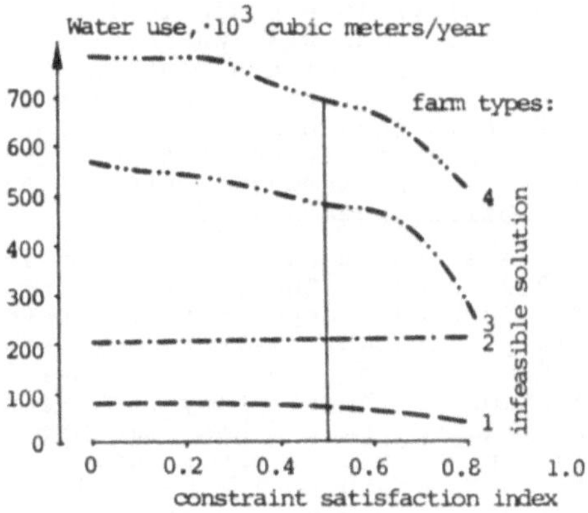


Fig. 2.A.3.

also resource requirements diminish. Thus, more resources are available while they are less needed. This race of the left and right hand sides of resource constraints is not readily interpretable especially inasmuch as the two groups of constraints are closely interrelated.

Thus, turning to the results obtained, Fig. 3.1, which

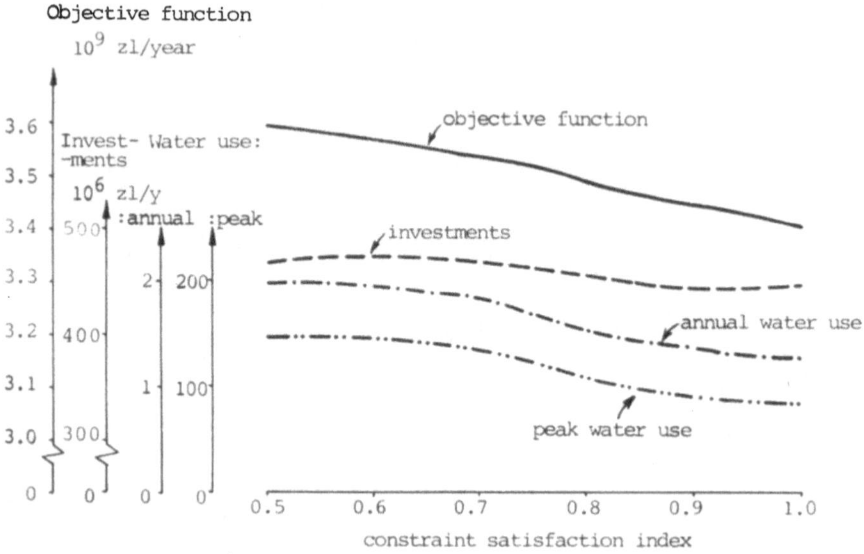


Fig. 3.1.

corresponds to Fig. 2.A.1 and partly to Fig. 2.A.3, shows the same sort of trends, deepened only by the addition of parallel trends in water constraints. This is especially clear from the curves of water use of Figs. 3.1 and 3.3 which are almost the same as in Fig. 2.A.3. The race phenomenon mentioned can be observed on the curve of total investment volumes. The shape of this curve indicated that it is reasonable to plan for α values approximating 0.6 - 0.7. As previously pointed out, stable economic and water use structures obtained for $\alpha \in [0.6, 0.7]$ must be checked for their feasibility when $\alpha \in (0.7, 1.0]$.

A more detailed analysis can be performed using information contained in Figs. 3.2 and 3.3. Thus, for instance, water use in farm types 3 and 4 is highly dependent upon both investment and water availability. Stronger influence is, obviously, exerted by investment since beyond $\alpha=0.7$ water duals are zero. This is, again, an indication of the strategy to be employed. Note that the situation of farm type p=5 has somewhat changed in that it now not only generates some water demand, but also increases this demand for α approaching 1.0 which is, presumably, a result of stricter conditions set on other farm types. Thus, it would be safe, if less paying, to anticipate produc-

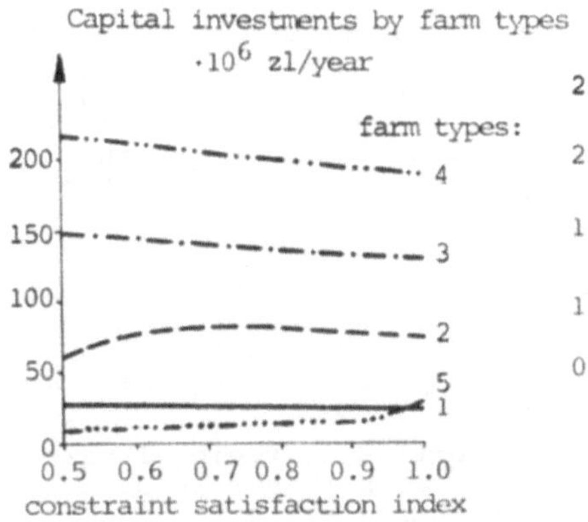


Fig. 3.2.a.

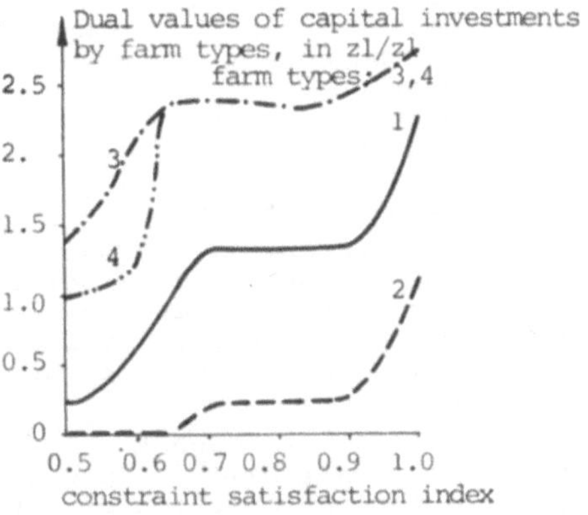


Fig. 3.2.b.

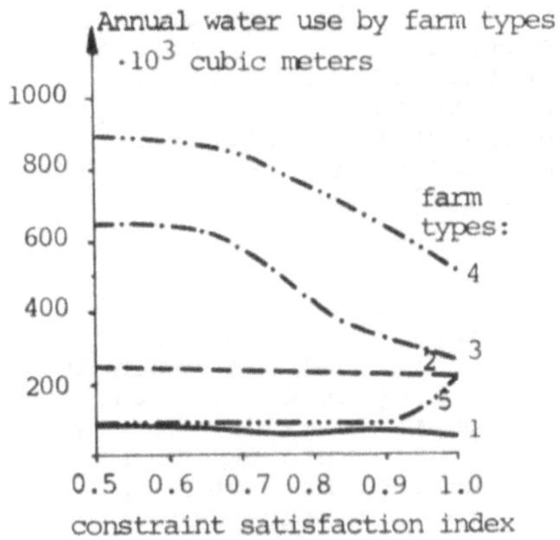


Fig. 3.3.a.

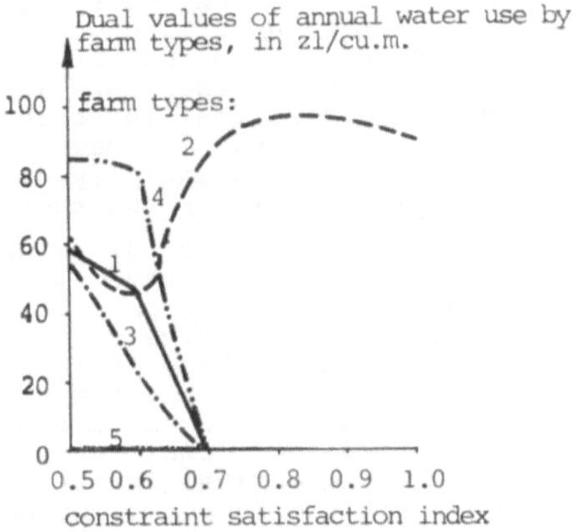


Fig. 3.3.b.

tion structures in which more water will be allocated to and used by farm type p=5.

## 5. POLICY INDICATIONS

Analyses which were exemplified here, performed with the regional agricultural model, are meant, in general, to yield:

* production and trade structures which, while being near to global optimality , could secure robustness and adequate income for particular farm types,
* conditions (economic, resource-wise,etc.) under which these structures could be realized, with special atten-

tion paid to possible bottlenecks, threats or setbacks, as revealed by sensitivity analyses,

* policies which would ensure attainment and maintenance of the conditions advantageous for the appearance of robust near-optimal structures, these policies being both oriented at the inside of the regional agricultural system, e.g., allocations or allocation rules and at the outside of it (points of negotiations with the central authorities).

On these three levels of resolution water resources and their use are given special consideration. The first of these levels is, unfortunately, beyond the possibility of presentation in this short paper, primarily because of the complexity of the model. Only some aspects of the two other levels could be somehow illustrated.

Thus, it was found that in this particular regional case investment availability and uncertainty with this respect is decisive for introduction of water-consuming activities and technologies - except for farm type p=2 where demand realized seems to be limited solely by supply.

Secondly, the analysis performed with regard to uncertainty, and illustrated here, indicates that it is sufficient to check the robustness of optimal structures obtained for satisfaction index $\alpha$=0.7-0.8 since that is where the major structural changes occur. To realize structures corresponding to $\alpha$, say, around 0.6 would not add much to the objective function but would lead to an essentially different economic situation, with a very low possibility and therefore fragility.

Having defined in this manner robust and rationally possible production and trade structures (the runs of the model given here as examples have led to structures with high livestock numbers, intensive grassland farming, irrigation of forage crops, etc., in case of implemented additional water supply), one is obliged to look for appropriate policy rules. For this particular case and area of consideration it turns out, on the basis of some additional data, that the rule could be: to allocate equally capital investment per hectare. Note that this rule is very simple but by no means obvious. In fact, there are some other equally simple and often advocated rules: allocation according to manpower, to existing assets, to productivity or efficiency. The rule formulated is oriented primarily at the economic rationality of water use. Hence, for instance, it can change only indirectly the situation of the farm type p=5 by attracting additional labour force. It is, on the other hand, economically justified to attract labour force to farm type p=5, even at a cost, but this question lies outside of the scope of this paper.

For a hint, however, of the limiting role played by labour force constraints, examine Figs. 4.1 and 4.2, obtained in the same formal setting as Figs. 3.1 - 3, but for significantly increased labour force RHS's in appropriate constraints for farm types p=2,4 and 5. Obviously, the effect is striking with respect to farm type p=5, i.e. bigger farms, which now consume much more capital, use 4 times as much water as in the previous case. At the magnitudes of $\alpha$ above 0.8 very high values of water availability duals appear. Obviously, the structures obtained for p=5 are now entirely different and production is much higher. Influence, although weaker, is also exerted on other farm types.

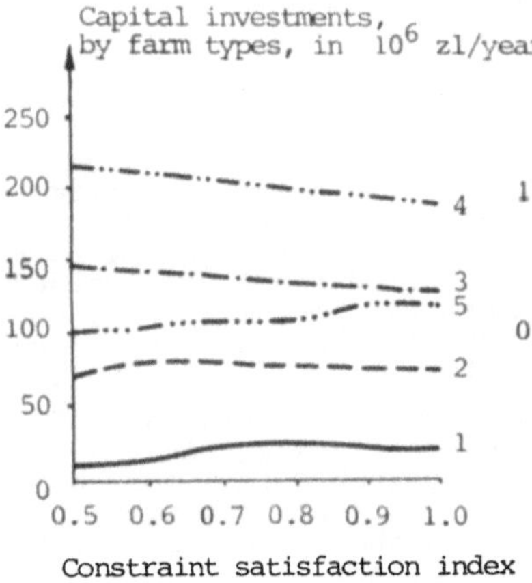


Figure 4.1.a.

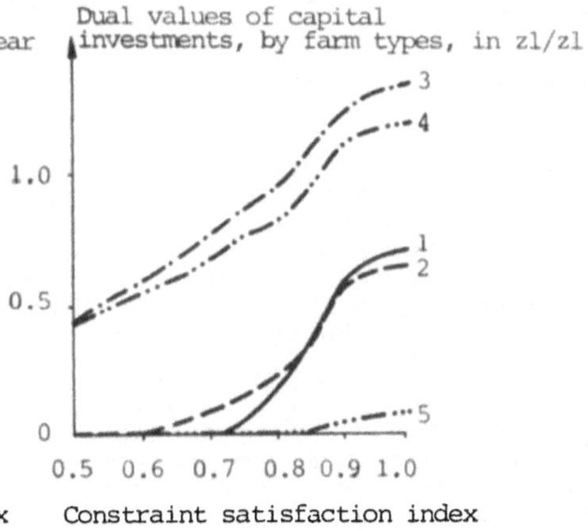


Figure 4.1.b.

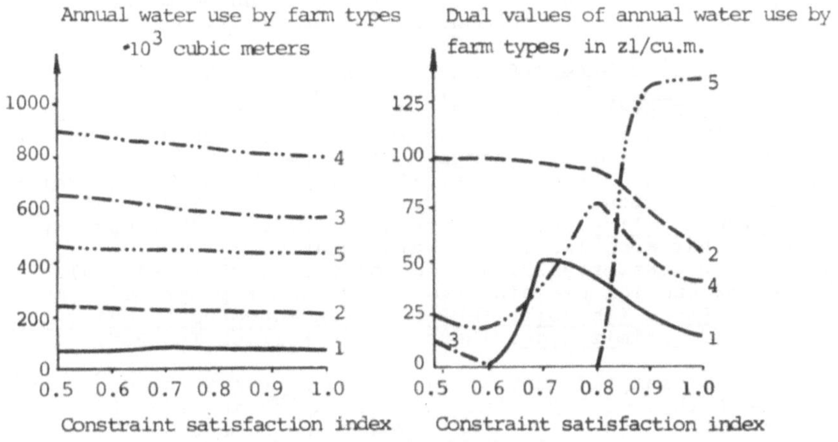


Fig. 4.2.

Thus, by representing some of the uncertainties in the model via the fuzzy-theoretic constructs, additional insight was gained into the potential conditions and courses of action. Certainly, similar conclusions could also be drawn through some other forms of the post-optimal analysis; in the case of fuzzy representations, however, interpretation, though not always straightforward, is deeper and information-richer.

## 6. CONCLUDING REMARKS

In this paper we have presented a summary of results obtained by using a linear programming model with fuzzy constraint coefficients and RHS's for the analysis of some agricultural water problems in a region. As it may be seen from the computational results obtained, and their thorough analysis presented in the tables and figures and commented upon, the fuzzy model assumed gives much insight into many issues of regional water problems in an agricultural context.

In general, the experience gained from this work, as well as some of our previous studies related to it, suggests that fuzzy mathematical programming may be a valuable tool for the analysis and solution of many real world agricultural problems in which there is a pervasive imprecision that cannot be dealt with by using probabilistic/statistical means.

## REFERENCES

Albegov M., Kacprzyk J., Orchard-Hays Wm., Owsiński J.W. and Straszak A. (1982) A general regional agricultural model (GRAM) applied to a region in Poland. Research Report RR-82-26, IIASA, Laxenburg, Austria.

Kacprzyk J. and Owsiński J.W. (1984). Nonstandard mathematical programming models accounting for imprecision as a planning tool for an agricultural enterprise (in Polish) Proc. Conf. on Organization of an Agricultural Enterprise in Varying Conditions. TNOiK, Szczecin.

Kacprzyk J., Owsiński J.W. and Zadrożny S. (1985). Fuzzy linear programming for regional agricultural policy analysis and design. First IFSA Congress, Palme de Mallorca (Spain).

Owsiński J.W. and Hołubowicz K. (1985). A case of energy-related development in an agricultural region. In: Strategic Regional Policy, A. Straszak and J.W. Owsiński, eds. Systems Research Institute, Warsaw, 1985.

Owsiński J.W., Kacprzyk J. and Zadrożny S. (1986). Agricultural production planning via fuzzy linear programming models. Proc. 4th Polish-GDR Symp. on Nonconventional Problems of Optimization. (Warsaw, 1984) Prace IBS PAN, Warsaw.

Owsiński J.W. and Zadrożny S. (1986). On a Practical Non-Iterative Method of Approximate Coordination in Large Scale Linear Systems. Syst. Anal. Model and Simul. 3, 171-181.

Tanaka H., Ichihashi H. and Asai K. (1984). A formulation of fuzzy linear programming problem based on comparison of fuzzy numbers. Control and Cybernetics 13, 185-194.

Verdegay J.L. (1984): A dual approach to solve the fuzzy linear programming problem. Fuzzy Sets and Systems 14, 131-141.

Wierzchoń S.T., Kindler J. and Tyszewski S. (1986). A class of solutions to the fuzzy linear programming (FLP) problems. Fuzzy Sets and Systems (to appear).

# AN INTERACTIVE METHOD FOR MULTIOBJECTIVE LINEAR PROGRAMMING WITH FUZZY PARAMETERS AND ITS APPLICATION TO WATER SUPPLY PLANNING

Roman Słowiński

Institute of Control Engineering
Technical University of Poznań
60-965 Poznań, Poland

Abstract. A multiobjective linear programming (MOLP) problem with fuzzy parameters in the constraints and objective functions in considered. Assuming fuzzy aspiration levels for the particular objectives and involving the comparison of fuzzy numbers, the problem is transformed into a multiobjective linear fractional program. For its solution an interactive method involving a linear programming procedure is proposed. This method has been developed for an application to long-term development planning of a water supply system. It has been first described in Słowiński (1986) and in this paper it is outlined and shown in the context of an updated review of recent proposals for multiobjective fuzzy linear programming.

Keywords: fuzzy multiobjective linear programming, fuzzy aspiration level, fuzzy number, interactive solution, multiobjective linear fractional programming, water supply system.

## 1. INTRODUCTION

A multiobjective linear programming (MOLP) problem with fuzzy parameters (MFLP) can be formulated as

$$\begin{bmatrix} f_1 = \underline{\tilde{c}}_1 \, \underline{x} \\ \cdots\cdots\cdots \\ f_k = \underline{\tilde{c}}_k \, \underline{x} \end{bmatrix} \longrightarrow \min_{\underline{x}} \tag{1}$$

s.t.:

$$\underline{\tilde{a}}_i \, \underline{x} \leqslant \tilde{b}_i \qquad i = 1,\dots,m_1 \tag{2}$$

$$\underline{\tilde{a}}_i \, \underline{x} \geqslant \tilde{b}_i \qquad i = m_1+1,\dots,m_2 \tag{3}$$

$$\underline{x} \geqslant \underline{0} \tag{4}$$

where $\underline{x}$ is a column vector of n decision variables, $\underline{\tilde{c}}_1,\dots,\underline{\tilde{c}}_k$, are row vectors of fuzzy cost coefficients corresponding to the objective functions $f_1,\ldots,f_k$; $\tilde{a}_i$ is the i-th line of the matrix of fuzzy coefficients $\tilde{A}$, and $\tilde{b}_i$ is its corresponding fuzzy right-hand-side. (Remark: the equality constraints $\tilde{a}_i x = \tilde{b}_i$ are not included in the above formulation since they can be obviously represented by pairs of inequality constraints). To complete this formulation we assume that for the particular criteria the decision maker (DM) is in a position to define fuzzy aspiration levels, thought of as goals, denoted by $\tilde{g}_1,\ldots,\tilde{g}_k$. All fuzzy parameters (marked with a tilde) are given as fuzzy numbers (normal and convex fuzzy subsets of the real line).

In Słowiński (1986) a survey of over 30 papers on multiobjective fuzzy linear programming published until 1983 is given. However, in 1984 a couple of new proposals appeared which we should like to mention below.

Most of the existing approaches follow Bellman and Zadeh's (1970) approach to decision making (see Kacprzyk and Orlovski's paper earlier in this volume) in which a fuzzy decision (solution) can be defined as the intersection of fuzzy constraints and fuzzy goals and the problem of finding a maximizing (optimal) decision can be reduced to a nonfuzzy mathematical programming problem. The classical approaches of this kind are due to Negoita and Sularia (1976), and Zimmermann (1978). Since then several extensions have been made. In particular, Leberling (1981) considered hyperbolic, instead of linear, membership functions; Sakawa (1983) considered five types of nonlinear membership functions and proposed a man-machine interactive procedure; Luhandjula (1982) used operators which allow some compensation between the aggregated membership functions; Chanas (1983) and Verdegay (1982) proposed the use of parametric programming for the identification of the complete fuzzy decision; finally, Nakamura (1984) considered piecewise linear membership functions. Another kind of approaches, but still in the framework of Bellman and Zadeh, is represented by fuzzy goal programming proposed by Narasimhan (1980) and then improved by Hannan (1981a, 1981b), Rubin and Narasimhan (1984), etc.

However, in the former approaches, only the goals and the right-hand-sides of the constraints, and in the latter approaches only the goals and their priorities, have been assumed fuzzy. The first approach dealing with fuzziness of parameters in a set of linear constraints is due to Dubois and Prade (1980a); they proposed an interpretation of fuzzy constraints as tolerance constraints involving the inclusion of fuzzy sets, or as approximate equality constraints involving a comparison of fuzzy numbers. By assuming the L-R representation of fuzzy numbers, they reduce the set of fuzzy constraints to a set of linear constraints. Then, Tanaka and Asai (1984), and Tanaka, Ichihashi and Asai (1984) considered fuzzy linear programming with all fuzzy parameters. Involving a comparison of fuzzy numbers, assuming the triangular membership functions, and using Bellman and Zadeh's framework, they reduced the MFLP problem to the conventional linear programming (LP) problem. Recently, Orlovski (1984) presented two approaches to multiobjective programming with fuzzy parameters where solutions are based on trade-offs between the greatest possible degree of nondominance and greatest possible degree of feasibility.

An interactive method for the MFLP problem proposed by Słowiński (1986) is related, in a certain sense, to those in Dubois and Prade (1980a), Orlovski (1984), Tanaka and Asai (1984), and Tanaka, Ichihashi and Asai (1984). Specifically, it uses the L-R type fuzzy numbers whose reference functions need not be linear; the transformation into a nonfuzzy mathematical programming problem is based on the comparison of fuzzy numbers in a different sense that in Dubois and Prade (1980a), Tanaka and Asai (1984), and Tanaka, Ichihashi and Asai (1984), and the solutions ensure the "best consistency" between the goals and the objective functions, and satisfy the constraints with a given "credibility". The nonfuzzy mathematical programming problem is linear fractional - it is solved using an interactive procedure involving linear programming. The method has been successfully applied to long-term development planning of a water supply system. In Section 2 we describe this method, and then we present its application as well as an example.

## 2. DESCRIPTION OF AN INTERACTIVE METHOD

We can easily see that the main question to be answered in MFLP consists in the comparison of the left-hand- and right-hand-sides of the objectives and constraints which are fuzzy numbers. This may be done in different ways. Let us notice first that inequality can be seen as the inclusion of fuzzy sets. Then, various "inclusion grades" can be used as comparison indices (see Dubois and Prade, 1980b; p. II.1.E). There are inclusion grades based on the intersection (or inclusion) and cardinality, and on the inclusion only. Another comparison index based on the cardinality is the relative Hamming distance between two fuzzy sets. All these indices, however, appear to be inefficient in fuzzy mathematical programming since they need the calculation of integrals or infima of special membership functions which, in turn, are functions of the decision variables. Instead, the comparison index in the sense of Zadeh (1965) can be used.

Let $\tilde{m}$ and $\tilde{n}$ be two fuzzy numbers. We want to evaluate the degree of possibility for $x \in R$, fuzzily restricted to belong to $\tilde{m}$, to be greater than $y \in R$, fuzzily restricted to belong to $\tilde{n}$, where R is the real line. Using Zadeh's extension principle, we get

$$\nu(\tilde{m} > \tilde{n}) = \sup_{x,y:\ x \geqslant y} \min(\mu_m(x), \mu_n(y))$$

where $\mu_m$ and $\mu_n$ are the respective membership functions of $\tilde{m}$ and $\tilde{n}$. $\nu(\tilde{m} > \tilde{n})$ is the truth value of "$\tilde{m}$ is greater than $\tilde{n}$". If there exists a pair $(x,y)$, such that $x \geqslant y$ and $\mu_m(x) = \mu_n(y) = 1$, then $\nu(\tilde{m} > \tilde{n}) = 1$.

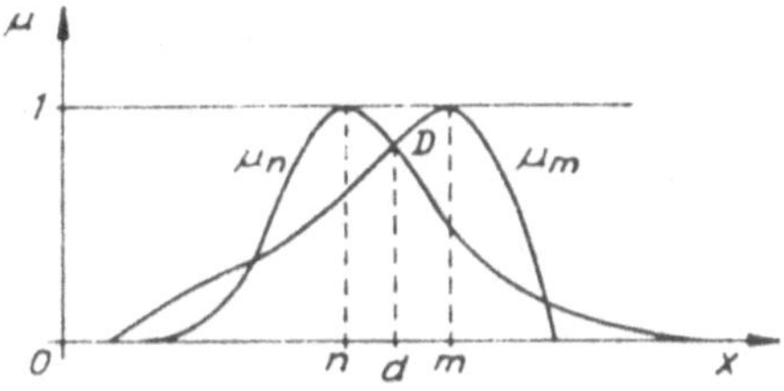


Fig. 1

In the case shown in Fig. 1, it can be easily checked that

$$\nu(\tilde{m} > \tilde{n}) = 1 \Leftrightarrow m \geqslant n$$

$$\nu(\tilde{n} > \tilde{m}) = \mathrm{hgt}(\tilde{m} \cap \tilde{n}) = \mu_m(d) = \mu_n(d), \text{ ordinate of } D$$

where $\mathrm{hgt}(\tilde{m} \cap \tilde{n})$ is the height of $\tilde{m} \cap \tilde{n}$ (see Dubois and Prade, 1980a).

It is shown in Dubois and Prade (1978) that a convenient representation for a fuzzy number $\tilde{m}$ is a triple of parameters $(m,\alpha,\beta)$ of its membership function

$$\mu_m(x) = \begin{cases} L((m-x)/\alpha) & \text{if} \quad x \leqslant m \\ R((x-m)/\beta) & \text{if} \quad x \geqslant m \end{cases}$$

where $m$ is the "mean" value, $\alpha$ and $\beta$ are nonnegative left and right "spreads" of $\tilde{m}$, respectively, and $L,R$ are symmetric bell-shaped reference functions that are non-increasing in $[0,\infty)$, and $L(0) = R(0) = 1$; $\tilde{m}$ is said to be of L-R type, written $\tilde{m} = (m,\alpha,\beta)_{LR}$. When the spreads are zero, $\tilde{m}$ is a nonfuzzy number, by convention.

For computational reasons we shall slightly change this definition. Let $L$ and $R$ be non-increasing function in $(-\infty,\infty)$ and $L(0) = R(0) = 1$, $L(1) = R(1) = 0$. For the same reasons we shall replace $\nu(\tilde{n}>\tilde{m})$ and $\nu(\tilde{m}>\tilde{n})$ by comparison indices $\sigma(\tilde{n}>\tilde{m})$ and $\sigma(\tilde{m}>\tilde{n})$, such that:

(i) $\sigma(\tilde{n}>\tilde{m}) = \nu(\tilde{n}>\tilde{m}) \leqslant 1$ and $\sigma(\tilde{m}>\tilde{n}) \geqslant \nu(\tilde{m}>\tilde{n}) = 1 \Leftrightarrow m \geqslant n$

(ii) $\sigma(\tilde{n}>\tilde{m}) \geqslant \nu(\tilde{n}>\tilde{m}) = 1$ and $\sigma(\tilde{m}>\tilde{n}) = \nu(\tilde{m}>\tilde{n}) \leqslant 1 \Leftrightarrow n \geqslant m$.

The closer $\sigma(\tilde{m}>\tilde{n})$ is to 0, the less true is the assertion "$\tilde{m}$ is greater than $\tilde{n}$". Assuming that $\tilde{m} = (m,\alpha,\beta)_{LR}$ and $\tilde{n} = (n,\alpha,\beta)_{LR}$ obey the modified definition of the reference functions, it can easily be checked that

$$\sigma(\tilde{n}>\tilde{m}) = \begin{cases} R^{-}((d_1-n)/\delta) = L((m-d_1)/\alpha) = \omega, & \text{if } n+\delta \geqslant m-\alpha \\ 0, & \text{otherwise} \end{cases} \tag{5}$$

$$\delta(\tilde{m}>\tilde{n}) = \begin{cases} L^{*}((n-d_2)/\gamma) = R((d_2-m)/\beta) = \Psi, & \text{if } m+\beta \geqslant n-\gamma \\ 0, & \text{otherwise} \end{cases} \qquad (6)$$

where $d_1$ and $d_2$ are abscissae or the intersection points of $R^{*}$ and L, and $L^{*}$ and R, respectively (see Fig. 2). This is equivalent to

$$n+\delta R^{*-1}(\omega) = m-\alpha L^{-1}(\omega)$$

$$n-\gamma L^{*-1}(\Psi) = m+\beta R^{-1}(\Psi)$$

and

$$\delta R^{*-1}(\omega)+\alpha L^{-1}(\omega) = m-n$$

$$\gamma L^{*-1}(\Psi)+\beta R^{-1}(\Psi) = n-m$$

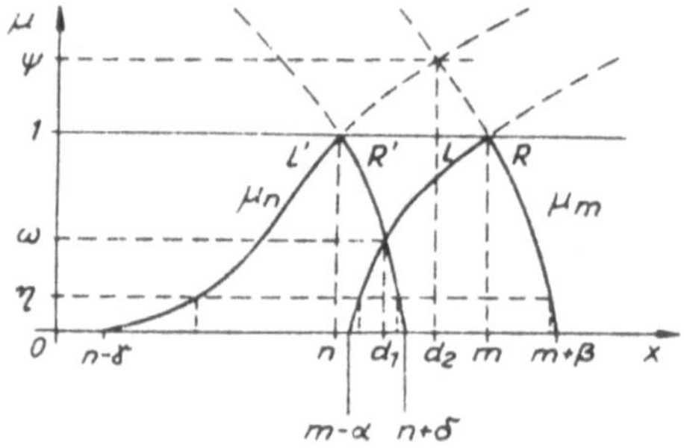


*Fig. 2.*

For $F(\omega) = (\delta R^{*-1}(\omega)+\alpha L^{-1}(\omega))^{-1}$ and $G(\Psi) = (\gamma L^{*-1}(\Psi)+\beta R^{-1}(\Psi))^{-1}$ we have

$$F^{-1}(\omega) = m-n$$

$$G^{-1}(\Psi) = n-m$$

which implies

$$\omega = F(m-n) \qquad (7)$$

$$\Psi = G(n-m) \qquad (8)$$

If $\tilde{m}=(m,\alpha,\beta)_{LR}$ and $\tilde{n}=(n,\gamma,\delta)_{RL}$, then (7) and (8) take the form

$$\omega = L((m-n)/(\alpha+\delta)) \qquad (7a)$$

$$\Psi = R((n-m)/(\beta+\gamma)) \qquad (8a)$$

In order to compare $\tilde{m}$ and $\tilde{n}$, we need both $\sigma(\tilde{m}>\tilde{n})$ and $\sigma(\tilde{n}>\tilde{m})$. If, e.g. $\sigma(\tilde{m}>\tilde{n}) \geqslant 1$, we know that either $\tilde{m}$ is greater than $\tilde{n}$, or both are too close to be separated. Then, we may choose a threshold $0 \leqslant \tau \leqslant 1$ and admit that $\tilde{m}$ is greater than $\tilde{n}$ at level $\tau$ as soon as $\sigma(\tilde{n}>\tilde{m}) \leqslant \tau$. If $\min(\sigma(\tilde{m}>\tilde{n}), \sigma(\tilde{n}>\tilde{m})) \geqslant \tau$, we say that $\tilde{m}$ and $\tilde{n}$ are approximately equal.

In case of the weak inequality, $\tilde{n} \geqslant \tilde{m}$, we only need $\sigma(\tilde{n}>\tilde{m})$. Indeed, for $\sigma(\tilde{n}>\tilde{m}) \geqslant 1$ the inequality is satisfied for any value of $\sigma(\tilde{m}>\tilde{n})$. Then, we may choose a credibility constant $0 \leqslant \tau \leqslant 1$ and admit that $\tilde{n}>\tilde{m}$ at credibility level $\tau$ as soon as $\sigma(\tilde{n}>\tilde{m}) \geqslant \tau$. We can see an important computational advantage from the use of $\sigma$ instead of $\nu$ : since $\tau \geqslant 0$, while calculating $\sigma(\tilde{n}>\tilde{m})$ we need not specify whether $\tilde{n}>\tilde{m}$ or $\tilde{m}>\tilde{n}$, as in case of $\nu(\tilde{n}>\tilde{m})$.

The comparison index $\sigma(\tilde{n}>\tilde{m})$ can be seen, however, as optimistic since it is based on the most favourable case - the intersection of the decreasing slope of $\tilde{m}$ with the increasing slope of $\tilde{n}$. To make the comparison more credible, we could use conjointly a pessimistic index based on an analysis of decreasing slopes of $\tilde{n}$ and $\tilde{m}$ for $\tilde{n} \geqslant \tilde{m}$, and increasing for $\tilde{m} \geqslant \tilde{n}$. For example, we may choose $0 \leqslant \eta \leqslant 1$ and define a pessimistic index as follows (see Fig. 2)

$$\pi(\tilde{n} >_{\eta} \tilde{m}) = n + \delta R^{*-1}(\eta) - m - \beta R^{-1}(\eta) \tag{9}$$

$$\pi(\tilde{m} >_{\eta} \tilde{n}) = m - \alpha L^{-1}(\eta) - n + \gamma L^{*-1}(\eta) \tag{10}$$

For $\eta = 0$ we have

$$\pi(\tilde{n} >_{0} \tilde{m}) = n + \delta - m - \beta$$

$$\pi(\tilde{m} >_{0} \tilde{n}) = m - \alpha - n + \gamma$$

Using the optimistic and the pessimistic index together, we may admit that $\tilde{n} \geqslant \tilde{m}$ at credibility levels $\tau$ and $\eta$ iff

$$\sigma(\tilde{n}>\tilde{m}) \geqslant \tau \tag{11}$$

and

$$\pi(\tilde{n} >_{\eta} m) \geqslant 0 \tag{12}$$

This result can be used directly to transform our MFLP problem into a deterministic one. Let us separately analyse the constraints (2) and (3) and the objective functions.

Fuzzy coefficients of the constraints are given as L-R fuzzy numbers with the modified definition of the reference functions:

$$\tilde{\underline{a}}_i = (\underline{a}_i, \underline{\alpha}_i, \underline{\beta}_i)_{LR}, \quad \tilde{b}_i = (b_i, 0, \delta_i)_{LL}, \quad i=1,\dots,m_1$$

$$\tilde{\underline{a}}_i = (\underline{a}_i, \underline{\alpha}_i, \underline{\beta}_i)_{LR}, \quad \tilde{b}_i = (b_i, \gamma_i, 0)_{RR}, \quad i=m_1+1,\dots,m_2$$

where $\underline{a}_i$, $b_i$ and $\underline{\alpha}_i$, $\underline{\beta}_i$, $\gamma_i, \delta_i$ are the mean values and (non-negative) left and right spreads, respectively; the left spread of $\tilde{b}_i$ for $i=1,\dots,m_1$, and its right spread for

$i=m_1+1,\ldots,m_2$ are zero because they are immaterial for the evaluation of satisfaction of the corresponding constraints. For clarity, we assume that the reference functions of all fuzzy coefficients are of two kinds only, as indicated above. It can be seen from the preceding considerations that this assumption is not necessary for the calculation of $\sigma$ and $\pi$ (cf. (7) and (8) and (9) and (10)).

For any $\underline{x} \geqslant \underline{0}$, $\underline{\tilde{a}}_i\underline{x} = (\underline{a}_i\underline{x},\ \underline{\alpha}_i\underline{x},\ \underline{\beta}_i\underline{x})_{LR}$, $i=1,\ldots,m_2$. For a given $\underline{x} \geqslant 0$ and some hypothetical data, the i-th constraint of type (2) is shown in Fig. 3, and the i-th constraint of type (3), in Fig. 4.

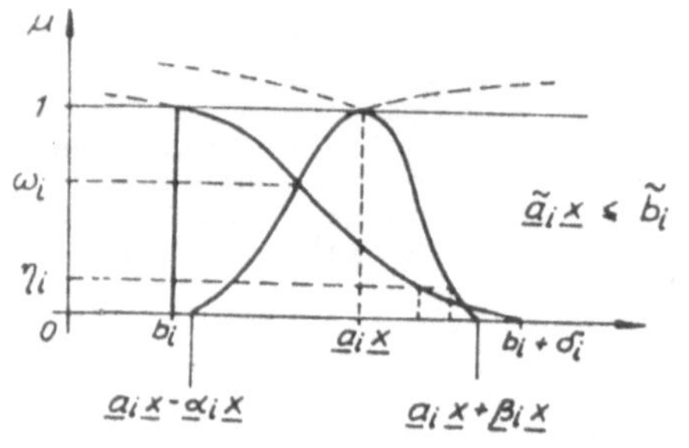


Fig. 3

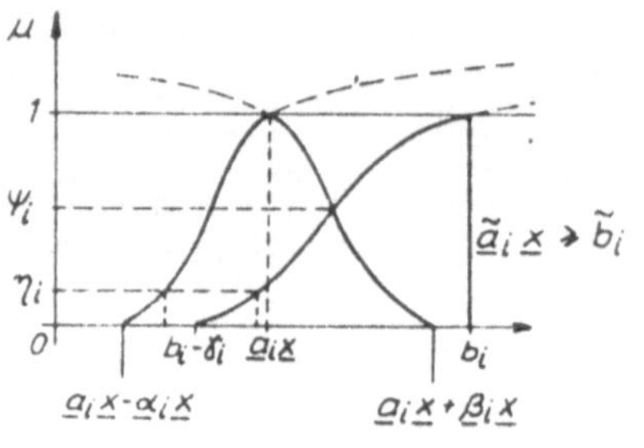


Fig. 4

It is easy to verify that for $i=1,\ldots,m_1$

$$\sigma(\tilde{b}_i > \underline{\tilde{a}}_i\underline{x}) = L((\underline{a}_i\underline{x}-b_i)/(\underline{\alpha}_i\underline{x}+\delta_i)) = \omega_i$$

$$\pi(\tilde{b}_i >_{\eta_i} \underline{\tilde{a}}_i\underline{x}) = b_i-\underline{a}_i\underline{x}+\delta_i L^{-1}(\eta_i)-\underline{\beta}_i\underline{x}R^{-1}(\eta_i)$$

and for $i=m_1+1,\ldots,m_2$

$$\sigma(\underline{\tilde{a}}_i\underline{x} > \tilde{b}_i) = R((b_i-a_i x)/(\underline{\beta}_i\underline{x}+\gamma_i)) = \Psi_i$$

$$\pi(\underline{\tilde{a}}_i\underline{x} >_{\eta_i} \tilde{b}_i) = \underline{a}_i\underline{x}-b_i - \underline{\alpha}_i\underline{x}L^{-1}(\eta_i)+\gamma_i R^{-1}(\eta_i)$$

Then, the DM must select some credibility constants $\eta_i$, $\tau_i \in [0,1]$, $i=1,\ldots,m_2$, and the constraints (2) and (3) may be expressed as:

$$L((\underline{a}_i\underline{x}-b_i)/(\underline{\alpha}_i\underline{x}+\delta_i)) \geqslant \tau_i \qquad i=1,\ldots,m_1 \tag{13}$$

$$R((b_i-\underline{a}_i\underline{x})/(\underline{\beta}_i\underline{x}+\gamma_i)) \geqslant \tau_i \qquad i=m_1+1,\ldots,m_2 \tag{14}$$

$$b_i-\underline{a}_i\underline{x}+\delta_i L^{-1}(\eta_i)-\underline{\beta}_i\underline{x}R^{-1}(\eta_i) \geqslant 0, \quad i=1,\ldots,m_1 \tag{15}$$

$$\underline{a}_i\underline{x}-b_i-\underline{\alpha}_i\underline{x}L^{-1}(\eta_i)+\gamma_i R^{-1}(\eta_i) \geqslant 0, \quad i=m_1+1,\ldots,m_2 \tag{16}$$

If L and R are strictly decreasing, we can transform (13) and (14) into the following equivalents

$$\underline{a}_i\underline{x} - b_i \leqslant L^{-1}(\tau_i)(\underline{\alpha}_i\underline{x} + \delta_i),\ i=1,\ldots,m_1 \tag{17}$$

$$b_i - \underline{a}_i x \leqslant R^{-1}(\tau_i)(\underline{\beta}_i\underline{x} + \gamma_i),\ i=m_1+1,\ldots,m_2 \tag{18}$$

In case of fuzzy objective functions, $\sigma$ and $\pi$ can be used to evaluate a degree of consistency between the fuzzy objectives and fuzzy goals. The fuzzy cost coefficients and goals are assumed to be L-R fuzzy numbers with the modified definition of reference functions:

$$\tilde{\underline{c}}_l = (\underline{c}_l,\ \underline{\varepsilon}_l,\ \underline{\varkappa}_l)_{LR},\quad \tilde{g}_l = (g_l,\ 0,\ \nu_l)_{LL},\ l=1,\ldots,k$$

where $\underline{c}_l$, $g_l$ and $\underline{\varepsilon}_l$, $\underline{\varkappa}_l$, $\nu_l$ are the mean values and (non-negative) left and right spreads, respectively; the left spread of $g_l$, $l=1,\ldots,k$, is zero because it is immaterial for the evaluation of consistency between the goals and objectives. Here again, the equality of the reference functions is not necessary.

For a given $\underline{x} \geqslant 0$ and some hypothetical data, the l-th condition $\tilde{\underline{c}}_l\underline{x} \leqslant \tilde{g}_l$ is shown in Fig. 5.

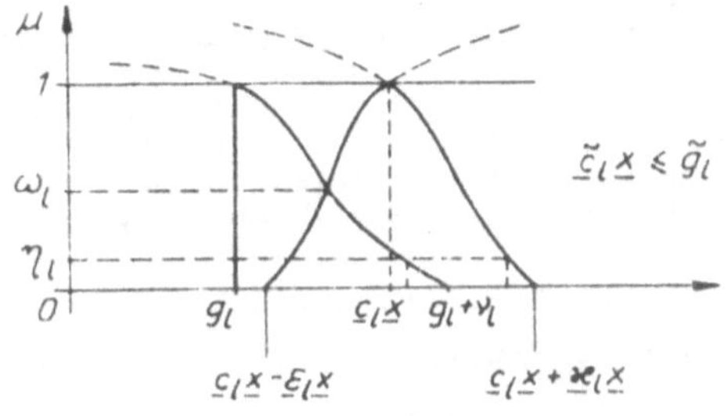


Fig. 5

It is easy to check that for $l=1,\ldots,k$:

$$\sigma(\tilde{g}_l > \tilde{\underline{c}}_l\underline{x}) = L((\underline{c}_l\underline{x} - g_l)/(\underline{\varepsilon}_l\underline{x} + \nu_l)) = \omega_l \tag{19}$$

$$\pi(\tilde{g}_l >_{\eta_l} \tilde{\underline{c}}_l\underline{x}) = g_l - \underline{c}_l\underline{x} + \nu_l L^{-1}(\eta_l) - \underline{\varkappa}_l x R^{-1}(\eta_l) \tag{20}$$

In order to ensure "the best consistency" between the goals and objectives, both these indices should be maximized for given $\eta_l$, $l=1,\ldots,k$. In consequence, we obtain the following non-fuzzy mathematical programming problem equivalent to (1) - (4):

$$\left[\begin{array}{l} z_1(\underline{x}) = L((\underline{c}_1\underline{x}-g_1)/(\underline{\varepsilon}_1\underline{x}+\nu_1)) \\ \cdots\cdots\cdots\cdots\cdots\cdots\cdots \\ z_k(\underline{x}) = L((\underline{c}_k\underline{x}-g_k)/(\underline{\varepsilon}_k\underline{x}+\nu_k)) \\ z_{k+1}(\underline{x}) = g_1-\underline{c}_1\underline{x}+\nu_1 L^{-1}(\eta_1)-\underline{\varkappa}_1\underline{x}R^{-1}(\eta_1) \\ \cdots\cdots\cdots\cdots\cdots\cdots\cdots\cdots\cdots \\ z_{2k}(\underline{x}) = g_k-c_k x+\nu_k L^{-1}(\eta_k)-\underline{\varkappa}_k\underline{x}R^{-1}(\eta_k) \end{array}\right] \rightarrow \max_{\underline{x}}$$

s.t.: (15), (16), (17), (18), and (4).

As is known, solving a multicriteria mathematical programming problem consists of finding a "best compromise" solution. The definition of the best compromise results from the DM's preferences concerning relationships among the criteria. Usually, the best compromise solution is selected from among at least weakly efficient solutions. If we reject an explicit comparison of all efficient solutions, we can use one of two possible approaches (Słowiński, 1984): either (i) to aggregate all the objective functions into a single function defining an overall utility and seek a compromise solution which maximizes this utility, or (ii) to progressively define a compromise through an exploration of the set of feasible solutions guided by the DM which results in a cluster of efficient solutions convergent to the best compromise.

The former has, however, some weak points. A traditionally acknowledged difficulty concerns the process involved in quantifying a relative importance the DM places a priori upon the different objectives. One of the ways to avoid the inherent difficulties of these "static" methods is to use an interactive method representing approach (ii). Our experience in multicriteria project scheduling (Słowiński, 1981) also shows that an interaction with the DM is very beneficial for the final decision.

Let us assume that L is linear, i.e.

$$(\tilde{g}_l > \tilde{\underline{c}}_l\underline{x}) = 1 - (\underline{c}_l\underline{x}-g_l)/(\underline{\varepsilon}_l\underline{x}+\nu_l), \quad l=1,\dots,k$$

Then, we have to solve the following multicriteria linear fractional programming (MLFP) problem:

$$\left[\begin{array}{l} z_1(\underline{x}) = (\underline{c}_1\underline{x}-g_l)/(\underline{\varepsilon}_l\underline{x}+\nu_l) \\ \cdots\cdots\cdots\cdots\cdots\cdots \\ z_k(\underline{x}) = (\underline{c}_k\underline{x}-g_k)/(\underline{\varepsilon}_k\underline{x}+\nu_k) \\ z_{k+1}(\underline{x}) = \underline{c}_1\underline{x}-g_1+\underline{\varkappa}_1\underline{x}R^{-1}(\eta_1) - \nu_1(1-\eta_1) \\ \cdots\cdots\cdots\cdots\cdots\cdots\cdots\cdots\cdots \\ z_{2k}(\underline{x}) = \underline{c}_k\underline{x}-g_k+\underline{\varkappa}_k\underline{x}R^{-1}(\eta_k)-\nu_k(1-\eta_k) \end{array}\right] \rightarrow \min_{\underline{x}}$$

s.t. $\underline{x}\in S$.

To be sure that the denominators of $z_1(\underline{x}),\dots,z_k(\underline{x})$ are positive for any $\underline{x} \in S$, we may admit that $\nu_1 > 0$ which is quite natural. To solve this problem we use the interactive algorithm of Choo and Atkins (1980). Let us describe informally its main idea.

In order to generate (weakly) efficient solutions, we aggregate the objective functions $z_1$ to $z_{2k}$ by the Chebyshev norm which is the maximum weighted deviation from some "ideal" point $\underline{u}^*$:

$$\varphi_l(z_l(\underline{x}) - u_l^*) \longrightarrow \min_{\underline{x} \in S} \max_l$$

or

$$y \rightarrow \min$$

$$\text{s.t.:} \quad y \geq \varphi_l(z_l(\underline{x}) - u_l^*) \quad l=1,\dots,2k$$

$$\underline{x} \in S$$

where $\varphi_l > 0$ are weights chosen a priori.

The point $\underline{u}^*$ is slightly smaller than the minimum of each criterion individually ensuring that no $u_l^*$ is in fact attainable. The isoquants for this norm are the "corners" lying along the line passing through $\underline{u}^*$ with direction $(1/\varphi_1,\dots,1/\varphi_{2k})$. The minimization of the Chebyshev norm chooses a "corner" closest to $\underline{u}^*$ and still in contact with the feasible region. This ensures weak efficiency. To start the algorithm, from the point $\underline{u}^*$ a search direction is chosen by selecting such a $\varphi_l$ which keeps the vertex of the "corner" in the central area of the feasible region. As is shown in Choo and Atkins (1980), a slightly modified form of the Chebyshev norm minimization is reduced in case of MLFP to a linear program with a single parameter. Thus we use any convenient univariate search method over this parameter to find a point as close to an efficient one as we wish. The "closeness" here is not critical and even a rough approximation is quite sufficient. This point is taken as a starting point for the iterative part of the algorithm. This crucially involves the DM. The search direction, and hence the choice of $\varphi_l$, is that from the "ideal" point to the starting point, which is then extended to include some more points.

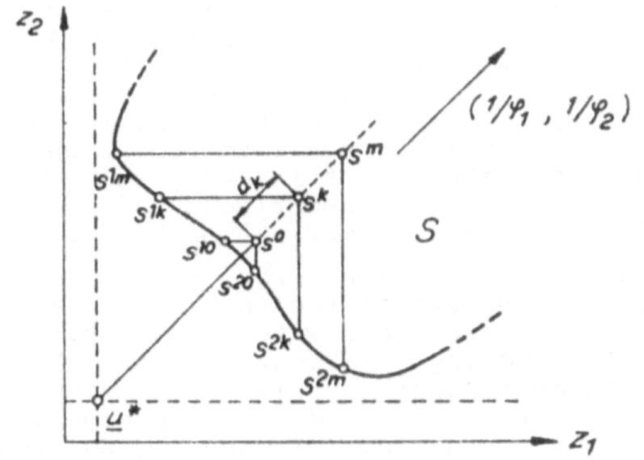


Fig. 6

Thus, in Fig. 6, $s^0$ is the starting point and $s^h$, with h=1,..., ...,m are the extra points. $d_h$ is the distance between $s^0$ and $s^h$. Then, by subsequently taking each criterion, say $z_1$ first, we minimize $z_1$ subject to all other criteria being at most equal to their value at $s^0$, then at $s^1, \ldots,$ until $s^m$. This gives a sequence of (weakly) efficient points $s^{10}, s^{11}, \ldots, s^{1m}$ for criterion 1 and $s^{20}$, $s^{21}$, etc. for criterion 2, etc. The minimization of $z_l(\underline{x})$ for $l=1,\ldots,2k$ is a linear fractional problem with linear constraints, so the Charnes and Cooper (1962) transformation into a linear program can be used. Thus, at each step a single-objective linear programming problem has to be solved. These (weakly) efficient solutions are then presented to the DM who is asked to select the one best fitting his needs.

This becomes the new starting point and the procedure is repeated but with a substantially reduced choice of the distance $d_m$ so that the search space may be focused on most interesting efficient points. The iteration stops when the most satisfactory efficient point is reached.

An important advantage of this algorithm is that the only optimization procedure to be used is linear programming. Moreover, it has a scheme very comprehensible to the DM, and allows reconsideration of points found uninteresting in previous iterations.

Of course, interaction with the DM can be extended to some changes of the right spreads of the goals and the credibility constants.

## 3. APPLICATION TO LONG-TERM DEVELOPMENT PLANNING OF A WATER SUPPLY SYSTEM

The method presented in section 2 has been successfully applied to long-term development planning of a jointly operated urban water supply and wastewater treatment system. The planning problem has been formulated first as a MOLP problem (Słowiński, Urbaniak and Węglarz, 1983) with deterministic coefficients. However, it turned out to be very difficult, if not impossible, to precisely estimate many data over the twenty-year planning horizon. Statistical estimation proved inefficient because of high subjectivity. Even if such statistical characteristics were defined, the obtained multicriteria stochastic LP problem would be too complicated to efficiently solve; anyway, precision of the stochastic model is superfluous and even misleading here. This compelled us to model imprecision by fuzzy sets.

The jointly operated urban water supply and wastewater treatment system consists of the following components: (i) water intakes, i.e. sources and water treatment plants with reservoirs; (ii) recycling treatment plants with reservoirs, i.e. wastewater treatment plants supplying reclaimed water; (iii) a distribution network, i.e. an aggregated pipeline network with pumps; (iv) water users, and (v) discharging treatment plants, i.e. wastewater treatment plants discharging effluents. A twenty-year planning horizon divided into T periods of equal length is assumed. We assume, moreover, that the characteristics of system components may only change stepwise at the border between two successive periods. This means that a new component scheduled for period t can be utilized from the very beginning of this period.

In Słowiński (1986) the following formulation of water supply system development planning in a fuzzy environment has been proposed. (Remark: The symbols used here have local meanings and should not be confused with similar ones in section 2).

Minimize:

$$f_1 = \sum_{t=1}^{T} \sum_{i=1}^{P(t)} (\tilde{c}_{it} x^{+}_{it} \tilde{q}_{t-1} + \tilde{k}_{it} L \sum_{j=1}^{M(t)} s_{ijt} x_{ijt} \tilde{q}_{it})$$

$$f_2 = \sum_{t=1}^{T} \sum_{i=P(t)+1}^{N(t)} (\tilde{c}_{it} x^{+}_{it} \tilde{q}_{t-1} + \tilde{k}_{it} L \sum_{j=1}^{M(t)} s_{ijt} x_{ijt} \tilde{q}_{it})$$

$$f_3 = \sum_{z=1}^{T} \sum_{l=1}^{Q(t)} (c_{lt} y^{+}_{lt} q_{t-1} + k_{lt} L y_{lt} q_{lt}$$

$$f_4 = \sum_{t=1}^{T} \sum_{j=1}^{M(t)} (\tilde{B}_{jt} - \sum_{i=1}^{N(t)} s_{ijt} \tilde{a}_{it} \tilde{a}_{ijt} x_{ijt}) \gamma_{jt} L \tilde{q}_t$$

$$f_5 = \sum_{t=1}^{T} (\tilde{\nu}(\tilde{\beta}_t \sum_{i=1}^{N(t)} x^{+}_{it} - \sum_{i=P(t)+1}^{N(t)} \tilde{\xi}_{it} x^{+}_{it} - \sum_{l=1}^{Q(t)} y^{+}_{lt}) + \tilde{\varkappa} \sum_{l=1}^{Q(t)} y^{+}_{lt})$$

subject to:

$$\sum_{j=1}^{M(t)} s_{ijt} x_{ijt} \leqslant \tilde{D}_{it}, \qquad i=1,\ldots,N(t),\ t=1,\ldots,T$$

$$\sum_{i=1}^{N(t)} s_{ijt} \tilde{a}_{it} \tilde{a}_{ijt} x_{ijt} \geqslant \alpha_{jt} B_{jt}, \quad j=1,\ldots,M(t),\ t=1,\ldots,T$$

$$\sum_{(i,j)\in V_{kt}} x_{ijt} \leqslant \tilde{X}_{kt}, \quad k=1,\ldots,V(t), \quad t=1,\ldots,T$$

$$\sum_{j=1}^{M(t)} (s_{ijt} x_{ijt} - s_{ij(t-1)} x_{ij(t-1)}) - x^{+}_{it} + x^{-}_{it} = 0$$

$$i=1,\ldots,N(t), \quad t=1,\ldots,T$$

$$y_{lt} - y_{l(t-1)} - y^{+}_{lt} + y^{-}_{lt} = 0 \quad l=1,\ldots,Q(t), \quad t=1,\ldots,T$$

where:

$f_1$ - expansion and operating cost of water intakes,
$f_2$ - expansion and operating cost of recycling treatment plants,
$f_3$ - expansion and operating cost of discharging treatment plants,

$f_4$ - loss to users resulting from shortfalls in water supply (reliability of supply),
$f_5$ - environmental quality,
$T$ - planning horizon divided into equal periods, $t=1,\ldots,T$,
$P(t)$ - number of existing and potential intakes in period $t$,
$M(t)$ - number of users in period $t$,
$N(t) - P(t)$ - number of existing and potential recycling treatment plants in period $t$,
$N(t)$ - total number of existing and potential water sources in period $t$,
$Q(t)$ - number of existing and potential discharging treatment plants in period $t$,
$x_{ijt}$ - intensity of water flow between source $i$ (i.e. water intake or recycling treatment plant) and user $j$ in period $t$,
$x_{it}^{+}, x_{it}^{-}$ - positive and negative capacity increment of source $i$ at the border of periods $t$ and $t-1$,
$y_{lt}$ - amount of wastewater delivered to plant $l$ in period $t$,
$y_{lt}^{+}, y_{lt}^{-}$ - positive and negative capacity increment of discharging treatment plant at the border of periods $t$ and $t-1$,
$\tilde{c}_{it}, \tilde{k}_{it}$ - expansion cost per unit capacity increment of source $i$ and its operating cost per unit capacity in period $t$,
$\tilde{c}_{lt}, \tilde{k}_{lt}$ - expansion cost per unit capacity increment of discharging treatment plant $l$ and its operating cost per unit capacity in period $t$,
$L$ - length of one period,
$\tilde{q}_t, \tilde{q}_{it}, \tilde{q}_{lt}$ - discount factor for expansion cost and operating cost of source $i$ and discharging treatment plant $l$ in period $t$, respectively,
$s_{ijt} \in \{0,1\}$ - parameter indicating whether user $j$ can be supplied (= 1) from source $i$ in period $t$, or not (= 0),
$\tilde{B}_{jt}$ - estimated demand of user $j$ in period $t$,
$\alpha_{jt}$ - minimum fraction of $B_{jt}$ which has to be supplied to user $j$ in period $t$ in order to maintain his activity at a tolerable level,
$\tilde{\gamma}_{jt}$ - unitary loss of user $j$ resulting from a shortfall in water supply by one unit of flow per day in period $t$,
$\tilde{D}_{it}$ - maximum capacity of source $i$ in period $t$,
$V(t)$ - number of existing and potential links of the aggregated distribution network in period $t$,
$V_{kt}$ - set of pairs $(i,j)$ such that $s_{ijt}=1$ and link $k$ belongs to the path connecting intake $i$ and user $j$,
$\tilde{X}_{kt}$ - maximum potential capacity of link $k$ in period $t$,
$\tilde{a}_{it}, \tilde{a}_{ijt}$ - reliability coefficients of source $i$ and potential connection of source $i$ and user $j$ in period $t$,
$\tilde{\beta}_t$ - average wastewater discharge rate in period $t$,

$\tilde{\gamma}_{it}$ - ratio of the amount of wastewater entering recycling treatment plant i to the amount of reclaimed water in period t,

$\tilde{\gamma}$ (resp. $\tilde{\varkappa}$) - average index of environmental water pollution following from a direct (resp. after treatment) discharge of a unit of wastewater per time unit.

It is asserted (Słowiński, Urbaniak and Węglarz, 1983) that for each feasible solution, positive and negative capacity increments are mutually exclusive. We assume that the cost and reliability coefficients, maximum capacities, water pollution indices, discount factors and the user´s demands are given as fuzzy numbers.

## 4. EXAMPLE

Let us reconsider a simple urban setting from Słowiński, Urbaniak and Węglarz (1983) presented schematically in Fig. 7, with water supply and wastewater treatment installations to be provided till the end of the planning horizon.

T=4, and the initial state of the system is as follows: users $U_1$ and $U_2$ are supplied from water intake $I_1$ with intensity $x_{110} = 4 \cdot 10^3$ and $x_{120} = 3 \cdot 10^3$, respectively. The

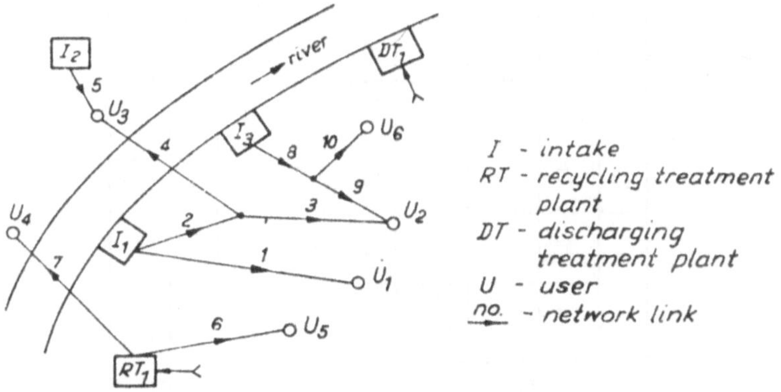


Fig. 7

characteristics of water sources, the distribution network and the water users are listed in Tables 1 - 3, respectively. For fuzzy parameters, only mean values are given in these tables. The mean values for fuzzy reliability coefficients vary from 0.9 to 0.95 for all components of the system. For simplicity, $\gamma_{it}=0$, $i=1,\ldots,P(t)$, and $\gamma_{it}=1$, $i=P(t)+1,\ldots,N(t)$, $t=1,\ldots,4$; $\beta_{jt} = \beta_t = 1$, for each j, t, and $\tilde{q}_{it} = \tilde{q}_t$, for each i, t, where the mean values are: $q_1 = 0.6$, $q_2 = 0.3$, $q_3 = 0.15$, $q_4 = 0.07$.

Table 1

| Parameter | t = 1 | | t = 2 | | | t = 3 | | | t = 4 | | | |
|---|---|---|---|---|---|---|---|---|---|---|---|---|
| | $I_1$ | $I_2$ | $I_1$ | $I_2$ | $RT_1$ | $I_1$ | $I_2$ | $RT_1$ | $I_1$ | $I_2$ | $I_3$ | $RT_1$ |
| $D_{it} \cdot 10^3$ | 10 | 7 | 13 | 7 | 11 | 18 | 9 | 13 | 18 | 9 | 17 | 17 |
| $c_{it} \cdot 10^2$ | 136 | 165 | 136 | 165 | 201 | 136 | 165 | 201 | 136 | 165 | 136 | 201 |
| $k_{it}$ | 1.2 | 1.0 | 1.2 | 1.0 | 1.5 | 1.4 | 1.1 | 1.6 | 1.6 | 1.2 | 1.4 | 1.7 |

Table 2

| No. of link | 1 | 2 | 3 | 4 | 5 | 6 | 7 | 8 | 9 | 10 |
|---|---|---|---|---|---|---|---|---|---|---|
| $x_{kt} \cdot 10^3$ | 12.5 | 10 | 7 | 2 | 7 | 7 | 9 | 11 | 3 | 8 |

Table 3

| t | $U_1$ | | | $U_2$ | | | $U_3$ | | | $U_4$ | | | $U_5$ | | | $U_6$ | | |
|---|---|---|---|---|---|---|---|---|---|---|---|---|---|---|---|---|---|---|
| | $B_{jt}\cdot 10^3$ | $\alpha_{jt}$ | $\gamma_{jt}$ | $B_{jt}\cdot 10^3$ | $\alpha_{jt}$ | $\gamma_{jt}$ | $B_{jt}\cdot 10^3$ | $\alpha_{jt}$ | $\gamma_{jt}$ | $B_{jt}\cdot 10^3$ | $\alpha_{jt}$ | $\gamma_{jt}$ | $B_{jt}\cdot 10^3$ | $\alpha_{jt}$ | $\gamma_{jt}$ | $B_{jt}\cdot 10^3$ | $\alpha_{jt}$ | $\gamma_{jt}$ |
| 1 | 4.5 | 0.9 | 1.2 | 2.5 | 0.8 | 0.3 | 5.0 | 0.8 | 1.3 | - | - | - | - | - | - | - | - | - |
| 2 | 6.0 | 0.9 | 1.2 | 3.5 | 0.8 | 0.3 | 5.5 | 0.8 | 1.3 | 3.5 | 0.85 | 1.2 | 4.0 | 0.9 | 1.1 | - | - | - |
| 3 | 9.0 | 0.9 | 1.2 | 6.0 | 0.86 | 0.3 | 6.0 | 0.9 | 1.3 | 5.0 | 0.9 | 1.3 | 5.0 | 0.9 | 1.2 | - | - | - |
| 4 | 11.0 | 0.9 | 1.3 | 7.0 | 0.85 | 0.4 | 7.0 | 0.95 | 1.4 | 8.0 | 0.9 | 1.4 | 6.0 | 0.95 | 1.2 | 7.0 | 0.85 | 0.3 |

As to the discharging treatment plant $DT_1$, the mean values are: $c_{13}=c_{14}=102\cdot 10^2$ and $k_{13}=1.3$, $k_{14}=1.4$; $\gamma=0.009$ and $\varkappa=0.001$.

The obtained MFLP problem has 84 variables and 116 constraints. All reference functions of fuzzy data are assumed linear. All the left and right spreads range from 5 to 10 per cent except for the goals. The mean values of goals are obtained from the individual minimization of each criterion with non-fuzzy cost coefficients fixed on their mean values:

$$g_1=238\cdot 10^6,\quad g_2=116\cdot 10^6,\quad g_3=0.0,\quad g_4=16\cdot 10^6,\quad g_5=12$$

The maximum values of $f_l$ obtained for the optimal solutions with respect to another criterion are:

$$f_1=381\cdot 10^6,\quad f_2=474\cdot 10^6,\quad f_3=92\cdot 10^6,\quad f_4=84\cdot 10^6,\quad f_5=259$$

Then, the right spreads of the goals are chosen to be 30%, 40%, 20%, and 20% of the respective difference $\bar{f}_l-g_l$, $l=1,\ldots,5$. The credibility constants $\tau_i$ and $\eta_i$ are equal to 0.6 and 0.2, respectively, for all the constraints. The objective functions $z_6(x),\ldots,z_{10}(x)$ are transposed to the constraints with $\eta_l=0$ and the least satisfactory value equal to the right spreads of the corresponding goals.

The results of using the interactive solution method are the following. The ideal solution to the MLFP problem is $\underline{u}^*=(0,0,0,0)$ and the starting point $s^0=(0.61, 0.52, 0.67, 0.7, 0.72)$ which corresponds to $\underline{f}=(264, 171, 24.6, 25.5, 47.4)$ (here and on $10^6$ standing beside the first four criteria is omitted). Assuming m=2 and $d_h=0.1$, we get 10 (weakly) efficient solutions as in matrix $S^1$, and translated into the original criteria in matrix $F^1$.

$S^1$ =

| | | | | |
|---|---|---|---|---|
| 0.45 | 0.6 | 0.73 | 0.76 | 0.8 |
| 0.67 | 0.41 | 0.73 | 0.76 | 0.8 |
| 0.67 | 0.6 | 0.47 | 0.76 | 0.8 |
| 0.67 | 0.6 | 0.73 | 0.52 | 0.8 |
| 0.67 | 0.6 | 0.73 | 0.76 | 0.43 |
| 0.37 | 0.68 | 0.79 | 0.82 | 0.88 |
| 0.73 | 0.31 | 0.79 | 0.82 | 0.88 |
| 0.73 | 0.68 | 0.37 | 0.82 | 0.88 |
| 0.73 | 0.68 | 0.79 | 0.4 | 0.88 |
| 0.73 | 0.68 | 0.79 | 0.82 | 0.32 |

$F^1$ =

| | | | | |
|---|---|---|---|---|
| 257 | 180 | 26.8 | 26.3 | 51.5 |
| 266 | 160 | 26.8 | 26.3 | 51.5 |
| 266 | 180 | 17.3 | 26.3 | 51.5 |
| 266 | 180 | 26.8 | 23 | 51.5 |
| 266 | 180 | 26.8 | 26.3 | 33.2 |
| 253 | 188 | 29 | 27.2 | 55.4 |
| 269 | 149 | 29 | 27.2 | 55.4 |
| 269 | 188 | 13.6 | 27.2 | 55.4 |
| 269 | 188 | 29 | 21.4 | 55.4 |
| 269 | 188 | 29 | 27.2 | 27.8 |

The DM selects the underlined point which becomes the new starting point. Taking m=1 and $d_h=0.05$, we get 5 (weakly) efficient solutions presented in the matrices $S^2$ and $F^2$.

$$S^2 = \begin{bmatrix} 0.59 & 9.63 & 0.76 & 0.79 & 0.45 \\ 0.7 & 0.55 & 0.76 & 0.79 & 0.45 \\ 0.7 & 0.63 & 0.62 & 0.79 & 0.45 \\ \hline 0.7 & 0.63 & 0.76 & 0.67 & 0.45 \\ 0.7 & 0.63 & 0.76 & 0.79 & 0.37 \end{bmatrix} \quad F^2 = \begin{bmatrix} 263 & 183 & 28 & 26.7 & 34.2 \\ 268 & 175 & 28 & 26.7 & 34.2 \\ 268 & 183 & 22.8 & 26.7 & 34.2 \\ \hline 268 & 183 & 28 & 25.1 & 34.2 \\ 268 & 183 & 28 & 26.7 & 30.3 \end{bmatrix}$$

In this illustrative example the DM finds the new underlined solution to be the best compromise.

REFERENCES


Bellman, R.E. and L.A. Zadeh (1970). Decision-making in a fuzzy environment. Mang. Sci. 17, 141-164.

Chanas, S. (1983). The use of parametric programming in fuzzy linear programming. Fuzzy Sets and Syst. 11, 243-251.

Charnes, A. and W.W. Cooper (1962). Programming with linear fractional functionals. Naval Res. Logistics Quart. 9, 181-196.

Choo, E.U. and D.R. Atkins (1980). An interactive algorithm for multicriteria programming. Comput. and Op. Res. 7, 81-87.

Dubois, D. and H. Prade (1978). Operations on fuzzy numbers. Int. J. Syst. Sci. 9, 613-626.

Dubois, D. and H. Prade (1980a). Systems of linear fuzzy constraints. Fuzzy Sets and Syst. 3, 37-48.

Dubois, D. and H. Prade (1980b). Fuzzy Sets and Systems - Theory and Applications. Academic Press, New York.

Hannan, E.L. (1981a). Linear programming with multiple fuzzy goals. Fuzzy Sets and Syst. 6, 235-248.

Hannan, E.L. (1981b). On fuzzy goal programming. Decision Sci. 12, 522-531.

Leberling, H. (1981). On finding compromise solutions in multicriteria problems using the fuzzy min-operator. Fuzzy Sets and Syst. 6, 105-118.

Luhandjula, M.K. (1982). Compensatory operators in fuzzy linear programming with multiple objectives. Fuzzy Sets and Syst. 8, 245-252.

Nakamura, K. (1984). Some extensions of fuzzy linear programming. Fuzzy Sets and Syst. 14, 211-229.

Narasimhan, R. (1980). Goal programming in a fuzzy environment. Decision Sci. 11, 325-336.

Negoita, C.V. and M. Sularia (1976). On fuzzy mathematical programming and tolerances in planning. Econ. Comp. and Econ. Cybern. Stud. and Res. 1, 3-15.

Orlovski, S.A. (1984). Multiobjective programming problems with fuzzy parameters. Control and Cyber. 13, 175-183.

Rubin, P.A. and R. Narasimhan (1984). Fuzzy goal programming with nested priorities. Fuzzy Sets and Syst. 14, 115-129.

Sakawa, M. (1983). Interactive fuzzy decision making for multiobjective linear programming problems and its application. In E. Sanchez and M.M. Gupta (eds.), Fuzzy Information, Knowledge Representation and Decision Analysis. Pergamon Press, Oxford, 293-298.

Słowiński, R. (1981). Multiobjective network scheduling with efficient use of renewable and nonrenewable resources. Eur. J. Op. Res. 7, 265-273.

Słowiński, R. (1984). A review of multiobjective linear programming methods (in Polish). Przegląd Statystyczny 31; Part I - no. 1-2, Part II - no, 3-4.

Słowiński, R. (1986). A multicriteria fuzzy linear programming method for water supply system development planning. Fuzzy Sets and Syst. To appear.

Słowiński, R., A. Urbaniak and J. Węglarz (1983). Multicriteria capacity expansion planning of a water supply and wastewater treatment system. In S.G. Tzafestas and M.H. Hamza (eds.), Advances in Modelling, Planning, Decision and Control of Energy, Power and Environmental Systems. Acta Press, Zürich, 275-278.

Tanaka, H. and K. Asai (1984). Fuzzy linear programming with fuzzy numbers. Fuzzy Sets and Syst. 13, 1-10.

Tanaka, H., H. Ichihashi and K. Asai (1984). A formulation of fuzzy linear programming problem based on comparison of fuzzy numbers. Control and Cyber. 13, 185-194.

Verdegay, J.L. (1982). Fuzzy mathematical programming. In M.M. Gupta and E. Sanchez (eds.), Fuzzy Information and Decision Analysis. North-Holland, Amsterdam.

Zadeh, L.A. (1965). Fuzzy sets. Information and Control 8, 338-353.

Zimmerman H.J. (1978). Fuzzy programming and linear programming with several objective functions. Fuzzy Sets and Syst. 1, 45-55.

# FUZZY EVALUATION OF PARETO POINTS AND ITS APPLICATION TO HYDROCRACKING PROCESSES

Michael Wagenknecht and Klaus Hartmann

Technische Hochschule "Carl Schorlemmer"
Leuna-Merseburg, German Democratic Republic

Abstract. In this paper we deal with polyoptimal decision making in chemical engineering. Using information on the pairwise importance of the performance criteria we construct a global rank-ordering which permits us to evaluate a given set of Pareto points. A modification of Saaty's ratio scale method is given. The case of fuzzy weights is discussed. The developed method has been applied to the investigation of hydrocracking processes as well as to magnetic tape production.

Keywords: polyoptimization, Pareto point, rank ordering, Saaty's ratio scale method, fuzzy weight.

## 1. INTRODUCTION

The method of polyoptimization has found during recent years increasing application for the solution of problems in chemical engineering. This is because, besides the increasing investment capital of the process systems and sales of feedstock and energy, by quantities, as well as the necessity of evaluation of various system properties as, e.g., the quality of utilization of raw materials and energy, costs, reliability and air and water pollution, of higher demands on technology and products which are subjected to essential dynamics. The changes of sales for raw materials, products, energy, equipments, new processes and automation systems influence the structure and control of the process under consideration in such a way that, for the determination of an optimal structure and control, respectively, we have to take into account a set of performance criteria. The demand for flexibility and global optimality includes further uncertainties being present in modelling and evaluation because of unknown parameters as well as necessary simplifications of the models.

The following two problems are typical for the optimization of process systems in chemical engineering:

1. Choice of an optimal structure of a process system (i.e. elements and their connections). This problem may include the design of a new process or the reconstruction of an existing one.

2. Choice of an optimal production strategy (i.e. optimal

planning and control).

It often happens that we have a great number of favoured variants in both cases. Nowadays we are able to get a certain set of Pareto-optimal points for a large number of problems. The user is faced with the problem of choosing some "best" variants. Since the Pareto-points are of equal "goodness" within the given partial order, we need further information for making this choice. Often, information is available on the importance of performance criteria (e.g., quality may be of greater significance than energy costs, etc.). Nevertheless, we may have to allow for difficulties when obtaining a global rank-ordering in particular in the case of a greater number of criteria. On the other hand, it often seems to be much easier to evaluate pairwise importance of two criteria. This is very often done verbally, e.g., in the form "the first criterion is much more important than the second one". Saaty (1978) has developed a ratio scale method to get a global rank-ordering from such information.

In the second section we present this method in short with a modification leading to numerical advantages. Moreover, we consider the case of fuzzy weights and deduce a mathematical apparatus. In the third section we show how to use these results of the evaluation of a set of Pareto-points. The following two sections are devoted to concrete applications. We consider various structures for hydrocracking processes in crude-oil distillation whereby the questionned experts are evaluated on their part by higher level authorities. In a second application we investigate performance criteria in the magnetic tape production.

## 2. METHODS FOR GLOBAL RANK-ORDERING

Assume that n performance criteria $Y_1,\dots,Y_n$ and a set of statements on the pairwise comparison of these criteria, e.g., in the form

"$Y_1$ is more important than $Y_2$"

"$Y_1$ is as important as $Y_3$"

....................

"$Y_{n-1}$ is much more important than $Y_n$"

are given.

There are several possibilities of grasping these statements quantitatively and we are going to consider two of them.

On the one hand, we can construct a "scale of importance" as has been done by Saaty (1978). The statement "$Y_1$ is related to $Y_j$" is evaluated by a positive number $a_{ij} \in (0,9]$ with the following graduation:

$a_{ij}=1$ - equal importance

$a_{ij}=3$ - weakly more important

$a_{ij}=5$ - more important

$a_{ij}=7$ - demonstratively more important

$a_{ij}=9$ - absolutely more important.

while numbers in-between denote intermediate cases.

This scale has been found the best in comparison with other ones. From the $a_{ij}$ we determine the weights $w_i$ (with a convenient normalization, e.g., $\Sigma\, w_i = 1$) expressing the rank -ordering of the $Y_i$. On the other hand, we can use fuzzy numbers instead of crisp ones. The resulting weights are also fuzzy then. This method is more convenient with fuzzy information of the decision maker but we are faced with an increased numerical effort. In practical application we achieved results using the first method which had been accepted by the user.

2.1. The crisp case

Let a pairwise comparison matrix $A = (a_{ij})$, with $a_{ij}$ as descrited above (we set $a_{ii} = 1$), be given. Since we use a ratio scale, we are seeking $w_i > 0$ with

$$a_{ij} = \frac{w_i}{w_j}\ ; \quad i,j = 1,\ldots,n \tag{1}$$

For an arbitrary given $a_{ij}$ there may not exist a $w_i$ fulfilling (1). Saaty proposed to determine the $w_i$ as the components of the eigenvector corresponding to the greatest eigen value of A which gives the exact $w_i$ in the case when (1) is solvable. Chu et al. (1979) considered the following problem:

$$(P_1)\quad \begin{cases} \sum\limits_{i,j} \left(a_{ij} - \dfrac{w_i}{w_j}\right)^2 \rightarrow \min \\ \sum\limits_{i} w_i = 1; \quad w_i > 0; \end{cases} \tag{2}$$

whereby $(P_1)$ is slightly modified for better numerical handling.

We will modify $(P_1)$ into $(P_2)$ aiming at an explicit solution, i.e.

$$(P_2)\quad \begin{array}{l} \sum\limits_{i,j} \left(\ln a_{ij} - \ln \dfrac{w_i}{w_j}\right)^2 \rightarrow \min \\ \prod\limits_{i} w_i = 1; \quad w_i > 0 \end{array} \tag{3}$$

where $\pi$ is the product operator.

$(P_2)$ may be deduced from the idea that $a_{ij} \approx \frac{w_i}{w_j}$ iff

$$\ln a_{ij} \approx \ln \frac{w_i}{w_j} .$$

Using the Lagrange multipliers we obtain

$$w_i = \left[\prod_{j \neq i} \frac{a_{ij}}{a_{ji}}\right]^{\frac{1}{2n}} \tag{4}$$

In the reciprocal case, i.e. $a_{ij} = \frac{1}{a_{ji}}$, we have

$$w_i = \left[\prod_{j \neq i} a_{ij}\right]^{\frac{1}{n}} \tag{5}$$

i.e. the i-th weight is equal to the geometric mean of the elements of the i-th row of A. For a better interpretation and comparability, we normalize the $w_i$ by $\sum_i w_i = 1$.

## 2.2. The fuzzy case

Now we assume the $a_{ij}$ to be positive fuzzy numbers of the (L,R)-type discussed by, e.g., Dubois and Prade (1980). We denote these numbers by $\tilde{A}_{ij}$, the corresponding weights by $\tilde{W}_i$. The latter we also assume to be of (L,R)-type, and positive. We replace relationship (1) by

$$\tilde{A}_{ij} \approx \tilde{C}_{ij} \tag{6}$$

with $\tilde{C}_{ij} = \frac{\tilde{W}_i}{\tilde{W}_j}$ (i.e. fuzzy division).

There are different ways to express (6) mathematically. We can use fuzzy intersection $\tilde{A}_{ij} \wedge \tilde{C}_{ij}$, and determine the $\tilde{W}_i$ by several demands on this intersection.

On the other hand, it is well-known that each (L,R)-number can be characterized by three parameters: the mean value, left spread and right spread. We can write:

$$\begin{aligned} \tilde{A}_{ij} &= (a_{ij}, \underline{a}_{ij}, \bar{a}_{ij}) \\ \tilde{C}_{ij} &= (c_{ij}, \underline{c}_{ij}, \bar{c}_{ij}) \\ \tilde{W}_i &= (w_i, \underline{w}_i, \bar{w}_i) \end{aligned} \tag{7}$$

Relationship (6) is grasped by the demand that the corresponding parameters of $\tilde{A}_{ij}$ and $\tilde{C}_{ij}$ should be as close as possible. Therefore we use the error sum:

$$\sum_{j=1}^{n} \sum_{\substack{i=1 \\ i \neq j}}^{n} \left[(c_{ij}-a_{ij})^2 + (\underline{c}_{ij}-\underline{a}_{ij})^2 + (\bar{c}_{ij}-\bar{a}_{ij})^2\right] \rightarrow \min$$

$$(P_3) \qquad \sum_i w_i = 1; \qquad w_i > 0 \tag{8}$$

When using the "supmin" - composition rule for Zadeh's extension principle,we get (cf. Dubois and Prade, 1980)

$$c_{ij} = \frac{w_i}{w_j}; \; \underline{c}_{ij} = \frac{\bar{w}_j w_i + \underline{w}_i w_i}{w_j^2}; \qquad \bar{c}_{ij} = \frac{\underline{w}_j w_i + \bar{w}_i w_j}{w_j^2} \tag{9}$$

The $a_{ij}$ and $w_i$ should not be confused with those from the previous section.

$(P_3)$ represents a nonlinear optimization task with 3n unknowns. There are good algorithms available for its solution.

## 3. THE EVALUATION OF PARETO-POINTS

Now we will deal with the problem how to use the weights $w_i$ (or $\tilde{W}_i$) to evaluate the set of Pareto-points. Again we start with the crisp case. The most simple way consists in the use of the well-known linear objective function $\sum_i w_i \times Y_i$, and to assign this value to each point. Nevertheless, we have to deal with the drawback that the $Y_i$ may be scalled in very different manners making this method useless since the performance criteria with small (absolute) values are discriminated. Therefore we have to use a convenient normalization of the criteria making them comparable. A normalization to [0,1] seems to be suitable, since the $w_i$ are of the same order of magnitude. Using the denotation $_i$ for the range of the i-th criterion and $\tau_i$ for a scaling function, we obtain $\tau_i : \mathcal{Y}_i \to [0,1]$. A widely used scaling method consists in the application of the individual extrema, e.g.

$$\tau_i(Y_i) = \left( \frac{Y_i - Y_i^{min}}{Y_i^{max} - Y_i^{min}} \right)^{\alpha_i} \tag{10}$$

with $\alpha_i > 0$, $Y_i^{min}$ is the minimum of $Y_i$ over $\mathcal{Y}_i$ and $Y_i^{max}$ is the maximum. We obviously have $\tau_i\,(Y^{min}) = 0$, $\tau_i(Y_i^{max}) = 1$.

There are several reasonable methods to evaluate a given Pareto-point. Denoting $Y^{(l)} = (Y_1^{(l)},\ldots,Y_n^{(l)})$, we have, e.g., the following evaluations $v_l$ of $Y^{(l)}$:

$$v_1 = v(Y^{(1)}) = \sum_{i=1}^{n} \tau_i(Y_i^{(1)})\ w_i \tag{11}$$

$$v_1 = \frac{1}{2}\,[\min_{1\leq i\leq n} \tau_i(Y_i^{(1)})\ w_i + \max_{1\leq i\leq n} \tau_i(Y_i^{(1)})] \tag{12}$$

$$v_1 = \max_{1\leq i\leq n}\ \min\ (\tau_i(Y_i^{(1)}), w_i) \tag{13}$$

Each approach has its characteristic properties. The choice depends on information and objectives of the decision maker.

We took the compensatory approach (11) since the results were sufficient. The methods (12) and (13) take into account merely two and one value of $Y^{(1)}$, respectively. This may be disadvantageous in practice.

In the fuzzy case we have fuzzy weights $\tilde{W}_i$. Here we construct a relation $\tilde{B}$ of the possible weights $w_i$ (non-normalized) and the set $\{Y^{(1)}\}$ of Pareto-points to be evaluated. With $\mu_{\tilde{B}}$ for the membership function of $B$ we define

$$\mu_B(w_1,\ldots,w_n,y_1,\ldots,y_n) = \frac{\sum_{i=1}^{n} w_i\ \tau_i(y_i)}{\sum_{i=1}^{n} w_i} \tag{14}$$

with $w_i \geq 0;\ \sum_{i=1}^{n} w_i > 0;\ y_i \in \mathcal{Y}_i$

($y_i$ stand for $Y_i^{(1)}$).

Denoting

$$\tilde{W} = \tilde{W}_1 \cap \ldots \cap \tilde{W}_n \tag{15}$$

("$\cap$" is the fuzzy intersection, e.g., the min operator), we get the following fuzzy set of evaluations

$$\tilde{V} = \tilde{B} \circ \tilde{W} \tag{16}$$

where "$\circ$" means a convenient composition rule of the fuzzy relation $\tilde{B}$ and the fuzzy set $\tilde{W}$. When using the sup-min composition we obtain

$$\mu_{\tilde{V}}(y_1,\ldots,y_n) = \sup_{w_1,\ldots,w_n} \min[\mu_{\tilde{B}}(w_1,\ldots,w_n,y_1,\ldots,y_n),\ \mu_{\tilde{W}}(w_1,\ldots,w_n)] \tag{17}$$

Finally we get

$$v_1 = \mu_{\tilde{V}}(Y_1^{(1)},\ldots,Y_n^{(1)}) \tag{18}$$

which is the desired evaluation in the fuzzy case.

4. DESIGN OF A PROCESS FOR CRUDE OIL MANUFACTURING

Due to increasing prices and decreasing availability of crude oil, the production of gasoline and diesel fuel only by distillation is not economic at the present time. On the other hand, we have to apply conversion processes (i.e. chemical reactions) to obtain the desired products (fuels) from higher boiling crude oil fractions. But this can be done only by using the expensive hydrogen. Cracking of higher boiling distillates is commonly realized catalytically with high temperatures and pressures. Moreover, we have to apply hydroraffination because of the presence of undesired sulfur, nitrogen and oxygen compounds in the raw material. These two processes can be carried out either in a reactor with a bifunctional catalyst or in separated reactors with a mixed catalyst. That is, the reactor system can be one-, two- or quasi-one-stage type. Since conversion of the raw material is not complete, we have to separate the non-converted raw material from the products for recycling. Moreover, we have to split the final products, i.e., the gasoline and diesel fuel, for removing undesired by-products as, e.g., cracking gas products. Depending on the separation of the reaction mixture, being realised after the first or second reactor, we differentiate the intermediate and final fractionation. Both kinds may occur combined in one stage as well. Now the problem consists in the determination of a structure of a process system which guarantees high conversion of raw material with low costs for the erecting and running of equipment. The analysis of alternate process structures leads to six main variants:

1. A double-stage reactor system with common intermediate and final fractionation (Fig. 1),
2. A quasi-one-stage reactor system with final fractionation (Fig. 2),
3. A double-stage reactor system without intermediate fractionation (Fig. 3).

In the first reaction stage a refining catalyst is used, in the second one a cracking catalyst is taken. The remaining three variants are analogous, only this time we use bifunctional and cracking catalysts.

As the performance criteria, reflecting different properties of the process system, we chose the following ones:

1. $Q_1$ - Fixed costs → min
2. $Q_2$ - Variable costs → min
3. $Q_3$ - Overall costs → min
4. $Q_4$ - Need for hydrogen → min
5. $Q_5$ - By-products (gas) → min

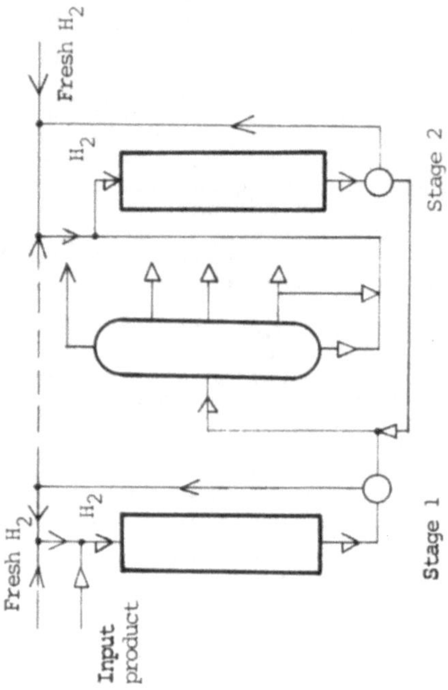


Fig. 1. Two stage cracking process with intermediate fractioning (a separation stage)

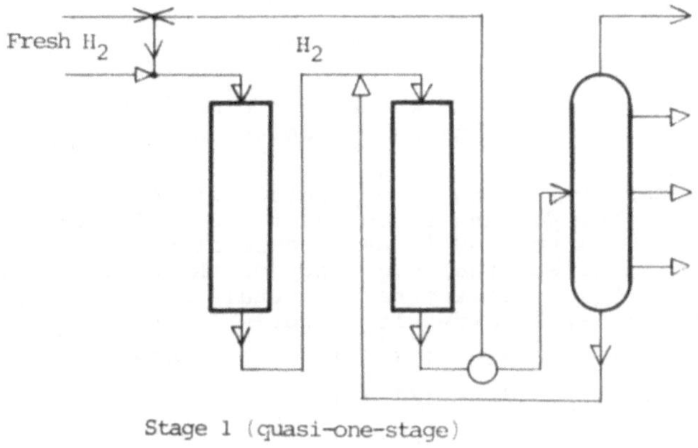


Fig. 2. Quasi-one-stage cracking process.

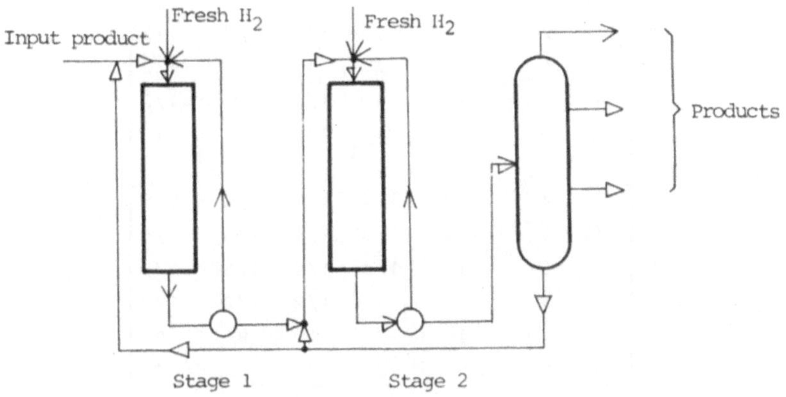


Fig. 3. Two-stage cracking process without intermediate fractioning.

6. $Q_6$ - Yield of gasoline → max
7. $Q_7$ - Yield of diesel fuel → max

The last two criteria include demands on the flexibility of the process system.

The control variables are the main dimensions of the process elements, technological variables, composition of raw material, etc.

The process of finding a compromise between the above criteria is difficult. We cannot construct a global overall objective function in the usual way because of the heterogeneity of the criteria. It was the complexity of the evaluations which led us to question three experts for obtaining the pairwise preference matrices. As a result we got:

$$A_1 = \begin{bmatrix} 1 & 1/5 & 1/3 & 1/5 & 1 & 1/7 & 1/7 \\ 5 & 1 & 3 & 1/7 & 1 & 1/3 & 1/3 \\ 3 & 1/3 & 1 & 1/3 & 1 & 1/5 & 1/5 \\ 5 & 7 & 3 & 1 & 5 & 1/5 & 1/5 \\ 1 & 1 & 1 & 1/5 & 1 & 1/5 & 1/5 \\ 7 & 3 & 5 & 5 & 5 & 1 & 1 \\ 7 & 3 & 5 & 5 & 5 & 1 & 1 \end{bmatrix}$$

$$A_2 = \begin{bmatrix} 1 & 2 & 2 & 3 & 1/3 & 1/4 & 1/5 \\ 1/2 & 1 & 3 & 1/2 & 1/5 & 1/5 & 1/5 \\ 1/2 & 1/3 & 1 & 1/4 & 1/4 & 1/4 & 1/5 \\ 1/3 & 2 & 4 & 1 & 1/3 & 1/4 & 1/5 \\ 3 & 5 & 4 & 3 & 1 & 3 & 1/2 \\ 4 & 5 & 4 & 4 & 1/3 & 1 & 1/2 \\ 5 & 5 & 4 & 5 & 1/3 & 2 & 1 \end{bmatrix}$$

$$A_3 = \begin{bmatrix} 1 & 2 & 1/5 & 1/3 & 1/4 & 1/5 & 1/5 \\ 1/2 & 1 & 1/5 & 1/3 & 1/4 & 1/5 & 1/5 \\ 5 & 5 & 1 & 1/3 & 1/4 & 1/5 & 1/5 \\ 3 & 3 & 3 & 1 & 1/2 & 1/2 & 1/2 \\ 4 & 4 & 4 & 2 & 1 & 9 & 9 \\ 5 & 5 & 5 & 2 & 1/9 & 1 & 1 \\ 5 & 5 & 5 & 2 & 1/9 & 1 & 1 \end{bmatrix}$$

The corresponding weights are, due to (5) :

$w^{(1)} = (0.031, 0.083, 0.055, 0.157, 0.051, 0,312, 0.312)$

$w^{(2)} = (0.086, 0.052, 0.038, 0.07, 0.312, 0.192, 0.249)$

Table 1

| $l_{ind}$ | $l_p$ | $l_{proc}$ | $Q_1$ | $Q_2$ | $Q_3$ | $Q_4$ | $Q_5$ | $Q_6$ | $Q_7$ | $v_I$ | $v_{II}$ |
|---|---|---|---|---|---|---|---|---|---|---|---|
| 1 | 1 | 5 | .632 | 5.61 | 6.24 | 644 | 69.5 | 524 | 692 | .444 | .448 |
| | 2 | 2 | .642 | 5.44 | 6.08 | 616 | 62.0 | 507 | 604 | .488 | .491 |
| | 3 | 4 | .732 | 5.95 | 6.68 | 696 | 96.7 | 482 | 619 | .321 | .336 |
| | 4 | 1 | .904 | 5.90 | 6.80 | 631 | 64.9 | 513 | 620 | .466 | .471 |
| | 5 | 6 | .995 | 5.49 | 6.49 | 577 | 70.7 | 544 | 574 | .438 | .429 |
| | 6 | 3 | 1.046 | 5.09 | 6.14 | 472 | 47.4 | 560 | 588 | .616 | .599 |
| 2 | 7 | 3 | 1.076 | 4.85 | 5.93 | 476 | 47.3 | 540 | 606 | .636 | .620 |
| | 8 | 2 | .643 | 5.41 | 6.06 | 613 | 64.1 | 495 | 607 | .476 | .479 |
| | 9 | 6 | 1.010 | 5.46 | 6.47 | 584 | 64.1 | 543 | 591 | .485 | .479 |
| | 10 | 5 | .635 | 5.54 | 6.18 | 640 | 75.9 | 501 | 599 | .417 | .423 |
| | 11 | 4 | .768 | 5.81 | 6.57 | 693 | 92.8 | 473 | 639 | .346 | .361 |
| 3 | 12 | 3 | 1.070 | 4.86 | 55.93 | 475 | 47.7 | 543 | 604 | .635 | .619 |
| | 13 | 2 | .644 | 5.41 | 6.05 | 613 | 64.4 | 494 | 607 | .474 | .477 |
| | 14 | 5 | .635 | 5.55 | 6.18 | 640 | 75.6 | 502 | 600 | .421 | .427 |
| | 15 | 6 | 1.009 | 5.46 | 6.47 | 584 | 65.1 | 544 | 591 | .484 | .478 |
| | 16 | 4 | .750 | 5.81 | 6.66 | 694 | 92.8 | 474 | 628 | .346 | .361 |
| | 17 | 1 | .921 | 5.58 | 6.61 | 624 | 57.4 | 497 | 635 | .509 | .514 |
| 4 | 18 | 3 | 1.057 | 5.01 | 6.07 | 457 | 58.2 | 553 | 579 | .571 | .552 |
| | 19 | 2 | .786 | 5.72 | 6.51 | 553 | 61.2 | 391 | 691 | .520 | .531 |
| | 20 | 6 | 1.157 | 5.88 | 7.03 | 572 | 46.4 | 545 | 586 | .494 | .485 |
| | 21 | 1 | 1.239 | 6.16 | 7.40 | 583 | 0.0 | 471 | 663 | .649 | .648 |
| | 22 | 5 | .799 | 6.08 | 6.88 | 607 | 14.9 | 400 | 670 | .563 | .572 |
| | 23 | 4 | 1.071 | 6.17 | 7.24 | 642 | 67.8 | 469 | 676 | .481 | .494 |
| 5 | 24 | 2 | .831 | 5.88 | 6.71 | 572 | 0.0 | 451 | 677 | .707 | .712 |
| | 25 | 1 | 1.234 | 6.18 | 7.41 | 587 | 0.0 | 479 | 569 | .648 | .648 |
| | 26 | 3 | 1.427 | 6.67 | 8.09 | 501 | 0.0 | 520 | 607 | .594 | .578 |
| | 27 | 4 | 1.092 | 6.31 | 7.40 | 650 | 5.6 | 487 | 675 | .651 | .660 |
| | 28 | 5 | .789 | 6.21 | 6.99 | 622 | 9.1 | 458 | 656 | .611 | .619 |
| 6 | 29 | 6 | 1.081 | 6,27 | 7.35 | 587 | 66.6 | 562 | 573 | .415 | .408 |
| | 30 | 3 | 1.212 | 5.62 | 6.83 | 484 | 87.4 | 561 | 584 | .456 | .441 |
| | 31 | 2 | .911 | 5.92 | 6.84 | 626 | 62.9 | 535 | 590 | .439 | .439 |
| | 32 | 5 | .686 | 5.84 | 6.53 | 647 | 71.9 | 534 | 584 | .415 | .419 |
| | 33 | 1 | .966 | 6.43 | 7.39 | 638 | 60.4 | 529 | 609 | .438 | .442 |
| | 34 | 4 | .958 | 6.39 | 7.35 | 687 | 58.9 | 528 | 613 | .428 | .439 |
| 7 | 35 | 2 | .823 | 5.75 | 6.57 | 555 | 5.5 | 389 | 692 | .659 | .666 |
| | 36 | 5 | .947 | 6.58 | 7.53 | 636 | 17.5 | 413 | 685 | .549 | .562 |
| | 37 | 4 | 1.001 | 6.07 | 7.07 | 652 | 7.0 | 488 | 680 | .679 | .690 |
| | 38 | 1 | 1.164 | 6.02 | 7.18 | 594 | .5 | 497 | 669 | .702 | .704 |
| | 39 | 6 | 1.484 | 6.71 | 8.19 | 592 | 7.9 | 481 | 655 | .589 | .584 |
| | 40 | 3 | 1.380 | 5.80 | 6.18 | 515 | 0.0 | 513 | 642 | .725 | .709 |

where: $l_{ind}$ - number of criterion the individual extremum of which has been determined

$l_p$ - enumeration of Pareto-point

$l_{proc}$ - number of the process whose best value is realized

Units of measurement: $[Q_1]=[Q_2]=[Q_3]$ = unity of costs, $[Q_4]=m^3H_2/m^3$ raw material, $[Q_5] = [Q_6] = [Q_7]$ = kilotons

$$w^{(3)} = (0,041,0.033,0.73,0.124,0.392,0.168,0.168)$$

For expert $E_1$ the production of gasoline and diesel fuel is most important, for expert $E_2$ the minimization of by-products and maximal gasoline output. Expert $E_3$ thinks that the minimization of by-product is very important. The questioned experts on their part were evaluated by an authorized board with regard to their knowledge.

We consider two cases:

a) The knowledge of the experts is of equal quality. That is, we have $E = (c_{ij})$; $i,j=1,2,3$, and $c_{ij}=1$ when E denotes the preference matrix for the experts.

b) We consider the following preference matrix

$$E = \begin{bmatrix} 1 & 1/3 & 5 \\ 3 & 1 & 7 \\ 1/5 & 1/7 & 1 \end{bmatrix}$$

Let $w^{(e)}$ be the vector of weights for the evaluation of the experts; then if $w^{(e)} = (w_1^{(e)}, w_2^{(e)}, w_3^{(e)})$, we have

$$w_i = \sum_{j=1}^{3} w_i^j \, w_j^{(e)} \qquad i = 1,\ldots,7,$$

This means that we take the average over all the preference matrices $A_j$. The numerical results are listed in Table 1. There, $v_I$ is for case a) and $v_{II}$ for case b). The parameters $\alpha_j$ are equal to 1. The results for $v_I$ and $v_{II}$ obtained by (11) do not differ as much as one could expect. This is explained by the situation that all the three experts consider the performance criteria $Q_5$-$Q_7$ very important. Therefore those Pareto-points have a high evaluation which is near to the individual extrema for $Q_5$-$Q_7$. Three process systems are therefore favoured (i.e. 40, 38, 24).

## 5.MAGNETIC TAPE PRODUCTION

Magnetic tapes belong to those kinds of information carriers which are being produced as audio, video and computer tapes in an increasing volume and qualities for quite different users.

A magnetic tape consists of a support and several layers applied to the support, e.g., an intermediate layer improving the adhesiveness of the magnetic film. The latter contains the ferric or chromic oxide. Various kinds of special layers may occur. In Fig. 4 the main production process stages are summarized. The production system consists of the following subsystems:

- production of support,
- production of intermediate layer,
- production of magnetic suspension,
- coating and drying,
- repulping and confectioning (cutting, testing, sizing, packaging).

The subsystem "production of magnetic suspension" and "coating and drying" are subject to polyoptimal control because of their special importance for the quality of the final products.

Homogenization of the suspension is the crucial operation of the first subsystem. In the second subsystem the crucial operations are coating, magnetization, and drying, with a special profile of temperature, and calendering.

High requirements are put on the magnetic tape with regard to its mechanic, magnetic and electro-acoustic properties. We have, e.g., the thickness of the film, oscillation, tensile strength, etc. as essential mechanic influence factors. The electro-acoustic properties and factors are determined as relative values to reference tapes. We can list, e.g., the limiting current interval, nominal current interval, sensitivity, noise voltage interval, etc. They may be different for different types of tapes.

Not all properties and influence factors are non-correlated and of equal importance; some of them can be treated as restrictions. The following performance criteria are of special importance for the optimal control: sensitivity ($D_s$), high frequency sensitivity ($D_h$), distortion damping measure ($D_d$), maximal modulation for 10 kHz ($D_{10max}$). They are all electro-acoustic factors.

The following significant influence factors turned out as a result of statistical modelling (the list is not complete): specific polarisation of saturation, stability and bulk density of the pigment, total reflexion and orientation quotient of the calendered web, coercive field intensity, etc. The models have been set up in the form of polynomial functions with a high significance level:

$$Q_i(x_1,\dots,x_{15}) = \sum_{j=1}^{15} (a_{ij}x_j^2 + b_{ij}x_j); \qquad i=1,\dots,4$$

The diversity of the desired quality set of the products and the pertinent production situation, involve flexible and quickly applicable decision aids. We are going to consider three different decision making situations:

1. Magnetic tapes are desired with high quality properties for the recording and reproducing of high frequencies. We obtain the following preference matrix

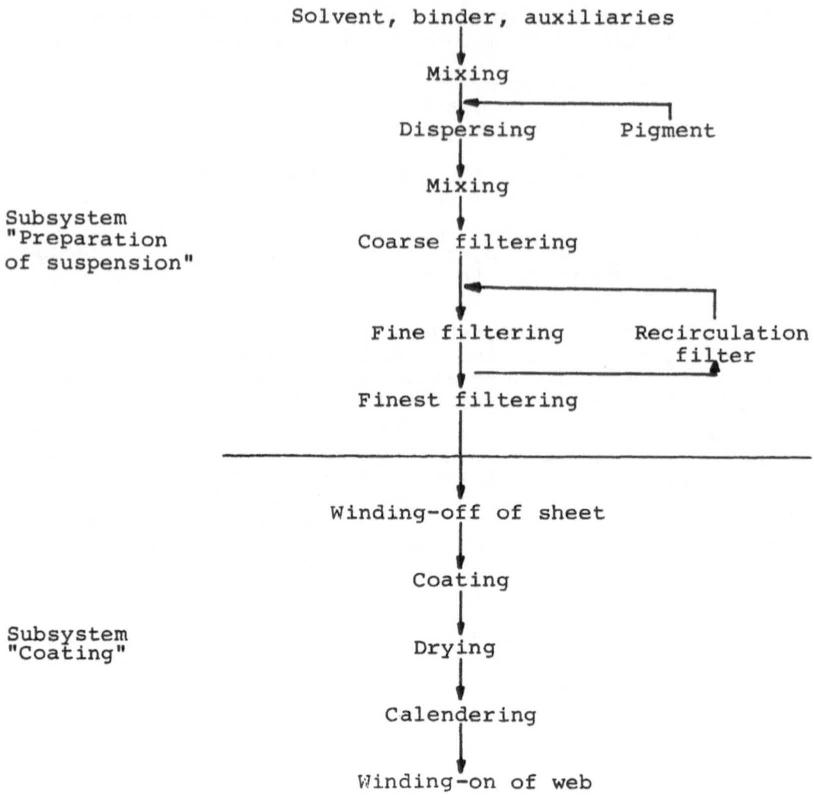


Figure 4: Main stages of magnetic tape production (Subsystem "Preparation of suspension" and "Coating")

$$A = \begin{array}{c} \\ D_s \\ D_h \\ D_d \\ D_{10max} \end{array} \begin{array}{c} \begin{matrix} D_s & D_h & D_d & D_{10max} \end{matrix} \\ \begin{bmatrix} 1 & 1/8 & 3/2 & 1/8 \\ 8 & 1 & 5 & 2 \\ 2/3 & 1/5 & 1 & 1/5 \\ 8 & 1/2 & 5 & 1 \end{bmatrix} \end{array}$$

2. There are tapes to be produced with sensitivity and high distortion damping measure. We use

$$A = \begin{bmatrix} 1 & 8 & 3/2 & 8 \\ 1/8 & 1 & 1/5 & 3/2 \\ 2/3 & 5 & 1 & 5 \\ 1/8 & 2/3 & 1/5 & 1 \end{bmatrix}$$

3. We are faced with the demand to improve the high frequency sensitivity without an essential deterioration of the relative sensitivity $D_s$. We used

$$A = \begin{bmatrix} 1 & 1/5 & 5 & 3/2 \\ 5 & 1 & 5 & 5 \\ 1/5 & 1/5 & 1 & 5 \\ 2/3 & 1/5 & 1/5 & 1 \end{bmatrix}$$

Applying formula (5), we obtain the following global weights

Table 2

| | case 1 | case 2 | case 3 |
|---|---|---|---|
| $D_s$ | 0.066 | 0.526 | 0.200 |
| $D_h$ | 0.508 | 0.074 | 0.590 |
| $D_d$ | 0.068 | 0.340 | 0.050 |
| $D_{10max}$ | 0.358 | 0.060 | 0.160 |

We investigate 40 situations of production efficient with regard to the four criteria introduced above. In Table 3 we list some of them to give an idea of the quantitative relationships

Table 3

| No | $D_s$ | $D_h$ | $D_d$ | $D_{10max}$ |
|---|---|---|---|---|
| 1 | 0.52 | 5.07 | 41.91 | 5.75 |
| 2 | 1.13 | 4.82 | 43.43 | 6.30 |
| 3 | 2.13 | 4.39 | 46.27 | 7.16 |
| 4 | 2.43 | 4.25 | 47.38 | 7.39 |
| 5 | 2.73 | 4.12 | 48.56 | 7.56 |
| 6 | 2.92 | 3.75 | 52.56 | 7.47 |
| 7 | 2.78 | 3.82 | 51.85 | 7.44 |
| 8 | 2.35 | 4.02 | 50.62 | 7.08 |
| 9 | 2.66 | 3.88 | 51.92 | 7.19 |
| Maximum | 5.23 | 6.60 | 67.95 | 8.31 |
| Minimum | -9.38 | -1.04 | -77.33 | -9.09 |

The units are in dB.

Table 4 contains the rank-ordering of the points, whereby we changed the exponents $\alpha_j$:

Table 4

| Case | Exponents $\alpha_j$ | Global rank-ordering (five best points) | | | | |
|---|---|---|---|---|---|---|
| 1 | $\alpha_1 - \alpha_4 = 1$ | 1 | 2 | 3 | 4 | 5 |
| | $\alpha_1 - \alpha_4 = 5$ | 5 | 4 | 3 | 6 | 7 |
| 2 | $\alpha_1 - \alpha_4 = 1$ | 6 | 7 | 9 | 5 | 8 |
| | $\alpha_1 - \alpha_4 = 5$ | 6 | 7 | 9 | 5 | 8 |
| 3 | $\alpha_1 - \alpha_4 = 1$ | 1 | 2 | 3 | 4 | 5 |
| | $\alpha_1 - \alpha_4 = 5$ | 1 | 5 | 4 | 3 | 2 |

From these results we derived valuable parameters for optimal production regimes as well as necessary improvements of quality. It is of great importance for the decision maker that he is enabled (by a dialogue system) to respond much more flexibly and quickly to the customer's demands and to special export conditions into areas with changeable requirements as to quality properties.

REFERENCES

Chu, A., et al. (1979). A comparison of two methods for determining the weights of belonging to fuzzy sets. J. Opt. Theory Appl. 27, 531-538.

Dubois, D., and H. Prade (1980). Fuzzy Sets and Systems. Academic Press, New York.

Saaty, T.L. (1978). Exploring the interface between hierarchies, multiple objectives and fuzzy sets. Fuzzy Sets and Syst. 1, 57-68.

Wagenknecht, M., and K. Hartmann (1983). On fuzzy rank ordering in polioptimization. Fuzzy Sets and Syst. 11, 253-264.

# OPTIMAL CLASSIFIER DESIGN USING FUZZY k-NEAREST NEIGHBOR RULES

Siew Chuah and James C. Bezdek

Computer Science Department
University of South Carolina
Columbia, SC 29208, USA

Abstract. A recently proposed classifier based on the fuzzy k-nearest neighbor rule (NNR) which is optimized over k with respect to predicted error rate is implemented. Optimal fuzzy partitions of (labelled) training sets produced by (i) Jozwik's method and (ii) Fuzzy c-Means (FCM) are used as inputs to fuzzy k-NNR's for classification of new objects. Performance of the resultant classifiers are compared to the conventional (hard) k-NNR. Numerical examples illustrating comparative performance of the three classifiers using samples from synthetic mixtures of normal densities and Anderson's Iris data are given. Our results indicate that partitioning labelled data with FCM, followed by selection of an optimal k using iteration on k, yield somewhat better fuzzy k-NNR performance on subsequent test sets than the method of Jozwik. Moreover, the fuzzy k-NNR classifier based on either method of fuzzy labelling seems superior to the conventional hard k-NNR.

Keywords: classifier design, fuzzy c-means, fuzzy k-NNR, nearest neighbor rules, pattern recognition.

## 1. INTRODUCTION

A statistical pattern recognition system is a decision rule that assigns test samples to one of c classes. The performance of such a system is measured by estimating the predicted probability of error of misclassification. The theoretically minimum error for all statistical decision rules is the Bayes error (which is called Bayes risk for non 0-1 loss matrices), $p_B$, given by $p_B = 1 - \int \max_i \left\{ p_i g(\underline{x}/i) \right\} \underline{dx}$, where $g(\underline{x}/i)$ and $p_i$ are respectively the class-conditional probability density function and the a priori probability of the i-th class. The mixture from which training and testing data are drawn is thus $f(\underline{x}) = \sum_i p_i g(\underline{x}/i)$, with feature vectors $\underline{x} \in R^q$. The "true" Bayes error $p_B$ is in practice unknown since it is a function of the prior probabilities and probability density functions of the classes, which in applications are themselves unknown. One of the most widely used approaches for classifier design when labelled samples are available is the k-NNR. This method is called "non-parametric" because it functions without knowledge of the unknown parameters of the $g(\underline{x}/i)$. k-NNR's can be used to produce estimates of bounds of $p_B$ which are asymptotically correct. Moreover, k-NNR's can also be used as the classifier of record for on line decision making. In this latter instance an estimate of $p_B$ is obtained by counting the number of samples misclassified by the k-NNR. In this paper a scheme for a fuzzy k-NNR developed by Jozwik (1983) will be implemented. In Jozwik's scheme, a "best" reference set (training set), along with an optimal k and optimal fuzzy partition for the training set, will be used to classify test sets. The results will be compared to those obtained when a fuzzy partition produced by the fuzzy c-means method described in Bezdek (1981) is used instead. Our numerical examples will show that fuzzy partitions derived from either method result in lower misclassification errors than hard partitions do.

In subsequent sections the following notation will be used. Let $X = \{\underline{x}_1, \underline{x}_2, \dots, \underline{x}_n\}$ be a sample of n observations $\underline{x}_k \in R^q$; $x_k$ is the k-th feature vector; $x_{kj}$ the j-th feature of $\underline{x}_k$. A conventional hard c-partition of X can be represented by a cxn matrix $W = [\underline{W}^{(1)} \underline{W}^{(2)} \dots \underline{W}^{(n)}]$ which satisfies three conditions:

$$w_{ik} \in \{0,1\}, \quad 1 \leq i \leq c, \quad 1 \leq k \leq n \tag{1a}$$

$$\sum_{i=1}^{c} w_{ik} = 1, \qquad 1 \leq k \leq n \tag{1b}$$

$$0 < \sum_{k=1}^{n} w_{ik} < n, \qquad 1 \leq i \leq c \tag{1c}$$

It is useful in the sequel to regard the i-th column $\underline{W}^{(i)}$ of W as a label vector attached to $\underline{x}_i$; e.g., $(\underline{W}^{(i)})^T = (0,0,1,0) \in R^4 \Rightarrow \underline{x}_i \in$ class 3. When the $w_{ik}$'s can take on values in the unit interval $[0,1]$ in (1a), W becomes a fuzzy c-partition of X, In this case the values $w_{ik}$ can be interpreted as grades of membership of the $\underline{x}_k$'s in c fuzzy subsets of X. We shall denote the set of all fuzzy c-partitions of X by $M_{fc}$. We partition X into a training (or design) set $X_d$ with $n_d$ elements; a test set $X_t$ with $n_t$ elements; and a third set $X_c$ with $n_c$ elements for testing the performance of the k-NNR classifier: $(n_d + n_t + n_c) = n$. The samples corresponding to $(n_t + n_c)$ are ordinarily pooled as a "test set". We shall reserve $n_c$ of the n samples for "final" testing, because Jozwik's method sometimes chooses an optimal k to be used with all $(n_d + n_t)$ samples. Kittler and Devijver (1981) suggest dividing $X_d \cup X_t$ with a partition ratio $(n_d:n_t) = (1:2)$. Our examples use $n_c = (n/10)$ and other partition ratios, namely: (1:1), (2:1), (1:9), and (9:1).

The traditional hard k-NNR, the fuzzy c-means method, and Jozwik's schemes are briefly described in sections 2, 3 and 4, respectively. Numerical results of the performance of all three methods are discussed in section 5.

## 2. HARD k-NNR

The 1-NNR decision rule assigns a test sample to the class of the nearest neighbor in a set of labelled points according to some distance metric. In the k-NNR case, the test sample is assigned to the class which has the highest representation amongst the k nearest neighbors. Ties are broken arbitrarily. This decision rule is well discussed in, e.g., Kittler and Devijver (1982).

## 3. FUZZY C-MEANS (FCM)

The fuzzy c-means algorithms can be used to obtain fuzzy partitions for the training set to be used in the fuzzy k-NNR. Optimal fuzzy partitions are found by minimizing the objective function (c.f. Bezdek, 1981)

$$J_m(W,\underline{z}) = \sum_{k=1}^{n} \sum_{i=1}^{c} (w_{ik})^m \| x_k - \underline{z}_i \|^2 \qquad (2)$$

where $1 \leq m < \infty$, $\| * \|$ is any inner product norm on $R^q$, $W \in M_{fc}$ and $\underline{z} = (\underline{z}_1, \underline{z}_2, \ldots, \underline{z}_c)$ is a vector of centers, $\underline{z}_i \in R^q$, $1 \leq i \leq c$. The exponent (m) controls the relative weights placed on each of the squared errors $\| \underline{x}_k - \underline{z}_i \|^2$. As m approaches 1, partitions that minimize $J_m$ become increasingly hard. (i.e., $w_{ik} \to 0$ or 1 as $m \overset{+}{\to} 1$). Briefly, we summarize the

FUZZY c-MEANS (FCM) ALGORITHM:

Step 1: Fix c,m, and $\| * \|$. Choose an initial $W_o \in M_{fc}$; then at step j, j = 1,2,... JMAX (some maximum number of iterations):

Step 2. Compute means $(\underline{z}_i)_j$, i = 1,2,...,c, by

$$(\underline{z}_i)_j = \sum_{k=1}^{n} (w_{ik})_j^m \underline{x}_k / \sum_{k=1}^{n} (w_{ik})_j^m$$

Step 3. Compute an updated membership array $W_{j+1}$ by

$$(w_{ik})_{j+1} = \left( \sum_{s=1}^{c} (d_{ik}/d_{sk})^{2/(m-1)} \right)^{-1} ; \quad 1 \leqslant i \leqslant c; \quad 1 \leqslant k \leqslant n$$

where $d_{ik} = \| \underline{x}_k - (\underline{z}_i)_j \| > 0 \forall i,k$, and $m \in (1,\infty)$.

Step 4. Compare $W_{j+1}$ to $W_j$. If $\| W_{j+1} - W_j \| < \epsilon$; Stop, and at termination call $W_{j+1} = W_{fcm}$; Otherwise, set $W_{j+1} = W$ and return to step 2.

Further details of the FCM algorithms are available in Bezdek (1981).

## 4. JOZWIK'S FUZZY k-NNR

In order to describe Jozwik's method, we digress to define the "standard" fuzzy k-NNR (it is, of course, non-unique - there are other ways to fuzzify the conventional k-NN decision strategy). Let W be a fuzzy c-partition of $X_d = \{ \underline{x}_1, \underline{x}_2, \ldots, \underline{x}_{nd} \}$, and let $\underline{x}$ be a test vector to be classified. ($\underline{x}$ may or may not be in $X_d$!). Find the k-NN's to $\underline{x}'$ in $X_d$, and let $K(x')$ denote indices of the columns in W corresponding to the k-NN's of $x'$. Form the fuzzy label vector

$$\underline{v}(\underline{x}') = \sum_j \underline{W}^{(j)}/k; \quad j \in K(\underline{x}')$$

The vector $\underline{v}(\underline{x}')$ has components between 0 and 1 that represent the membership of $\underline{x}'$ in each of the c classes as determined by the k-NN's to $\underline{x}'$. If the columns of W are all hard, $\underline{v}(\underline{x}')$ reduces to an estimate of $\underline{p}(*/\underline{x}')$, i.e.

$$\underline{v}(\underline{x}') \approx \underline{p}(*/\underline{x}') = \begin{bmatrix} p(1/\underline{x}') \\ p(2/\underline{x}') \\ \cdot \\ \cdot \\ \cdot \\ p(c/\underline{x}') \end{bmatrix} , \qquad (4)$$

a non-parametric estimate of the posterior probabilities that, given $\underline{x}'$, it came from the various classes. Thus, a strategy that is formally analogous to Bayesian decision theory is to assign $\underline{x}'$ to the class in which it achieves maximum membership. This is the fuzzy k-NNR alluded to below. Note that when all of the $W^{(j)}$'s in (3) are hard, labelling $\underline{x}'$ by maximum membership is exactly the k-NN strategy: assign $\underline{x}'$ to the class receiving the maximum number of votes from neighbors in $K(\underline{x}')$. Now we describe Jozwik's implementation of this fuzzy decision strategy. Jozwik's scheme involves two stages:

Stage 1: Start with a labelled training data set $X_d$. The class labels of the $n_d$ points in $X_d$ correspond to a unique hard c-partition of $X_d$ (say $W_o$). Estimate the probability of error using the "leave one out" idea introduced by Lachenbruch (1965) as follows. Let k be the number of neighbors nearest to $\underline{x}_j$. Then:

(1.1): For $k = 1,2,\ldots,n_d-1$; for $j = 1,2,\ldots,n_d$:

Take $\underline{x}_j$ from $X_d$. Classify $\underline{x}_j$ by forming the fuzzy decision vector $\underline{v}(\underline{x}_j) = \sum_s \underline{W}_o^{(s)} / k$; $s \in K(\underline{x}_j)$. Then assign $\underline{x}_j$ to the class associated with its largest membership. If $\underline{x}_j$ is wrongly classified, then set $e_{jk} = 1$, otherwise set $e_{jk} = 0$. ($e_{jk}$ is an error counter); next k.

(1.2) For each $k = 1,2,\ldots,n_d-1$, estimate the Bayes error as $q_k = (\sum_{j=1}^{n_d} e_{jk})/n_d$.

(1.3) The k which minimizes $q_k$ is the optimal number of initial nearest neighbors. Set $k_o$ = the index which minimizes $q_k$ and $p_o = q_{k_o}$. Now we have $(W_o, k_o, p_o)$, where here the initial $W_o$ is the supplied hard c-partition of $X_d$ (cf. Step 1 of FCM, where $W_o$ might be fuzzy; the return from Step 2.1 below also has $W_o$ fuzzy).

(1.4) For $h = 0,1,2,\ldots$: define $\underline{W}_{h+1}^{(i)} = (k_h \underline{v}(\underline{x}_i) + W_h^{(i)})/(k_h+1)$, $i = 1,2,\ldots,n_d$. $\underline{W}_{h+1}^{(i)}$ is the i-th column of a (new) fuzzy c-partition $W_{h+1}$ of $X_d$, and $\underline{v}(\underline{x}_i)$ is calculated as in (1.1) but with $W_h$.

(1.5) Repeat (1.1) - (1.4) to obtain a sequence $(W_o,k_o,p_o)$, $(W_1,k_1,p_1),\ldots,(W_h,k_h,p_h)$. Iteration on h is terminated at the smallest index h such that $p_h \leqslant p_{h+1}$. At this h set $(W_1^*,k_1^*,p_1^*) = (W_h,k_h,p_h)$, where $W_1^*$ is the "optimal" fuzzy partition of $X_d$, $k_1^*$ is the optimal number of nearest neighbors, and $p_1{}^*$ is the optimal estimate of the empirical error. We emphasize that in this process $W_o$ is hard (the initial labels for $X_d$) and $W_1, W_2, \ldots, W_{h+1}$ is a sequence of h fuzzy c-partitions of $X_d$.

Stage 2:

(2.1) With $W_1{}^*$ use the fuzzy $k_1{}^*$ labels at (3) to produce, with $X_t = \left\{ \underline{x}_1, \ldots, \underline{x}_{nt} \right\}$ the "test" set, $n_t$ fuzzy column vectors $\underline{v}(\underline{x}_1), \ldots, \underline{v}(\underline{x}_{n_t})$. Let $V_1{}^*$ denote the $(c \times n_t)$ matrix of these column vectors. Append $V_1{}^*$ to $W_1{}^*$, and call the $c \times (n_d + n_t)$ matrix $[W_1{}^*, V_1{}^*] = W_o$. This $W_o$ is a fuzzy c-partition of $X_d \cup X_t$. Repeat (1.1) - (1.5) by iteration on h until from (1.5) we obtain:

(2.2) $(W_2{}^*, k_2{}^*, p_2{}^*)$ corresponding to the smallest index h such that $p_h \leqslant p_{h+1}$.

(2.3) Let $(X_{df}, W_f, k_f, p_f)$ denote the final reference (design) set, fuzzy c-partition, number of nearest neighbors, and estimate for $p_B$. The "optimal" choice for this 4-tuple is selected by minimizing $p_k{}^*$:

$$\text{If } p_1{}^* \leqslant p_2{}^*: \ (X_{df}, W_f, k_f, p_f) = (X_d, W_1{}^*, k_1{}^*, p_1{}^*) \tag{5a}$$

$$\text{If } p_1{}^* > p_2{}^*: \ (X_{df}, W_f, k_f, p_f) = (X_d \cup X_t, W_2{}^*, k_2{}^*, p_2{}^*) \tag{5b}$$

(2.4) Apply $(X_{df}, W_f, k_f, p_f)$ to $X_c$ as a check against the predicted performance $p_f$. This results in a second estimate (say $p_c$) of $p_B$.

Jozwik's algorithm is admittedly complex. However, the examples below seem to justify it in terms of better performance than the hard k-NNR. Several theoretical and computational questions concerning the method invite attention. For example, the termination scheme for (1.5) stops at the first h where $p_{h+1} \geqslant p_h$. One wonders: (i) is $p_h$ monotone decreasing with h until this point? (ii) is there a pair (h+m, h+m+1) with m > 1 so that $p_{h+m} \leqslant p_{h+m+1}$ and $p_{h+m} < p_h$? (i.e. does it stop too soon?). Finally, one wonders whether the computations of both stages could be circumvented by starting with an optimal $W_f$ obtained by a different fuzzy partitioning scheme. The experiments described below are designed with these questions in mind.

## 5. NUMERICAL RESULTS

The Euclidean norm was used in all calculations involving nearest neighbors for the data sets A,B,C,C' described below. In Jozwik's scheme, classification is tried with k nearest neighbors for k = 1 to $n_d-1$. Our first observation is that at each h, there is a threshold value for k beyond which the probability of error $p_h$ never decreased for larger values of k. Consequently, we limited k to kmax = floor $(\sqrt{n_d})$, the limit suggested by Kittler and Devijver (1981). Jozwik's method picks a "best" reference set $X_{df} = X_d$ or $(X_d \cup X_t)$ to be used to classify new data. However, empirical evidence suggests that stage 2 of the algorithm is often unnecessary. Table 1 exhibits the results of processing the 100 samples $X_c$ from data set C using stage 1 and then stage 2 of Jozwik's method: $p_c$ is the observed error rate achieved by applying, respectively, $(W_1{*}, k_1{*})$ from $X_d$, and $(W_2{*}, k_2{*})$ from $X_d \cup X_t$, to $X_c$. The mean absolute difference in observed errors is exactly 1%, with Stage 1 achieving the smaller average error rate. From this one may infer that - for this sample at least - the stage 2 calculations added nothing to future classifier performance. Secondly, stage 1 of Jozwik's method was (for data sets A and B) executed for h = 50 iterations to investigate the stopping criterion for (1.5). Once the h so that $p_h \leqslant p_{h+1}$ was found, no smaller error could be found for larger h.

Table 1. Effect of stage 2 calculations on $p_c$ for $X_c$ from data set C*

| Partition ratio $(n_d:n_t)$ | $p_c$ using $W_1{*},k_1{*}$ $X_R = X_d$ | $p_c$ using $W_2{*},k_2{*}$ $X_R = X_d \cup X_t$ |
|---|---|---|
| (1:2)) | .26 | .30 |
| (1:1) | .28 | .29 |
| (2:1) | .32 | .31 |
| (1:9) | .28 | .28 |
| (9:1) | .29 | .30 |

$^*p_B$ = 0.23 for data set C

This argues well for the proposed rule in (1.5). The three data sets used to illustrate our results are:

A: Anderson's Iris data (c.f. Fisher, 1936)
$c=3$ classes; $q = 4$ features;
50 samples from each class: $(n_d+n_t+n_c) = 150$

B: Artifically generated univariate normal mixtures
$c=2$ classes; $q = 1$ feature;
$$\left.\begin{array}{ll}\text{class 1:} & N(1,1)\\ \text{class 2:} & N(2,1)\end{array}\right\} \Rightarrow p_B = 0.31 \sim \text{True (optimal) error rate}$$
500 samples from each class; $(n_d+n_t+n_c) = 1000$.

C: Artifically generated bivariate normal mixtures
$c=2$ classes; $q=2$ features; $I = [\delta_{ij}]$ the $2\times2$ identity matrix
$$\left.\begin{array}{ll}\text{class 1:} & N(\mu_1,I);\ \mu_1^T = (1,1)\\ \text{class 2:} & N(\mu_2,I);\ \mu_2^T = (2,2)\end{array}\right\} \Rightarrow p_B = 0.24 \sim \text{True error rate}$$
500 samples from each class; $(n_d+n_t+n_c) = 1000$.

For each of the three data sets and each partition ratio $(n_d:n_t)$ Jozwik's scheme was executed for $k = 1$ to $\sqrt{n}_d$, and the optimal outputs $(X_{df}, W_f, k_f, p_f)$ were used to classify $X_c$. Observed probabilities of error using the fuzzy $k_f$-NNR on $X_c$ are called $p_c$. Using fuzzy c-means (for m=2) a fuzzy c-partition was obtained on $X_{df}$ and was also used to classify $X_c$ with the fuzzy $k_f$-NRR. Different values of m did not produce significantly different results so the outputs shown below are based on m = 2. Similarly, the hard k-NNR was implemented for k=1 to $\sqrt{n}_d$ and an optimal $k_{nn}$ to minimize observed $p_{nn}$ was obtained. Finally, all three methods were used to classify a new data set from mixture C to see which method's observed probability of error appeared least sensitive to sample variations. For the Iris data $p_1$* was first obtained at the 2nd iteration, and no smaller $p_1$* could be found for larger values of h. Similarly, for the univariate mixture, $p_1$* was obtained at h=2 and remained constant for all h>2. From these results we infer that Jozwik's stopping rule (1.5) is probably "optimal" in the sense that $p_1$* achieves a global minimum at the first h such that $p_1$* is found.

IRIS Data: A

Table 2 shows the results obtained using all three methods on the Iris data. In general, there is not much difference in the probability of error obtained using Jozwik's method and the

Table 2. Iris data set A

a) Jozwik

| Partition Ratio $(n_d:n_t)$ | $n_d$ | $n_t$ | $n_c$ | Stage 1 $k_1$* | Stage 1 $p_1$* | Stage 2 $k_2$* | Stage 2 $p_2$* | $p_c$ (Using $W_f,k_f$) |
|---|---|---|---|---|---|---|---|---|
| (1:2) | 36 | 78 | 36 | 1 | .056 | 1 | .044 | .028 |
| (1:1) | 57 | 57 | 36 | 1 | .053 | 1 | .018 | .028 |
| (2:1) | 78 | 36 | 36 | 8 | .026 | 10 | .018 | .028 |
| (1:9) | 12 | 102 | 36 | 1 | .000 | 5 | .070 | .000 |
| (9:1) | 102 | 12 | 36 | 8 | .019 | 10 | .017 | .056 |

b) fuzzy c-means

| | $X_R = X_d$ | | | | $X_R = X_d \cup X_t$ | | | |
|---|---|---|---|---|---|---|---|---|
| Partition ratio $(n_d:n_t)$ | $k_{fcm}$ | $p_{fcm}$ | $k_f$ | $p_c$ (Using $W_{fcm},k_f$) | $k_{fcm}$ | $p_{fcm}$ | $k_f$ | $p_c$ (Using $W_{fcm},k_f$) |
| (1:2) | 6 | .333 | 1 | .361 | 1 | .056 | 1 | .056 |
| (1:1) | 1 | .056 | 1 | .056 | 1 | 0.56 | 10 | .056 |
| (2:1) | 1 | .056 | 8 | .111 | 1 | .056 | 10 | .083 |
| (1:9) | 1 | .389 | 1 | .389 | 1 | .056 | 1 | .056 |
| (9:1) | 1 | .056 | 10 | .083 | 1 | .056 | 10 | .083 |

c) hard k-NNR

| | $X_R = X_d$ | | | | $X_R = X_d \quad X_t$ | | | |
|---|---|---|---|---|---|---|---|---|
| Partition ratio $(n_d:n_t)$ | $k_{nn}$ | $p_{nn}$ | $k_f$ | $p_c$ (using $W_R,k_f$) | $k_{nn}$ | $p_{nn}$ | $k_f$ | $p_c$ (using $W_R,k_f$) |
| (1:2) | 1 | .028 | 1 | .028 | 1 | .028 | 1 | .028 |
| (1:1) | 3 | .000 | 1 | .056 | 1 | .028 | 1 | .028 |
| (2:1) | 4 | .000 | 10 | .028 | 1 | .028 | 10 | .056 |
| (1:9) | 1 | .000 | 1 | .000 | 1 | .028 | 1 | .028 |
| (9:1) | 1 | .028 | 10 | .056 | 1 | .028 | 1.0 | .056 |

Notes

1. $k_f$ = optimal number of NN's using $W_f$ from Józwik's Method
2. $k_{fcm}$ = optimal number of NN's using $W_{fcm}$ from FCM
3. $p_{fcm}$ = minimal error rate achieved using $(W_{fcm},k_{fcm})$ on $X_R$.
4. $k_{nn}$ = optimal number of NN's using the hard NN rule with $W_R$
5. $p_{nn}$ = minimal error rate achieved using $(W_R,k_{nn})$ on $X_R$
6. $W_R$ = given hard c-partition of $X_R$
7. $p_c$ = observed error rate attained on $X_c$ using 1 of 3 rules.

hard k-NNR. The error ranges from 0.028 to 0.056. The optimal $k_{nn}$'s using the hard k-NNR alone are in general smaller than Jozwik's $k_f$, and resulted in lower probabilities of observed error on $X_c$. Using Jozwik's optimal $k_f$ with the hard k-NNR results in higher error only in the (1:1) case. This example also shows that fuzzy c-means does not perform well with small training data sets. Note, e.g., that $p_{fcm}$ for $X_d$ at (1:9) is 38.9%, as opposed to 5.6% at (9:1). The probability of error increases as the size of the training set decreases. Due to the small number of samples in the Iris data, all of these results must be interpreted with some caution.

Univariate normal mixtures: B

As seen in Table 3, the hard k-NNR using Jozwik's optimal $k_f$ with the smaller reference set yields an observed probability of error greater than that obtained by Jozwik's method in all cases except for the (2:1) partition ratio. Using the enlarged reference set, the error is smaller in 2 out of the 5 cases for the hard $k_f$-NNR, but the optimal $p_{nn}$ achieved by the hard k-NNR is smaller than Jozwik's in all cases except for the (1:9) partition ratio when using the smaller reference set. With the larger reference set, the optimal $p_{nn}$ attained by the hard k-NNR is 0.29, which is smaller than that of Jozwik's, although this is achieved at very high k values. Finally, fuzzy c-means performs better than either of the others overall, with errors between 0.29 and 0.30. FCM-NNR errors seem more stable, since they are not affected greatly by the number of nearest neighbors or different partition ratios. The "true" Bayes error rate for data set B is 0.3085; the FCM-NNR method gives very close estimates to this Bayes error.

Bivariate normal mixtures: C

Basically, the same conclusions can be derived here as for the univariate case. In Table 4 it can be seen that the hard k-NNR using the optimal $k_f$ found by Jozwik's method resulted in a greater error than in Jozwik's method in all cases except the (1:9) partition ratio. As in the univariate case, the minimum error $p_{nn}$ achieved by the hard k-NNR is slightly smaller

Table 3. Univariate normal mixture data set B*

a) Jozwik

| Partition ratio $(n_d:n_t)$ | $n_d$ | $n_t$ | $n_c$ | Stage 1 $k_1^*$ | Stage 1 $p_1^*$ | Stage 2 $k_2^*$ | Stage 2 $p_2^*$ | $p_c$ (using $W_f, k_f$) |
|---|---|---|---|---|---|---|---|---|
| (1:2) | 300 | 600 | 100 | 9 | .28 | 21 | .32 | .32 |
| (1:1) | 450 | 450 | 100 | 8 | .28 | 5 | .32 | .36 |
| (2:1) | 600 | 300 | 100 | 3 | .28 | 2 | .30 | .42 |
| (1:9) | 90 | 810 | 100 | 1 | .25 | 3 | .32 | .32 |
| (9:1) | 810 | 90 | 100 | 5 | .29 | 5 | .29 | .31 |

b) fuzzy c-means

| | $X_R = X_d$ | | | | $X_R = X_d \cup X_t$ | | | |
|---|---|---|---|---|---|---|---|---|
| Partition ratio $(n_d:n_t)$ | $k_{fcm}$ | $p_{fcm}$ | $k_f$ | $p_c$ (using $W_{fcm}, k_f$) | $k_{fcm}$ | $p_{fcm}$ | $k_f$ | $p_c$ (using $W_{fcm}, k_f$) |
| (1:2) | 2 | .29 | 9 | .30 | 11 | .29 | 9 | .30 |
| (1:1) | 1 | .29 | 8 | .30 | 11 | .29 | 8 | .30 |
| (2:1) | 1 | .29 | 3 | .29 | 11 | .29 | 3 | .30 |
| (1:9) | 8 | .31 | 1 | .33 | 11 | .29 | 1 | .30 |
| (9:1) | 1 | .29 | 5 | .29 | 11 | .29 | 5 | .30 |

c) hard k-NNR

| | $X_R = X_d$ | | | | $X_R = X_d \cup X_t$ | | | |
|---|---|---|---|---|---|---|---|---|
| Partition ratio $(n_d:n_t)$ | $k_{nn}$ | $p_{nn}$ | $k_f$ | $p_c$ (using $W_R, k_f$) | $k_{fcm}$ | $p_{fcm}$ | $k_f$ | $p_c$ (using $W_R, k_f$) |
| (1:2) | 15 | .29 | 9 | .38 | 25 | .29 | 9 | .32 |
| (1:1) | 15 | .34 | 8 | .40 | 25 | .29 | 8 | .33 |
| (2:1) | 18 | .34 | 3 | .39 | 25 | .29 | 3 | .36 |
| (1:9) | 9 | .40 | 1 | .52 | 25 | .29 | 1 | .41 |
| (9:1) | 16 | .27 | 5 | .38 | 25 | .29 | 5 | .35 |

$^{*}{}_{PB}$ = 0.31 for data set B

Table 4. Bivariate normal mixture data set $C^*$

a) Jozwik

| Partition ratio $(n_d:n_t)$ | $n_d$ | $n_t$ | $n_c$ | Stage 1 $k_1^*$ | Stage 1 $p_1^*$ | Stage 2 $k_2^*$ | Stage 2 $p_2^*$ | $p_c$ (using $W_f, k_f$) |
|---|---|---|---|---|---|---|---|---|
| (1:2) | 300 | 600 | 100 | 1 | .160 | 2 | .221 | .26 |
| (1:1) | 450 | 450 | 100 | 1 | .191 | 1 | .219 | .28 |
| (2:1) | 600 | 300 | 100 | 2 | .158 | 4 | .210 | .32 |
| (1:9) | 90 | 810 | 100 | 3 | .20 | 4 | .261 | .28 |
| (9:1) | 810 | 90 | 100 | 5 | .199 | 7 | .207 | .29 |

b) fuzzy c-means

| | $X_R = X_d$ | | | | $X_R = X_d \cup X_t$ | | | |
|---|---|---|---|---|---|---|---|---|
| Partition ratio $(n_d:n_t)$ | $k_{fcm}$ | $p_{fcm}$ | $k_f$ | $p_c$(using $W_{fcm}, k_f$) | $k_{fcm}$ | $p_{fcm}$ | $k_f$ | $p_c$(using $W_{fcm}, k_f$) |
| (1:2) | 9 | .25 | 1 | .27 | 7 | .26 | 1 | .27 |
| (1:1) | 13 | .25 | 1 | .27 | 7 | .26 | 1 | .27 |
| (2:1) | 22 | .25 | 2 | .27 | 7 | .26 | 2 | .27 |
| (1:9) | 7 | .23 | 3 | .24 | 7 | .26 | 3 | .27 |
| (9:1) | 2 | .26 | 5 | .27 | 7 | .26 | 5 | .27 |

c) hard k-NNR

| | $X_R = X_d$ | | | | $X_R = X_d \cup X_t$ | | | |
|---|---|---|---|---|---|---|---|---|
| Partition ratio $(n_d:n_t)$ | $k_{nn}$ | $p_{nn}$ | $k_f$ | $p_c$(using $W_R, k_f$) | $k_{nn}$ | $pp_{nn}$ | $k_f$ | $p_c$(using $W_R, k_f$) |
| (1:2) | 8 | .24 | 1 | .33 | 12 | .27 | 1 | .34 |
| (1:1) | 16 | .27 | 1 | .33 | 12 | .27 | 1 | .34 |
| (2:1) | 17 | .27 | 2 | .33 | 12 | .27 | 2 | .35 |
| (1:9) | 6 | .24 | 3 | .25 | 12 | .27 | 3 | .32 |
| (9:1) | 11 | .27 | 5 | .34 | 12 | .27 | 5 | .34 |

$^*p_B$ = 0.23 for data set C

than Jozwik's, but this is achieved at a high value of k. Fuzzy c-means again resulted in an even smaller probability of error in all cases, with an average error of about 0.27. The "true" Bayes error for mixture C is 0.2398.

Fresh bivariate normal mixture: C'

Table 5 lists the results of processing a new sample C' from bivariate mixture C of size n = 1000 (500 vectors each from $n(\underline{\mu}_1, I), n(\underline{\mu}_2, I)$). The entire data set C' is regarded as a "final" test set (in our previous notation, $C' = X_c'$). For each of the five partition ratios of the previous examples, the $n_d$ design samples (with $(n_d/2)$ from each class), together with their algorithmic outputs, were used to test classify data set C'. For Jozwik's method, $p_c'$ using $(W_f, k_f)$ is the error achieved on C' using $k_f$ from processing C as in Table 4; $W_f$ is from Stage 1 of Table 4 processing. The desire in Table 5 is to see whether the performance of rules (W,k) predicted by $p_c$ in Table 4 on $X_c$ holds up on a new sample $X_c$. Comparing corresponding columns shows fairly good agreement between $p_c$ and $p_c'$. The values $k''1$ shown at the right, however, are the optimal numbers of Stage 1 neighbors in $X_d$ with $W_f$ fixed found by applying $(W_f, k)$ to C' for k from 1 to $\sqrt{n_d}$. Evidently one can secure a somewhat better performance from $(X_d, W_f, k_f'')$ than is realized by $(X_d, W_f, k_f)$. Note, however, that the average decrease in error rate is only 1.6%, achieved at the expense of a greatly increased computing burden (average $k_f$ = 2.4 neighbors; average $k_f''$ = 17.4 neighbors). This result supports Jozwik's assertion that $(W_f, k_f, p_f)$ is the "best" combination to use for classification on fresh data from the same process.

Part 2 of Table 5 reports the same comparison for the FCM/k-NNR. The average difference between $p_{fcm}$ for $X_c$ using $(W_{fcm}, k_{fcm})$ and $p_c$ for C' using the same classifier is negligible. Note again that $k_{fcm}$, the optimal number of neighbors with $W_{fcm}$ found by direct calculation on C', is generally higher than $k_{fcm}$ - and that a slight reduction in error rates again results.

Finally, we may compare $p_{nn}$ and $p_c'$ for the hard $(W_R, k_{nn})$-NNR on $X_d$ and C': the result is again that $p_{nn}$ and $p_c'$ are quite close, but that $k_{nn}''$ can be found that improves the $(W_R, k_{nn})$ performance.

Table 5. Observed error rate $p'_c$ on data set c'*

a) Jozwik

| $X_d \subset C$ $n_d$ | $X'_c = C'$ $n_c$ | $k_f$ | $p'_c$ (using $W_f, k_f$) | $p''_c$ (using $W_f, k''_1$) | $k''_f$ |
|---|---|---|---|---|---|
| 300 | 1000 | 1 | .27 | .25 | 11 |
| 450 | 1000 | 1 | .27 | .25 | 21 |
| 600 | 1000 | 2 | .28 | .25 | 23 |
| 90 | 1000 | 3 | .27 | .27 | 6 |
| 810 | 1000 | 5 | .26 | .25 | 27 |

b) FCM/k-NNR

| $X_d \subset C$ $n_d$ | $X'_c = C'$ $n_c$ | $k_{fcm}$ | $p'_c$ (using $W_{fcm}, k_{fcm}$) | $p''_c$ (using $W_{fcm}, k''_{fcm}$) | $k''_{fcm}$ |
|---|---|---|---|---|---|
| 300 | 1000 | 9 | .25 | .25 | 7 |
| 450 | 1000 | 13 | .25 | .25 | 20 |
| 600 | 1000 | 22 | .26 | .25 | 12 |
| 90 | 1000 | 7 | .27 | .27 | 2 |
| 810 | 1000 | 2 | .26 | .25 | 27 |

c) hard k-NNR

| $X_d \subset C$ $n_d$ | $X'_c = C'$ $n_c$ | $k_{nn}$ | $p'_c$ (using $W_R, k_{nn}$) | $p''_c$ (using $W_R, k''_{nn}$) | $k''_{nn}$ |
|---|---|---|---|---|---|
| 300 | 1000 | 8 | .26 | .26 | 16 |
| 450 | 1000 | 16 | .26 | .26 | 21 |
| 600 | 1000 | 17 | .26 | .26 | 20 |
| 90 | 1000 | 6 | .32 | .28 | 7 |
| 810 | 1000 | 11 | .27 | .25 | 27 |

*$_{PB}$ = 0.23 for data set C'

Table 5 seems to support two conclusions: first, predicted error rates using "optimal" pairs (W, k) from all three rules to be rather dependable, but (not unexpectedly!) slightly better performance can be realized by "retraining" on new data; and secondly, the FCM/k-NNR seems to yield slightly better performance than either of the other two rules.

## 6. CONCLUSIONS

The results obtained by the FCM/k-NNR appear to produce the best observed classifier performance among the three k-NNR's studied. Depending on the data set and partition ratio, each of the three methods result in fairly realistic estimates of the Bayes error $p_B$. In all three methods, Kittler and Devijver's partition ratio of (1:2) does provide a good rule of thumb for data division since the error associated with this ratio is usually comparable to that achieved for other partition ratios. Finally, it appears that fuzzy k-NNR's in general perform slightly better than the conventional hard k-NNR. It may be, however, that a hard $(k,l_i)$-NNR is in some sense equivalent to fuzzy k-NNR's: this will be the target of a future investigation.

## ACKNOWLEDGEMENT

This research was supported by NSF Grant IST-8407860.

## REFERENCES

Bezdek, J.C. (1981). Pattern Recognition with Fuzzy Objective Function Algorithms. Plenum, New York.
Fisher, R.A. (1936). The use of multiple measurements in taxonomic problems. Ann. Eugenics 7, 179-188.
Jozwik, A. (1983). A learning scheme for a fuzzy k-NN rule. Pattern Recogn. Letters 1, 287-289.
Kittler, J. and P.A. Devijver (1981). An efficient estimator of pattern recognition system error probability. Pattern Recogn. 13, 245-249.
Kittler, J. and P.A. Devijver (1982). Pattern Recognition: A Statistical Approach. Prentice-Hall, London.
Lachenbruch, P.A. (1965). Estimation of Error Rates in Discriminant Analysis. Ph.D. Dissertation (unpublished), University of California, Los Angeles.

# GREY DECISION MAKING AND ITS USE FOR THE DETERMINATION OF IRRIGATION STRATEGIES

Deng Julong

Department of Automation
Huazhong University of Science and Technology
Wuhan, People's Republic of China

Abstract. Using the concepts of a topological space and state, an approach to multitarget decision making - called grey decision making - has been developed. Some definitions and properties are given in this paper. As a successful example of grey decision making, the derivation of an irrigation strategy for the People's Victory Channel is outlined. Experience of recent years proves that the effectiveness of this strategy is satisfactory.

Keywords: grey decision making, grey systems, systems theory, decision making.

## 1. OUTLINE OF GREY DECISION MAKING

Open sets of a topological space play an important role in the topology of space. An open set means that its bounds are uncertain. A number whose real value cannot be determined for lack of information, but when an open interval where this number is located is known, is called a grey number. Thus we say that an open set is a grey number. In the topology of space, a neighborhood of a point is the set of points which lie "close" enough to that point. Usually, a neighborhood contains some open set, thus the neighborhood is an extension of a grey number, or, in other words, neighborhoods imply that some close elements are located around a key element. According to the theory of grey systems, a key element is one of a pointed whitening value of a grey number.

A state (situation) composed of an event and a game (a counter-measure) is an essential concept of grey systems, while a pair consisting of an event(s) and a game(s) is an essential element of decision making.

As a general rule, the effects of a state dealt with by different games for the same event are different. The ultimate aim of decision making is to obtain a class of satisfactory states according to the effects: in this class a key state has to be included. In addition to the key state, all of the satisfactory states abut on the key state according to the given targets. To accept the key state as a kernel, a neighborhood of the satisfactory states should be composed; we call that neighborhood a grey butt of decision making. For irrigation decision making we have to arrange an irrigation procedure month by month or day by day. Thus the irrigation time intervals are events, and different manners of irrigation - such as no irrigation, channel irrigation and well irrigation - are games.

We know that the season of wheat sowing starts on about October 15 and that a great quantity of water is necessary, hence

(October 15, channel irrigation)

is a key state. About December 15 there is the season of wheat sowing to keep a full stand of seed, and warm water is necessary, thus

(December 15, well irrigation)

is a key state.

The targets (criteria) to make out a butt of irrigation are as follows:

1. to control the underground water below a critical level in order to prevent the soil from alkalisalination;
2. to rationally irrigate the plants;
3. to obtain maximum benefits.

The key state can provide the plants with good growing conditions because it is obtained from the agricultural requirements. Channel irrigation can decrease the level of underground water and well irrigation can icrease it, thus the underground water can be controlled.

## 2. STATE TOPOLOGICAL SPACES

Definition 1. Let a be a qualified event, b be a qualified game, A be a set of qualified events, and B be a set of qualified games. Let for each $a \in A$, $\underline{U}_a = \{U(a)\}$ be a non-empty family of subsets of A associated with a, such that:

1° $a \in U(a)$, for each $U(a) \in \underline{U}_a$

2° If $V \supseteq U(a)$ for some $U(a)$, then $V \in \underline{U}_a$

3° If $U$ and $V \in \underline{U}_a$, then $U \cap V \in \underline{U}_a$

4° If $U \in \underline{U}_a$, then there exists a $V \in \underline{U}_a$ such that if $\alpha \in V$, then $U \in \underline{U}_\alpha$.

Then, $\underline{U}_a$ is called a system of neighborhood at event a and a neighborhood U(a) is called a grey number of event, denoted by $\otimes(a)$ and a is its whitening value. Similarly, we have a grey number $\otimes(b)$, with b being its whitening value.

We call

$$(A, J_a), \quad J_a = \{U_a \mid a \in A\}$$
$$(B, J_b), \quad J_b = \{U_b \mid b \in B\}$$

the topological space of event and game, respectively. Usually, we call A and B the topological spaces.

Remark 1. For an irrigation strategy, the irrigation time interval is a qualified event, the ratio of irrigation area between channel and well irrigation is a qualified game.

Definition 2. Let $X_i$, i=1,2, be topological spaces and $\theta_i$ be the associated families of open sets for each i,

$$X = \mathop{\times}_{i=1}^{2} X_i$$

be the Cartesian product of $X_1$ and $X_2$ and let

$$\underline{S} = \left\{ S \mid S = \mathop{\times}_{i=1}^{s} Y_i, \quad \text{when} \quad y_i = x_i, \quad \text{for all} \quad i \neq j, \quad \text{and} \quad y_j = u_h \in \theta_j, \quad j=1,2 \right\}$$

Then $\underline{S}$ is a subbase for the usual topological space for X.

The topology so defined is called the state topology. And we call

$$\otimes(a) = \otimes(a,b), \quad \text{or} \quad \otimes(s) = (\otimes(a), \otimes(b))$$

the grey state or the neighborhood of $\underline{S}$; we call $\tilde{x}(s) = s = (a,b)$ the whitening state associated with a whitening event a and a whitening game b; we call $S_{ij} = (a_i, b_j)$ the whitening sub--state associated with a subevent $a_i$ and a subgame $b_j$.

Definition 3. Let $\sigma_{ij}^{(p)}$ be an effect measure of a whitening state for the p-th target; thus

$$\sigma : \{S_{ij}\} \to \{\sigma_{ij}^{(p)}\}$$

is a mapping from the states to the effects of the p-th target.

## 3. AXIOMS OF DECISION MAKING FOR THE STATE

We have the following four basic axioms:

1) If $\sigma_{ij}^{(p)} \in \emptyset$, then $S_{ij}$ is not a state;

2) If $S_{ij} \in \emptyset$, then $\sigma_{ij}^{(p)}$ is not an effect;

3) Let $\sigma_{ij}^{(p)}$, $\sigma_{kj}^{(p)}$, $\sigma_{ik}^{(p)}$ be effects of the states $S_{ij}$, $S_{kj}$, $S_{ik}$ of the p-th target, let I be an index set of the events, K be an index set of the games, for $i \in I$, $j,k \in K$, and there exist $\sigma_{ij}^{(p)} = \sigma_{ik}^{(p)}$, $\sigma_{ij}^{(p)} = \sigma_{kj}^{(p)}$;

4) Let $\sigma_{ik}$, $\sigma_{kj}$, $\sigma_{ij}$ be effect sets for all targets, i.e.

$$\sigma_{ik} = \sigma_{ik}^{(p)} \qquad p=1,2,\ldots,m$$

$$\sigma_{kj} = \sigma_{kj}^{(p)} \qquad p=1,2,\ldots,m$$

$$\sigma_{ij} = \sigma_{ij}^{(p)} \qquad p=1,2,\ldots,m$$

If p is sufficiently large, then

$$\sigma_{ij} = \sigma_{ik}, \quad \sigma_{ij} = \sigma_{kj},$$

cannot occur.

Definition 4. 1) If $\sigma_{ij}^{(p)} = \sigma_{ik}^{(p)}$, then we say that games $b_j$ and $b_k$ are equivalent with respect to an associated event $a_i$ in the p-th target. We call the set

$$\{\sigma_{ij}^{(p)}\} = \{\sigma_{ij}^{(p)} \mid \sigma_{ij}^{(p)} = \sigma_{ik}^{(p)},\ k \in K,\ K \text{ is the index set of games}\}$$

the equivalence classes of effects associated with an event $a_i$.

2) If $\sigma_{ij}^{(p)} = \sigma_{kj}^{(p)}$, then we say that $a_i$ and $a_k$ are equivalent with respects to an associated game $b_j$. We call

$$\{\sigma_{kj}^{(p)}\} = \{\sigma_{kj}^{(p)} \mid \sigma_{kj}^{(p)} = \sigma_{ij}^{(p)},\ i \in I,\ I \text{ is the index set of events}\}$$

the equivalence classes of effects associated with a game $b_j$.

3) We call

$$\{s_{ij}\} = \{s_{ij} \mid \sigma(s_{ij}) = \sigma_{ij}^{(p)},\ \forall \sigma_{ij}^{(p)} \in \{\sigma_{ij}^{(p)}\}\}$$

the equivalence set. Similarly, $\{s_{kj}\}$ can be defined.

4) We call

$$\underline{S} = \{\{s_{ij}\} \mid \forall \sigma_{ij}^{(p)} \in \{\sigma_{ij}^{(p)}\},\quad \forall \sigma_{kj}^{(p)} \in \{\sigma_{kj}^{(p)}\}\}$$

the equivalent family of states.

Remark 2. The power of $\{s_{ij}\}$ in $\underline{S}$ may be 1 or more.

Proposition 1. Let $\underline{S}$ be an equivalent family of states and $E^{(p)}$ be an effect set of the p-th target. Thus the mapping

$$\sigma^{(p)} : \underline{S} \to E^{(p)}$$

is 1 - to -1 and onto.

Proposition 2. Let E be a p-dimensional effect space such that

$$e_{ij} \in E,\quad e_{ij} = (\sigma_{ij}^{(p)}, \sigma_{ij}^{(i)}, \ldots, \sigma_{ij}^{(n)}),\quad e_{ij} \in R^n$$

Let $\underline{S}$ be a state set such that $\sigma^{-1} : E \to \underline{S}$. Then $\sigma^{-1}$ is

1- to -1 and onto if and only if $n$ is suffciently large.

Definition 5. Let $X$ be a topological state space

$$X = \left\{ U_x(\otimes), U_y(\otimes) \right\}, \quad U_x(\otimes) \in a, \quad U_y(\otimes) \in b,$$

$$\text{or} \quad X \triangleq (\otimes(a,b))$$

where $a$ is a set of events and $b$ is a set of games, $a_i$ is a whitening event, and $a_i \in a$; let

$$Y = \left\{ (a_i), U_y(\otimes) \right\} \subseteq X$$

for each $\gamma \in Y$, $V_\gamma = Y \cap \underline{U}_\gamma$ where $V_\gamma$ is a neighborhood system. Then, for

$$V \in V_\gamma, \quad U \in \underline{U}_\gamma, \quad V = Y \cap U$$

we have a topology $J' = \left\{ V_\gamma | \gamma \in Y \right\}$ which is said to be induced on $Y$ by the topology $J$ of $X$.

Usually, we call $J'$ the topology associated with a whitening event $a_i$. Similarly, the topology associated with a game $b_i$ can be obtained.

Definition 6. Let $X$ be a partially ordered set under some ordering "$\leqslant$", i.e.

1) $x \leqslant y$ and $y \leqslant \gamma$ imply $x \leqslant \gamma$,
2) $x \leqslant x$, for all $x \in X$,
3) $x \leqslant y$ and $y \leqslant x$ imply $x = y$,

If we define

$$P_r(x) = \left\{ y | x_r \leqslant y \right\}, \quad r \in R$$

$$\underline{U}_x = \left\{ U | U \supseteq P_r(x) \right\}$$

then $\underline{U}_x$ is a grey number or a neighborhood and $J = \left\{ \underline{U}_x | x \in X \right\}$ is called the right order topology for $X$.

Definition 7. Let $\left\{ (X,J) \right\}$ be a right order topology of states' effects, $\left\{ S_n | n=1,2,\ldots \right\}$ be a state sequence, and $\left\{ A_k | k=1,2,\ldots,n \right\}$ be a sequence of states' effects such that

$$A_i = \left\{ \sigma_i^{(1)}, \sigma_i^{(2)}, \ldots, \sigma_i^{(p)} \right\}, \quad A_i \in \left\{ A_k \right\}$$

$U_i \subseteq X$, and $\left\{ D_\alpha | \alpha \in I \right\}$ be families of subsets of $X$. Then, $\left\{ D_\alpha \right\}$ is called a cover or covering for $U_i$ provided

$\bigcup_{\alpha\in I} D_\alpha \supseteq U_i$, when $\bigcup_{\beta\in I} D_\beta \supseteq U_i$, $\forall \bigcup_{\beta\in I} D_\beta \supseteq \bigcup_{\alpha\in I} D_\alpha$.
We call $\{D_\alpha\}$ a compact cover, and we have

$$A_i \subseteq U_i \subseteq \underline{U}_i$$

$$U_i = \{U_i | U_i \supseteq \{y | x_i \leqslant y\},\ i \in I\},\ J = \{U_i\}$$

Then, $U_i$ is called a grey neighborhood system of $A_i$.

For a sequence $\{A_k\}$, a sequence of neighborhoods

$$\{U_n\} = \{U_k | k=1,2,\ldots\}$$

exists. If there is a right order topology $J$, a neighborhood of $U_o$ is

$$U_o \supseteq \{y | x_o \leqslant y\},\quad U_o \subseteq X,\quad x_o \geqslant x_i,\quad \forall i \in I$$

and $U_o$ belongs to a compact subspace, then $U_o$ is said to be a grey butt associated with a state sequence $\{S_n\}$, if and only if for each $U_n \in \{U_o\}$ there exists an integer $N$ such that $n \geqslant N$ implies $U_o \supseteq U_n$.

Definition 8. A state $S_n$ associated with $U_n$ is called a suboptimal state. If $S_n = (a_i, b_{j^*})$, then a game $b_{j^*}$ is called a suboptimal game.

Theorem 2. Let $X$ be a partially ordered set of multitarget effects and let each simply ordered subset of $X$ have an upper bound in $X$. Then, $X$ has a maximal element, i.e. if for each $Y \subseteq X$ such that $Y$ is simply ordered, there exists $z \in X$ such that $y \in Y$ implies $y \leqslant z$, then there exists $m \in X$ such that for each $x \in X$ either $x$ or $m$ are incomparable, i.e. neither $x \leqslant m$ nor $m \leqslant x$ is true. The proof proceeds according to Zorn's lemma.

Definition 9. A maximal element of a partially ordering set of multitarget effects is a whitening optimal state.

## 3. EXAMPLE

The People's Victory Channel of He Nan Province is a huge irrigating system. Its planning has been done by grey decision making.

In order to simplify the calculations, let us assume that all states during a month are equal. Then a neighborhood degenerates to one point. To calculate the effect, some measures such as: a synthesis measure for the target preventing the soil from alkalisalination,a centre measure for the target of rational irrigation, and a bound measure for the target of benefits, are used.

Let $U_{ij}$ be effect samples of states $S_{ij} = (a_i, b_j)$, where $a_i$ in the i-th month and $b_j$ is the j-th irrigation procedure. Let $\sigma_{ij,o}$ be a centre measure and $\sigma_{ij,m}$ be a bound measure such that:

$$\sigma_{ij,o} = \frac{\min(U_{ij}, U_o)}{\max(U_{ij}, U_o)}, \quad \forall j$$

$$\sigma_{ij,m} = \frac{U_{ij}}{U_m}, \quad \forall j$$

where $U_o$ is a pointed effect and $U_m$ is

$$U_m = \max_j U_{ij}$$

Based on statistics and an effect function, a synthesis measure can be obtained. First, some grey numbers have to be listed:

1. for rain intensity:

   heavy rain(a), $\otimes(a) > 80$ mm

   medium rain(b), $20 \leqslant \otimes(b) \leqslant 80$ mm

   low rain(c), $x(c) < 20$ mm

2. for burying depth of underground water (in m):

   high burying depth(α), $\otimes(\alpha) > 2.2$

   medium burying depth(β), $1.8 \leqslant \otimes(\beta) \leqslant 2.2$

   shallow burying depth(γ), $x(\gamma) < 1.8$

Let us denote:

$H_{n+1}, H_n$ - burying depth of underground water at time n+1 and n, respectively;

$\Delta H_e$ - increment of burying depth due to evaporation;

$\Delta H_\gamma$ - increment of burying depth due to rain

$\Delta H_i$, i=1,2,1 - increments of burying depth resulting from different irrigation procedures:

- $\Delta H_1$ - no irrigation,
- $\Delta H_2$ - channel irrigation,
- $\Delta H_3$ - well irrigation;

$\Delta H_\Sigma$ - total increment of burying depth;

$P_i$, $i=\alpha,\beta,\gamma$ - probability of burying depth:

- $P_\alpha$ - probability of high burying depth,

$P_\beta$ - probability of medium burying depth,
$P_\gamma$ - probability of shallow burying depth;

$P_i$, i=a,b,c, - probability of rain:
$P_a$ - probability of heavy rain,
$P_b$ - probability of medium rain,
$P_c$ - probability of low rain;

$a_i$, i=1,2,...,12 - intervals of irrigation time
$a_1$ - January,
$a_2$ - February,
...............
$a_{12}$ - December.

$b_i$, i=1,2,3, - irrigation procedure:
$b_1$ - no irrigation,
$b_2$ - channel irrigation,
$b_3$ - well irrigation;

$H_i$, i=α,β,γ - burying depth of underground water:
$H_\alpha$ - high,
$H_\beta$ - medium ,
$H_\gamma$ - shallow.

To calculate the synthesis measure, the data are given in Tables 1-4.

Table 1. Rain intensities

| rain | H (m/month) |
|---|---|
| heavy(a) | $\geq 0.6$ |
| medium(b) | [0.3 - 0.6) |
| low(c) | [0 - 0.3) |

Table 2. Burying depth of underground water
(monthly distribution)

| H<br>month | $H_\alpha$<br>$H_e$(m/month) | $H_\beta$<br>$H_e$(m/month) | $H_\gamma$<br>$H_e$(m/month) |
|---|---|---|---|
| January-April | 0.1 | 0.36 | 0.45 |
| May-August,<br>November,<br>December | 0.2 | 0.45 | 0.55 |
| September,<br>October | 0.3 | 0.54 | 0.65 |

Table 3. Increments of burying depth for different
irrigation procedures

| $b_i$ | $H_i$ |
|---|---|
| $b_1$ | $H_1(k) = 0$ |
| $b_2$ | $H_2(k) = -0.4$ |
| $b_3$ | $H_3(k) = 0.3$ |

Table 4. Probabilities of burying depths and rain
for the consecutive months

| Month | $P_\alpha$ | $P_\beta$ | $P_\gamma$ | $P_a$ | $P_b$ | $P_c$ |
|---|---|---|---|---|---|---|
| 1 | 0.46 | 0.54 | 0 | 0 | 0 | |
| 2 | 0.54 | 0.37 | 0.09 | 0 | 0.08 | 0.92 |
| 3 | 0.28 | 0.72 | 0 | 0 | 0.23 | 0.77 |
| 4 | 0.18 | 0.82 | 0 | 0 | 0.85 | 0.15 |
| 5 | 0.46 | 0.54 | 0 | 0.08 | 0.38 | 0.54 |
| 6 | 0.28 | 0.72 | 0 | 0.31 | 0.54 | 0.15 |
| 7 | 0.18 | 0.54 | 0.28 | 0.85 | 0.15 | 0 |
| 8 | 0.18 | 0.18 | 0.64 | 0.54 | 0.46 | 0 |
| 9 | 0.22 | 0.33 | 0.45 | 0.38 | 0.54 | 0.08 |
| 10 | 0.22 | 0.67 | 0.11 | 0.08 | 0.69 | 0.23 |
| 11 | 0.44 | 0.45 | 0.11 | 0 | 0.31 | 0.69 |
| 12 | 0.45 | 0.44 | 0.11 | 0 | 0.15 | 0.85 |

By using the mentioned data, the synthesis measure of October can be obtained. Denote the total increment of burying
depth in October by $\Delta H_\Sigma(10)$. Then:

- from table 2, when it is a high depth, we have $\Delta H_e(10,\alpha) = 0.3$;
- from table 1, when it is a heavy rain, we have $\Delta H_\gamma(10,\alpha) = -0.6$;
- from table 3, when the channel irrigation is used, we have $\Delta H_2(10,b_2) = -0.4$.

Thus, the total increment of burying depth, $\Delta H_\Sigma$, for October, a high burying depth($\alpha$), a heavy rain(a), the channel irrigation($b_2$), is

$$\begin{aligned}\Delta H_\Sigma(10,\alpha,a,b_2) &= \Delta H_e(10,\alpha)+\Delta H_\gamma(10,a)+\Delta H_2(10,b_2) = \\ &= 0.3 + (-0.6) + (-0.4) = \\ &= -0.7\end{aligned}$$

Similarly, we have :

$$\begin{aligned}\Delta H_\Sigma(10,\alpha,b,b_2) &= \Delta H_e(10,\alpha)+\Delta H_\gamma(10,b)+\Delta H_2(10,b_2) = \\ &= 0.3 + (-0.3) + (-0.4) = -0.4\end{aligned}$$

$$\begin{aligned}\Delta H_\Sigma(10,\alpha,c,b_2) &= \Delta H_e(10,\alpha)+\Delta H_\gamma(10,c)+\Delta H_2(10,b_2) = \\ &= 0.3 + 0 + (-0.4) = \\ &= -0.1\end{aligned}$$

Then, the burying depths in October for the different cases are:

$$\begin{aligned}H(10,a) &= 2.4+\Delta H_\Sigma(10,\alpha,a,b_2) = 2.4 - 0.7 = 1.7\\ H(10,b) &= 2.4+\Delta H_\Sigma(10,\alpha,b,b_2) = 2.4 - 0.4 = 2\\ H(10,c) &= 2.4+\Delta H_\Sigma(10,\alpha,c,b_2) = 2.4 - 0.1 = 2.3\end{aligned}$$

Let the effect function of H be given as shown in Fig. 1.

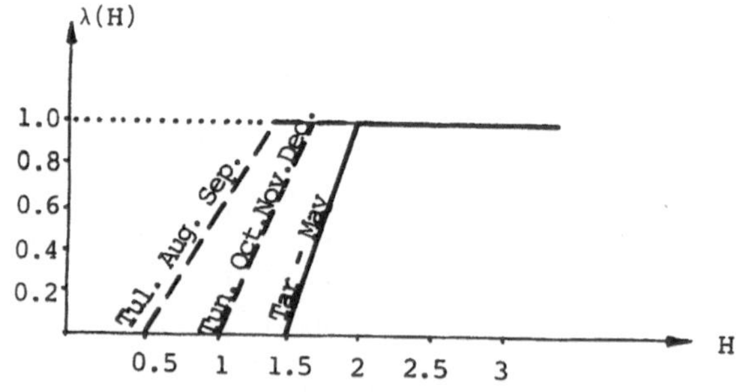


Fig. 1

Therefore, from Fig. 1 the effect values are:

$$\lambda_a = \lambda(H) = \lambda(1.7) = 0.75$$
$$\lambda_b = \lambda(H) = \lambda(2) = 0.95$$
$$\lambda_c = \lambda(H) = \lambda(2.3) = 1$$

The synthesis measure of a high burying depth is therefore as follows

$$\sigma^{(1)}_{10,2} = \lambda_a P_a + \lambda_b P_b + \lambda_c P_c =$$
$$= 0.75 \cdot 0.08 + 0.95 \cdot 0.69 + 1 \cdot 0.23 = 0.94$$

Similarly, we have :

$$\sigma^{(1)}_{10,2}(\beta) = 0.82$$
$$\sigma^{(1)}_{10,2}(\gamma) = 0.33$$

Thus the synthesis measure for October is

$$\sigma^{(1)}_{10,2} = P_\alpha \sigma^{(1)}_{10,2}(\alpha) + P_\beta \sigma^{(1)}_{10,2}(\beta) + P_\gamma \sigma^{(1)}_{10,2}(\gamma) =$$
$$= 0.22 \cdot 0.94 + 0.67 \cdot 0.82 + 0.11 \cdot 0.33 = 0.8$$

and the effect of state

$$S_{10,2} = (\text{October, channel})$$

is 0.8 for the first target.

A partially ordered set of multitarget effects can be obtained by using Theorem 2, and the following satisfactory states are obtained:

$S_{11}=(a_1,b_1)$, $S_{23}=(a_2,b_3)$, $S_{31}=(a_3,b_1)$ or $(a_3,b_2)$

$S_{42}=(a_4,b_2)$, $S_{51}=(a_5,b_1)$, $S_{62}=(a_6,b_2)$

$S_{71}=(a_7,b_1)$, $S_{83}=(a_8,b_3)$, $S_{91}=(a_9,b_1)$

$S_{10,2}=(a_{10},b_2)$, $S_{11,1}=(a_{11},b_1)$, $S_{12,3}=(a_{12},b_3)$

The satisfactory irrigation strategies are:

| | |
|---|---|
| January: no | July: no |
| February: well | August: well |
| March: no or well | September: no |
| April: channel | October: channel |
| May: no | November: no |
| June: channel | December: well |

## 4. CONCLUDING REMARKS

The experience of recent years has proved that the effectiveness of the irrigation strategy developed using the method presented is satisfactory. Mainly the expected level of underground water has been controlled and harvests have taken place in the channel valley.

## REFERENCES

Deng Julong (1984). Grey decision making and topological space. Contributions to Grey Systems and Agriculture, 12, Taivuan, Shanxin (in Chinese).

Deng Julong (1984). The theory and methods of socio-economic grey systems. Social Science in China, 6 (in Chinese).

Deng Julong et al. (1984). Some models of grey systems and their applications. Exploration of Nature 3 (in Chinese).

INDEX

**THEORY AND DECISION LIBRARY**

General Editors: W. Leinfellner and G. Eberlein

---

SERIES B: **MATHEMATICAL AND STATISTICAL METHODS**

Editor: H. Skala (Paderborn)

*Already Published:*

ROBUSTNESS OF STATISTICAL METHODS AND NONPARAMETRIC STATISTICS, *Edited by* DIETER RASCH and MOTI LAL TIKU.
ISBN 90-277-2076-2. TDLB 1

STOCHASTIC OPTIMIZATION AND ECONOMIC MODELS,
*by* JATI K. SENGUPTA.
ISBN 90-277-2301-X. TDLB 2

A SHORT COURSE ON FUNCTIONAL EQUATIONS,
BASED UPON RECENT APPLICATIONS TO THE SOCIAL AND BEHAVIORAL SCIENCES,
*by* J. ACZEL.
ISBN 90-277-2376-1 (HB), ISBN 90-277-2377-X (PB). TDLB 3